Landscape of 21st Century Mathematics

Bogdan Grechuk

Landscape of 21st Century Mathematics

Selected Advances, 2001–2020

 Springer

Bogdan Grechuk (iD)
Department of Mathematics
University of Leicester
Leicester, UK

ISBN 978-3-030-80629-3   ISBN 978-3-030-80627-9   (eBook)
https://doi.org/10.1007/978-3-030-80627-9

Mathematics Subject Classification: 00-02, 00A05, 00A09, 01A61

This Springer imprint is published by the registered company Springer Nature Switzerland AG.
The registered company address is: Gewerbestrasse 11, 6330 Cham, Switzerland

Preface

Great mathematicians of the seventeenth–eighteenth century like Euler and Gauss were experts in essentially all areas of mathematics that existed at the time. This level of broad expertise is impossible in modern mathematics, where over 100 mathematical papers are posted online on arXiv every day. The best one may hope for is to have deep expertise in some selected areas of mathematics and a "landscape view" of what is happening in mathematics as a whole.

Most mathematical papers, surveys and books are devoted to the in-depth exploration of some specific area of mathematics. This book has a completely opposite aim, namely to give an overview of the landscape of a broad spectrum of twenty-first century mathematics, without going deeply into any particular subject.

To achieve this aim, we first compiled a big list of *great recent theorems* from all areas of mathematics. By "great theorems" we mean theorems recommended by experts in the corresponding areas, the main results of papers published in leading mathematical journals, or those receiving a large number of citations in the corresponding subject category, theorems that were awarded a prestigious prize, theorems which resolved important long-standing open problems in their fields, or recent preprints with clear potential to satisfy the above criteria in the near future. From this big list, we then selected those theorems that have a reasonably *elementary statement*. This means that the theorem's formulation can be rigorously explained in a few paragraphs to mathematicians and Ph.D. students working in a different area, and even to graduate and undergraduate students with a good background in standard disciplines like analysis, algebra, geometry and probability theory. We then grouped the selected theorems by subject category.

In the present book, for each theorem chosen as explained above, we give a brief background and history, list some related previous results, and finally give the rigorous formulation of the theorem itself, including all necessary definitions. For some theorems we also say a few words about the proof's main ideas, but, unlike the theorem formulations, the proofs are discussed only at a non-rigorous, intuitive level, often without formal definitions of the involved concepts, and referring to the original papers for further details. For many theorems, the proofs are not discussed at all.

The theorems included in this book are at the same time great and beautiful. In many cases, the theorem formulations are so easy that the reader may be surprised that such elementary-looking statements could have remained open for many years and only recently be resolved. For Ph.D. students and mathematicians, this book may be a good place to learn developments outside of their areas of expertise, which can be read for pleasure and without much effort. For undergraduate and graduate students, the book may serve as a gentle introduction to modern mathematics. The students will learn *what* mathematicians are doing, and, we hope, may become interested in *how* they are doing this. The "how" question requires learning much deeper mathematics, and the bibliography of this book contains over 1000 references for further reading.

This book does not aim to explain difficult-to-understand mathematical theorems in a simple way. Instead, it collects great theorems with easy-to-understand statements from all areas of mathematics. Unfortunately, there are some relatively technical areas, like Hodge theory, in which we were not able to find any recent great theorem whose formulation is sufficiently elementary to be included here. However, all *broad* areas of mathematics, from number theory and combinatorics to traditionally more advanced topics like abstract algebra or topology, are presented.

The present book bears some similarities to my previous book "Theorems of the 21st Century" [520], which presents *some* easy-to-state great recent theorems and aims to explain them to an undergraduate or even an able high-school student. However, only a small portion of interesting theorems are covered there. In contrast, this book assumes some minimal undergraduate background and does not provide explanations of what 'integral', 'derivative', or 'prime number' means, for example. This allows the inclusion of more theorems. In fact, the aim was to not miss a single twenty-first century theorem which is both great and easy to understand. Readers who see any clear omission are encouraged to contact me; any great theorem with an elementary statement which we missed in this book will be included in the next edition.

This book collects theorems published in the period from 2001 to 2020, as well as some theorems that are not published in journals and exist only as arXiv preprints. The most notable example of the latter is the celebrated proof of the Poincaré conjecture by Perelman. For theorems published in journals, we use the publication date as a chronological marker. So for example, if we write "In 2008, Green and Tao proved that..." we mean that the publication date is 2008, while the date of the first submission to arXiv may have been much earlier. Some recent arXiv submissions with great potential are also included. We hope that the readers will enjoy having all such theorems collected in one place. Moreover, the plan is to add new theorems once they are published and include them in further editions.

Leicester, UK   Bogdan Grechuk
May 2021

Acknowledgements

I would like to thank Dr Tetiana Grechuk, my wife, for her continued support, encouragement, and patience while I was writing this book.

I also thank Vasyl Grechuk, my father, for many useful discussions, suggestions, and help with the figures.

I thank my MSc student, Adam Smith, for doing an MSc project on this topic under my supervision and finding some theorems for this book on the way.

I thank the mathematicians, leading experts in their areas, who found time to look at the preliminary version of this book and suggested the inclusion of some theorems which were missing. I have received very valuable suggestions from Professors Karim Adiprasito, Peter Bűhlmann, Yuansi Chen, William Gasarch, Andrew Granville, Martin Hairer, Michael Larsen, Alex Lubotzky, Gunter Malle, James Maynard and Benjamin Sudakov.

I thank my colleagues at the University of Leicester for reading parts of this book and providing suggestions and corrections.

I thank Springer, and especially Rémi Lodh, for their interest in publishing this book. I also thank the referees for their useful suggestions.

Contents

Acronyms

$\mathbb{N}$	the set of positive integers.
$\mathbb{Z}$	the set of integers.
$\mathbb{Q}$	the set of rational numbers.
$\mathbb{R}$	the set of all real numbers.
$\mathbb{R}^d$	standard (Euclidean) d-dimensional space.
$\lvert \cdot \rvert$	if x is a real number, $\lvert x \rvert$ denotes the absolute value of x.
$\exp(x)$	denotes e^x, where $e = 2.71828\ldots$ is the base of the natural logarithm.
$\log$	denotes the natural logarithm, unless specifically stated otherwise.
$\in$	denotes membership of a set.
$X \subset Y$	indicates that the set X is a subset of the set Y.
$\forall$	for every. For example, "$\forall n \in \mathbb{Z}\ldots$" means "For every integer $n\ldots$".
$\exists$	exists. For example, "$\exists n \in \mathbb{Z}\ldots$" means "There exists an integer $n\ldots$".
$\{\ldots \mid \ldots : \ldots\}$	is a notation used to describe a set. For example, the set A of all even integers can be written as $A = \{n \in \mathbb{Z} \mid \exists k \in \mathbb{Z} : n = 2k\}$.
$\lvert \cdot \rvert$	if S is a finite set, $\lvert S \rvert$ denotes the number of elements in S.
$\sum$	denotes summation, while $\prod$ denotes a product.
$\cup$	denotes the union of sets. Also, $\bigcup\limits_{i=1}^{n} A_i$ denotes the union $A_1 \cup A_2 \cup \cdots \cup A_n$.
$\cap$	denotes the intersection of sets. Also, $\bigcap\limits_{i=1}^{n} A_i$ denotes the intersection $A_1 \cap A_2 \cap \cdots \cap A_n$.
$!$	for an integer $n \geq 0$, $n!$ denotes $\prod_{i=1}^{n} i$, with the convention that $0! = 1$.
$:=$	denotes "equal by definition". For example, $n! := \prod_{i=1}^{n} i$ for every positive integer n.
$\inf$	denotes the infimum, while $\sup$ denotes the supremum.

$a \equiv b \pmod{c}$	means that $a - b$ is divisible by c.
$\langle x, y \rangle$	denotes the inner product of vectors x, y.
$\lvert \cdot \rvert$ or $\lVert \cdot \rVert$	if x is an element of a normed space, its norm may be denoted by $\lvert x \rvert$ or $\lVert x \rVert$.
$\mathbb{P}$	denotes the probability.
$f(x) = O(g(x))$	means that $\lvert f(x) \rvert \leq C g(x)$ for some constant C.
$f(x) = o(g(x))$	means that $\lim_{x \to \infty} \frac{\lvert f(x) \rvert}{g(x)} = 0$.

Chapter 1
Number Theory

1.1 Gaps Between Primes

Small Gaps Between Primes

The theory of prime numbers is full of conjectures which are very easy to state but extremely difficult to prove. A famous example is the *twin prime conjecture* predicting that $p_{n+1} - p_n = 2$ for infinitely many values of n, where p_n denotes the n-th prime number. By the prime number theorem, $p_{n+1} - p_n$ is asymptotically $\log n$ on average. In 1940, Erdős [426] proved the existence of an $\epsilon < 1$ such that

$$p_{n+1} - p_n < \epsilon \log n \qquad (1.1)$$

infinitely often. Then many authors proved (1.1) for smaller and smaller values of ϵ. The value $\epsilon = 0.2484...$, achieved by Maier [832] in 1988, was the best one for several decades. A long-standing goal was to prove that (1.1) remains true for every $\epsilon > 0$. In 2009, Goldston et al. [503] achieved this milestone.

Theorem 1.1

$$\liminf_{n \to \infty} \frac{p_{n+1} - p_n}{\log p_n} = 0.$$

Theorem 1.1, in the authors' words, proves that "there exist consecutive primes which are closer than any arbitrarily small multiple of the average spacing". While this result was already a breakthrough, the methods developed for its proof opened the road for further progress. By refining these methods, in 2010 the same authors [504] dramatically improved Theorem 1.1, and proved that there exist pairs of primes nearly within the square root of the average spacing.

© The Author(s), under exclusive license to Springer Nature Switzerland AG 2021
B. Grechuk, *Landscape of 21st Century Mathematics*,
https://doi.org/10.1007/978-3-030-80627-9_1

Theorem 1.2

$$\liminf_{n \to \infty} \frac{p_{n+1} - p_n}{\sqrt{\log p_n} (\log \log p_n)^2} < \infty.$$

Goldston et al. [503, 504] proved Theorems 1.1 and 1.2 by connecting the theory of small gaps between primes to the theory of the distribution of primes in arithmetic progressions. In 1837, Dirichlet proved that if positive integers a, q are relatively prime, then the arithmetic progression $a + qk$, $k \in \mathbb{Z}$, contains infinitely many primes. Intuitively, one would expect that if N is much larger than q, then the prime numbers up to N would be approximately equally distributed in arithmetic progressions $a_1 + qk, a_2 + qk, \ldots, a_{\phi(q)} + qk$, where $0 < a_1 < a_2 < \cdots < a_{\phi(q)} < q$ are all positive integers less than q which are relatively prime with q, and $\phi(q)$ denotes the number of such integers. Formally, let $\mathcal{P}(N, q, a)$ be the set of primes less than N congruent to a modulo q, and let

$$h(N, q) = \max_{1 \leq m \leq \phi(q)} \left| \theta(N, q, a_m) - \frac{N}{\phi(q)} \right|, \quad \text{where } \theta(N, q, a) = \sum_{p \in \mathcal{P}(N, q, a)} \log p.$$

The quantity $h(N, q)$ gives a measure of how the primes up to N deviate from being equally distributed in arithmetic progressions modulo q. In 1965, Bombieri [163] and Vinogradov [1235] proved that the deviations $h(N, q)$ are uniformly small for all $1 \leq q \leq N^{1/2-\epsilon}$, in the sense that, for any $v \leq 1/2$, any $\epsilon > 0$, and any $A > 0$, there is a constant C such that

$$\sum_{q \leq N^{v-\epsilon}} h(N, q) \leq C \frac{N}{(\log N)^A}, \quad \forall N. \tag{1.2}$$

In 1968, Elliott and Halberstam [387] conjectured that (1.2) remains correct for any $v \leq 1$.

Goldston et al. [503, 504] derived Theorems 1.1 and 1.2 from the Bombieri–Vinogradov theorem. Moreover, they proved that if (1.2) is true for $v = 1/2 + \delta$ for any $\delta > 0$, then there is a constant C, depending on δ, such that

$$p_{n+1} - p_n \leq C \tag{1.3}$$

for infinitely many values of n. Because δ can be arbitrarily small, the authors say that they "appear to be within a hair's breadth" of obtaining (1.3). Moreover, they proved that (1.2) with $v \geq 0.971$ would imply (1.3) with $C = 16$. However, improving v beyond $1/2$ in the Bombieri–Vinogradov theorem was a long-standing open question with no progress for over 40 years, and, for this reason, proving (1.3) with any C had been considered a hopeless task by many experts. In was therefore a big surprise when Zhang [1291] confirmed (1.3) in 2014.

Theorem 1.3 *There exists a constant $C_1 < \infty$ such that $p_{n+1} - p_n \leq C_1$ for infinitely many values of n. In fact, one can take $C_1 = 7 \cdot 10^7$.*

To prove Theorem 1.3, Zhang did not prove (1.2) for $v = 1/2 + \delta$. Instead, he proved that a weaker version of (1.2) suffices to establish (1.3), and then proved this weaker version.

The constant $C_1 = 7 \cdot 10^7$ in Theorem 1.3 arises from the theory of admissible k-tuples. By a *k-tuple* we mean a set $H = \{h_1, \ldots, h_k\}$ of k distinct integers, and we may assume that $h_1 < h_2 < \cdots < h_k$. H is called *admissible* if for any prime p there exists an integer n such that none of the integers

$$n + h_1, n + h_2, \ldots, n + h_k \tag{1.4}$$

are divisible by p. The *prime tuples conjecture*, formulated by Hardy and Littlewood [566] in 1923, predicts that for every admissible k-tuple H, all integers (1.4) are simultaneously prime for infinitely many values of n. The *twin prime conjecture* is the special case of this conjecture with $k = 2$ and $H = \{0, 2\}$. Zhang proved that if $k \geq k_0 = 3.5 \cdot 10^6$, then, for any admissible k-tuple H, there are infinitely many values of n such that at least 2 of the integers (1.4) are prime. This implies (1.3) with $C = h_k - h_1$. He then observed that there exists an admissible k_0-tuple with $h_{k_0} - h_1 \leq 7 \cdot 10^7$, and Theorem 1.3 follows.

In 2009, Gowers introduced the idea of *polymath projects*, a way of collaboration where a large group of mathematicians tries to solve a problem collectively by discussing it in a comments section of a blog. When the proof of Theorem 1.3 was announced in 2013, Tao organized a polymath project whose aim was to reduce the constant C_1 in the theorem as much as possible. This could be done by (i) proving Zhang's k-tuple result for a smaller value of k, or (ii) for a given k, finding an admissible k-tuple with a smaller value of $h_k - h_1$. While part (i) required understanding the details of Zhang's proof, part (ii) was computational in nature and required only a minimal background in number theory. This allowed mathematicians with different backgrounds to successfully cooperate, and the value of C_1 was reduced on a weekly, and sometimes on a daily basis. In a few months, the participants of the polymath project proved that Theorem 1.3 is valid with $C_1 = 4680$.

After this, Maynard announced that he, working alone, achieved $C_1 = 600$. Even more importantly, Maynard's theorem [860] proves the existence of infinitely many bounded intervals containing m primes for all m. The same result was proved independently by Tao.

Theorem 1.4 *For any positive integer m, there exists a constant $C_m < \infty$ such that $p_{n+m} - p_n \leq C_m$ for infinitely many values of n. Moreover, one can take $C_1 = 600$, and $C_m = Cm^3 e^{4m}$ for some absolute constant C.*

Maynard and Tao proved Theorem 1.4 by showing that for any m, there is a constant $k = k(m)$ such that for every admissible k-tuple H at least m of the integers (1.4) are prime for infinitely many values of n. Maynard also proved that the Elliott–Halberstam conjecture implies Theorem 1.4 with $C_1 = 12$ and $C_2 = 600$.

Then Maynard joined the polymath project and, with collective efforts, mathematicians [997] proved Theorem 1.4 with $C_1 = 246$ and $C_m = Cm \exp((4 - \frac{28}{157})m)$ for some absolute constant C. Also, the authors of [997] formulated a plausible conjecture (a slightly modified version of the generalized Elliott–Halberstam conjecture formulated by Bombieri et al. [165] in 1986) which would imply Theorem 1.3 with $C_1 = 6$. On the other hand, they showed that $C_1 = 6$ is the limit of the current techniques, and substantially new ideas are required to reach $C_1 = 2$ and prove the twin prime conjecture.

Large Gaps Between Primes

In the previous section we discussed how small the gaps between consecutive primes can be. In this section we will discuss how large they can be.

Euclid famously proved that there are infinitely many primes. Legendre (who lived in 1752–1833) made a stronger conjecture that, for every positive integer k, there is a prime number between k^2 and $(k + 1)^2$. This conjecture remains open. If we denote by

$$G(X) = \sup_{p_{n+1} \leq X} (p_{n+1} - p_n)$$

the largest gap between primes up to X, Legendre's conjecture is essentially equivalent to the upper bound $G(X) \leq \sqrt{X}$. In 1852, Chebyshev, confirming a conjecture of Bertrand, proved that, for every $n > 1$, there exists a prime p such that $n < p < 2n$. This shows that $G(X) < X/2$ for all $X \geq 3$. In 1930, Hoheisel [616] showed the existence of constants X_0 and $\theta < 1$ such that $G(X) < X^\theta$ for all $X \geq X_0$. In fact, Hoheisel showed that one can take $\theta = \frac{32999}{33000}$. Then many authors improved the constant θ. Before 2001, the best result was $\theta = 0.535$, proved by Baker and Harman [82] in 1996. In 2001, Baker et al. [83] improved this to $\theta = 0.525$.

Theorem 1.5 *There exists a constant X_0 such that for all $x > X_0$ the interval $[x - x^{0.525}, x]$ contains at least one prime number.*

A version of Theorem 1.5 with exponent 0.5 instead of 0.525 would prove Legendre's conjecture for all sufficiently large k, but this remains open. To gain some intuition as to why Theorem 1.5 and Legendre's conjecture should be true, consider a simple probabilistic model of the primes. Let us form a random set A of positive integers into which each integer $k \geq 3$ is included with probability $\frac{1}{\log k}$. By the prime number theorem, the integers in the set A are about "as dense" as the primes. In 1936, Cramér [309] computed that the largest gap between elements of A up to X grows proportionally to $(\log X)^2$ as $X \to \infty$ with probability 1, and, based on this, conjectured that the same is true for $G(X)$. Because $\sqrt{X}$ grows much faster than $(\log X)^2$, Legendre's conjecture is expected to be true with a large margin.

What about the lower bounds for $G(X)$? The prime number theorem implies that $G(X)$ grows at least as $\log(X)$. In 1931, Westzynthius [1253] proved that $G(X) \geq c \frac{\log X \log_3 X}{\log_4 X}$ for some constant $c > 0$, where $\log_r(X)$ denotes the r-th fold logarithm.

This was improved to $G(X) \geq c \frac{\log X \log_2 X}{(\log_3 X)^2}$ by Erdős [394] in 1935, and to

$$G(X) \geq cf(X), \quad \text{where} \quad f(X) = \frac{\log X \log_2 X \log_4 X}{(\log_3 X)^2},$$

by Rankin [1025] in 1938. This strange-looking bound, surprisingly, stood for almost 80 years, except for improvements in the value of the constant c. Only in 2016, Ford et al. [460] proved that $G(X)$ grows faster than $f(X)$ by more than a constant factor.

Theorem 1.6 $\lim\limits_{X \to \infty} \frac{G(X)}{f(X)} = \infty$.

Theorem 1.6 was independently proved by Maynard [861]. Then the authors of [460] and [861] joined their efforts and in 2018 improved the bound further [459]:

Theorem 1.7   *There exists a constant $c > 0$ such that*

$$G(X) \geq c \frac{\log X \log_2 X \log_4 X}{\log_3 X}.$$

Theorems 1.6 and 1.7 were proved by constructing a sequence of consecutive integers in $(X/2, X]$, each of which has a "very small" prime factor. More formally, let $j(n)$ be the maximal gap between consecutive integers coprime to n, and let $J(X) = \max_{n \leq x} j(n)$. Then it is easy to see that $G(X) \geq J(X)$ for all $X \geq 7$. Theorems 1.6 and 1.7 were proved by obtaining the corresponding lower bounds for $J(X)$. However, this method has a serious limitation, because $J(X)$ is conjectured to grow much more slowly than $G(X)$. Specifically, Maier and Pomerance [830] conjectured in 1990 that $J(X)$ grows as $\log X (\log \log X)^{2+o(1)}$, and this conjecture is well-believed. If it is true, then the lower bound in the form $G(X) \geq \log X (\log \log X)^{2+o(1)}$ is the absolute limit of the current techniques. This bound is very far from matching Cramer's prediction that $G(X) \approx (\log X)^2$. In fact, the authors of [459] argue that even improving the lower bound in Theorem 1.7 should require a substantially new idea.

1.2   Prime Values of Polynomials

Prime Values of One Polynomial
Many famous open questions about prime numbers, including the *twin prime conjecture* (are p and $p + 2$ both primes infinitely often?), the *Sophie Germain conjecture* (are p and $2p + 1$ both primes infinitely often?), and *Landau's 4th problem* (is $x^2 + 1$ a prime for infinitely many integers x?) are special cases of the following general question.

Question 1.8 *Let $P_i(x_1, \ldots, x_n)$, $i = 1, \ldots, k$, be k polynomials in n variables with integer coefficients. Under what conditions on P_i does there exist infinitely many integer values of $x_1, \ldots, x_n$ such that all $P_i(x_1, \ldots, x_n)$, $i = 1, \ldots, k$, are simultaneously primes?*

Question 1.8 is highly non-trivial even for $k = 1$ polynomial in $n = 1$ variable. In 1837, Dirichlet proved that $ax + b$ is prime for infinitely many integers x if and only if the integers a, b are relatively prime. More generally, Bunyakovsky conjectured in 1857 that a polynomial $P(x)$ takes prime values infinitely often if and only if

(i) $P(x)$ is irreducible over the integers,
(ii) $P(x)$ is positive for infinitely many values of x, and
(iii) the values $P(1), P(2), P(3), \ldots$ have no common factor.

Dirichlet's theorem proves this conjecture for linear polynomials, but it is open for every single polynomial of degree at least 2.

Question 1.8 becomes easier for polynomials in several variables. Fermat stated and Euler proved that every prime p of the form $4k + 1$ can be written as $x^2 + y^2$ for integers x, y. By Dirichlet's theorem, there are infinitely many such primes. Euler also proved that there are infinitely many primes of the form $x^2 + 2y^2$ and $x^2 + 3y^2$. Later, Dirichlet solved the $k = 1$ case of Question 1.8 for any polynomial of the form $P(x, y) = ax^2 + bxy + cy^2$, and Iwaniec [643] extended this result to all quadratic polynomials in two variables which depend essentially on both variables.

For polynomials of higher degree, progress was limited for a long time. For example, Hardy and Littlewood [566] asked in 1923 whether $x^3 + y^3 + z^3$ is a prime for infinitely many triples of non-negative integers (x, y, z), and this remained open for almost 80 years. The first breakthrough for higher-degree polynomials happened in 1998, when Friedlander and Iwaniec [476] proved that there are infinitely many primes of the form $x^2 + y^4$ with integer x, y. In 2001, Heath-Brown [584] resolved this question for the polynomial $P(x, y) = x^3 + 2y^3$.

Theorem 1.9 *There are infinitely many primes of the form $x^3 + 2y^3$ with integer x, y.*

In particular, Theorem 1.9 gives a positive answer to the Hardy–Littlewood question in the strong form: $x^3 + y^3 + z^3$ is a prime infinitely often even if we require that $y = z$.

Note that, among the positive integers up to N, about $N^{3/4}$ of them are representable as $x^2 + y^4$, while only about $N^{2/3}$ are representable as $x^3 + 2y^3$. In this sense, Theorem 1.9 is harder than the Friedlander–Iwaniec theorem, and in fact the hardest result in this direction we currently have. Landau's 4th problem is even harder, because there are only $N^{1/2}$ positive integers up to N of the form $x^2 + 1$.

Instead of considering polynomials one by one, in 2020 Maynard [863] developed a method for resolving this problem for many infinite families of polynomials. The following theorem is one of the corollaries of his method.

Theorem 1.10 *For any integer a that is not a perfect square, there are infinitely many prime numbers of the form* $x^4 - ay^4 + 4axy^2z - 2ax^2z^2 + a^2z^4$ *with integer* x, y, z.

The proofs of Theorems 1.9 and 1.10 crucially rely on the theory of number fields and their field norms. Recall that a *number field* is a field F that contains $\mathbb{Q}$ and has finite dimension n when considered as a vector space over $\mathbb{Q}$. Then the operation of multiplication on any $\alpha \in F$ can be considered as a linear transformation of this vector space which is defined by some $n \times n$ matrix. The *field norm* $N_{F/\mathbb{Q}}(\alpha)$ is defined as the determinant of this matrix.

For example, let $d > 0$ be any square-free[1] integer and let us add $\sqrt{d}$ to the field $\mathbb{Q}$ of rational numbers. Then we must also add all numbers of the form $a + b\sqrt{d}$, $a, b \in \mathbb{Q}$, to ensure that the operations of addition and multiplication are well-defined. The resulting set forms a field F, and each element $a + b\sqrt{d} \in F$ can be associated to the vector $(a, b) \in \mathbb{Q}^2$. Hence, F is a number field of dimension $n = 2$, which is usually denoted by $\mathbb{Q}(\sqrt{d})$. Now, multiplication by $\alpha = \sqrt{d}$ sends each $a + b\sqrt{d} \in F$ to $(a + b\sqrt{d})\sqrt{d} = bd + a\sqrt{d}$. In terms of vectors, (a, b) is transformed into (bd, a). This can be written as a linear transformation

$$\begin{bmatrix} 0 & d \\ 1 & 0 \end{bmatrix} \cdot \begin{bmatrix} a \\ b \end{bmatrix} = \begin{bmatrix} bd \\ a \end{bmatrix}$$

and the determinant of the matrix $\begin{bmatrix} 0 & d \\ 1 & 0 \end{bmatrix}$ is $-d$. Hence, the field norm $N_{\mathbb{Q}(\sqrt{a})/\mathbb{Q}}(\sqrt{d}) = -d$.

Let us now adjoin $\sqrt[3]{2}$ to $\mathbb{Q}$. The resulting number field is denoted $\mathbb{Q}(\sqrt[3]{2})$, its elements have the form $a + b\sqrt[3]{2} + c\sqrt[3]{4}$ for some $a, b, c \in \mathbb{Q}$, and each such element can be associated to the vector $(a, b, c) \in \mathbb{Q}^3$. Let us consider the operation of multiplication by the element $x + y\sqrt[3]{2}$, where x and y are variables. We have

$$(x + y\sqrt[3]{2})(a + b\sqrt[3]{2} + c\sqrt[3]{4}) = (xa + 2yc) + (ya + xb)\sqrt[3]{2} + (yb + xc)\sqrt[3]{4}.$$

This corresponds to the linear transformation

$$\begin{bmatrix} x & 0 & 2y \\ y & x & 0 \\ 0 & y & x \end{bmatrix} \cdot \begin{bmatrix} a \\ b \\ c \end{bmatrix} = \begin{bmatrix} xa + 2yc \\ ya + xb \\ yb + xc \end{bmatrix},$$

and the corresponding matrix has determinant $x^3 + 2y^3$. Hence, $N_{\mathbb{Q}(\sqrt[3]{2})/\mathbb{Q}}(x + y\sqrt[3]{2}) = x^3 + 2y^3$, which is exactly the polynomial in Theorem 1.9. This fact is used crucially in the proof of Theorem 1.9, because there are established methods

---

[1] An integer is called square-free if none of its divisors is a square of an integer greater than 1.

for proving that field norms of linear expressions $N_{F/\mathbb{Q}}(\beta_1 x_1 + \cdots + \beta_n x_n)$ represent infinitely many primes.

Maynard [863] developed a powerful technique which allows one to set $k \le n/4$ variables in $P(x_1, \ldots, x_n) = N_{F/\mathbb{Q}}(\beta_1 x_1 + \cdots + \beta_n x_n)$ equal to 0 without affecting the conclusion that infinitely many primes can be represented. For example, let a be an integer which is not a perfect square, $\omega = \sqrt[4]{a}$, and $K = \mathbb{Q}(w)$ be the number field resulting from adjoining w to $\mathbb{Q}$. If we then compute the polynomial

$$P(x, y, z, t) = N_{K/\mathbb{Q}}(x + yw + zw^2 + tw^3),$$

and set $t = 0$, we get $P(x, y, z, 0) = x^4 - ay^4 + 4axy^2z - 2ax^2z^2 + a^2z^4$, and Theorem 1.10 follows.

Prime Values of Linear Polynomials, and Arithmetic Progressions
Dirichlet's theorem resolves Question 1.8 for one linear polynomial in one variable. What if we consider several linear polynomials? A nice special case is provided by the polynomials $P_i(x, y) = x + (i - 1) \cdot y, i = 1, \ldots, k$, with the additional requirement $y \neq 0$. In this case Question 1.8 reduces to the question of whether prime numbers contain infinitely many k-term arithmetic progressions.

In 1939, Van der Corput [304] proved that the set $\mathcal{P}$ of primes contains infinitely many 3-term arithmetic progressions (3APs for short). In 1953, Roth [1061] proved that any subset of the integers of positive upper density contains a 3AP. In 2005, Ben Green [522] proved a common generalization of these two results.

Theorem 1.11 *Every subset of primes of positive upper density[2] contains a 3-term arithmetic progression.*

In 1975, Szemerédi [1174] generalized Roth's theorem to longer arithmetic progressions and proved that every subset of integers of positive upper density contains infinitely many k-term arithmetic progressions for all k. However, this result is not directly applicable to prime numbers because their density is 0. For almost 70 years, no-one could prove the analogue of Van der Corput's theorem even for arithmetic progressions of length $k = 4$. In 2008, Green and Tao [523] resolved this question for all k.

Theorem 1.12 *For every positive integer k, there exist an arithmetic progression of length k consisting only of prime numbers. Moreover, every subset of primes of positive upper density contains infinitely many arithmetic progressions of length k for all k.*

To prove Theorem 1.12, Green and Tao first proved a version of Szemerédi's theorem for *pseudorandom sets*, that is, sets which resemble random sets of similar density in terms of certain statistical properties. This version is now known as

---

[2] A subset $A \subset \mathcal{P}$ of primes has *positive upper density* if $\limsup\limits_{n \to \infty} \frac{|A \cap \mathcal{P}_n|}{|\mathcal{P}_n|} > 0$, where $\mathcal{P}_n$ is the set of primes less than n.

the *relative Szemerédi theorem*. Informally, it states that if S is a suitably defined pseudorandom set of integers and A is a subset of S with positive relative density in S, then A contains arbitrarily long arithmetic progressions. As a final step, Green and Tao constructed a set S^* which satisfies their pseudorandomness conditions such that a positive fraction of elements of S^* are prime numbers.

One may also ask *how many* arithmetic progressions of length k are contained in the set of prime numbers up to N. The proof of Theorem 1.12 implies that there are at least $(1 + o(1))c_k \frac{N^2}{\log^k N}$ of them, where $c_k > 0$ are some small constants. In 2010, Green and Tao [524] provided the exact asymptotic count of the 4-term progressions.

Theorem 1.13 *The number of 4-tuples of primes $p_1 < p_2 < p_3 < p_4 \le N$ which lie in arithmetic progression is*

$$(1 + o(1))C \frac{N^2}{\log^4 N}, \quad \text{with} \quad C = \frac{3}{4} \prod_{p \ge 5} \left(1 - \frac{3p - 1}{(p - 1)^3}\right) \approx 0.4764,$$

where the product is over all primes $p \ge 5$.

Theorem 1.12 is a great breakthrough, but it resolves only a very special case of Question 1.8. More generally, let us consider a sequence $P = \{P_1, \ldots, P_k\}$ of linear polynomials of the form

$$P_i(x_1, \ldots, x_n) = c_{i1}x_1 + \cdots + c_{in}x_n + b_i, \quad i = 1, 2, \ldots, k, \tag{1.5}$$

where c_{ij} and b_i are integer coefficients. Let $g_P(N)$ be the number of choices of $x_1, \ldots, x_n$ in the range $-N \le x_i \le N, i = 1, \ldots, n$, such that $P_j, j = 1 \ldots, k$, are all primes. Similarly, let $f_P(N)$ be the number of choices of $x_1, \ldots, x_n$ in the same range such that P_j are all positive. Let

$$C_\infty(P) = \lim_{N \to \infty} \frac{f_P(N)}{(2N + 1)^n}.$$

The sequence P is called *admissible* if for every fixed prime p there exist integer values of $x_1, \ldots, x_n$ such that all $P_i(x_1, \ldots, x_n), i = 1, \ldots, k$, are *not* divisible by p. Also, we say that polynomials in P are *pairwise affine-independent* if there are no P_i, P_j in P such that $P_i(\cdot) = a P_j(\cdot) + b$ for some constants a, b.

In 2012, Green et al.[526] proved the following result.

Theorem 1.14 *Let $P = \{P_1, \ldots, P_k\}$ be a sequence of non-constant pairwise affine-independent linear polynomials in the form (1.5). Assume that P is admissible and $C_\infty(P) > 0$. Then there are infinitely many values of $x_1, \ldots, x_n$ such that all*

$P_i(x_1, \ldots, x_n)$, $i = 1, \ldots, k$, are simultaneously prime. Moreover

$$g_P(N) = (C_P + o(1)) \frac{N^n}{\log^k N},$$

where $C_P > 0$ is a positive constant depending on P.

The conditions that P is admissible and $C_\infty(P) > 0$ are necessary for the conclusion. Hence, Theorem 1.14 resolves Question 1.8 for all linear pairwise affine-independent polynomials. The additional condition of pairwise affine-independence excludes the hardest cases corresponding to the twin prime and Sophie Germain conjectures. Theorem 1.14 also counts how often all P_i are simultaneously primes. For example, applying it to the arithmetic progressions, the authors derived that, for any fixed integer $k \geq 2$, the number of k-tuples of primes $p_1 < p_2 < \cdots < p_k \leq N$ which lie in an arithmetic progression is

$$(C_k + o(1)) \frac{N^2}{\log^k N}, \quad \text{with} \quad C_k = \frac{1}{2(k-1)} \prod_p \beta_p,$$

where $\beta_p = \frac{1}{p} \left(\frac{p}{p-1} \right)^{k-1}$ if $p \leq k$ and $\beta_p = \left(1 - \frac{k-1}{p} \right) \left(\frac{p}{p-1} \right)^{k-1}$ if $p \geq k$. Theorem 1.13 is the special case of this statement with $k = 4$.

The proof of Theorem 1.14 has an interesting history. First, Green and Tao [524] presented a conditional proof in 2010, assuming two unproven conjectures, called the *Möbius and nilsequences conjecture* (MN) and the *inverse Gowers-norm conjecture* (GI). These conjectures are too technical to state here formally, but we will give a very informal introduction. Given a bounded sequence $\{a_n\}_{n=1}^\infty$ of complex numbers, how can we "test" if $\{a_n\}$ "looks like a random sequence"? For example, we expect that, if $\{a_n\}$ is a random sequence, then the sub-sequence $a_1, a_3, a_5, \ldots$ of its odd terms is "similar" to the sub-sequence $a_2, a_4, a_6, \ldots$ of its even terms. To formalize this, let us define the sequence $\{b_n\}$ as $b_n = 1$ for odd n and $b_n = -1$ for even n, and postulate that if $\{a_n\}_{n=1}^\infty$ is "random" then the average

$$\frac{1}{N} \sum_{n=1}^N a_n b_n$$

should converge reasonably quickly to 0 as $N \to \infty$. If this is the case, we say that the sequences $\{a_n\}$ and $\{b_n\}$ are *asymptotically orthogonal*. Of course, the odd-even test was just an example, and we can "test the randomness" of $\{a_n\}$ by checking its orthogonality to different "test sequences" $\{b_n\}$. That is, we can fix some set B of test sequences and call the sequence $\{a_n\}$ *B-pseudorandom* if it is asymptotically orthogonal to all sequences in B.

The MN conjecture predicted that the *Möbius function* $\mu(n)$, defined as $(-1)^k$ when n is the product of k distinct primes and 0 otherwise, is B-pseudorandom if B is a certain set of sequences called "polynomial nilsequences". The GI conjecture gives an explicit criterion for a bounded sequence of complex numbers $\{a_n\}$ to be B-pseudorandom with the same choice of B.

In 2010, Green and Tao [524] proved Theorem 1.14 assuming the MN and GI conjectures. From the known special cases of these conjectures, they unconditionally deduced some special cases of Theorem 1.14, including Theorem 1.13. In 2012, the same authors first proved the MN conjecture [525], and then, together with Ziegler, proved the GI conjecture in [526], thus finishing the unconditional proof of Theorem 1.14.

The methods developed to prove Theorem 1.14 are also applicable to linear polynomials with parameters, such as $P_1 = x_1$, $P_2 = x_2$, and $P_3 = n - x_1 - x_2$, where n is a parameter. In this case, P_1, P_2, P_3 can be all primes if and only if n can be represented as sum of three primes. Theorem 1.14 counts the number of such representations asymptotically if n is odd and goes to infinity.

In 1742, Goldbach conjectured that every odd integer $n \geq 7$ is the sum of three primes, while every even integer $n \geq 4$ is the sum of two primes. Goldbach's conjecture became one of the best-known unsolved problems in number theory and in all of mathematics. In 1937, Vinogradov [1236] proved the conjecture for all odd integers greater than some (large) constant n_0. In 1956, Borozdkin [173] proved that one can take $n_0 = e^{e^{16038}}$. In 2002, Liu and Wang [808] improved this to $n_0 = e^{3100}$. Finally, Helfgott [593] proved the conjecture for all odd n. He did this by improving the Liu–Wang bound on n_0 from $n_0 = e^{3100}$ to $n_0 = 10^{27}$, and verifying (together with Platt) the conjecture for smaller values of n by direct computation [594].

Theorem 1.15 *Every odd integer $n \geq 7$ is the sum of three primes.*

Note that Theorem 1.14 is asymptotic in nature, it has no control on what happens for small n, hence it implies Vinogradov's theorem but not Theorem 1.15. Also, Theorem 1.14 is not applicable to Goldbach's conjecture for even n, because the polynomials $P_1 = x$, $P_2 = n - x$ are not affine-independent. For even n, Goldbach's conjecture remains wide open, even for large n.

Prime Values in Polynomial Progressions

Let $P_1(m), \ldots, P_k(m)$ be a sequence of polynomials in one variable m with integer coefficients. For any integers x and m, the sequence $x + P_1(m), \ldots, x + P_k(m)$ is called a *polynomial progression*. Note that an arithmetic progression is a special case of a polynomial progression with $P_j(m) = (j - 1)m$, $j = 1, \ldots, k$.

In 1996, Bergelson and Leibman [134] proved that if $P_1(0) = \cdots = P_k(0) = 0$, then any set A of integers with positive upper density contains infinitely many progressions $x + P_1(m), \ldots, x + P_k(m)$ with $m > 0$. This is a generalization of Szemerédi's theorem to polynomial progressions.

In 2008, Tao and Ziegler [1190] proved the corresponding generalization of Theorem 1.12.

Theorem 1.16 *Given any polynomials $P_1(m), \ldots, P_k(m)$ with integer coefficients such that $P_1(0) = \cdots = P_k(0) = 0$, there are infinitely many pairs of positive integers x and m such that the numbers $x + P_1(m), \ldots, x + P_k(m)$ are simultaneously prime.*

Moreover, the authors proved that for any $\epsilon > 0$, there are infinitely many pairs (x, m) as in Theorem 1.16 with $1 \leq m \leq x^\epsilon$. In addition, they proved that even any positive proportion of primes contains infinitely many polynomial progressions.

The proof of Theorem 1.16 follows from the Bergelson–Leibman theorem (BLT for short) in roughly the same way as Theorem 1.12 is deduced from the Szemerédi theorem. More precisely, the proof of Theorem 1.16 relies on a certain multidimensional generalization of the BLT, also established in [134]. In 2018, Tao and Ziegler [1191] used this generalization of the BLT to prove a version of Theorem 1.16 for multivariate polynomials. Let $\mathbb{Z}[m_1, \ldots, m_r]$ denote the set of polynomials in r variables with integer coefficients.

Theorem 1.17 *Let d, r be natural numbers and $P_1, \ldots, P_k \in \mathbb{Z}[m_1, \ldots, m_r]$ be polynomials of degree at most d such that $P_i - P_j$ has degree exactly d for all $1 \leq i < j \leq k$. Suppose that for each prime p there exist $n \in \mathbb{Z}$ and $\mathbf{m} \in \mathbb{Z}^r$ such that none of $n + P_1(\mathbf{m}), \ldots, n + P_k(\mathbf{m})$ are divisible by p. Then there exist infinitely many natural numbers $n, m_1, \ldots, m_r$ such that $n + P_1(m_1, \ldots, m_r), \ldots, n + P_k(m_1, \ldots, m_r)$ are simultaneously prime.*

Theorem 1.17 not only generalizes Theorem 1.16 to the multivariate case, but also removes the condition $P_1(0) = \cdots = P_k(0) = 0$ in it. However, it introduces a new condition that "$P_i - P_j$ has degree exactly d". Theorem 1.17 solves Question 1.8 for polynomials of the form $x_1 + P(x_2, \ldots, x_n)$ under this additional condition. Proving any result of this kind without any additional condition is difficult, because the case $k = 2$ with $P_1 - P_2 = 2$ in Theorem 1.17 would imply the twin prime conjecture.

Square-Free Values of Polynomials
Many variants of Question 1.8 can be studied. For example, one can relax the requirement for values of polynomials from being primes to just being square-free. For one polynomial f in one variable, this problem is easy for linear polynomials, and it was solved for quadratic polynomials by Estermann [433] in 1931, and for cubic polynomials by Erdős [400] in 1953. Erdős further posed the question: for which cubic polynomials f is $f(p)$ square-free for infinitely many *primes* p? In 2014, Helfgott [596] answered this question.

For an integer $m > 0$, let $\mathcal{P}(m)$ be the set of integers $0 < k < m$ such that $\gcd(k, m) = 1$. For a function $f : \mathbb{Z} \to \mathbb{Z}$, let $\mathcal{P}_f(m)$ be the set of $k \in \mathcal{P}(m)$ such that $f(k)$ is divisible by m. Let

$$c_f = \prod_{p \in \mathcal{P}} \left(1 - \frac{|\mathcal{P}_f(p^2)|}{|\mathcal{P}(p^2)|} \right),$$

where $\mathcal{P}$ is the set of all primes.

Theorem 1.18 *Let f be a cubic polynomial with integer coefficients without repeated roots. Then the number of prime numbers $p \leq N$ such that $f(p)$ is square-free is*

$$(1 + o(1))c_f \frac{N}{\log N} + O(1).$$

In particular, Theorem 1.18 implies that $f(p)$ is square-free for infinitely many primes p if and only if $c_f > 0$. It is not difficult to check that $c_f > 0$ if and only if

(*) for every prime q there exists an integer x coprime to q such that $f(x)$ is not divisible by q^2.

Thus, Theorem 1.18 implies that the obvious necessary condition (*) is also sufficient for $f(p)$ to be square-free infinitely often. In fact, it is known that for every f there is a constant $B = B(f)$ such that if (*) holds for all $q < B$ then it holds for all q. Hence, (*) can be checked in finite time for any given f.

The proof of Theorem 1.18 proceeds as follows. If $f(p)$ is not square-free, then it is divisible by q^2 for some prime q. We next proceed to counting, for every q, the number of primes $p \leq N$ such that $f(p)$ is divisible by q^2. For small q, this problem reduces to counting primes in arithmetic progressions to small moduli and we can apply the Bombieri–Vinogradov theorem (1.2). For medium q, it can be shown that the number of such p is relatively small. For large q, we need to count prime solutions $p, q \leq N$ to the equation $f(p) = mq^2$, where m is small compared to N. For each individual m, the equation $f(x) = my^2$ defines an elliptic curve, and one can use the rich theory of such curves to count the prime points on it. In particular, the proof crucially uses the modularity theorem, see Theorem 1.64 below.

1.3 Primes in Sets of Integers

Digits of Prime Numbers

Theorem 1.9 states that there are infinitely many primes in the set A of integers which can be represented in the form $x^3 + 2y^3$ with integer x, y. One may ask the same question for different sets A of integers, not necessary defined by polynomials. For example, are there infinitely many primes in the set of integers whose sum of decimal digits is even?

More generally, let $S_q(n)$ be the sum of digits of the integer n in the q-ary number system. Let $\pi(n)$ be the number of primes less than n, and let $\pi_{q,m,a}(n)$ be the number of primes p less than n such that $S_q(p) - a$ is a multiple of m. In 1968, Gelfond [494] asked whether $\pi_{q,m,a}(n) \approx \frac{\pi(n)}{m}$. In 2010, Mauduit and Rivat [859] answered this question affirmatively.

Theorem 1.19 *For all integers $m \geq 2$ and $q \geq 2$ such that $q - 1$ and m are relatively prime, there exist constants $C_{q,m} > 0$ and $\sigma_{q,m} > 0$ such that the inequality $\left| \pi_{q,m,a}(n) - \frac{\pi(n)}{m} \right| \leq C_{q,m} n^{1-\sigma_{q,m}}$ holds for all integers $n > 0$ and all integers a.*

In particular, the $q = 10$ and $m = 2$ case of Theorem 1.19 implies that (asymptotically) half the prime numbers have an even sum of decimal digits.

Alternatively, one may ask whether there are infinitely many primes in a set of integers whose decimal expansion does not have a specific digit d. Note that, among the integers up to N, about $N/2$ of them have an even sum of digits, but only about $N^{9/10}$ of them miss a given digit. Hence, this question looks much more difficult than the previous one. Nevertheless, Maynard [862] gave a complete answer in 2019.

Theorem 1.20 *Given any digit $d \in \{0, 1, \ldots, 9\}$, there are exist infinitely many primes p which do not have the digit d in their decimal expansion.*

Theorem 1.20 is a rare example of proved statement that there are infinitely many primes in a sparse set with no clear algebraic structure. Moreover, Maynard proved that the number $P_d(X)$ of primes $p \leq X$ without the digit d is in the range

$$C_1 \frac{X^{\log 9/\log 10}}{\log X} \leq P_d(X) \leq C_2 \frac{X^{\log 9/\log 10}}{\log X},$$

where $0 < C_1 < C_2 < \infty$ are some constants. One may also consider the more general problem of excluding s digits in the base q expansion. In this more general setting, Maynard [862] proved the analogue of Theorem 1.20 provided that q is sufficiently large and $s \leq q^{23/80}$. Under these conditions, the number of primes up to X with s excluded digits grows as $\frac{X^{\log(q-s)/\log q}}{\log X}$, again up to constant factors.

We now say a few words about the proof of Theorem 1.20. Fix a digit d, and let $\mathcal{A}$ be the set of positive integers less than X with no digit in their decimal expansion equal to d. Then we need to count the number $P(\mathcal{A})$ of primes in $\mathcal{A}$, or, in other words, the number of solutions of the equation $p - a = 0$ over primes p and integers $a \in \mathcal{A}$. It is known that this number can be written as

$$P(\mathcal{A}) = \frac{1}{X} \sum_{0 < a < X} S_{\mathcal{A}}\left(\frac{a}{X}\right) S_P\left(\frac{-a}{X}\right),$$

where

$$S_{\mathcal{A}}(\theta) = \sum_{a \in \mathcal{A}} \exp(2\pi i a\theta) \quad \text{and} \quad S_P(\theta) = \sum_{p < X} \exp(2\pi i p\theta).$$

The function $S_{\mathcal{A}}(\theta)$ is called the *Fourier transform* of the set $\mathcal{A}$. It turns out that this specific set $\mathcal{A}$ has a Fourier transform with a particularly nice structure, which allows us to prove a strong upper bound $S_{\mathcal{A}}(a/X) \leq X^{0.32}$ for "generic" a. This can then be used to bound $P(\mathcal{A})$. This method is known as the Hardy–Littlewood circle method.

A Surprising Number of Primes in Certain Intervals

Results like Theorem 1.19 may be difficult to prove but are hardly surprising: half of the integers have an even sum of digits, and so half of the primes. In particular, if $\mathcal{R}$ is a random set of integers with the same density as the primes, then, with probability 1, (asymptotically) half of the integers in $\mathcal{R}$ has an even sum of digits. Also, there will be infinitely many pairs in $\mathcal{R}$ with bounded gaps between them (as in Theorem 1.3), infinitely many integers in $\mathcal{R}$ in the form $x^3 + 2y^3$ (as in Theorem 1.9), infinitely many arbitrarily long arithmetic progressions in $\mathcal{R}$ (as in Theorem 1.12), and so on. Of course, there will be some properties of $\mathcal{R}$ not shared by primes (for example, there will be infinitely many even integers in $\mathcal{R}$), but the general impression one gets is that if a natural property is not trivially false for prime numbers and is true for $\mathcal{R}$ then it must be true for primes. Thus, even without proof, we can guess which properties are true for primes just by checking them for $\mathcal{R}$.

In 1985, Maier [831] gave a famous and sensational example demonstrating that this intuition may be fundamentally wrong. He proved that the expression

$$\frac{\pi(x + (\log x)^\lambda) - \pi(x)}{(\log x)^{\lambda - 1}},$$

where π is the prime-counting function, does not have a limit as x tends to infinity: the limit superior of this expression is greater than 1, and the limit inferior is less than 1. In contrast, the corresponding expression for $\mathcal{R}$ tends to 1 as $x \to \infty$ with probability 1 for every $\lambda \geq 2$. In other words, Maier proved that there are values of x for which the interval $(x, x + (\log x)^\lambda)$ contains significantly more (or significantly less) primes than predicted by a random model. This shows that other random model predictions (such as the twin prime conjecture) may be wrong as well, and we cannot be confident in such predictions until we get a rigorous proof.

Maier proved his theorem by constructing a matrix of integers such that its rows are intervals of consecutive integers, and its columns are arithmetic progressions with carefully selected moduli. Then known results on the number of primes in arithmetic progressions are used to estimate the number of primes in the columns, and hence in the matrix. Maier demonstrated how to construct such a matrix with a large/small number of primes. He then concluded that there must exist a row with a large/small number of primes. This method became known as *Maier's matrix method*.

By combining Maier's matrix method with more modern tools from the theory of multiplicative functions and integral equations, Granville and Soundararajan [517] proved in 2007 a much stronger result of this type, which they call "perhaps the best possible result in this context". As usual, let $\mathcal{P}$ be the set of primes. For $x > 0$, let

$\mathcal{P}(x) = \{p \in \mathcal{P} : p < x\}$ be the set of primes less than x, and let

$$\theta(x) = \sum_{p \in \mathcal{P}(x)} \log p, \quad \text{and} \quad \Delta(x, y) = \frac{\theta(x + y) - \theta(x) - y}{y}.$$

The function $\theta(x)$ counts primes up to x, each prime p with weight $\log p$. The prime number theorem states that $\theta(x) \approx x$ for large x, hence the number of weighted primes in the interval $(x, x + y]$ is about $\theta(x + y) - \theta(x) \approx y$. The function $\Delta(x, y)$ measures the quality of this approximation. Granville and Soundararajan [517] proved that there are intervals with much more and much less primes than average. In the authors' words, primes are not very well distributed.

Theorem 1.21 *Let x be large and y be such that $\log x \leq y \leq \exp\left(\frac{\beta\sqrt{\log x}}{2\sqrt{\log\log x}}\right)$, where $\beta > 0$ is an absolute constant. Then there exist numbers x_+ and x_- in $(x, 2x)$ such that $\Delta(x_+, y) \geq y^{-\delta(x,y)}$ and $\Delta(x_-, y) \leq -y^{-\delta(x,y)}$, where $\delta(x, y) = \frac{1}{\log\log x}\left(\log\left(\frac{\log y}{\log\log x}\right) + \log\log\left(\frac{\log y}{\log\log x}\right) + O(1)\right).$*

Similar to Maier's theorem, Theorem 1.21 would be wrong with probability 1 for a sequence of randomly selected numbers with the same density as the primes. Hence, one should be very careful when making conjectures about prime numbers based on probabilistic arguments. In fact, the authors proved the much more general (and surprising) result that a version of Theorem 1.21 holds for any "arithmetic sequence" in place of primes, see [517] for details.

1.4   Finding Integer Solutions to Equations and Systems

A Question of Solution Existence

Many famous problems in number theory can be formulated in terms of finding integer solutions to some equation or system of equations. The first question one may ask is whether any integer solutions exist at all. Or, if an equation has some known easy-to-find solutions, one may ask if there are any other solutions.

For example, Catalan [255] conjectured in 1844 that $8 = 2^3$ and $9 = 3^2$ are the only consecutive positive integers which are perfect powers. This is equivalent to the statement that $3^2 - 2^3 = 1$ is the only positive integer solution to the equation

$$x^a - y^b = 1 \tag{1.6}$$

such that $x, y > 0$, $a, b > 1$. If x, y, a, b is a solution, and p, q are any prime factors of a, b, respectively, then

$$(x^{a/p})^p - (y^{b/q})^q = 1,$$

hence it suffices to consider the case when a and b in (1.6) are primes. In 1850, Lebesgue [774] proved that there are no positive integers x, y and $a > 1$ such that $x^a = y^2 + 1$, thus solving the case $b = 2$ of (1.6). In 1960, Chao [263] resolved the case $a = 2$. This reduces the problem to the case when a, b are both odd primes.

In 1976, Tijdeman [1203] proved that if (1.6) has a solution in positive integers (other than $3^2 - 2^3 = 1$), then it must have one with $a, b < C$, where C is some absolute constant. Later the same year, Langevin [762] proved that one can take $C = 10^{110}$. Because it is known that for any fixed a, b Eq. (1.6) can be solved in finite time, this reduces the Catalan's conjecture to checking a finite but unfeasibly large number of cases.

In 2003, Mihăilescu [890] proved that if a, b are prime solutions to (1.6), then $a^{b-1} \equiv 1 \pmod{b^2}$ and $b^{a-1} \equiv 1 \pmod{a^2}$. Any pair of prime numbers satisfying these conditions is called a *double Wieferich prime pair*. This necessary condition greatly narrowed the search for possible exponents, and played a crucial role in the full resolution of Catalan's conjecture, which was done by Mihăilescu [889] in 2004.

Theorem 1.22 *The integers* $8 = 2^3$ *and* $9 = 3^2$ *are the only consecutive positive integers which are perfect powers.*

An example of a question about the existence of integer solutions to a system of equations is a famous problem about Diophantine quintuples. More than 2000 years ago, Diophantus noticed that the set of rational numbers $\{\frac{1}{16}, \frac{33}{16}, \frac{17}{4}, \frac{105}{16}\}$ has the property that the product of any two of them plus one is the square of a rational number. Later, Fermat found positive integers $\{1, 3, 8, 120\}$ with this property.

In general, a set of m distinct positive integers $\{a_1, a_2, \ldots, a_m\}$ is called a *Diophantine m-tuple* if $a_i a_j + 1$ is a perfect square for all $1 \leq i < j \leq m$. In particular, any pair of positive integers a, b such that $ab + 1 = r^2$ is a perfect square is called a *Diophantine pair*. Euler noticed that, to every such pair, we can add the third number $c = a + b + 2\sqrt{ab + 1} = a + b + 2r$ and form a *Diophantine triple* which is called an *Euler triple*. Moreover, we can also add a fourth number $d = 4r(a + r)(b + r)$ to form a *Diophantine quadruple*. It follows that there are infinitely many Diophantine quadruples.

A long-standing folklore conjecture predicts that no five positive integers with this property (a *Diophantine quintuple*) exist. In other words, the conjecture predicted that the system of equations $a_i a_j + 1 = (b_{ij})^2$, $1 \leq i < j \leq 5$, has no solutions in positive integers. As partial progress, Dujella [370] proved in 2004 that there is no Diophantine sextuple and that there can be at most finitely many Diophantine quintuples. In 2019, He et al.[581] proved the conjecture in full. The key idea of the proof is to introduce a special "∂-operator" which is defined for non-Euler Diophantine triples and has the property that if we apply it repeatedly, we always arrive at an Euler triple after a finite number of iterations. Then the author introduces the *degree* of a Diophantine triple as the number of times we need to apply the ∂-operator to get an Euler triple. Given a potential Diophantine quintuple, the author first proves that it cannot contain an Euler triple, and then excludes the cases that it contains a Diophantine triple of (a) degree 1 and (b) degree at least 2.

Theorem 1.23 *There is no Diophantine quintuple.*

If an equation has some parameters, one may ask for which values of the parameters it is solvable. For example, a famous equation arises from an attempt to approximate the irrational number $\sqrt{d}$, where $d > 0$ is a square-free integer, by a rational number $\frac{x}{y}$. We have $\frac{x}{y} \approx \sqrt{d}$ if $x^2 \approx dy^2$, or $x^2 - dy^2 \approx 0$. Because $x^2 - dy^2 \neq 0$, the best possibilities are $x^2 - dy^2 = 1$ and

$$x^2 - dy^2 = -1, \tag{1.7}$$

which are called the *positive* and *negative Pell equation*, respectively. While the positive Pell equation always has integer solutions, the negative equation has no integer solutions for some values of d, such as $d = 3$, or, more generally, for any d which has a divisor of the form $4k + 3$. Moreover, if S is the set of positive square-free integers with no such divisors, there are $d \in S$, such as $d = 34$, for which Eq. (1.7) is not solvable in integers. In 1993, Stevenhagen [1161] conjectured that (1.7) is solvable for about 58% of values $d \in S$.

More formally, let $f(N)$ be the number of positive square-free integers $d \leq N$ such that (1.7) has an integer solution, let $g(N)$ be the cardinality of the set $S \cap \{1, \ldots, N\}$, and let $\alpha = \prod_{j=1}^{\infty}(1 + 2^{-j})^{-1} = 0.419\ldots$. Stevenhagen's conjecture predicts that $\lim_{N \to \infty} \frac{f(N)}{g(N)} = 1 - \alpha = 0.580\ldots$ In 2010, Fouvry and Klüners [461] made dramatic progress in this direction.

Theorem 1.24 *There exist constants $0 < C_1 \leq C_2 < 1$ such that*

$$(C_1 - o(1))g(N) \leq f(N) \leq (C_2 + o(1))g(N), \quad \text{as} \quad N \to \infty.$$

In fact, this is true for $C_1 = \alpha$ and $C_2 = \frac{2}{3}$.

In a later work, Fouvry and Klüners [462] showed that Theorem 1.24 remains true with $C_1 = \frac{5\alpha}{4} = 0.524\ldots$ To prove Stevenhagen's conjecture, one needs to achieve $C_1 = C_2 = 1 - \alpha$.

Zeros of Quadratic and Cubic Forms
A classical 1884 Theorem of Meyer [884] states that any indefinite quadratic form over the integers in $n \geq 5$ variables has a non-trivial integer zero.[3] The bound $n \geq 5$ is the best possible, because the equation

$$(x_1^2 + x_2^2) - 3(x_3^2 + x_4^2) = 0$$

has no non-trivial integer solutions.

---

[3] A *quadratic form* is a polynomial $P(x_1, \ldots, x_n) = \sum_{i=1}^{n} \sum_{j=1}^{n} c_{ij} x_i x_j$ with integer coefficients c_{ij}. It is *indefinite* if it is less than 0 for some values of x_i and greater than 0 for others. A *zero* of P is a solution to the equation $P(x_1, \ldots, x_n) = 0$. The solution $x_1 = \cdots = x_n = 0$ is a *trivial zero*, other zeros are *non-trivial*.

All cubic forms are indefinite, and it is conjectured that they must have a non-trivial integer zero if $n \geq 10$. In 1963, Davenport [321] proved this for $n \geq 16$. After more than 40 years with no further progress, in 2007 Heath-Brown [586] proved the same result for $n \geq 14$.

Theorem 1.25 *For every cubic form*

$$P(x_1, \ldots, x_n) = \sum_{i=1}^{n} \sum_{j=1}^{n} \sum_{k=1}^{n} c_{ijk} x_i x_j x_k$$

with integer coefficients in $n \geq 14$ variables, there exists integers $x_1^*, \ldots, x_n^*$, not all zero, such that $P(x_1^*, \ldots, x_n^*) = 0$.

For the special case when P has a diagonal form $P(x_1, \ldots, x_n) = \sum_{i=1}^{n} a_i x_i^3$, the conclusion of Theorem 1.25 is true for all $n \geq 7$, see [81]. In 1977, Vaughan [1227] proved that for $n \geq 16$ the system

$$\sum_{i=1}^{n} a_i x_i^3 = \sum_{i=1}^{n} b_i x_i^3 = 0 \tag{1.8}$$

always has a non-trivial integer solution. Here, the bound $n \geq 16$ is the best possible, because there is a system (1.8) in $n = 15$ variables which has no non-trivial solutions in the 7-adic field[4] $\mathbb{Q}_7$, and hence no non-trivial integer solutions. In 2007, Brüdern and Wooley [215] showed that, if $n \geq 13$, then this type of counterexample in the only possible one.

Theorem 1.26 *For any integer $n \geq 13$, and any integer coefficients a_i, b_i, $1 \leq i \leq n$, the system (1.8) has a non-trivial solution in integers if and only if it has a non-trivial solution in $\mathbb{Q}_7$.*

We say that an equation (or system) satisfies the *Hasse principle* if it either has a non-trivial solution in rational numbers, or has no non-trivial solutions in either real numbers or in $\mathbb{Q}_p$ for some prime p. The famous Hasse–Minkowski theorem [574] states that this principle holds for all quadratic forms. For cubic forms, the Hasse principle fails in general, but Theorem 1.26 implies that it holds for the system (1.8) in $n \geq 13$ variables. The bound $n \geq 13$ is the best possible.

Is the Number of Solutions Finite?

If an equation has some solutions (or we cannot prove that it does not) the next question to investigate is whether the solution set is finite or infinite.

We start with the following example. In 1857, Liouville asked whether the product of two or more consecutive non-zero integers can be a perfect power. In

---

[4] Informally, "no non-trivial solutions in the p-adic field $\mathbb{Q}_p$" means that the absence of non-trivial integer solutions can be proved using divisibility analysis modulo p in combination with the method of infinite descent.

1975, Erdős and Selfridge [418] answered this question negatively. In turn, they conjectured that a similar result is true for the products of terms of sufficiently large arithmetic progressions. Specifically, they conjectured the existence of a constant k_0 such that, for every $k \geq k_0$, the equation

$$n(n + d) \ldots (n + (k - 1)d) = y^l \tag{1.9}$$

does not have solutions in positive integers n, d, y and l, with $\gcd(n, d) = 1$ and $l \geq 2$. It is known (see Laishram and Shorey [760]) that the Erdős–Selfridge conjecture is a corollary of the famous *abc conjecture* of Masser and Oesterlé [946]. This conjecture states that if a, b, c are coprime positive integers such that $a + b = c$, then the product of the distinct prime factors of abc is not much smaller than c. More precisely, if $\mathrm{rad}(n)$ denotes the product of the distinct prime factors of n, then the abc conjecture predicts that for every real number $\epsilon > 0$, there exists a constant $K_\epsilon < \infty$ such that for all triples (a, b, c) of coprime positive integers such that $a + b = c$, we have

$$c \leq K_\epsilon \mathrm{rad}(abc)^{1+\epsilon}.$$

The abc conjecture has a number of important consequences but remains open.

Unconditionally, the Erdős–Selfridge conjecture has only been proved under some restrictive assumptions. For example, in 1985 Marszalek [848] proved the conjecture for every fixed d, while in 1996 Shorey [1124] proved the conjecture if n is fixed and $l \geq 7$. A major breakthrough came in 2020, when Bennett and Siksek [124] proved that, for every fixed large k, there may be at most finitely many arithmetic progressions of length k that violate this conjecture.

Theorem 1.27 *There exists a constant k_0 such that, for every integer $k \geq k_0$, Eq. (1.9) has at most finitely many solutions in positive integers n, d, y and l, with $\gcd(n, d) = 1$ and $l \geq 2$.*

The proof of Theorem 1.27 is quite involved and relies on some deep results in number theory, including (an explicit version of) Roth's theorem on 3-term arithmetic progressions (see the discussion before Theorem 1.11) and the modularity of elliptic curves (see Theorem 1.64 below).

Theorem 1.27 is partial progress towards a conjecture that an equation has no solutions at all. However, there are types of equations for which the solution set is non-empty but still finite.

For example, in a 1925 letter to Mordell, Siegel proved that if $f(x)$ is a polynomial of degree $k \geq 3$ with integer coefficients and no repeated roots, then the equation $f(x) = y^2$ has at most finitely many solutions in integers x and y. In 1964, LeVeque [787] proved the same result for the equation

$$f(x) = y^l$$

for any fixed $l \geq \max\{2, 5-k\}$, while in 1976 Schinzel and Tijdeman [1087] proved that this equation has at most finitely many solutions even if $l \geq \max\{2, 5-k\}$ is a variable.

What about this type of equation in more variables? A special case of a 1995 theorem of Darmon and Granville [318] states that, for any irreducible binary cubic form $F(x, y)$ with integer coefficients[5] and for any fixed $l \geq 4$, the equation

$$F(x, y) = z^l \tag{1.10}$$

has at most finitely many solutions in co-prime integers (x, y, z). In 2013, Bennett and Dahmen [123] proved that, for "almost all" binary cubic forms F, this remains true even if l is also a variable.

To formulate this result, we need some notation. Let $\Delta_F = 18abcd - 4b^3d + b^2c^2 - 4ac^3 - 27a^2d^2$ denote the discriminant of F. Let S_F be the set of primes dividing $2\Delta_F$, and let U_F be the set of non-zero integers u with the property that all prime factors of u belong to S_F. Let $\mathcal{F}$ be the set of binary cubic forms F such that $F(x, y) \notin U_F$ for all integers x and y.

Theorem 1.28 *For any $F \in \mathcal{F}$, Eq. (1.10) has at most finitely many solutions in integers (x, y, z, l) such that $\gcd(x, y) = 1$ and $l \geq 4$.*

For example, Theorem 1.28 is applicable to $F(x, y) = 3x^2 + 2x^2y + 5xy^2 + 3y^3$. As a more general example, the authors show that $F(x, y) = 3x^3 - ax^2y - (a + 9)xy^2 - 3y^3$ belongs to $\mathcal{F}$ for infinitely many values of a, which provides an infinite family of forms F for which Theorem 1.28 is applicable. Moreover, the authors provide a heuristic argument which suggests that "almost all" binary cubic forms belong to $\mathcal{F}$.

Presenting Infinitely Many Solutions

If an equation has finitely many solutions, one can in principle list them all. However, what do we mean by "solving" an equation if it has infinitely many solutions? Because in this case we cannot list all solutions explicitly, the best we can hope for is to present some formulas with parameters which represent all solutions.

We say that a set $A \subset \mathbb{Z}^n$ is a *polynomial family with k parameters* if there exist n polynomials $P_1, P_2, \ldots, P_n$ in k variables with integer coefficients such that $x = (x_1, x_2, \ldots, x_n)$ belongs to A if and only if there exists integers $y_1, y_2, \ldots, y_k$ such that $x_i = P_i(y_1, y_2, \ldots, y_k)$, $i = 1, 2, \ldots, n$. An old problem, which goes back to a 1938 paper of Skolem [1133], asks if the solution set of the equation

$$x_1x_4 - x_2x_3 = 1 \tag{1.11}$$

is a polynomial family. Despite the simplicity of the equation, this problem was open for over 70 years, and was solved only in 2010 by Vaserstein [1226].

---

[5] That is, $F(x, y) = ax^3 + bx^2y + cxy^2 + dy^3$ with integer a, b, c, d. F is called *irreducible* if it cannot be written as $(a'x + b'y)(c'x^2 + d'xy + e'y^2)$ for some integers a', b', c', d', e'.

Theorem 1.29 *The set of all integer solutions of Eq.* (1.11) *is a polynomial family with* 46 *parameters.*

By known reductions, Theorem 1.29 immediately implies the existence of polynomial families describing the solutions of some other, more complicated equations, see [1226] for details.

1.5 Counting Solutions to Equations and Inequalities

Counting Solutions to Equations

If an equation (or system of equations) has infinitely many solutions, describing all of them may be a difficult task even for the simplest equations (see Theorem 1.29) and is often a hopeless task for more difficult equations. In such cases, one may ask if we can at least estimate the number of solutions with absolute value up to some bound.

We start to investigate this problem with the equation

$$F(x_1, x_2, \ldots, x_n) = 0 \tag{1.12}$$

where F is a form (that is, homogeneous polynomial) of degree d with integer coefficients. We assume that F is absolutely irreducible, otherwise (1.12) reduces to equations of lower degree. To count multiples of the same solution only once, we call a solution $\mathbf{x} = (x_1, x_2, \ldots, x_n)$ to (1.12) *simple* if $\gcd(x_1, x_2, \ldots, x_n) = 1$ and the first non-zero component of the vector $(x_1, x_2, \ldots x_n)$ is positive. Let $N_F(B)$ be the number of simple solutions to (1.12) such that $\max_{1 \le i \le n} |x_i| \le B$.

Counting integer solutions to (1.12) is easy for forms in $n = 1$ and $n = 2$ variables. For (irreducible) forms in $n = 3$ variables, Pila [984] showed in 1995 that $N_F(B) \le C(d, \epsilon)B^{1+1/d+\epsilon}$ for every $\epsilon > 0$. The beauty of this result is that the bound is completely independent of F. In 2002, Heath-Brown [585] improved the exponent in this bound from $1 + 1/d + \epsilon$ to $2/d + \epsilon$, which is in fact optimal.

Theorem 1.30 *For any absolutely irreducible form $F(x, y, z)$ of degree d in $n = 3$ variables, and any $\epsilon > 0$, we have $N_F(B) \le C(d, \epsilon)B^{2/d+\epsilon}$, where $C(d, \epsilon)$ is a constant depending only on d and ϵ.*

Of course, many equations of interest are not of the form (1.12). For example, the famous Markoff equation is $x_1^2 + x_2^2 + x_3^2 = 3x_1x_2x_3$. Its solutions in positive integers are known as *Markoff triples*, and have applications in number theory, group theory, and geometry. One may study a more general equation

$$x_1^2 + x_2^2 + \cdots + x_n^2 = ax_1x_2 \ldots x_n + k, \tag{1.13}$$

which is called the *generalized Markoff–Hurwitz equation*. For $B > 0$, let $M_{n,a,k}(B)$ be the number of integer solutions $(x_1, \ldots, x_n)$ to (1.13) with $|x_i| \le B$

for $1 \leq i \leq n$. In 2019, Gamburd et al. [490] established an asymptotic formula describing how $M_{n,a,k}(B)$ grows with B.

Theorem 1.31 *If $n \geq 3$, $a \geq 1$, and $k \in \mathbb{Z}$ are such that $k - n + 2$ and $k - n - 1$ are not perfect squares, then either $\lim\limits_{B \to \infty} M_{n,a,k}(B) < \infty$ or*

$$M_{n,a,k}(B) = (c + o(1))(\log B)^{\beta},$$

where $c = c(n, a, k) > 0$ and $\beta = \beta(n) > 0$ are positive constants.

The proof of Theorem 1.31 crucially relies on the earlier result of Baragar [96], which provides the weaker asymptotic

$$M_{n,a,k}(B) = (\log B)^{\beta + o(1)},$$

where $\beta = \beta(n) > 0$ are the same constants as in Theorem 1.31. The proof also uses the following phenomenon, which was first observed for the original Markoff equation $x_1^2 + x_2^2 + x_3^2 = 3x_1 x_2 x_3$: if (x_1, x_2, x_3) is a solution to this equation, then $(x_2 x_3 - x_1, x_2, x_3)$, $(x_1, x_1 x_3 - x_2, x_3)$, and $(x_1, x_2, -x_1 x_2 - x_3)$ are also solutions. It turns out that a similar idea works for the general Eq. (1.13) and is very useful for counting solutions.

So far we have discussed integer solutions only. For some equations, one may also be interested in counting real solutions. In 1968, Littlewood [806] studied the equation

$$\sum_{i=1}^{N} \cos(n_i x) = 0$$

where $n_1, n_2, \ldots, n_N$ are integers which are all different. For an integer $N > 0$, let $f(N)$ be the minimal number of the equation's real solutions in $[0, 2\pi)$. Littlewood asked to estimate $f(N)$ and guessed that the answer is "Possibly $f(N) = N - 1$, or not much less". In 2008, Borwein et al. [177] proved that this prediction is wrong, and in fact $f(N)$ can be much less than $N - 1$.

Theorem 1.32 *There exist a constant $C > 0$, and a sequence of integers $N_1 < N_2 < \cdots < N_m < \ldots$, such that $f(N_m) \leq C N_m^{5/6} \log N_m$ for all m.*

Counting Solutions to Systems of Equations

One may also be interested in counting solutions to systems of equations. One of the most famous theorems of this type is *Vinogradov's mean value theorem*. For positive integers s, k, and X, let $J_{s,k}(X)$ be the number of integral solutions of the system of equations

$$x_1^j + \cdots + x_s^j = y_1^j + \cdots + y_s^j, \quad 1 \leq j \leq k, \tag{1.14}$$

such that $1 \leq x_i, y_i \leq X$ for $1 \leq i \leq s$. Strong bounds on $J_{s,k}(X)$ have applications in many areas of number theory. In 1935, Vinogradov proved that if $s \geq k^2 \log(k^2 + k) + \frac{1}{4}k^2 + \frac{5}{4}k + 1$, then

$$J_{s,k}(X) \leq C_{s,k}(\log X)^{2s} X^{2s - \frac{1}{2}k(k+1) + \frac{1}{2}},$$

where $C_{s,k}$ is a constant depending on s and k. A conjecture known as "the main conjecture in Vinogradov's mean value theorem" predicted that a stronger estimate

$$J_{s,k}(X) \leq C_{s,k,\epsilon}(X^{s+\epsilon} + X^{2s - \frac{1}{2}k(k+1) + \epsilon}) \tag{1.15}$$

should hold for any $s \geq 1$, $k \geq 2$, and any fixed $\epsilon > 0$. If true, this estimate would be optimal up to ϵ. This conjecture attracted a lot of attention. In particular, Wooley [1274] proved the conjecture if $s \geq k(k + 1)$.

Theorem 1.33 *For any natural numbers $k \geq 2$ and $s \geq k(k + 1)$, and any real $\epsilon > 0$, there exists a constant $C = C(s, k, \epsilon)$ such that $J_{s,k}(X) \leq C X^{2s - \frac{1}{2}k(k+1) + \epsilon}$.*

The proof of Theorem 1.33 is based on the novel efficient congruencing method, which, in the author's words, "iteratively exploits the translation invariance of associated systems of Diophantine equations to derive powerful congruence constraints on the underlying variables". In a subsequent work, Wooley [1275] used the same method to prove (1.15) in the case when $k = 3$ and s is arbitrary. Finally, Bourgain, Demeter, and Guth [185] proved the conjecture in full in 2016.

Theorem 1.34 *For any natural numbers $s \geq 1$ and $k \geq 2$, and any real $\epsilon > 0$, there exists a constant $C = C(s, k, \epsilon)$ such that (1.15) is true.*

While all the previous work on Vinogradov's mean value theorem used number-theoretic methods, the proof of Theorem 1.34 in [185] uses completely different, harmonic analysis techniques. As a consequence of this new approach, the proven result is in fact much more general than Theorem 1.34. For example, instead of counting integer solutions to (1.14), we may count solutions in S, where $S \subset \mathbb{R}$ is an arbitrary set of well-separated real numbers.

Counting Integer Solutions to Inequalities
A related research direction is counting integer solutions to an inequality or a system of inequalities. Geometrically, a system of inequalities in n real variables describes a region in $\mathbb{R}^n$, and the problem reduces to estimating the number of integer points in this region. Intuitively, this number should be approximately equal to the volume of the region, and many researchers study under which conditions this approximation works, and how well.

A well-studied special case of this problem is to estimate the number of integer solutions to the inequality

$$|F(x_1, \ldots, x_n)| \leq m, \tag{1.16}$$

where F is a homogeneous polynomial in n variables. In 2001, Thunder [1200] solved this problem for the case when F is a *decomposable form*, that is, a polynomial $F(x_1, \ldots x_n) = \prod_{i=1}^{d}(a_{i1}x_1 + \cdots + a_{in}x_n)$, where the coefficients a_{ij} are non-zero complex numbers.

Theorem 1.35 *For every decomposable form F of degree d in n variables with integer coefficients, the number $N_F(m)$ of integer solutions to (1.16) is either infinite or at most $c(n, d)m^{n/d}$, where $c(n, d)$ is an effectively computable constant depending only on n and d.*

The case $d = 2$ of the theorem was proved by Mahler [829] in 1933, but progress in estimating $N_F(m)$ for $d > 2$ was limited. Theorem 1.35 estimates $N_F(m)$ for all d, confirming a 1989 conjecture of Schmidt. It implies that volume approximation works reasonably well in this case.

A more precise connection between the number of integer points in a region and its volume is available for some specific shapes such as ellipsoids. Let Q be an invertible $d \times d$ matrix with real entries such that (a) $(Qx, y) = (x, Qy)$ for all $x, y \in \mathbb{R}^d$, and (b) $(Qx, x) > 0$ for every non-zero $x \in \mathbb{R}^d$. For $s > 0$, let E_s be the set of points $x \in \mathbb{R}^d$ such that $(Qx, x) \leq s$. Any set of the form $rE_1 + a$ (for $a \in \mathbb{R}^d$, and $r > 0$) is called an *ellipsoid*. For any ellipsoid $S \subset \mathbb{R}^d$, let $\mathrm{Vol}(S)$ and $\mathrm{Vol}_{\mathbb{Z}}(S)$ be its volume and the number of integer points in it, respectively. In 1999, Bentkus and Götze [126] proved that, in dimensions $d \geq 9$, $\mathrm{Vol}(rE_1 + a) = \mathrm{Vol}(rE_1)$ approximates $\mathrm{Vol}_{\mathbb{Z}}(rE_1 + a)$ up to a relative error which decreases at a rate $O(r^{-2})$ as $r \to \infty$. In 2004, Götze [507] proved that this is the case in all dimensions $d \geq 5$.

Theorem 1.36 *For any $d \geq 5$ and any Q as above,*

$$\sup_{a \in \mathbb{R}^d} \left| \frac{\mathrm{Vol}_{\mathbb{Z}}(rE_1 + a) - \mathrm{Vol}(rE_1)}{\mathrm{Vol}(rE_1)} \right| = O(r^{-2})$$

as $r \to \infty$.

The condition $d \geq 5$ in Theorem 1.36 is the best possible, because, for $d \leq 4$, the error rate is known to decrease more slowly than $O(r^{-2})$, even if E_1 is the unit sphere.

1.6 Special Integer Sequences

One of the most famous and well-studied sequences of integers in mathematics is the *Fibonacci sequence $F_0, F_1, \ldots, F_n, \ldots$* defined by $F_0 = 0$, $F_1 = 1$, and $F_{n+2} = F_{n+1} + F_n$ for $n \geq 0$. This sequence is studied so intensively that there is an entire journal, *The Fibonacci Quarterly*, which publishes only papers dedicated to its properties. Despite this, many basic questions about $\{F_n\}$ either remain open or have been resolved only recently. One such question is: what are the perfect powers in this sequence?

It follows from a 1951 paper of Ljunggren [809] that the only perfect squares in the Fibonacci sequence are $0 = 0^2, 1 = 1^2$, and $144 = 12^2$. This result was rediscovered by Cohn [291] in 1964. In 1969, London and Finkelstein [811] proved that the only perfect cubes are 0, 1 and $8 = 2^3$. Are there any other perfect powers in the sequence? By 2006, it was known that there are no more p-th powers with $p \leq 17$ or with prime $p \geq 5.1 \cdot 10^{17}$. In 2006, Bugeaud et al. [225] resolved this question for all p.

Theorem 1.37 *The only perfect powers in the Fibonacci sequence are $F_0 = 0$, $F_1 = 1$, $F_2 = 1$, $F_6 = 8$, and $F_{12} = 144$.*

In fact, the choice of initial values $F_0 = 0$, $F_1 = 1$ is rather arbitrary. Another famous sequence $L_0, L_1, \ldots, L_n, \ldots$ starts with $L_0 = 2$, $L_1 = 1$, and then applies the same recurrence $L_{n+2} = L_{n+1} + L_n$ for $n \geq 0$. In [225], it is also proved that $L_1 = 1$ and $L_3 = 4$ are the only perfect powers is this sequence. The proof combines the classical theory of exponential Diophantine equations with a modular approach developed to prove Fermat's Last Theorem, see the discussion before Theorem 1.64 below.

Of course, one may start with any pair of initial values or modify the recurrence relation. Here is a way to define a broad class of Fibonacci-like sequences. The n-th term of the Fibonacci sequence is given by

$$F_n = \frac{\alpha^n - \beta^n}{\alpha - \beta},$$

where $\alpha = \frac{1+\sqrt{5}}{2}$ and $\beta = \frac{1-\sqrt{5}}{2}$. More generally, for any pair of (possibly complex) algebraic integers (α, β) such that $\alpha + \beta$ and $\alpha\beta$ are coprime integers and α/β is not a root of unity,[6] the sequence

$$\frac{\alpha^n - \beta^n}{\alpha - \beta}$$

consists only of integers and is called the sequence of *Lucas numbers*. Also, a pair of algebraic integers (α, β) is called a *Lehmer pair* if $(\alpha + \beta)^2$ and $\alpha\beta$ are non-zero coprime integers and α/β is not a root of unity. For every Lehmer pair (α, β), the sequence $u_n = u_n(\alpha, \beta)$ defined by the formulas $u_n = \frac{\alpha^n - \beta^n}{\alpha - \beta}$ for odd n and $u_n = \frac{\alpha^n - \beta^n}{\alpha^2 - \beta^2}$ for even n is a sequence of integers which is called the sequence of *Lehmer numbers*. One of the main themes in studying general Lucas and Lehmer numbers is the study of primitive divisors. A prime number p is called a *primitive divisor* of u_n if p divides u_n but does not divide $(\alpha^2 - \beta^2)^2 u_1 \ldots u_{n-1}$. An old and difficult problem, which goes back to at least the beginning of the twentieth century,

---

[6] An *algebraic integer* is a complex number that is a root of some polynomial with integer coefficients and leading coefficient 1. A *root of unity* is a complex number z such that $z^n = 1$ for some positive integer n.

is to find all triples (α, β, n) such that (α, β) is a Lehmer pair and $u_n(\alpha, \beta)$ has no primitive divisors. In 2001, this problem was solved by Bilu et al. [149].

Theorem 1.38 *For every Lehmer pair (α, β) and every $n > 30$, $u_n(\alpha, \beta)$ has a primitive divisor*

The condition $n > 30$ in Theorem 1.38 is the best possible because, for example, $\alpha = (1 - \sqrt{-7})/2$ and $\beta = (1 + \sqrt{-7})/2$ is a Lehmer pair for which u_{30} has no primitive divisors. With the help of a computer, the authors found all Lehmer numbers with no primitive divisors for $n \leq 30$, while the theorem states that there are no such numbers with $n > 30$. This completely solves the problem. As a special case, the theorem also solves the corresponding problem for Lucas numbers.

The proof of Theorem 1.38 combines mathematical arguments with long and involved (but rigorous) electronic computations. In particular, computers have been used to solve many high degree *Thue equations*, that is, equations of the form $f(x, y) = r$, where f is an irreducible bivariate form over the rational numbers, and $r \neq 0$ is a rational number.

One interesting application of Theorem 1.38 is a lower bound for the greatest prime factors of Lehmer numbers. For any integer m let $P(m)$ denote the greatest prime factor of m, with the convention that $P(m) = 1$ when m is 1, 0, or -1. In 1886, Bang [94] proved that $P(a^n - 1) \geq n + 1$ for any integers $a > 1$ and $n > 2$. More generally, Zsigmondy [1293] proved in 1892 that $P(a^n - b^n) \geq n + 1$ for integers $n > 2$ and $a > b > 0$. Theorem 1.38 implies that a similar bound $P(u_n) \geq n - 1$ holds for any Lehmer number u_n and any $n > 30$.

It has long been conjectured that the bounds in the previous paragraph are far from optimal. For example, Erdős [404] conjectured in 1965 that

$$\lim_{n \to \infty} \frac{P(2^n - 1)}{n} = +\infty. \tag{1.17}$$

However, this remained open, and even Bang's original bound remained unimproved for over 125 years. Finally, Stewart [1162] proved in 2013 a strong form of conjecture (1.17) in the much more general context of Lehmer numbers.

Theorem 1.39 *For every Lehmer pair (α, β), there is a constant $C = C(\alpha, \beta)$ such that for every $n > C$,*

$$P(u_n(\alpha, \beta)) > n \exp\left(\frac{\log n}{104 \log \log n}\right).$$

Stewart [1162] also proved the same lower bound for greatest prime factors of Lucas numbers, and for $P(a^n - b^n)$ for any integers $a > b > 0$. In particular, Erdős' conjecture (1.17) is true, even with $P(a^n - b^n)$ in place of $P(2^n - 1)$.

1.7 Multiplicative Functions

General Estimates

Some functions of central importance in number theory (the most prominent example is the *Liouville function* $\lambda(n)$, equal to $+1$ or -1 depending on whether the number of prime factors of n is even or odd) are completely multiplicative. A function $f : \mathbb{N} \to \mathbb{R}$ is called *completely multiplicative* if $f(x \cdot y) = f(x) \cdot f(y)$ for all positive integers x, y. A central topic in the theory of such functions is to estimate sums in the form $\sum_{n=a}^{b} f(n)$. For example, if f takes values in $[-1, 1]$, what are the possible values of the mean $\frac{1}{N} \sum_{n=1}^{N} f(n)$? The maximum value is clearly 1, but determining the minimum is non-trivial. In 1996, Hall [559] proved the existence of a constant $c > -1$ such that $\frac{1}{N} \sum_{n=1}^{N} f(n) \geq c + o(1)$. He observed that c in his theorem cannot exceed

$$\delta = 1 - 2\log(1 + \sqrt{e}) + 4 \int_{1}^{\sqrt{e}} \frac{\log t}{t + 1} dt = -0.656999\ldots,$$

and conjectured that in fact the statement is true with $c = \delta$. In 2001, Granville and Soundararajan [515] proved this conjecture.

Theorem 1.40 *For every completely multiplicative function $f : \mathbb{N} \to [-1, 1]$,* $\frac{1}{N} \sum_{n=1}^{N} f(n) \geq \delta + o(1)$.

More generally, a completely multiplicative function $f : \mathbb{N} \to \mathbb{C}$ may take complex values, and Granville and Soundararajan investigated possible values of $\lim_{N \to \infty} \frac{1}{N} \sum_{n=1}^{N} f(n)$, see [515] for details.

Further, one may ask what are the "typical" values of the sum $\sum_{n=1}^{N} f(n)$ if the function f is selected at random. To define a completely multiplicative function f, it suffices to define the values $f(p)$ for primes p. We say that a completely multiplicative function f is *(Steinhaus) random* if the values $f(p)$ are selected independently, uniformly at random from the unit circle $\{z \in \mathbb{C} : |z| = 1\}$. In 2010, Helson [597] predicted that $\mathbb{E} \left| \sum_{n \leq x} f(n) \right| = o(\sqrt{x})$. In words, the conjecture predicts "better than square root cancellation" for expectation. In 2020, Harper [568] proved a much stronger result. For $0 \leq q \leq 1$, define $g_q(x) = \left(\frac{x}{1 + (1-q)\sqrt{\log\log x}} \right)^{q}$.

Theorem 1.41 *If f is a Steinhaus random multiplicative function, then there exist positive constants c, C, and x_0 such that, for all $0 \leq q \leq 1$ and all $x \geq x_0$,*

$$c g_q(x) \leq \mathbb{E} \left| \sum_{n \leq x} f(n) \right|^{2q} \leq C g_q(x).$$

With $q = 1/2$, Theorem 1.41 implies that

$$c\frac{\sqrt{x}}{(\log\log x)^{1/4}} \le \mathbb{E}\left|\sum_{n\le x} f(n)\right| \le C\frac{\sqrt{x}}{(\log\log x)^{1/4}}.$$

This not only confirms Helson's conjecture, but establishes the rate of growth of $\mathbb{E}\left|\sum_{n\le x} f(n)\right|$ up to a constant factor.

The proof of Theorem 1.41 is based on the *Euler product formula*

$$F_x(s) = \prod_{p<x}\left(1 - \frac{f(p)}{p^s}\right)^{-1} = \sum_{n\in S(x)} \frac{f(n)}{n^s}, \tag{1.18}$$

where the product is over all primes p less than x, and $S(x)$ is the set of all positive integers n whose all prime factors are less than x. This formula is particularly useful for the study of Steinhaus random multiplicative functions because the factors $\left(1 - \frac{f(p)}{p^s}\right)^{-1}$ in the product are independent. Harper [568] first expressed the moments $\mathbb{E}\left|\sum_{n\le x} f(n)\right|^{2q}$ as expectations involving $F_x(s)$, and then applied (1.18).

A function $f : \mathbb{N} \to [-1, 1]$ is called *multiplicative* if $f(1) = 1$ and $f(x \cdot y) = f(x) \cdot f(y)$ holds for all *coprime* $x, y \in \mathbb{N}$. Some important functions, such as the *Möbius function*,[7] are multiplicative but not completely multiplicative. Theorems 1.40 and 1.41 estimate sums of values of f over "long" intervals. Similar results are available for multiplicative functions as well. However, in many applications, it is important to estimate the average values $\frac{1}{h}\sum_{x\le n\le x+h} f(n)$ of f in "short" intervals, that is, in the case when h is much smaller than x. In 2016, Matomäki and Radziwiłł [854] proved that such averages are approximately equal to the averages of the form $\frac{1}{X}\sum_{X\le n\le 2X} f(n)$ over a "long" interval $[X, 2X]$, which are much easier to estimate.

Theorem 1.42 *There exist absolute constants B and C (one can take $B = 20\,000$) such that for any multiplicative function $f : \mathbb{N} \to [-1, 1]$, for any $2 \le h \le X$, and any $\delta > 0$, the inequality*

$$\left|\frac{1}{h}\sum_{x\le n\le x+h} f(n) - \frac{1}{X}\sum_{X\le n\le 2X} f(n)\right| \le \delta + B\frac{\log\log h}{\log h}$$

holds for all but at most $CX\left(\frac{(\log h)^{1/3}}{\delta^2 h^{\delta/25}} + \frac{1}{\delta^2(\log X)^{1/50}}\right)$ *integers $x \in [X, 2X]$.*

---

[7] For an integer $n > 0$, let $\mu(n) = 0$ if n has a squared prime factor, and $\mu(n) = (-1)^{\Omega(n)}$ for square-free n, where $\Omega(n)$ is the number of prime factors of n.

Theorem 1.42 has many applications. For example, it was used to prove formula (1.19) and Theorem 2.30 discussed later in this book.

The Liouville Function and the Chowla Conjecture

Maybe the most important and famous example of a completely multiplicative function is the *Liouville function* $\lambda(n) = (-1)^{\Omega(n)}$, where $\Omega(n)$ is the number of prime factors of the positive integer n, counted with multiplicity. Our lack of understanding of $\lambda(n)$ is demonstrated by the following fact. Let $P_\lambda(m)$ denote the *block complexity* of $\lambda(n)$, that is, the number of sign patterns of size m that are taken by consecutive values of $\lambda(n)$. Then, intuitively, it is "completely obvious" that $P_\lambda(m)$ should be equal to 2^m, that is, all possible sign patterns of size m should be realized as consecutive values of the Liouville function. However, the best lower bound for $P_\lambda(m)$ known before 2018 was only $P_\lambda(m) \geq m + 5$ for $m \geq 3$, see [855]. In 2018, Frantzikinakis and Host [470], using the deep theory of dynamical systems, made a significant improvement and proved that $P_\lambda(m)$ grows superlinearly in m.

Theorem 1.43 $\lim\limits_{m \to \infty} \dfrac{P_\lambda(m)}{m} = \infty.$

A number of important conjectures in number theory are formulated in terms of the Liouville function. One of the most famous and important of them is the *Chowla conjecture* [276]. It states that if $k \geq 1$, $a_1, \ldots, a_k$ are natural numbers, and $b_1, \ldots, b_k$ are distinct non-negative integers such that $a_i b_j - a_j b_i \neq 0$ for $1 \leq i < j \leq k$, then

$$\sum_{n \leq x} \lambda(a_1 n + b_1) \cdots \lambda(a_k n + b_k) = o(x) \quad \text{as } x \to \infty.$$

The $k = 1$ case of the conjecture is equivalent to the prime number theorem, and already the $k = 2$ case remains open. In 2016, Tao [1182] proved a theorem which he calls the "logarithmically averaged version" of the $k = 2$ case of the Chowla conjecture.

Theorem 1.44 *Let a_1, a_2 be natural numbers, and let b_1, b_2 be integers such that $a_1 b_2 - a_2 b_1 \neq 0$. Let $1 \leq w(x) \leq x$ be a quantity depending on x that goes to infinity as $x \to \infty$. Then one has*

$$\sum_{x/w(x) < n \leq x} \frac{\lambda(a_1 n + b_1)\lambda(a_2 n + b_2)}{n} = o(\log w(x)) \quad \text{as } x \to \infty.$$

The proof of Theorem 1.44 crucially relies on the estimate

$$\sup_\alpha \int_X^{2X} \left| \sum_{x < n \leq x+H} \lambda(n) e(-\alpha n) \right| dx = o(XH) \quad \text{as } X \to \infty, \tag{1.19}$$

where $e(t) = \exp(2\pi i t)$, and $H = H(X)$ is any function such that $H \to \infty$ as $X \to \infty$. Estimate (1.19) was proved in [857] using Theorem 1.42. To obtain a logarithmically averaged version of the Chowla conjecture for all k, it is important to have an estimate similar to (1.19), but with supremum and integral exchanged. Before 2020, this was known to hold only for $H \geq X^\theta$ with $\theta > \frac{5}{8}$, see [1192]. In 2020, Matomäki et al. [856] proved such an estimate for $H = X^\theta$ with $\theta > 0$ arbitrarily small.

Theorem 1.45 *Let $\theta \in (0, 1)$ be given and set $H = X^\theta$. Then*

$$\int_X^{2X} \sup_\alpha \left| \sum_{x < n \leq x+H} \lambda(n) e(-\alpha n) \right| dx = o(XH) \quad \text{as } X \to \infty.$$

The authors of [856] expressed a hope that their methods can be extended to reach $H = \exp((\log X)^{1-\delta})$ for some $\delta > 0$. However, it is still unclear how to prove Theorem 1.45 for all functions H such that $H \to \infty$ as $X \to \infty$.

1.8 Diophantine Approximation

Approximating Irrational Numbers by Fractions and Irreducible Fractions

A fundamental research direction in number theory is to understand how well an irrational number α can be approximated by rational numbers $\frac{m}{n}$. We may (and will) always assume that $n > 0$. There is a trade-off between the quality of the approximation and the value n of the denominator. Dirichlet's famous *approximation theorem* states that, for any α, the inequality $\left| \alpha - \frac{m}{n} \right| < \frac{1}{n^2}$ is satisfied for infinitely many pairs of integers m and n. In 1891, Hurwitz [631] proved the stronger statement that, for $k = \sqrt{5}$, and every $\alpha \in \mathbb{R}$, there are infinitely many $m, n \in \mathbb{Z}$ satisfying

$$\left| \alpha - \frac{m}{n} \right| < \frac{1}{kn^2}. \tag{1.20}$$

This result is the best possible, because for the golden ratio $\alpha = \frac{1+\sqrt{5}}{2}$, and any $k > \sqrt{5}$, inequality (1.20) has only finitely many solutions in integers m, n. However, (1.20) can be improved if we ignore exceptional numbers like the golden ratio, and ask what is the best quality of approximation achievable for a "typical" real number. For any function $\psi : \mathbb{N} \to \mathbb{R}^+$, where $\mathbb{R}^+ = [0, \infty)$, we say that $\alpha \in \mathbb{R}$ is ψ-*approximable* if the inequality

$$\left| \alpha - \frac{m}{n} \right| < \frac{\psi(n)}{n} \tag{1.21}$$

has infinitely many solutions in integers m, n. In 1924, Khintchine [709] proved a fundamental theorem, stating that

(i) if $\sum_{n=1}^{\infty} \psi(n) = \infty$, then (Lebesgue) almost all $\alpha \in \mathbb{R}$ are ψ-approximable;
(ii) conversely, if $\sum_{n=1}^{\infty} \psi(n) < \infty$, then almost all $\alpha \in \mathbb{R}$ are not ψ-approximable.

For example, by applying part (i) to the function $\psi(n) = \frac{1}{kn}$, we conclude that, for any $k > 0$, and almost all $\alpha \in \mathbb{R}$, inequality (1.20) has infinitely many solutions in integers m, n.

In 1941, Duffin and Schaeffer [368] studied approximation of α by *irreducible* fractions $\frac{m}{n}$. Let $S(\psi)$ be the set of real numbers $\alpha \in \mathbb{R}$ such that inequality (1.21) has infinitely many *coprime* solutions m, n. Duffin and Schaeffer conjectured that set $S(\psi)$ has zero or full Lebesgue measure depending on whether the series $\sum_{n=1}^{\infty} \phi(n) \frac{\psi(n)}{n}$ converges, where $\phi(n)$ denotes the number of positive integers which are less than n and co-prime with it. The Duffin–Schaeffer conjecture became one of the fundamental conjectures in rational approximation, attracted a lot of attention, and has been proved in a number of special cases. For example, Khinchin's theorem implies the Duffin–Schaeffer conjecture when $n\psi(n)$ is decreasing. In the 1970s, Erdős [405] and Vaaler [1222] proved the conjecture for $\psi(n) = O(1/n)$. However, despite all efforts, the general case of the conjecture remained open for almost 80 years. In 2020, Koukoulopoulos and Maynard [741] proved the conjecture in full.

Theorem 1.46 *For any function $\psi : \mathbb{N} \to \mathbb{R}^+$,*

(i) *if $\sum_{n=1}^{\infty} \phi(n) \frac{\psi(n)}{n} = \infty$, then the set $S(\psi)$ has full Lebesgue measure in $\mathbb{R}$;*
(ii) *if $\sum_{n=1}^{\infty} \phi(n) \frac{\psi(n)}{n} < \infty$, then the set $S(\psi)$ has zero Lebesgue measure.*

In fact, Koukoulopoulos and Maynard only needed to prove part (i) of Theorem 1.46. Part (ii) was known before, but we state it here for completeness.

Theorem 1.46 is formulated in terms Lebesgue measure. However, the sets of zero Lebesgue measure can be classified much more precisely in terms of Hausdorff measure and Hausdorff dimension. Let $f : [0, \infty) \to [0, \infty)$ be a continuous, non-decreasing function, with $f(0) = 0$. Let F be any subset of an arbitrary metric space (X, d). For $\delta > 0$, a δ-*cover* of the set F is a sequence of balls $\{B_i\}_{i=1}^{\infty}$ in X with radii $r(B_i) \leq \delta$ for all i such that $F \subset \bigcup_{i=1}^{\infty} B_i$. Let $H_\delta^f(F) = \inf \sum_{i=1}^{\infty} f(r(B_i))$, where the infimum is taken over all δ-covers of F. The number

$$H^f(F) = \lim_{\delta \to 0} H_\delta^f(F) \tag{1.22}$$

is called the *Hausdorff f-measure* of F. In particular, if $f(r) = r^d$, $d > 0$, then $H^f(F)$ is denoted by $H^d(F)$ and is called the *d-dimensional Hausdorff measure* of F. For example, if d is an integer and $X = \mathbb{R}^d$, then H^d coincides with d-dimensional Lebesgue measure, up to a constant factor. However, H^d is defined for all real $d > 0$. The number

$$\dim(F) = \inf\left\{d > 0 : H^d(S) = 0\right\} \tag{1.23}$$

is called the *Hausdorff dimension* of the set F. For example, subsets $F \subset \mathbb{R}$ of Lebesgue measure zero may have any dimension from 0 to 1.

In 2006, Beresnevich and Velani [130] proved that the Duffin–Schaeffer conjecture implies the following much more general statement.

Theorem 1.47 *Let* $f : [0, \infty) \rightarrow [0, \infty)$ *be a continuous, non-decreasing function, with* $f(0) = 0$*, and such that* $f(r)/r$ *is monotonic on* $(0, \infty)$*. Let* $\psi : \mathbb{N} \rightarrow \mathbb{R}^+$ *be arbitrary and let* $S(\psi)$ *be as in Theorem 1.46. Then*

(i) *if* $\sum_{n=1}^{\infty} \phi(n) f\left(\frac{\psi(n)}{n}\right) = \infty$*, then* $H^f(\mathbb{R} \setminus S(\psi)) = 0$*;*

(ii) *if* $\sum_{n=1}^{\infty} \phi(n) f\left(\frac{\psi(n)}{n}\right) < \infty$*, then* $H^f(S(\psi)) = 0$*.*

To prove that the Duffin–Schaeffer conjecture implies Theorem 1.47, Beresnevich and Velani developed a new powerful method, which they called the *mass transference principle*, which allows one to transfer certain Lebesgue measure-theoretic statements to Hausdorff measure-theoretic statements.

Theorem 1.46 proves the Duffin–Schaeffer conjecture, and thus establishes Theorem 1.47 unconditionally. For example, with $f(r) = r^d, d \in (0, 1)$, Theorem 1.47 implies that if $\sum_{n=1}^{\infty} \phi(n) \frac{\psi(n)^d}{n^d} = \infty$, then the set $\mathbb{R} \setminus S(\psi)$ of points which cannot be approximated by irreducible fractions with accuracy (1.21) not only has Lebesgue measure zero (which follows from Theorem 1.46), but in fact has Hausdorff dimension at most d.

Simultaneous Approximation: Curves and Manifolds

In the previous subsection we discussed approximation of one irrational number by rational numbers. One may also ask with what accuracy we can approximate *several* irrational numbers $y_1, \ldots, y_d$ by rational numbers $\frac{m_1}{n}, \ldots, \frac{m_d}{n}$ with the same denominator $n > 0$.

Khintchine's theorem (see the discussion before Theorem 1.46) has a multidimensional version concerning simultaneous approximation. Let $\psi : \mathbb{R}^+ \rightarrow \mathbb{R}^+$ be a real positive decreasing function. A set of d real numbers $\mathbf{y} = (y_1, \ldots, y_d) \in \mathbb{R}^d$ is called *simultaneously ψ-approximable* if the inequality

$$\max_{1 \leq i \leq d} \left| y_i - \frac{m_i}{n} \right| \leq \frac{\psi(n)}{n}$$

has infinitely many solutions in integers $m_1, \ldots, m_d, n$. The (d-dimensional version of) Dirichlet's approximation theorem implies that all $\mathbf{y} \in \mathbb{R}^n$ are simultaneously ψ-approximable for $\psi(n) = n^{-1/d}$. In general, let $S_d(\psi)$ be the set of simultaneously ψ-approximable points in $\mathbb{R}^d$. The multidimensional version of Khintchine's theorem [709] states that

(i) if $\sum_{n=1}^{\infty} \psi(n)^d = \infty$, then the set $S_d(\psi)$ has full Lebesgue measure in $\mathbb{R}^d$;
(ii) conversely, if $\sum_{n=1}^{\infty} \psi(n)^d < \infty$, then the set $S_d(\psi)$ has zero Lebesgue measure in $\mathbb{R}^d$.

   In particular, part (i) with $d = 2$ states that if $\sum_{n=1}^{\infty} \psi(n)^2 = \infty$, then the set $\mathbb{R}^2 \setminus S_2(\psi)$ has measure zero in $\mathbb{R}^2$. For example, applying this to $\psi(n) = \frac{\epsilon}{\sqrt{n}}$, we conclude that for any $\epsilon > 0$, and almost all pairs $(\alpha, \beta) \in \mathbb{R}^2$, the inequality

$$\max \left\{ \left| \alpha - \frac{m_1}{n} \right|, \left| \beta - \frac{m_2}{n} \right| \right\} \leq \frac{\epsilon}{n\sqrt{n}} \tag{1.24}$$

has infinitely many solutions in integers m_1, m_2, n.

   However, this does not imply that, for example, the pair $\alpha = x$, $\beta = x^2$ is simultaneously ψ-approximable for almost all x. The set of such pairs has measure zero in $\mathbb{R}^2$, and Khintchine's theorem says nothing about such sets. We say that a curve C in $\mathbb{R}^2$ is of *Khintchine type*:

(i) *for divergence*, if almost all points on C belong to $S_2(\psi)$ whenever $\sum_{n=1}^{\infty} \psi(n)^2 = \infty$.
(ii) *for convergence*, if almost all points on C does not belong to $S_2(\psi)$ whenever $\sum_{n=1}^{\infty} \psi(n)^2 < \infty$.

   In 2006, Vaughan and Velani [1228] proved that a broad class of smooth planar curves are of Khintchine type for convergence.

Theorem 1.48 *Let $\psi : \mathbb{R}^+ \to \mathbb{R}^+$ be a decreasing function with $\sum_{n=1}^{\infty} \psi(n)^2 < \infty$. Let f be a twice continuously differentiable function on some interval (a, b), such that $f''(x) \neq 0$ for almost all $x \in (a, b)$. Then for almost all $x \in (a, b)$ the pair $(x, f(x))$ is not simultaneously ψ-approximable.*

   Complementing Theorem 1.48, Beresnevich et al. [129] proved in 2007 that a similar result is true for Khintchine type for divergence.

Theorem 1.49 *Let $\psi : \mathbb{R}^+ \to \mathbb{R}^+$ be a decreasing function with $\sum_{n=1}^{\infty} \psi(n)^2 = \infty$. Let f be a three times continuously differentiable function on some interval (a, b), such that $f''(x) \neq 0$ for almost all $x \in (a, b)$. Then for almost all $x \in (a, b)$ the pair $(x, f(x))$ is simultaneously ψ-approximable.*

   The notions of Khintchine type for divergence and convergence can be straight-forwardly generalized from curves in $\mathbb{R}^2$ to manifolds in $\mathbb{R}^d$. Let

$$M = \{(x_1, \ldots, x_k, f_1(x), \ldots, f_m(x)) \in \mathbb{R}^d : x = (x_1, \ldots, x_k) \in U)\},$$

where U is an open subset of $\mathbb{R}^k$ and $f = (f_1, \ldots, f_m) : U \to \mathbb{R}^m$ is an analytic map. We will call M an *analytic submanifold* of $\mathbb{R}^d$. M is called *non-degenerate* if it is not contained in a proper affine subspace of $\mathbb{R}^d$. M is called of *Khintchine type for divergence* if for any ψ such that $\sum_{n=1}^{\infty} \psi(n)^d = \infty$, almost all points on

M belong to $S_d(\psi)$. In 2012, Beresnevich [127] generalized Theorem 1.49 from curves to manifolds.

Theorem 1.50 *Any non-degenerate analytic submanifold in $\mathbb{R}^d$ is of Khintchine type for divergence.*

The proof of Theorem 1.50 follows from the study of the distribution of rational points near manifolds. The authors obtained sharp lower bounds for the number of rational points near non-degenerate manifolds in dimensions $d > 2$, showed that the distribution of such points is close to uniform, and deduced Theorem 1.50 from these facts.

Simultaneous Approximation: Littlewood's Conjecture

The previous section discusses simultaneous approximability of *almost all* pairs (or d-tuples) of points. In this section, we discuss a famous conjecture concerning simultaneous approximability of *all* pairs (α, β) of real numbers.

It follows from (1.24) that for almost all pairs $(\alpha, \beta) \in \mathbb{R}^2$, and any $\epsilon > 0$, the inequality

$$\left| \alpha - \frac{m_1}{n} \right| \cdot \left| \beta - \frac{m_2}{n} \right| \le \frac{\epsilon^2}{n^3}$$

has infinitely many solutions in integers m_1, m_2, n. Multiplying both sides by n^3, and denoting by $||x|| = \min\{|x - m|, m \in \mathbb{Z}\}$ the distance from x to the nearest integer, we conclude that the inequality $n \, ||\alpha n|| \, ||\beta n|| \le \epsilon^2$ is solvable (in n) for any $\epsilon > 0$, or, equivalently,

$$\liminf_{n \to \infty} n \, ||\alpha n|| \, ||\beta n|| = 0. \tag{1.25}$$

Recall that the argument above works for almost all pairs $(\alpha, \beta) \in \mathbb{R}^2$. In the 1930s Littlewood conjectured that in fact (1.25) remains correct for *all* pairs of real numbers α, β.

Let $T \subset \mathbb{R}^2$ be the set of all pairs (α, β) for which (1.25) fails. Littlewood's conjecture predicts that T is the empty set. From the argument above, we know that T has Lebesgue measure 0 in $\mathbb{R}^2$. In 2006, Einsiedler, Katok, and Lindenstrauss [382] used deep methods from the theory of dynamical systems to prove the much stronger result that in fact T is a countable union of sets of box dimension 0. We say that set $S \subset \mathbb{R}^2$ has *box dimension* 0 if $\lim_{\epsilon \to 0+} \frac{\log N_S(\epsilon)}{\log(1/\epsilon)} = 0$, where $N_S(\epsilon)$ is the minimal number of squares of side length ϵ needed to cover S.

Theorem 1.51 *The set T of exceptions to Littlewood's conjecture can be written as a union of sets $S_1 \cup S_2 \cup S_3 \cup \ldots$, where each S_i has box dimension 0. In particular, T has Hausdorff dimension 0.*

Littlewood's conjecture predicts that, for some values of n, αn and βn are "close" to being integers. More generally, one may ask whether, given any real numbers γ

and δ, we can find n such that the fractional parts of αn and βn are close to γ and δ, respectively. More formally, for which pairs $(\alpha, \beta) \in \mathbb{R}^2$ do we have

$$\forall \gamma, \delta \in \mathbb{R}, \quad \liminf_{|n| \to \infty} |n| \, ||n\alpha - \gamma|| \, ||n\beta - \delta|| = 0 \ ? \tag{1.26}$$

Before 2011, the existence of even *one* pair $(\alpha, \beta) \in \mathbb{R}^2$ satisfying (1.26) was an open question, raised independently by several researchers, including, for example, Bugeaud [224]. In 2011, Shapira [1114] proved that in fact (1.26) is true for almost all pairs (α, β).

Theorem 1.52 *Statement (1.26) is true for (Lebesgue) almost every pair of real numbers $(\alpha, \beta) \in \mathbb{R}^2$.*

Simultaneous Approximation: Exceptionally Good and Exceptionally Bad Pairs

The two-dimensional version of Dirichlet's approximation theorem states that for any pair of real numbers $(\alpha, \beta) \in \mathbb{R}^2$ and any integer $N > 0$, there exist integers m_1, m_2 and $1 \le n \le N$ such that

$$\max \left\{ \left| \alpha - \frac{m_1}{n} \right|, \left| \beta - \frac{m_2}{n} \right| \right\} \le \frac{1}{n\sqrt{N}}.$$

A vector $(\alpha, \beta) \in \mathbb{R}^2$ is called a *singular pair* if for every $\delta > 0$ there exists an N_0 such that for all $N > N_0$, there exist integers m_1, m_2 and $1 \le n \le N$ such that

$$\max \left\{ \left| \alpha - \frac{m_1}{n} \right|, \left| \beta - \frac{m_2}{n} \right| \right\} \le \frac{\delta}{n\sqrt{N}}.$$

In other words, singular pairs are pairs of numbers which can be simultaneously approximated better than guaranteed by Dirichlet's approximation theorem by an arbitrarily large factor. All points on any line $c_1\alpha + c_2\beta + c_3 = 0$ with rational coefficients c_1, c_2, c_3 are known to be singular pairs. The set of points on such lines has Hausdorff dimension 1. By translating the problem into the language of symbolic dynamics, Cheung [274] proved in 2011 that the set of singular pairs is in fact much larger.

Theorem 1.53 *The set of all singular pairs in $\mathbb{R}^2$ has Hausdorff dimension $\frac{4}{3}$.*

Theorem 1.53 studies pairs of numbers which are simultaneously approximable unusually well. In the opposite direction, one may study numbers (or pairs of numbers) which are approximable unusually badly. A real number α is said to be *badly approximable* if there exists a constant $c(\alpha) > 0$ such that $||n\alpha|| > \frac{c(\alpha)}{n}$ for all $n \in \mathbb{N}$. Khintchine's theorem implies that the set of badly approximable numbers has Lebesgue measure zero.

This notion can be generalized to pairs of numbers. It follows from (1.24) that, for any $\epsilon > 0$, and almost all $(\alpha, \beta) \in \mathbb{R}^2$, we have $\max\{||n\alpha||, ||n\beta||\} \leq \frac{\epsilon}{\sqrt{n}}$ for infinitely many n. Equivalently, $\max\{||n\alpha||^2, ||n\beta||^2\} \leq \frac{\epsilon^2}{n}$ infinitely often. Hence, there is only a measure zero set of "badly approximable" pairs for which this condition is not satisfied.

More generally, let S be the set of pairs of real numbers (i, j) such that $0 \leq i, j \leq 1$ and $i + j = 1$. For $(i; j) \in S$, let $\mathrm{Bad}(i; j)$ denote the set of pairs $(\alpha; \beta) \in \mathbb{R}^2$ for which there exists a positive constant $c = c(\alpha, \beta)$ such that

$$\max\{||n\alpha||^{1/i}, ||n\beta||^{1/j}\} \geq \frac{c}{n} \quad \text{for all } n \in \mathbb{N}.$$

We have observed above that the set $\mathrm{Bad}(1/2; 1/2)$ has (two-dimensional) Lebesgue measure 0. By a similar argument, one can show that set $\mathrm{Bad}(i; j)$ has measure 0 for any $(i, j) \in S$.

In 1983, Schmidt [1089] conjectured that the set $\mathrm{Bad}(1/3; 2/3) \cap \mathrm{Bad}(2/3; 1/3)$ is non-empty. He noted that if someone could prove that this conjecture is false, or, more generally, could find any $(i_1, j_1) \in S$ and $(i_2, j_2) \in S$ such that $\mathrm{Bad}(i_1; j_1) \cap \mathrm{Bad}(i_2; j_2) = \emptyset$, this would imply the truth of Littlewood's conjecture (see Theorem 1.51 and the discussion before it). However, Badziahin, Pollington and Velani [78] proved in 2011 that Schmidt's conjecture is true, even if we allow intersections of more than two sets $\mathrm{Bad}(i_t; j_t)$. Moreover, the intersection is not only non-empty, but in fact has Hausdorff dimension 2, the maximal possible.

Theorem 1.54 *For any finite number $(i_1; j_1); \ldots; (i_k; j_k)$ of pairs from S, the set $\bigcap_{t=1}^{k} \mathrm{Bad}(i_t; j_t)$ is non-empty. Moreover, this set has Hausdorff dimension 2.*

The notion of badly approximable pairs can be generalized to triples, quadruples, and so on. Let $\mathcal{R}_d \subset \mathbb{R}^d$ be the set of $\mathbf{r} = (r_1, \ldots, r_d)$ such that $r_i \geq 0$, $i = 1, \ldots, d$ and $\sum_{i=1}^{d} r_i = 1$. Given $\mathbf{r} \in \mathcal{R}_d$, we say that a point $\mathbf{y} = (y_1, \ldots, y_d) \in \mathbb{R}^d$ is $\mathbf{r}$-*badly approximable* if there exists a $c = c(\mathbf{y}) > 0$ such that

$$\max_{1 \leq i \leq d} ||ny_i||^{1/r_i} \geq \frac{c}{n} \quad \text{for all } n \in \mathbb{N},$$

with the convention that $||ny_i||^{1/0} = 0$. Let $\mathrm{Bad}(\mathbf{r})$ be the set of all $\mathbf{r}$-badly approximable points in $\mathbb{R}^d$.

In 2015, Beresnevich [128] generalized Theorem 1.54 to an arbitrary dimension. The proof uses a lattice-based reformulation of $\mathrm{Bad}(\mathbf{r})$ similar to the one suggested by Dani [315] in 1985, in combination with a novel technique for the efficient counting of lattice points in convex bodies.

Theorem 1.55 *For any finite subset $W \subset \mathcal{R}_d$, the set $\bigcap_{\mathbf{r} \in W} \mathrm{Bad}(\mathbf{r})$ is non-empty. Moreover, this set has Hausdorff dimension d.*

The Structure of the Lagrange Spectrum

As discussed above, the best constant k in inequality (1.20) which works for *all* α is $k = \sqrt{5}$. However, approximability with a better constant k may be achievable for individual irrational numbers. For any $\alpha \in \mathbb{R} \setminus \mathbb{Q}$, let $k(\alpha)$ be the supremum of all $k > 0$ for which inequality (1.20) holds for infinitely many rational numbers $\frac{m}{n}$. The value $k(\alpha)$ measures "how well" the specific irrational number α can be approximated by rationals.

The set $L = \{k(\alpha) \mid \alpha \in \mathbb{R} \setminus \mathbb{Q}, k(\alpha) < +\infty\}$ of all possible finite values of $k(\alpha)$ is called the *Lagrange spectrum*. The set L encodes many Diophantine properties of real numbers. Hurwitz's theorem, which states that $k(\alpha) \geq \sqrt{5}$ for all $\alpha \in \mathbb{R} \setminus \mathbb{Q}$ and this is the best possible, implies that $\sqrt{5}$ is the smallest element of L. It is known that there are only countably many elements of L less than 3, but, on the other hand, L contains all real numbers greater than $c = 4.5728...$ [474]. These facts leave open the intriguing question of understanding the set L between 3 and c. In 2008, Bugeaud [223] studied the function $d(t) = \dim(L \cap (-\infty, t)), t \in \mathbb{R}$, and, in particular, asked if $d(t)$ is continuous. This question was answered by Moreira [907] in 2018. This deeply enriched our understanding of L.

Theorem 1.56 *The function $d(t)$ is a continuous non-decreasing surjective[8] function from $\mathbb{R}$ to $[0, 1]$, such that $\max\{t \in \mathbb{R} \mid d(t) = 0\} = 3$ and $d(\sqrt{12} - \delta) = 1$ for some $\delta > 0$.*

The proof of Theorem 1.56 is based on the idea of approximating parts of the Lagrange spectrum from inside and from outside by sums of regular Cantor sets, see the discussion before Theorem 3.7 below for relevant definitions. In fact, the proof of Theorem 1.56 uses techniques developed by Moreira and Yoccoz in [908] to prove Theorem 3.7.

Approximation by Algebraic Integers

In addition to approximating irrational numbers by rationals with bounded denominator, one may study approximation of transcendental numbers by algebraic integers of given degree with bounded height.[9] In 1969, Davenport and Schmidt [323] proved that there is a constant C such that for any irrational number ξ there are infinitely many algebraic integers α of degree at most 2 such that

$$0 < |\xi - \alpha| < \frac{C}{H(\alpha)^2},$$

---

[8] A function $f : \mathbb{R} \to [0, 1]$ is called *surjective* if for every $x \in [0, 1]$ there exists a $t \in \mathbb{R}$ such that $f(t) = x$.

[9] Recall that a real number α is called an *algebraic number* if there exists a polynomial P with integer coefficients such that $P(\alpha) = 0$. Real numbers which are not algebraic are called *transcendental*. We may and will assume that P is irreducible over the integers. The largest absolute value of the coefficients of P is called the *height* of α. If the leading coefficient of P is 1, α is called an *algebraic integer*. The *degree* of an algebraic integer α is the degree of P.

where $H(\alpha)$ is the height of α, and that this result is the best possible up to a constant factor. The same authors also proved that if ξ cannot be written exactly in the form $\frac{a+b\sqrt{c}}{d}$ for some integers a, b, c, d, then there are infinitely many algebraic integers α of degree at most 3 such that

$$|\xi - \alpha| < \frac{C}{H(\alpha)^{\gamma^2}} \qquad (1.27)$$

for some $C \in \mathbb{R}$, where $\gamma = \frac{1+\sqrt{5}}{2}$ is the golden ratio. Some researchers believed that the exponent $\gamma^2 \approx 2.618$ in (1.27) can be improved to 3. More generally, it has been conjectured that $\tau_n = n$ for all integers $n \geq 2$, where τ_n is the supremum of all $\tau \in \mathbb{R}$ such that for any transcendental $\xi \in \mathbb{R}$ there exist infinitely many algebraic integers α of degree at most n such that $|\xi - \alpha| \leq H(\alpha)^{-\tau}$. The value τ_n measures the quality of approximation of real numbers by algebraic integers of degree n. It is known that $\tau_2 = 2$ and (1.27) implies that $\tau_3 \geq \gamma^2$.

Somewhat surprisingly, Roy [1065] proved in 2003 that the exponent γ^2 in (1.27) cannot be improved.

Theorem 1.57 *There exist a transcendental real number ξ and constant $c > 0$ such that, for any algebraic integer α of degree at most 3, we have $|\xi - \alpha| \geq cH(\alpha)^{-\gamma^2}$.*

Theorem 1.57 is proved by providing explicit examples of such numbers ξ based on the Fibonacci continued fractions. The theorem states that $\tau_3 \leq \gamma^2$. Together with (1.27), this implies that $\tau_3 = \gamma^2 = \frac{3+\sqrt{5}}{2} < 3$. In particular, the $\tau_n = n$ conjecture is false even for $n = 3$.

The quality of approximation can be improved if we approximate transcendental real numbers by algebraic numbers (as opposed to algebraic integers) of bounded degree. In this case, the long-standing Wirsing–Schmidt conjecture (see, for example, [1092]) predicts that for any real transcendental number ξ and any positive integer n there is a constant $C = C(\xi, n) > 0$ such that the inequality

$$0 < |\xi - \alpha| < \frac{C}{H(\alpha)^{n+1}} \qquad (1.28)$$

holds for infinitely many algebraic numbers α of degree at most n. The conjecture is easy for $n = 1$ and was proved by Davenport and Schmidt [324] in 1967 for $n = 2$, but remains open for $n > 2$. The upper bound in (1.28), if true, is optimal up to a constant factor. In 2015, Beresnevich [128] proved that, for every $N > 0$, there exists a transcendental number ξ satisfying (1.28) and a constant $C'(\xi, N) > 0$ such that $|\xi - \alpha| \geq \frac{C'}{H(\alpha)^{n+1}}$ for all algebraic numbers α of degree $n \leq N$. Moreover, the Hausdorff dimension of such ξ in any interval I is equal to 1.

1.9   Elliptic Curves

Integer Points on an Elliptic Curve

A fundamental problem in number theory is finding integer and rational solutions
to polynomial equations. This problem is easy for equations in one variable, as well
as for linear and quadratic equations in two variables. Hence, the first interesting
case is cubic equations in two variables with rational coefficients. The set of pairs
of real numbers satisfying such an equation is called a *general cubic curve*, and the
problem is to analyse integer or rational points on such a curve. It is known that if
such a curve is non-singular[10] and contains at least one rational point, then, after an
appropriate change of variables, it can be written in a unique way in the form

$$y^2 = x^3 + ax + b, \tag{1.29}$$

where a, b are integers such that (i) if $p^4 | a$ for some prime p, then $p^6 \nmid b$, and
(ii) $\Delta_{a,b} = -16(4a^3 + 27b^2) \neq 0$. The curve (1.29) is called a *non-singular
elliptic curve over $\mathbb{Q}$ in Weierstrass form*, or just an "elliptic curve" for short. The
quantity $\Delta_{a,b}$ is called the *discriminant* of the elliptic curve (1.29). The quantity
$h = \max\{4|a|^3, 27b^2\}$ is called its *height*.

A fundamental theorem of Siegel [1127] implies that Eq. (1.29) has only finitely
many integer solutions $x, y \in \mathbb{Z}$. In 1992, Schmidt [1090] proved that the number
$N_{a,b}$ of integer solutions to (1.29) is at most $C_\epsilon |\Delta_{a,b}|^{1/2+\epsilon}$, for any $\epsilon > 0$. In
2006, Helfgott and Venkatesh [591] improved the constant $1/2$ in the exponent to
$0.2007....$

Theorem 1.58 *For every $\epsilon > 0$ there is a constant $C_\epsilon < \infty$ such that*

$$N_{a,b} \leq C_\epsilon |\Delta_{a,b}|^{\beta+\epsilon},$$

where $\beta = \dfrac{4\sqrt{3}\log(2+\sqrt{3}) - 6\log 2 - 3\log 3}{12\log 2} = 0.2007....$

The bound in Theorem 1.58 remained unimproved for over a decade. In 2020,
Bhargava et al. [144] improved the constant in the exponent by a factor of almost 2.

Theorem 1.59 *For every $\epsilon > 0$ there is a constant $C_\epsilon < \infty$ such that*

$$N_{a,b} \leq C_\epsilon |\Delta_{a,b}|^{\alpha+\epsilon},$$

where $\alpha = 0.1117...$ is an explicit constant.

---

[10] A curve is *non-singular* if it has no singular points. Intuitively, singular points are "cusps" (like
the point $(0, 0)$ on the curve $x^2 = y^3$) or self-intersections.

The Rank of an Elliptic Curve

While the number of integer solutions to (1.29) is finite, the structure of *rational* solutions to (1.29) is much more interesting. It is known that if $P = (x_1, y_1)$ and $Q = (x_2, y_2)$ are rational points on (1.29) such that $x_1 \neq x_2$, then the line $l = PQ$ intersects the curve (1.29) in a third rational point (x_3, y_3). The point $R = (x_3, -y_3)$ is called the "sum" of the rational points P and Q, and we write

$$P + Q = R. \tag{1.30}$$

If $P = Q$, then the sum is defined in the same way, except that l is now the line tangent to the curve at P. If we also define $P + Q = \infty$ when $x_1 = x_2$ but $y_1 \neq y_2$, and $\infty + P = P + \infty = P$, then the set

$$E(\mathbb{Q}) = \{\text{rational sotutions to (1.29)}\} \cup \{\infty\} \tag{1.31}$$

forms a group with operation "+" and identity element $e = \infty$. A fundamental theorem of Mordell [906] states that this group is isomorphic to the direct product $\mathbb{Z}^r \oplus E(\mathbb{Q})_{\text{tors}}$, where $r \geq 0$ is an integer, and $E(\mathbb{Q})_{\text{tors}}$ is the subgroup of elements of finite order[11] in $E(\mathbb{Q})$. The integer r is called the *rank* of the elliptic curve E and is written as rank(E).

An old theorem of Nagell [920] and Lutz [822] states that if $(x, y) \in E(\mathbb{Q})_{\text{tors}}$ and $(x, y) \neq \infty$, then x, y are integers, and either $y = 0$ or y^2 is a divisor of $4a^3 + 27b^2$. Given the coefficients a and b in (1.29), this theorem gives an easy method for computing $E(\mathbb{Q})_{\text{tors}}$. If we could also compute rank r, we would determine the group $E(\mathbb{Q})$, up to isomorphism. However, no algorithm for computing the rank of an elliptic curve is known. A famous conjecture of Birch and Swinnerton-Dyer [150] proposes a formula for r, but it remains open, and is one of the problems for which a million dollars is offered for a solution [252].

Many other basic questions about rank remain open. For example, it is conjectured but remains unproven that elliptic curves with rank $r \geq 2$ are rare. There is an old conjecture that if the elliptic curves are ordered by height, then 50% of all elliptic curves have rank 0, and 50% have rank 1, which leaves 0% for the curves of rank $r \geq 2$. In particular, this conjecture would imply that the average rank of all elliptic curves is 0.5. However, before 2015, no one could prove that the average rank is bounded from above by any finite constant whatsoever. In 2015, Bhargava and Shankar [142] proved that the average rank is at most 1.5.

Theorem 1.60 *Let $S(h)$ be the set of elliptic curves of height at most h. Then*

$$\limsup_{h \to \infty} \left(\frac{1}{|S(h)|} \sum_{E \in S(h)} \text{rank}(E) \right) \leq 1.5.$$

---

[11] The *order* of a non-identity element a in a group G is the lowest positive integer n such that $a^n = e$, with the convention that the order is infinite if no such n exists.

The proof of Theorem 1.60 relies on counting the asymptotic number of binary quartic forms having bounded invariants. The idea goes back to Gauss, who studied integral binary quadratic forms $f(x, y) = ax^2 + bxy + cy^2$ with integer coefficients. Gauss considered two such forms *equivalent* if they can be transformed into each other by a linear invertible change of variables $x = px' + qy'$, $y = rx' + sy'$, where p, q, r, s are integers such that $ps - qr = 1$. He observed that any two equivalent quadratic forms have the same *discriminant* $\Delta(f) = b^2 - 4ac$, and posed the question of counting the number of equivalence classes of irreducible integral binary quadratic forms having discriminant D. This question was answered by Siegel [1128] in 1944. In 1951, Davenport [320] solved the corresponding question for binary cubic forms. The next case is that of binary quartic forms

$$f(x, y) = ax^4 + bx^3y + cx^2y^2 + dxy^3 + ey^4$$

with integer coefficients a, b, c, d, e. In this case, there are two independent invariants

$$I(f) = 12ae - 3bd + c^2 \quad \text{and} \quad J(f) = 72ace + 9bcd - 27ad^2 - 27eb^2 - 2c^3$$

preserved by linear changes of variables as above, and the question becomes to count the number of equivalence classes of integral binary quartic forms with bounded values of I and J. Bhargava and Shankar [142] answered this question and deduced Theorem 1.60 from there.

Even with Theorem 1.60, it was still possible that only 0% of elliptic curves have rank 0. In a later work [143], the same authors improved the bound in Theorem 1.60 from 1.5 to $\frac{7}{6} < 1.17$, and also proved that a positive proportion of elliptic curves have rank 0.

Theorem 1.61 *Let $S(h)$ be the set of elliptic curves of height at most h. Also, let R_0 be the set of all elliptic curves having rank 0. Then*

(i) $\limsup\limits_{h \to \infty} \left(\frac{1}{|S(h)|} \sum\limits_{E \in S(h)} \text{rank}(E) \right) \leq \frac{7}{6}$, *and*

(ii) $\liminf\limits_{h \to \infty} \frac{|S(h) \cap R_0|}{|S(h)|} > 0$.

The proof method of Theorem 1.61 is similar to that for Theorem 1.60, except this time the authors needed to solve the more complicated problem of counting the equivalence classes of integral ternary (that is, in three variables x, y, z) cubic forms having bounded invariants.

Another fascinating open question about rank is whether it can be arbitrarily large. In 2006, Elkies discovered an elliptic curve (1.29) with rank $r \geq 28$, and this is the current record. In 1967, Shafarevich and Tate [1193] solved a version of this problem for elliptic curves over function fields. Let $\mathbb{F}_p$ be the field of integers with operations modulo a prime p, and let $\mathbb{F}_p(t)$ be the field of rational functions of the form

$$f(t) = \frac{a_m t^m + a_{m-1} t^{m-1} + \cdots + a_1 t + a_0}{b_n t^n + b_{n-1} t^{n-1} + \cdots + b_1 t + b_0},$$

where $a_m, \ldots, a_1, a_0, b_n, \ldots, b_1, b_0 \in \mathbb{F}_p$. An *elliptic curve E over the field* $\mathbb{F}_p(t)$ is a curve (1.29) with coefficients $a, b \in \mathbb{F}_p(t)$ with $\Delta_{a,b} = -16(4a^3 + 27b^2) \neq 0$, and we are looking for solutions $x, y \in \mathbb{F}_p(t)$. Mordell's theorem [906] still holds in this generality, and the rank $r = \text{rank}(E)$ is defined exactly as above. In 1967 Shafarevich and Tate [1193] produced elliptic curves over $\mathbb{F}_p(t)$ of arbitrarily large rank.

There was a hope that this result could give an insight into how to produce elliptic curves of large rank over $\mathbb{Q}$, but this did not happen. A possible reason is that the curves produced by Shafarevitch and Tate have a special property, called isotriviality, which has no analogue for elliptic curves over $\mathbb{Q}$. The *j-invariant* of an elliptic curve E given by (1.29) is $j(E) = 1728 \frac{4a^3}{4a^3 + 27b^2}$. An elliptic curve E over the field $\mathbb{F}_p(t)$ is called *isotrivial* if $j(E) \in \mathbb{F}_p$, and *non-isotrivial* otherwise. Since 1967, there was an open question which asked whether there exist non-isotrivial examples of elliptic curves over $\mathbb{F}_p(t)$ of arbitrarily large rank. In 2002, Ulmer [1221] constructed such examples.

Theorem 1.62 *The elliptic curve E defined over the field $\mathbb{F}_p(t)$ by the equation $y^2 + xy = x^3 - t^d$, where $d = p^n + 1$ for some positive integer n, is non-isotrivial and has rank at least $(p^n - 1)/2n$.*

The proof of Theorem 1.62 is based on establishing a version of the Birch and Swinnerton-Dyer conjecture for this class of curves, which makes it possible to use zeta functions to determine the rank.

The Sato–Tate Conjecture
A central theme in the theory of elliptic curves is to estimate the number N_p of solutions to (1.29) in the p-element field $\mathbb{F}_p$. In 1936, Hasse [575], confirming a 1924 conjecture of Artin, proved that

$$p + 1 - 2\sqrt{p} \leq N_p \leq p + 1 + 2\sqrt{p} \tag{1.32}$$

for every prime p which is not a divisor of the discriminant $\Delta_{a,b}$. In other words, $N_p \approx p + 1$ with error at most $2\sqrt{p}$. There are some elliptic curves for which N_p is exactly equal to $p + 1$ for half of the primes p, that is,

$$\lim_{N \to \infty} \frac{|\{p \in \mathcal{P} \mid p \leq N, N_p = p + 1\}|}{|\{p \in \mathcal{P} \mid p \leq N\}|} = \frac{1}{2}, \tag{1.33}$$

where $\mathcal{P}$ is the set of prime numbers not dividing $\Delta_{a,b}$, and $|\cdot|$ denotes the number of elements in a finite set. The elliptic curves for which (1.33) holds are known as curves "with complex multiplication". However, it is known that if elliptic curves (1.29) are ordered, by, for example, height $h = \max\{4|a|^3, 27b^2\}$, then the proportion of curves with complex multiplication tends to 0 as $h \to \infty$. In other words, (1.33) fails for the vast majority of elliptic curves. Such curves are known as curves "without complex multiplication", and understanding how N_p is distributed within the interval $[p + 1 - 2\sqrt{p}, p + 1 + 2\sqrt{p}]$ for such curves has been a major research direction.

It follows from (1.32) that $-1 \leq \frac{p+1-N_p}{2\sqrt{p}} \leq 1$, which implies the existence of an angle $\theta_p \in [0, \pi]$ such that $\cos \theta_p = \frac{p+1-N_p}{2\sqrt{p}}$. In 1960, Sato and Tate independently conjectured that for every elliptic curve over $\mathbb{Q}$ without complex multiplication and for every two real numbers $0 \leq \alpha < \beta \leq 1$, one has

$$\lim_{N \to \infty} \frac{|\{p \in \mathcal{P} \mid p \leq N, \, \alpha \leq \theta_p \leq \beta\}|}{|\{p \in \mathcal{P} \mid p \leq N\}|} = \frac{2}{\pi} \int_\alpha^\beta \sin^2 \theta \, d\theta. \tag{1.34}$$

This conjecture gives much more precise information about N_p than Hasse's theorem (1.32). The conjecture attracted a lot of attention and has been proved by many authors under various additional conditions. Following a large chain of partial results, Barnet-Lamb et al. [100] finally proved the conjecture in full in 2011.

Theorem 1.63 *Equality* (1.34) *holds for every elliptic curve over* $\mathbb{Q}$ *without complex multiplication.*

The Modularity Theorem
The theory of elliptic curves has been a central instrument in the resolution of many important problems whose formulation had nothing to do with elliptic curves. The most famous example of such a story is the proof of Fermat's Last Theorem. Around 1637, Fermat stated without proof that, for any integer $n \geq 3$, there are no positive integers x, y, z such that $x^n + y^n = z^n$. This statement became known as *Fermat's Last Theorem*, and was the most famous open problem in mathematics for over 300 years.

In 1955, Taniyama and Shimura studied the connection of elliptic curves to the theory of functions f defined on the complex upper half-plane $\mathbb{H} = \{z \in \mathbb{C}, \text{Im}(z) > 0\}$, satisfying the symmetries

$$f(z) = f\left(\frac{az+b}{cz+d}\right), \tag{1.35}$$

where a, b, c, d are integers such that $ad - bc = 1$. For example, all such symmetries are satisfied by the function

$$j(z) = 1728 \frac{20G_4(z)^3}{20G_4(z)^3 - 49G_6(z)^2}, \quad \text{where} \quad G_k(z) = \sum_{(m,n) \neq (0,0)} (m + nz)^{-k},$$

and, more generally, by any function $f : H \to \mathbb{C}$ of the form

$$f(z) = \frac{P(j(z))}{Q(j(z))}, \tag{1.36}$$

where P and Q are some polynomials with complex coefficients. Functions of the form (1.36) are called *modular functions of level 1*, see [46, Theorem 2.8]. More generally, for any integer N, one may define *modular functions of level N*, which satisfy all symmetries (1.35) such that $ad - bc = 1$ and c is a multiple of N.

An elliptic curve (1.29) is called *modular*[12] if there exist non-constant modular functions $f(z)$, $g(z)$ of the same level N such that $f(z)^2 = g(z)^3 + ag(z) + b$. Taniyama and Shimura conjectured that in fact *every elliptic curve over $\mathbb{Q}$ is modular*. This statement became known as Taniyama–Shimura conjecture, or the modularity conjecture.

In 1986, Ribet [1043], confirming a conjecture of Serre, proved that if Fermat's Last Theorem is wrong, then there exists an elliptic curve (now known as the *Frey curve*) which is not modular. In 1995, Wiles [1262] proved the modularity conjecture for a class of elliptic curves which includes the Frey curve. The combination of these results confirms Fermat's Last Theorem.

While Wiles' theorem is, without doubt, one of the greatest theorems of the twentieth century, it leaves open whether the modularity conjecture is true for all elliptic curves over $\mathbb{Q}$. In 2001, Breuil et al. [208] confirmed the modularity conjecture in full.

Theorem 1.64 *Every elliptic curve over $\mathbb{Q}$ is modular.*

The proof of Theorem 1.64 builds on Wiles's work, but enriches and deepens his ideas to treat the remaining cases. Since 2001, Theorem 1.64 and the methods developed for its proof have been used to prove many other results of central importance in number theory, including, for example, Theorems 1.18, 1.27 and 1.37 discussed earlier in this book.

The Congruent Number Problem
This section describes another example of application of the theory of elliptic curves to a nice and old problem in number theory. A positive integer n is called a *congruent number* if it is the area of some right triangle with rational side lengths. In other words, $n = \frac{1}{2}ab$ for some rational numbers a, b such that $c = \sqrt{a^2 + b^2}$ is also a rational number. Determining whether or not a given number is congruent is called the *congruent number problem* and is one of the oldest unsolved problems in mathematics with over 1000 years of history.

If n is a congruent number, then $x = \frac{n(a+c)}{b}$ and $y = \frac{2n^2(a+c)}{b^2}$ is a non-zero rational point on the elliptic curve

$$y^2 = x^3 - n^2 x. \tag{1.37}$$

---

[12] There are many equivalent definitions of this concept, we follow the one given in [307].

Conversely, if (x, y) is a non-zero rational point on (1.37), then n is the area of a right triangle with side lengths $a = \frac{x^2-n^2}{y}$, $b = \frac{2nx}{y}$, and $c = \frac{x^2+n^2}{y}$. Hence, the congruent number problem is equivalent to determining whether (1.37) has a non-zero rational solution. Using this equivalence, Tunnell [1215] solved the congruent number problem assuming that the Birch and Swinnerton-Dyer conjecture is true for (1.37).

A related problem is to determine what proportion of positive integers are congruent numbers. In 2016, Smith posted online a preprint [1138] in which he proved that this proportion is strictly positive. Moreover, this is true even if we require that n is equal to 5, 6, or 7 mod 8. In a later work, Smith [1139] proved that this is not true if n is equal to 1, 2, or 3 mod 8.

Theorem 1.65 *The set of congruent numbers equal to 1, 2, or 3 mod 8 has zero natural density in* $\mathbb{N}$.

The proof of Theorem 1.65 uses the equivalence of the congruent number problem with (1.37) and is based on the deep theory of elliptic curves.

Hyperelliptic Curves
Elliptic curves (1.29) are special cases of a larger family of curves, defined by polynomial equations of higher degree. Let $\mathcal{P}$ be the set of odd-degree polynomials $P(x) = \sum_{i=0}^{2g+1} a_i x^{2g+1-i}$ with rational coefficients a_i and leading coefficient $a_0 = 1$. The set of real solutions to the equation

$$y^2 = P(x) \tag{1.38}$$

for $P \in \mathcal{P}$ is known as an odd degree *hyperelliptic curve*, and rational solutions to (1.38) are called finite rational points on this curve. With the change of variables $x' = u^2 x$, $y' = u^{2g+1} y$, we obtain new coefficients $a_i' = u^{2i} a_i$, $i = 1, 2, \ldots, 2g+1$, and, by selecting u to be the least common denominator of $a_1, a_2, \ldots, a_{2g+1}$, we can make all coefficients integers. After this, define the *height* of $P \in \mathcal{P}$ by $H(P) = \max\{|a_1|, |a_2|^{1/2}, \ldots, |a_{2g+1}|^{1/(2g+1)}\}$. In 2014, Poonen and Stoll [1003] proved that if odd degree hyperelliptic curves are ordered by height, then most of them have no finite rational points.

Theorem 1.66 *For a fixed integer* $g > 0$ *and real* $X > 0$, *let* $\mu(X, g)$ *be the fraction of the polynomials* $P \in \mathcal{P}$ *of degree* $2g + 1$ *and height less than* X *for which the equation* $y^2 = P(x)$ *has no rational solutions. Then*

$$\lim_{g \to \infty} (\lim_{X \to \infty} \mu(X, g)) = 1.$$

Poonen and Stoll also proved in [1003] that for "almost all" polynomials $P \in \mathcal{P}$ (in the same sense as in Theorem 1.66), there is an explicit algorithm, with the coefficients of P as input, and output certifying that there are indeed no rational solutions to $y^2 = P(x)$. In other words, there exists a universal method which is able to solve almost all equations in the form $y^2 = P(x)$, $P \in \mathcal{P}$, at once.

The proof of Theorem 1.66 depends crucially on the methods developed by Bhargava and Shankar in [142] to prove Theorem 1.60, which was later extended to hyperelliptic curves by Bhargava and Gross in [141].

1.10   Other Topics in Number Theory

Properties of Expansions of Real Numbers

The most standard way to write a real number is its (finite or infinite) decimal expansion. Some very basic properties about digits in this expansion remain open. For example, we cannot even prove that every digit occurs infinitely often in the decimal expansion of $\sqrt{2}$.

More generally, the *complexity function* (or just "complexity") of a real number α is the number $p(n)$ of distinct blocks of digits of length n occurring in its decimal expansion. It is natural to conjecture that the decimal expansion of every irrational algebraic number[13] contains all possible digit patterns, and $p(n) = 10^n$. However, only much weaker lower bounds on $p(n)$ for algebraic α are known. The best result before 2007 was the 1997 theorem of Ferenczi and Mauduit [451] stating that

$$\liminf_{n \to \infty}(p(n) - n) = +\infty. \tag{1.39}$$

In 2007, Adamczewski and Bugeaud [5] proved a significantly better lower bound for $p(n)$.

Theorem 1.67 *The complexity function $p(n)$ of every irrational algebraic number α satisfies*

$$\liminf_{n \to \infty} \frac{p(n)}{n} = +\infty. \tag{1.40}$$

The main tool in the proof of Theorem 1.67 is a new combinatorial criterion for the real number to be transcendental. This criterion is a refinement of an earlier transcendence criterion developed by Ferenczi and Mauduit in [451] to prove (1.39). Using the refined criterion, Adamczewski and Bugeaud showed that if an irrational number α does not satisfy (1.40), then it must be transcendental. Moreover, Theorem 1.67 remains true in any b-ary expansion with any base $b \geq 2$. The reader may compare it with Theorem 1.43, stating a similar result for the complexity of the Liouville function.

Another useful way to write real numbers is as a continued fraction. For a real number x, define the sequence $\{x_n\}$ by the rules: (i) $x_0 = x$, and (ii) for $n \geq 0$, if x_n is an integer then stop, otherwise let $x_{n+1} = \frac{1}{x_n - \lfloor x_n \rfloor}$, where $\lfloor z \rfloor$ denotes the largest integer not exceeding z. This sequence may be finite or infinite depending on x. Let

---

[13] A real number α is called an *algebraic number* if there exists a polynomial P with integer coefficients such that $P(\alpha) = 0$.

$a_n = \lfloor x_n \rfloor$ for all n. The sequence $\{a_i\}$ gives the *continued fraction expansion* of x, which is notated as $x = [a_0; a_1, a_2, \ldots, a_n, \ldots]$. It has many applications, for example, the continued fraction expansion of $\sqrt{d}$ can be used to find a solution to the negative Pell equation (1.7), provided that it exists.

If we fix a bound A and study the rational numbers which have a finite continued fraction expansion $[a_0; a_1, a_2, \ldots, a_n]$ such that $a_i \leq A$, $\forall i$, what positive integers can occur as a denominator of such a number in reduced form? In 1972, Zaremba [1288] conjectured that in fact every positive integer can. More formally, for $A > 0$, let $D(A)$ be the set of $d \in \mathbb{N}$ for which there exist an integer b such that $0 < b < d$, $\gcd(b, d) = 1$, and $\frac{b}{d} = [0; a_1, \ldots a_n]$ with $a_k \leq A$, $k = 1, \ldots, n$. Zaremba conjectured that there exists an $A > 0$ such that $D(A) = \mathbb{N}$, and even suggested that $A = 5$ may work. Zaremba's conjecture has important applications to pseudorandom number generation and numerical integration, but remains open. However, Bourgain and Kontorovich [190] proved in 2014 that, for $A = 50$, "almost every" positive integer belongs to $D(A)$.

Theorem 1.68 *Let* $K_A(N)$ *be the number of integers less than N belonging to* $D(A)$. *Then there exists a constant A such that* $\lim_{N \to \infty} \frac{K_A(N)}{N} = 1$. *In fact, this is true for $A = 50$.*

In 2015, Huang [625] proved that Theorem 1.68 remains true for $A = 5$.

Walking on a Circle with Irrational Step Size
Let us select any point A_0 on a circle with centre O, and let $A_0, A_1, A_2 \ldots$ be a sequence of points on the circle such that the angle $A_n O A_0$ is $n\beta$, where β is some fixed angle. If $\alpha = \frac{\beta}{2\pi}$ is rational, this sequence is periodic. Otherwise, for each n, let us select a point Z on the circle such that $\prod_{k=0}^{n} |Z A_k|$ is maximal. In 1959, Erdős and Szekeres [421] asked if this product goes to infinity with n. In the language of complex numbers, it asks whether $\lim_{n \to \infty} \max_{|z|=1} \prod_{k=1}^{n} \left| z - e^{2\pi i k \alpha} \right| = \infty$. In 2017, Avila, Jitomirskaya, and Marx [62] answered this question negatively.

Theorem 1.69 *For any irrational α, one has*

$$\liminf_{n \to \infty} \max_{|z|=1} \prod_{k=1}^{n} \left| z - e^{2\pi i k \alpha} \right| < \infty.$$

Interestingly, the authors needed to solve this elementary-looking number theoretic problem on the way to deriving deep results related to the spectral theory of Harper's model used in statistical mechanics, see [62] for details.

Minkowski's Conjecture About Unimodular Lattices
A *lattice* L in $\mathbb{R}^n$ is a set of the form

$$L = \left\{ \sum_{i=1}^{n} a_i v_i \;\middle|\; a_i \in \mathbb{Z} \right\},$$

where $v_1, \ldots, v_n$ is a basis for $\mathbb{R}^n$. A lattice L is called *unimodular* if the determinant of the $n \times n$ matrix with entries $\langle v_i, v_j \rangle$ is 1 or -1, where $\langle \cdot, \cdot \rangle$ denotes the scalar product. This property does not depend on the choice of basis for L.

For $x = (x_1, \ldots, x_n) \in \mathbb{R}^n$, let $N(x) = |x_1 \cdot x_2 \cdots \cdot x_n|$. *Minkowski's conjecture* states that for any unimodular lattice $L \subset \mathbb{R}^n$, and any $x \in \mathbb{R}^n$, there is a $y \in L$ such that $N(x - y) \leq 2^{-n}$. The conjecture is interesting on its own, and also has number-theoretic consequences. In 1976, Skubenko [1134] proved it for $n \leq 5$, but no further progress happened for almost 30 years, until 2005, when McMullen [866] resolved the conjecture for $n = 6$.

Theorem 1.70 *For any unimodular lattice $L \subset \mathbb{R}^6$, and any $x \in \mathbb{R}^6$, there is a $y \in L$ such that $N(x - y) \leq 2^{-6}$.*

The proof of Theorem 1.70 follows from reduction of Minkowski's conjecture to an easier conjecture about well-rounded lattices. Given a lattice $L \subset \mathbb{R}^n$, let $|L|$ be the infimum of $|y|$ over all non-zero $y \in L$. A vector $y \in L$ is called *minimal* if $|y| = |L|$, and the lattice L is called *well-rounded* if its minimal vectors span $\mathbb{R}^n$. The *covering conjecture* (C_n) of Woods [1273] states that for any well-rounded unimodular lattice $L \subset \mathbb{R}^n$ we have

$$\sup_{x \in \mathbb{R}^n} \inf_{y \in L} |x - y| \leq \frac{\sqrt{n}}{2}.$$

McMullen [866] proved that, for all integers $n_0 > 0$, if the covering conjecture (C_n) holds for all $n \leq n_0$, then Minkowski's conjecture is also true for all lattices in $\mathbb{R}^n$, $n \leq n_0$. Because the covering conjecture (C_n) was proved for all $n \leq 6$ by Woods [1273] in 1972, Theorem 1.70 follows. Moreover, McMullen's result opened the road for further progress, allowing Minkowski's conjecture to be proved by establishing the covering conjecture in the corresponding dimensions. Using this method, Hans-Gill, Raka and Sehmi proved Minkowski's conjecture for $n = 7$ in 2009 [561] and for $n = 8$ in 2011 [562]. Then Kathuria and Raka resolved the case $n = 9$ in 2016 [690] and $n = 10$ in 2020 [691]. However, Regev, Shapira and Weiss [1035] in 2017 constructed a counterexample to the covering conjecture (C_n) in dimensions $n \geq 30$. Hence, fundamentally new ideas are needed to prove Minkowski's conjecture in high dimensions.

Chapter 2
Combinatorics

2.1 Ramsey Theory on the Integers

We start the combinatorics chapter by discussing topics in combinatorial number theory. These topics are on the border of number theory and combinatorics, and would fit equally well in the number theory chapter. We start with Ramsey theory on the integers.

In 1916, Schur [1098] proved that, for any integer $r > 0$, and for any partition of the set $\mathbb{N}$ of positive integers into r parts $C_1, \ldots, C_r$ (called a "finite colouring"), we can always find integers x, y, z from the same part C_i, such that $x + y = z$. In fact, the same is true if we partition only the integers $\{1, 2, \ldots, N\}$, where N is a constant depending on r. The smallest N with this property is denoted $S(r)$ and called the *r-th Schur number*. It is known that $S(1) = 2$, $S(2) = 5$, $S(3) = 14$, $S(4) = 45$, and $S(5) = 161$. The value of $S(5)$ was computed by Heule [601] in 2018, and the (computer-assisted) proof took up 2 petabytes of space.

In 1927, van der Waerden [1240] proved that, for any partition of $\mathbb{N}$ into r parts, and for any k, we can find an arithmetic progression of length k contained in the same part C_i. Schur's and van der Waerden's theorems originated the Ramsey theory, which aims to understand which patterns can be found in one colour of any finite colouring of $\mathbb{N}$. A set $S \subset \mathbb{N}$ is called *monochromatic* if $S \subset C_i$ for some i. A set of k polynomials $P_1, \ldots, P_k$ in s variables $x_1, \ldots, x_s$ with integer coefficients is called a *Ramsey family* if for any finite colouring of $\mathbb{N}$ there exist $x_1, \ldots, x_s \in \mathbb{N}$ such that the set $\{P_1(x_1, \ldots, x_s), \ldots, P_k(x_1, \ldots, x_s)\}$ is monochromatic. In this language, Schur's theorem states that the set $\{x, y, x + y\}$ is Ramsey, while van der Waerden's theorem states that the same is true for the set $\{x, x+y, \ldots, x+(k-1)y\}$. More generally, in 1933 Rado [1015] developed a method which allows one to determine whether a family of linear polynomials is Ramsey or not. A classical problem asks to extend Rado's result to general polynomials. However, before 2017, it was not even known if the simple family $\{xy, x + y\}$ is Ramsey. In 2017, Moreira [909] proved that this is the case, even for the larger family $\{x, xy, x + y\}$.

© The Author(s), under exclusive license to Springer Nature Switzerland AG 2021
B. Grechuk, *Landscape of 21st Century Mathematics*,
https://doi.org/10.1007/978-3-030-80627-9_2

Theorem 2.1 *For any finite colouring of* $\mathbb{N}$ *there exist a pair (in fact, infinitely many pairs)* x, $y \in \mathbb{N}$ *such that the set* $\{x, xy, x + y\}$ *is monochromatic.*

Theorem 2.1 was proved by first developing a "correspondence principle" which allowed the problem to be translated to the language of topological dynamics, and then solving the dynamical problem using the methods available in that area.

A notion closely related to "Ramsey family" is "partition regularity". A polynomial equation $P(x_1, \ldots, x_s) = 0$ is called *partition regular* if any finite colouring of $\mathbb{N}$ contains a monochromatic set $\{x_1, \ldots, x_s\}$ of distinct integers such that $P(x_1, \ldots, x_s) = 0$. In this language, Schur's theorem states that the equation $x + y = z$ is partition regular, and Rado's Theorem [1015] can be used to decide partition regularity for any linear equation. Using the methods developed to prove Theorem 2.1, Moreira [909] was able to establish partition regularity for some new non-linear equations, for example, $x^2 - y^2 = z$ and $x^2 + 2y^2 - 3z^2 = w$. However, in general this problem is notoriously hard even for simple quadratic equations. A famous question of Erdős and Graham [409] asks if the equation $x^2 + y^2 = z^2$ is partition regular. In fact, before 2016, this question was open even for $r = 2$ colours. In 2016, Heule et al. [602] solved the two-colour case.

Theorem 2.2 *The set* $\{1, \ldots, 7824\}$ *can be partitioned into two parts, such that no part contains a Pythagorean triple (set of three positive integers* x, y, *and* z, *such that* $x^2 + y^2 = z^2$), *while this is impossible for* $\{1, \ldots, 7825\}$.

The proof of Theorem 2.2 proceeds by reformulating the problem as a satisfiability (SAT) problem,[1] and then uses special computer programs, SAT solvers, to solve the latter. The SAT solvers can either find a satisfying assignment of variables, or produce a certificate (proof) confirming that such an assignment does not exist. In the case of Theorem 2.2, the resulting certificate is almost 200 terabytes in size.

The problem remains open for more colours. More generally, there are no integers a, b, c for which the equation $ax^2 + by^2 = cz^2$ is known to be partition regular. The problem becomes somewhat easier if we allow one variable, which we denote by λ, to vary freely. We say that an equation $p(x, y, \lambda) = 0$ is *partition λ-regular* if for every finite colouring of $\mathbb{N}$, one can find distinct x, y of the same colour that satisfy the equation for some integer λ. A classical theorem of Furstenberg and Sárközy [484, 1081] states that we can always find a pair of same-coloured integers whose difference is a perfect square, or, in other words, that the equation $x - y = \lambda^2$ is partition λ-regular. In 2006, Khalfalah and Szemerédi [708] proved the same for the equation $x + y = \lambda^2$. In 2017, Frantzikinakis and Host [469] established partition λ-regularity for an infinite family of quadratic equations.

---

[1] Let $x_1, x_2, \ldots, x_n$ be boolean variables. A *clause* is any expression of the form $y_1 \vee y_2 \vee \cdots \vee y_k$, where each y_i is either x_j or $\neg x_j$ for some j. The *SAT problem* asks, given m clauses, to decide if it is possible to satisfy them all.

Theorem 2.3 *Let $p(x, y, z) = ax^2 + by^2 + cz^2 + dxy + exz + fyz$, where a, b, c are non-zero and d, e, f are arbitrary integers. Suppose that the three integers $e^2 - 4ac$, $f^2 - 4bc$ and $(e + f)^2 - 4c(a + b + d)$ are all non-zero perfect squares. Then the equation $p(x, y, \lambda) = 0$ is partition λ-regular.*

In particular, Theorem 2.3 implies that the equation $ax^2 + by^2 = c\lambda^2$ is partition λ-regular whenever the integers a, b, c, and $a + b$ are non-zero perfect squares. For example, this holds for the equation $16x^2 + 9y^2 = \lambda^2$. Theorem 2.3 is proved using a surprising connection between partition regularity and the problem of establishing the asymptotic behaviour of averages of multiplicative functions, which is a central topic in analytic number theory, see, for example, Theorems 1.40 and 1.42.

One may also look for different monochromatic patterns in finite colourings of $\mathbb{N}$, not related to polynomials and polynomial equations. For example, Erdős and Graham [409] asked in 1980 whether any finite colouring of the integers $2, 3, 4, \ldots$ contains a monochromatic finite set S such that

$$\sum_{n \in S} \frac{1}{n} = 1. \tag{2.1}$$

In 2003, Croot [312] answered this question affirmatively.

Theorem 2.4 *There exists a constant b such that for every partition of the integers in $[2, b^r]$ into r classes, there is always one class containing a subset S satisfying (2.1).*

Theorem 2.4 not only answers the Erdős–Graham question positively, but also states that the set S can be composed only of integers from 2 to b^r. Croot also proved that, for sufficiently large r, one may choose $b = e^{167000}$. However, Theorem 2.4 fails for any $b < e$, hence the bound b^r is the best possible up to the value of b.

2.2 Patterns in Sets of Integers

3-Term Arithmetic Progressions
For integer $N > 0$ and $k \geq 3$, let $r_k(N)$ be the cardinality of the largest subset of $\{1, 2, \ldots, N\}$ which contains no nontrivial k-term arithmetic progressions. Szemerédi's celebrated theorem [1174] states that, for any $k \geq 3$,

$$\lim_{N \to \infty} \frac{r_k(N)}{N} = 0. \tag{2.2}$$

Equivalently, every subset of integers of positive upper density[2] contains infinitely many k-term arithmetic progressions for all k. A famous conjecture of Erdős [407] states that the same is true for any set A of positive integers such that $\sum_{n \in A} \frac{1}{n}$ diverges. This conjecture would follow from the upper bound

$$r_k(N) \leq C_k \frac{N}{(\log N)^{1+c_k}}, \tag{2.3}$$

where C_k and c_k are positive constants.

For $k = 3$, (2.2) was conjectured by Erdős and Turán [425] in 1936, and proved by Roth [1061] in 1953. In fact, Roth proved that

$$r_3(N) \leq C \frac{N}{\log \log N}$$

for some constant C. In 1999, Bourgain [183] improved this to

$$r_3(N) \leq C \frac{N \sqrt{\log \log N}}{\sqrt{\log N}}.$$

In 2011, Sanders [1077] obtained a bound very close to (2.3).

Theorem 2.5 *There exists a constant C such that*

$$r_3(N) \leq C \frac{N (\log \log N)^6}{\log N}.$$

The exponent 6 in Theorem 2.5 has been reduced to 4 by Bloom [157], and to 3 by Schoen [1094]. This sequence of results made the bound closer and closer to (2.3), but never achieved it. Finally, this milestone was reached by Bloom and Sisask [158] in 2020.

Theorem 2.6 *There exist positive constants C_3 and c_3 such that (2.3) holds for $k = 3$. In particular, any set A of positive integers such that $\sum_{n \in A} \frac{1}{n}$ diverges contains infinitely many 3-term arithmetic progressions.*

In 2011, Bateman and Katz [106] proved a version of Theorem 2.6 for cap sets, see Theorem 2.16 below. One of the motivations to study this problem for cap sets was the hope that (i) the problem may be easier in that setting and (ii) some of the ideas developed for solving the easier problems may be used to solve the harder one. This is exactly what happened! Many ideas introduced by Bateman and Katz to prove Theorem 2.16 were indeed used by Bloom and Sisask in the proof of Theorem 2.6.

---

[2] A set A of integers has *positive upper density* if $\limsup\limits_{N \to \infty} \frac{|A \cap \{1,2,\dots,N\}|}{N} > 0$.

Theorem 2.6 is a big milestone with many corollaries. In particular, it proves that Theorem 1.11 is valid not only for the primes, but for any set as dense as the primes. However, even with Theorem 2.6, we are still far from understanding how fast $r_3(N)$ actually grows. A construction by Behrend [110], dating back to 1946, implies that

$$r_3(N) \geq N e^{-C\sqrt{\log N}}$$

for some $C > 0$, which leaves a substantial gap between the best-known upper and lower bounds.

Longer Arithmetic Progressions

For $k \geq 4$, the question of whether (2.3) holds remains open. Szemerédi's original proof [1174] of (2.2) implies only a bound of the form

$$r_k(N) \leq CN(\log\log\ldots\log N)^{-1},$$

where log is repeated $f(k)$ times. In 2001, Gowers [508] presented a new proof of Szemerédi's theorem, which gives the much better bound $r_k(N) < N(\log\log N)^{-c(k)}$.

Theorem 2.7 *For every positive integer k, every subset of $\{1, 2, \ldots, N\}$ of size at least $N(\log\log N)^{-c}$, where $c = 2^{-2^{k+9}}$, contains an arithmetic progression of length k.*

Roth's 1953 proof of the $k = 3$ case of (2.2) introduced a beautiful "density increment argument". For any $\delta > 0$, let $P(\delta)$ denote the assertion that for N sufficiently large, all subsets of $\{1, \ldots, N\}$ of density at least δ contains a 3-term progression. Note that $P(\delta)$ is clearly true for $\delta \geq 1$. Roth proved that

(*)   for any $0 < \delta_0 \leq 1$, $P(\delta_0 + f(\delta_0))$ implies $P(\delta_0)$,

where f is some strictly increasing function with $f(0) = 0$. From (*), the infimum of all $\delta > 0$ for which $P(\delta)$ holds must be 0, and the $k = 3$ case of (2.2) follows. The proof of (*) is based on certain exponential-sum estimates, and seemed to not generalize to $k > 3$. Szemerédi's proof of (2.2) is completely different, and is full of intricate combinatorial arguments. Gowers showed that Roth's density increment argument can be generalized to all k, which makes it possible to obtain Szemerédi's theorem with a much better bound on $r_k(N)$.

Szemerédi's theorem (2.2) can be generalized to hold in sparse and random sets. We say that a finite set I of integers is $(\delta; k)$-*Szemerédi* if every subset of I of cardinality at least $\delta|I|$ contains a nontrivial k-term arithmetic progression. In this language, (2.2) states that the set $[n] = \{1, 2, \ldots, n\}$ is $(\delta; k)$-Szemerédi for every sufficiently large n.

Let $[n]_p$ be a random subset of $[n]$ in which each element of $[n]$ is chosen independently with probability $p \in [0, 1]$. In 2008, Green and Tao [523] observed that Szemerédi's theorem implies the existence of a function $p = p(n)$ with

$\lim\limits_{n \to \infty} p(n) = 0$ such that

$$\lim_{n \to \infty} \mathbb{P}([n]_p \text{ is } (\delta; k)\text{-Szemerédi}) = 1, \qquad (2.4)$$

and used this in the proof of Theorem 1.12. In 2016, Conlon and Gowers [298] proved that (2.4) remains true for every function $p = p(n)$ such that $p \geq Cn^{-1/(k-1)}$.

Theorem 2.8 *For any $\delta > 0$ and any integer $k \geq 3$, there exists a constant $C = C(k, \delta)$ such that (2.4) is true whenever $p \geq \min\{1, Cn^{-1/(k-1)}\}$.*

The bound on p in Theorem 2.8 is optimal up to a constant factor because it is known that, for some constant $c = c(k, \delta) > 0$, $\lim\limits_{n \to \infty} \mathbb{P}([n]_p \text{ is } (\delta; k)\text{-Szemerédi}) = 0$ whenever $p \leq cn^{-1/(k-1)}$.

Theorem 2.8 is just one of many corollaries of the powerful "transference principle" developed in [298] which allows the extension of many well-known combinatorial theorems to the setting of sparse random sets. Informally, the basic idea behind this transference principle is the following statement. If X is a set (such as $\{1, 2, \ldots, n\}$), and U is a sparse random subset of X, then, for every subset $A \subset U$ there is a subset $B \subset X$ with properties similar to A. In particular, the relative density of B in X is approximately the same as the relative density of A in U, and the number of substructures (such as arithmetic progressions) in A is an appropriate multiple of the number of such substructures in B. Now, Szemerédi's theorem controls the number of arithmetic progressions in B, which, by the transference principle, controls the number of such progressions in A.

To formulate an even more powerful generalization of Szemerédi's theorem, we need some notation. For an integer $m > 0$ and real number $x \geq m$, define $\binom{x}{m} = \frac{1}{m!} \prod_{i=0}^{m-1} (x - i)$. By convention, we define $\binom{x}{m} = 0$ if $x < m$. Also, by an *m-subset* of $[n]$ we mean a subset which has m elements. In 2015, Balogh et al. [90] proved the following result, which they called a sparse analogue of Szemerédi's theorem.

Theorem 2.9 *For every real $\beta > 0$ and every integer $k \geq 3$, there exist constants C and $n_0 > 0$ such that for all integers $n \geq n_0$ and $m \geq Cn^{1-1/(k-1)}$, there are at most $\binom{\beta n}{m}$ m-subsets of $[n]$ that contain no k-term arithmetic progression.*

Theorem 2.9 easily implies both Szemerédi's original theorem and Theorem 2.8. In turn, Theorem 2.9 itself is just one of many corollaries of a powerful theorem in graph theory proved by the authors, see [90] for details.

There are many other generalizations of Szemerédi's theorem (2.2). In particular, Bergelson and Leibman [134] extended this theorem to polynomial progressions. Let $\mathbb{Z}[y]$ be the set of polynomials in one variable with integer coefficients. Let $P = (P_1, \ldots, P_m)$ be a finite set of polynomials $P_i \in \mathbb{Z}[y]$. Let $r_P(N)$ be the size of the largest subset of $\{1, 2, \ldots, N\}$ containing no subset of the form $x, x +$

$P_1(y), \ldots, x + P_m(y)$ with $y \neq 0$. Bergelson and Leibman [134] proved in 1996 that

$$\lim_{N \to \infty} \frac{r_P(N)}{N} = 0,$$

provided that all polynomials P_i have zero constant term. In the special case $P_i(y) = iy$, $i = 1, \ldots, m$, this theorem reduces to the Szemerédi's theorem (2.2) with $k = m + 1$. In this special case, we have an explicit upper bound $r_P(N) < N(\log \log N)^{-c}$, established in Theorem 2.7, and it was an open question to obtain a similar upper bound in general. In 2020, Peluse [970] achieved this in the case when all polynomials P_i have distinct degrees.

Theorem 2.10 *Let* $P = (P_1, \ldots, P_m)$ *be a finite set of polynomials* $P_i \in \mathbb{Z}[y]$ *with distinct degrees, each having zero constant term. Then there exists a constant* $c_P = c(P_1, \ldots, P_m) > 0$ *such that*

$$r_P(N) \leq \frac{N}{(\log \log N)^{c_P}}.$$

The proof of Theorem 2.10 uses a density increment argument as in Gowers' proof of Theorem 2.7. However, Gowers used in a substantial way that arithmetic progressions are preserved under translation and dilation. This is not the case for polynomial progressions, so several new ideas were required to make the argument work.

Additive Patterns in Integers

A set B of integers is called *sum-free* if there are no distinct $x, y, z \in B$ such that $x + y = z$. The set $\{1, 2, \ldots, n\}$ has a sum-free subset (the set of odd integers) of size $n/2$. In 1965, Erdős [403] proved that every set A of n integers has a sum-free subset B of size at least $n/3$. Erdős further asked if the constant $1/3$ in this theorem can be improved. In 2010, Lewko [789] found a 28-element set of integers with maximal sum-free subset of size 11, showing that the constant cannot be greater than $\frac{11}{28} \approx 0.39$. In 2014, Eberhard et al. [379] proved that no constant greater than $\frac{1}{3}$ can work, so Erdős' theorem is essentially the best possible.

Theorem 2.11 *There is a set of n positive integers with no sum-free subset of size greater than* $\frac{1}{3}n + o(n)$.

The proof of Theorem 2.11 uses the following well-known fact. Let $f(n)$ be the largest k such that every set of n non-zero integers contains a sum-free subset of size k. If A and B are sets of integers with no sum-free subsets of sizes larger than $f(|A|)$ and $f(|B|)$, respectively, and $N > 0$ is a sufficiently large integer, then the set $A \cup NB$ has no sum-free subset of size larger than $f(m) + f(n)$. This implies that $f(m + n) \leq f(m) + f(n)$. From this, it is easy to deduce that to show $f(n) \leq cn + o(n)$, it is sufficient to find a single set A with no sum-free subset of size larger than $c|A|$. The authors constructed, for each $\epsilon > 0$, a set A with no sum-free subset of size larger than $(1/3 + \epsilon)|A|$, and Theorem 2.11 followed.

In 1985, Babai and Sós [75] asked for a generalization of this result: does every finite group G contain a product-free subset of size $c|G|$ for some $c > 0$? Here, a subset S of a finite group G is called *product-free* if it does not contain three elements x, y and z with $x \cdot y = z$. As partial progress, Babai and Sós [75] managed to prove the existence of product-free subsets of size $c|G|^{4/7}$, which was improved to $c|G|^{11/14}$ by Kedlaya [698] in 1997. However, Gowers [511] proved in 2008 that the answer to the Babai and Sós question is negative.

Theorem 2.12 *There exists infinitely many finite groups G which have no product-free subsets of cardinality greater than $2|G|^{8/9}$.*

Theorem 2.12 is proved by providing an explicit example: Gowers proved that the statement is true for $G = \mathrm{PSL}_2(q)$, provided that q is sufficiently large. Recall that $\mathrm{PSL}_2(q)$ is the group of 2×2 matrices whose entries belong to the field $\mathbb{F}_q$ (the field of integers $\{0, 1, \ldots, q-1\}$ with arithmetic operations defined modulo q), the determinant is equal to 1, and matrices A and $-A$ are "identified".

Another example of additive patterns in integers is the Erdős sumset conjecture. Schur's theorem [1098] states that the set $\mathbb{N}$ of positive integers cannot be partitioned into a finite number of sum-free sets. In 1974, Hindman [607] proved a generalization of Schur's theorem, stating that, for any finite colouring of integers, there is an infinite set A such that all finite sums

$$S_A = \left\{ \sum_{x \in B} x \mid B \subset A, |B| < \infty \right\} \tag{2.5}$$

of integers from the set A have the same colour. In particular, this implies the existence of infinite sets $B \subset \mathbb{N}$ and $C \subset \mathbb{N}$, such that the set $B + C = \{b + c : b \in B, c \in C\}$ is monochromatic. A conjecture of Erdős [409], known as the *Erdős sumset conjecture*, predicted that any set A of positive upper density contains a set $B + C$ as above. Nathanson [925] proved in 1980 that this is true if the set B is infinite and C is arbitrarily large but finite. In 2015, Di Nasso et al. [349] proved that the set $B + C$ with infinite B, C can be found in any set A of upper density at least $1/2$. Finally, in 2019, Moreira et al. [910] proved the Erdős sumset conjecture in full.

Theorem 2.13 *Any set $A \subset \mathbb{N}$ of positive upper density contains $B + C$, where B and C are infinite subsets of $\mathbb{N}$.*

The set S_A defined in (2.5) has been studied in many other contexts. For example, we say that the set A is *complete* if S_A contains all sufficiently large integers. This notion, introduced by Erdős in the sixties, is well-studied, and the central question is to find sufficient conditions for completeness. In 1962, Erdős [402] conjectured the existence of $c > 0$ such that if (i) A has density[3] at least $c\sqrt{n}$ and (ii) S_A contains an element of every infinite arithmetic progression, then A is complete.

We say that A is *subcomplete* if S_A contains an infinite arithmetic progression. In 1966, Folkman [457] conjectured that every A satisfying (i) is subcomplete, and noted that this would easily imply the Erdős conjecture. In 2006, Szemerédi and Vu [1176] confirmed Folkman's conjecture and hence the Erdős conjecture.

Theorem 2.14 *There exist a constant $c > 0$ such that every set $A \subset \mathbb{N}$ with density at least $c\sqrt{n}$ is subcomplete.*

A key step in the proof of Theorem 2.14 was establishing the following finite analogue of Folkman's conjecture: there is a constant C such that if $A \subset \{1, \ldots, n\}$ has $|A| \geq C\sqrt{n}$ elements, then S_A contains an arithmetic progression of length n. This result is the best possible up to the value of C, in the sense that a weaker lower bound on $|A|$ is not sufficient for the conclusion.

Another additive pattern in the set of integers is a pattern of the form

$$a, a + n_1, a + n_2, a + n_1 + n_2,$$

or its generalization with k parameters $n_1, \ldots, n_k$. Formally, let $\mathbb{Z}^k$ be the set of vectors $x = (x_1, \ldots, x_k)$ with integer coordinates, $V_k \subset \mathbb{Z}^k$ be the set of vectors $\epsilon = (\epsilon_1, \ldots, \epsilon_k)$ with all coordinates 0 or 1, and, for any fixed parameters $n = (n_1, n_2, \ldots, n_k) \in \mathbb{Z}^k$, consider the pattern $a + \sum_{i=1}^{k} \epsilon_i n_i$, $\epsilon \in V_k$. A set $A \subset \mathbb{N}$ of positive integers contains such a pattern if the set $\bigcap_{\epsilon \in V_k} \left(A + \sum_{i=1}^{k} \epsilon_i n_i\right)$ is non-empty. If the set $A \subset \mathbb{N}$ has positive upper density $d(A) > \delta > 0$, a stronger conclusion would be that

$$d\left(\bigcap_{\epsilon \in V_k} \left(A + \sum_{i=1}^{k} \epsilon_i n_i\right)\right) \geq \delta^{2^k}, \tag{2.6}$$

which would imply that such patterns not only exist but are "as dense as expected".

In 2005, Host and Kra [620] proved that this stronger conclusion holds for "many" choices of parameters $n = (n_1, n_2, \ldots, n_k) \in \mathbb{Z}^k$. Specifically, for any set $B \subset \mathbb{Z}^k$ and $x \in \mathbb{Z}^k$ the *translate* $B + x$ is $\{y : y = b + x, b \in B\}$. Let $t(B)$ be the minimal number of translates of B needed to fully cover $\mathbb{Z}^k$. A set B is called *syndetic* if $t(B) < \infty$.

Theorem 2.15 *For any integer $k \geq 1$, and any $A \subset \mathbb{N}$ with upper density $d(A) > \delta > 0$, the set of $n = (n_1, n_2, \ldots, n_k) \in \mathbb{Z}^k$ such that (2.6) holds is syndetic.*

---

[3] We say that A has *density* at least $f(n)$ if $|A \cap \{1, 2, \ldots, n\}| \geq f(n)$ for all sufficiently large n.

In fact, Theorem 2.15 is a combinatorial reformulation of a deep theorem in the field of ergodic theory, which establishes L^2-convergence of so-called "ergodic averages taken along cubes whose sizes tend to infinity", see [620] for details.

2.3 Patterns in Lattices

The Cap Set Problem

Patterns like arithmetic progressions can also be studied in multidimensional grids (lattices). Let $\mathbb{F}_3$ be the 3-element field (that is, $\mathbb{F}_3 = \{0, 1, 2\}$ with operations defined modulo 3), and let $\mathbb{F}_3^n$ be the n-dimensional vector space over $\mathbb{F}_3$. We say that three different points $x, y, z \in \mathbb{F}_3^n$ form a *line* if $x + z = 2y$. With $d = y - x$, we have $y = x + d$, and $z = x + 2d$, so the line is also called a 3-*term arithmetic progression*. A set $A \subseteq \mathbb{F}_3^n$ is called a *cap set* if it contains no lines. The *cap set problem* is the problem of finding the size of the largest possible cap set, as a function of n. This problem is a finite field analogue of the problem of finding large sets of integers with no 3-term arithmetic progression, discussed in Sect. 2.2. For example, the bound in the form

$$\frac{|A|}{3^n} \leq \frac{C}{n^{1+\epsilon}} \tag{2.7}$$

is a finite field analogue of (2.3) with $k = 3$, and it was hoped that the proof of (2.7) can shed some light on how to prove (2.3). In 1995, Meshulam [883] came ϵ-close to (2.7) by proving that $\frac{|A|}{3^n} \leq \frac{C}{n}$ for some constant C and every cap set $A \subseteq \mathbb{F}_3^n$. In 2012, Bateman and Katz [106] proved (2.7).

Theorem 2.16 *There exist constants* $\epsilon > 0$ *and* $C < \infty$ *such that if* $A \subseteq \mathbb{F}_3^n$ *is a cap set, then* (2.7) *holds.*

When Bloom and Sisask proved the $k = 3$ case of (2.3) in a later work [1077] (see Theorem 2.5), their proof indeed built upon many of the ideas developed in [106] to prove Theorem 2.16. However, the bound (2.7) is far from being optimal. The famous *cap set conjecture* predicted that the much stronger bound

$$|A| \leq c^n, \quad \text{for some} \quad c < 3 \tag{2.8}$$

should hold. There are known constructions [380] of cap sets of size approximately $|A| \approx (2.2)^n$, so the cap set conjecture, if true, would give an optimal bound up to the value of c.

In 2017, Croot et al. [311] proved the analogue of (2.8) for progression-free subsets of $\mathbb{Z}_4^n$, where $\mathbb{Z}_4 = \{0, 1, 2, 3\}$ with addition defined modulo 4. A set $A \subseteq \mathbb{Z}_4^n$ is called *progression-free* if it contains no 3-term arithmetic progressions.

For $x \in (0, 1)$, let $H(x) = -x \log_2 x - (1 - x) \log_2 (1 - x)$. Let $\gamma = \max\limits_{0 < \epsilon < 0.25} \left(\frac{1}{2}(H(0.5 - \epsilon) + H(2\epsilon)) \right)$. Note that $\gamma \approx 0.926$. In particular, $\gamma < 1$.

Theorem 2.17 *If $n \geq 1$ and $A \subseteq \mathbb{Z}_4^n$ is progression-free, then $|A| \leq 4^{\gamma n}$.*

The proof of Theorem 2.17 is short and uses the polynomial method. The idea is to construct a low-degree multivariate polynomial that is zero on a certain set, and then proceed by arguing about algebraic properties of this polynomial. The method is known for producing "magically" short proofs of long-standing open problems in combinatorics.

Just a few weeks after the proof of Theorem 2.17 was published online, Ellenberg and Gijswijt [385] found a way to adapt the same method to prove the cap set conjecture.

Theorem 2.18 *If $A \subseteq \mathbb{F}_3^n$ is a cap set, then $|A| = o(2.756^n)$.*

Longer Lines and Other Patterns
The cap set problem studies multidimensional sets with no "lines" of length 3. What about longer lines? The famous *Hales–Jewett theorem* [554], states, informally, that for any positive integers k and m, there is a positive integer n such that if the n-dimensional $k \times k \times \cdots \times k$ cube is coloured in m colours, there must always be a monochromatic row, column, or diagonal. More formally, for positive integers k, n, let $[k] = \{1, 2, \ldots, k\}$ and let $[k]^n$ be the set of vectors $x = (x_1, \ldots, x_n)$ such that each $x_i \in [k]$. Let $y = (y_1, \ldots, y_n)$ be a vector such that each y_i either belongs to $[k]$ or is equal to the wildcard value $*$. Assume that at least one coordinate of y takes the wildcard value $*$. For $j = 1, 2, \ldots, k$, let $y^j \in [k]^n$ be a vector obtained from y by setting all the wildcards $*$ equal to j. The set $y^1, y^2, \ldots, y^k$ is called a *combinatorial line*. The Hales–Jewett theorem states that if k and m are fixed and n is sufficiently large, then any m-colouring of $[k]^n$ contains a monochromatic combinatorial line.

In 1991, Furstenberg and Katznelson [486] proved a stronger density version of this theorem: if an integer $k > 0$ and real $\delta > 0$ are fixed and n is sufficiently large, then any subset $A \subset [k]^n$ with at least δk^n elements contains a combinatorial line. This theorem generalizes (and easily implies) Szemerédi's theorem (2.2). However, Furstenberg and Katznelson's proof does not supply an explicit bound on the size of n in terms of k and δ. In 2009, a large group of mathematicians organized an online discussion, called a *polymath project*, and, combining their efforts, proved the density Hales–Jewett theorem with an explicit bound on the dimension [996].

Let $f : \mathbb{N} \times \mathbb{N} \to \mathbb{N}$ be a function such that $f(k, 1) = 2$ for all k, $f(1, n) = 2n$ for all n, and $f(k, n) = f(k - 1, f(k, n - 1))$ for all $k, n > 1$.

Theorem 2.19 *For every integer $k > 0$ and real number $\delta > 0$, there exists a positive integer $N = N(k, \delta)$ such that if $n \geq N$ and A is any subset of $[k]^n$ with at least δk^n elements, then A contains a combinatorial line. Moreover, one may take $N(3, \delta) = f(3, O(1/\delta^2))$ and $N(k, \delta) = f(k + 1, O(1/\delta))$ for $k \geq 4$.*

The proof of Theorem 2.19 is based on the density increment method. The key lemma states that if a subset A of $[k]^n$ of density δ does not contain a combinatorial line, then A must have density $\delta + \gamma$ inside some large subspace of $[k]^n$, where γ is a positive constant that depends only on δ. Then Theorem 2.19 follows from an iterative application of the lemma. The overall proof is purely combinatorial and surprisingly simple—in fact, it gives the simplest known proof of Szemerédi's theorem (2.2).

Szemerédi's theorem can also be generalized to multidimensional patterns which are more complicated than lines. Let $\mathbb{Z}^n$ be the set of vectors with n integer components. For $a \in \mathbb{Z}^n$, $X \subset \mathbb{Z}^n$, and integer $d > 0$, denote by $a + dX$ the set $\{y \in \mathbb{Z}^n : y = a + dx, x \in X\}$. In 1978, Furstenberg and Katznelson [485] proved that if an integer $n > 0$, real $\delta > 0$, and finite subset $X \subset \mathbb{Z}^n$ are fixed and k is sufficiently large, then any subset $A \subset [k]^n$ with at least δk^n elements has a subset of the form $a + dX$ for some $a \in \mathbb{Z}^n$ and integer $d > 0$. This is a deep result containing Szemerédi's theorem (2.2) as a very special case. However, the proof in [485] does not provide an explicit algorithm to compute how large k should be in terms of δ, n, and X. The existence of such an algorithm was proved by Gowers [510] in 2007. We say that a number $K = K(x_1, \ldots, x_m)$ is *explicitly computable* if there is an algorithm which takes $x_1, \ldots x_m$ as an input and returns K as an output.

Theorem 2.20 *For every integer $n > 0$, real $\delta > 0$, and finite subset $X \subset \mathbb{Z}^n$, there is an explicitly computable integer $K = K(\delta, n, X) > 0$ such that if $k \geq K$ and A is any subset of $[k]^n$ with at least δk^n elements, then A contains a subset of the form $a + dX$ for some $a \in \mathbb{Z}^n$ and integer $d > 0$.*

The proof of Theorem 2.20 is combinatorial and is based on establishing the hypergraph analogues for the key graph-theoretic lemmas in Szemerédi's original proof of (2.2).

2.4 Sum-Product Phenomena

For $A, B \subset \mathbb{Z}$, define $A + B = \{a + b, a \in A, b \in B\}$ and $A \times B = \{a \cdot b, a \in A, b \in B\}$. In 1983, Erdős and Szemerédi [423] proved the existence of positive constants c and ϵ such that the inequality

$$\max(|A + A|, |A \times A|) \geq c|A|^{1+\epsilon} \tag{2.9}$$

holds for any finite non-empty set A of integers (or real numbers). In words, this theorem states that the set $A + A$ of pairwise sums and the set $A \times A$ of pairwise products cannot both be small. In fact, they conjectured that the much stronger estimate

$$\max(|A + A|, |A \times A|) \geq c_\epsilon|A|^{2-\epsilon} \tag{2.10}$$

should hold for any $\epsilon > 0$, where $c_\epsilon > 0$ is a constant depending on ϵ. This conjecture remains open. In 2009, Solymosi [1143] proved (2.10) with the constant $\frac{4}{3}$ in place of 2. The constant $\frac{4}{3}$ has been improved only slightly. For example, Shakan [1105] in 2019 proved the same result with the constant $\frac{4}{3} + \frac{5}{5277}$.

More generally, for $A \subset \mathbb{Z}$ and positive integer k, define $kA = A + A + \cdots + A$ and $A^k = A \times A \times \cdots \times A$, where A in the sum and in the product is repeated k times. Erdős and Szemerédi [423] conjectured that

$$\max(|kA|, |A^k|) \geq c_\epsilon |A|^{k-\epsilon}. \tag{2.11}$$

As partial progress towards conjecture (2.11), Bourgain and Chang [184] proved in 2004 that, for any constant b, (2.11) holds with b in place of $k - \epsilon$, provided that k is sufficiently large.

Theorem 2.21 *For any integer $b > 0$ there is an integer $k = k(b) > 0$ such that* $\max(|kA|, |A^k|) \geq |A|^b$ *holds for any non-empty finite set $A \subset \mathbb{Z}$.*

Erdős and Szemerédi [423] also made a related conjecture about *simple* sums and products. Let $A = \{a_1, a_2, \ldots, a_k\}$ be a set of k distinct integers, and let $S(A) = \left\{\sum_{i=1}^k \epsilon_i a_i \mid \epsilon_i = 0 \text{ or } 1\right\}$ and $\Pi(A) = \left\{\prod_{i=1}^k a_i^{\epsilon_i} \mid \epsilon_i = 0 \text{ or } 1\right\}$ be the set of all possible simple sums and products of elements of A, respectively. Let

$$g(k) = \min_{A:|A|=k} (|S(A)| + |\Pi(A)|).$$

Erdős and Szemerédi [423] conjectured that the set of sums $S(A)$ and set of products $\Pi(A)$ cannot both be small. Specifically, they conjectured that $g(k)$ grows faster than any polynomial in k, that is, for any d, there exists a constant $k_0 = k_0(d)$ such that $g(k) > k^d$ for any $k \geq k_0(d)$. In 2003, Chang [262] proved a stronger lower bound on $g(k)$.

Theorem 2.22 *For any $\epsilon > 0$ there is a $k_0 = k_0(\epsilon)$ such that, for all $k \geq k_0$,*

$$g(k) > k^{(1/2-\epsilon)\frac{\log k}{\log(\log k)}}.$$

Theorem 2.22 implies that $g(k) > k^d$ for any $k \geq k_0(d)$. In fact, because it is known that $g(k) < k^{c\frac{\log k}{\log(\log k)}}$ for some constant c, the lower bound for $g(k)$ in Theorem 2.22 is the best possible up to a constant factor in the exponent.

While a common approach to proving results in this area is to assume that the sum set is small and then derive that the product set is large, the proof of Theorem 2.22 follows the reverse route: the authors assume that the product set is small and derive that in this case the sum set must be large.

In 2004, Bourgain et al. [188] proved a version of the Erdős–Szemerédi theorem (2.9) in finite fields. Let q be a prime, and let F_q be a finite field with q elements. For $A \subset F_q$, define $A + A = \{a + b, a \in A, b \in A\}$ and $A \cdot A = \{a \cdot b, a \in A, b \in A\}$.

Theorem 2.23 *For any $\delta > 0$ there exist positive constants c and ϵ such that for any prime q, and any $A \subset F_q$ such that $q^\delta < |A| < q^{1-\delta}$, we have*

$$\max(|A + A|, |A \cdot A|) \geq c|A|^{1+\epsilon}.$$

Sum-product theorems turn out to be very useful and have many applications. For example, the authors of [188] used Theorem 2.23 to prove the finite fields analogues of the Szemerédi–Trotter theorem (see Eq. (5.6) below), Erdős' distinct distances problem (see Theorem 5.18 below), and the $n = 3$ case of the Kakeya problem (later proved for all n in Theorem 3.88 below).

After Theorem 2.23 had been proved, many authors obtained versions of it with explicit parameters. A notable achievement in this direction is the 2016 theorem of Roche-Newton et al. [1053], which states that Theorem 2.23 is true with $\epsilon = 1/5$ if $\delta = 3/8$.

Theorem 2.24 *Let F be a field with positive odd characteristic[4] p, i.e. $F = F_q$ is a finite field with $q = p^k$ elements, where $k \geq 1$ is an integer. Let $A \subset F$ be a subset of cardinality $|A| < p^{5/8}$. Then*

$$\max(|A + A|, |A \cdot A|) \geq c|A|^{6/5}$$

for some absolute constant $c > 0$.

2.5 Other topics in Combinatorial Number Theory

Erdős' Multiplication Table

For positive integer N, let $M(N)$ denote the number of distinct integers n which can be written as $n = a \cdot b$, where a and b are positive integers not exceeding N. In other words, $M(N)$ is the number of distinct products in the $N \times N$ multiplication table. In 1965, Erdős [404] proved that $\lim\limits_{N \to \infty} \frac{M(N)}{N^2} = 0$, that is, most integers do not appear in the table. However, the question of precisely how few integers appear, or, in other words, how fast $M(N)$ grows with N, remained open. In 2008, Ford [458] resolved this question, up to a constant factor.

---

[4] Recall that the *characteristic* of a field F with additive identity 0 and multiplicative identity 1 is the smallest positive integer p such that $1 + 1 + \cdots + 1 = 0$, where the summand 1 is repeated p times. We say that F has characteristic 0 if no such p exists.

Theorem 2.25 *There exist positive constants c_1 and c_2 such that inequality*

$$c_1 \frac{N^2}{(\log N)^\delta (\log \log N)^{3/2}} \leq M(N) \leq c_2 \frac{N^2}{(\log N)^\delta (\log \log N)^{3/2}}$$

holds for all N, where $\delta = 1 - \frac{1+\log\log 2}{\log 2} = 0.08607....$

In fact, Theorem 2.25 is only one out of many corollaries of a deep theory developed by Ford for estimating the number $H(x, y, z)$ of positive integers $n \leq x$ having a divisor in the interval $(y, z]$ and the number $H_r(x, y, z)$ of positive integers $n \leq x$ having exactly r such divisors, see [458] for details.

Distinct Covering Systems

In 1950, Erdős [399] observed that every integer n satisfies either $n \equiv 0 \pmod 2$, or $n \equiv 0 \pmod 3$, or $n \equiv 1 \pmod 4$, or $n \equiv 3 \pmod 8$, or $n \equiv 7 \pmod{12}$, or $n \equiv 23 \pmod{24}$. More generally, a finite set S of distinct positive integers $0 < s_1 < s_2 < \cdots < s_k$ is called a *distinct covering system* if there exist integers $m_1, m_2, \ldots, m_k$ such that each integer n satisfies at least one of the congruences

$$n \equiv m_i \pmod{s_i}, \quad i = 1, 2, \ldots, k. \tag{2.12}$$

A famous problem of Erdős [399] from 1950, known as the *minimum modulus problem for covering systems*, asks whether for every N there is a distinct covering system S with all s_i greater than N.

In 1975, Erdős and Selfridge [406] conjectured that the answer to this problem is negative if we also require that the sum of reciprocals $\sum_{i=1}^{k} \frac{1}{s_k}$ of elements of S is bounded by some constant B. In 2007, Filaseta, Ford, Konyagin, Pomerance, and Yu [453] proved this conjecture.

Theorem 2.26 *For any number B, there is a number N_B such that inequality $\sum_{i=1}^{k} \frac{1}{s_k} > B$ holds for any distinct covering system S with $s_1 > N_B$.*

If $s_1, \ldots, s_k$ are pairwise coprime, then the Chinese remainder theorem implies that the density δ of the integers that are not in the union of the residue classes (2.12) is given by

$$\delta = \prod_{i=1}^{k} (1 - 1/s_i).$$

In particular, $\delta > 0$ if each $s_i > 1$. The authors of [453] developed a methodology to estimate δ when the moduli s_i are not necessarily pairwise coprime, and deduced Theorem 2.26 from there.

Theorem 2.26 has a number of interesting corollaries. In particular, it implies that for any number $K > 1$ and N sufficiently large, depending on K, there is no distinct covering system S with all s_i in the interval $(N, KN]$. However, it leaves the original minimum modulus problem unresolved. In 2009, Nielsen [937] constructed a distinct covering system with minimum modulus $s_1 = 40$. However, he conjectured that this cannot be done for arbitrarily large N, and the answer to the minimum modulus problem is negative. In 2015, Hough [621] confirmed this conjecture.

Theorem 2.27 *In every distinct covering system* $0 < s_1 < s_2 < \cdots < s_k$ *we have* $s_1 \leq 10^{16}$.

The proof of Theorem 2.27 uses ideas from the proof of Theorem 2.26 but extends them in a substantial way. The proof uses a probabilistic method and makes crucial use of the Lovász local lemma [413] (see Theorem 7.10).

Another famous question of Erdős and Selfridge is whether there exists a distinct covering system $1 < s_1 < s_2 < \cdots < s_k$ such that all s_i are odd. This problem remains open, despite all efforts.

The Threshold for Making Squares

One of the most famous problems in mathematics is the problem of efficiently determining the prime factorization of a given integer n. Currently, the fastest known algorithms use the following idea. They generate a random sequence of integers $a_1, a_2, \ldots$ such that $a_i = b_i^2 \pmod{n}$ for some integers b_i, and try to find a subsequence whose product is a perfect square: $a_{i_1} a_{i_2} \ldots a_{i_k} = y^2$. Then n is a divisor of $y^2 - z^2 = (y - z)(y + z)$, where $z = b_{i_1} b_{i_2} \ldots b_{i_k}$, and there is a chance that $\gcd(n, y - z)$ is a non-trivial factor of n.

To understand the performance of such algorithms, one needs to answer the following crucial question, asked by Pomerance [999] in 1994: *how many random integers between 1 and x should we select such that the product of some selected integers is a perfect square?* More formally, we say that a finite sequence S of positive integers has a *square dependence* if it has a subset $A \subset S$ such that the product $\prod_{n \in A} n$ is a perfect square. For $x > 1$, let us select integers $a_1, a_2, \ldots$ independently uniformly at random from $[1, x]$, and let $T = T(x)$ be the smallest integer such that the sequence $a_1, a_2, \ldots, a_T$ has a square dependence. Pomerance [999] conjectured that $T(x)$ exhibits a sharp threshold, that is,

$$\lim_{x \to \infty} \mathbb{P}\left[(1 - \epsilon)f(x) \leq T(x) \leq (1 + \epsilon)f(x)\right] = 1 \qquad (2.13)$$

for some function $f(x)$ and any $\epsilon > 0$.

In 2012, Croot, Granville, Pemantle and Tetali [310] conjectured the exact form of the function $f(x)$. To define it, we need some notation. An integer m is called y-*smooth* if all of its prime factors are at most y. Let $\Psi(x, y)$ be the number of y-smooth integers up to x, $\pi(y)$ be the number of primes up to y, and let $J(x) = \min_{2 \leq y \leq x} \frac{\pi(y)x}{\Psi(x,y)}$. Croot et al. [310] conjectured that (2.13) holds with $f(x) = e^{-\gamma} J(x)$, where $\gamma = \lim_{n \to \infty} \left(-\log n + \sum_{k=1}^n \frac{1}{k}\right) = 0.577\ldots$ is the Euler–Mascheroni constant. In fact, they proved the upper bound of this conjecture, and established the lower bound up to the factor $\frac{\pi}{4}$.

Theorem 2.28 *For any $\epsilon > 0$,*

$$\lim_{x \to \infty} \mathbb{P}\left[\frac{\pi}{4}(e^{-\gamma} - \epsilon)J(x) \le T(x) \le (e^{-\gamma} + \epsilon)J(x)\right] = 1.$$

The proof of Theorem 2.28 uses a "first moment method", and proceeds by counting the expected number of non-empty subsets $I \subset \{1, \dots, N\}$ such that $\prod_{i \in I} a_i$ is a perfect square. This approach proves the sharp upper bound for $T(x)$, but fails to provide the sharp lower bound, because there exists a constant $c > 0$ such that this expected number blows up when $N > (e^{-\gamma} - c)J(x)$. In 2018, Balister et al. [86] used a different approach to remove the factor $\frac{\pi}{4}$ from the lower bound and confirm the conjecture in full.

Theorem 2.29 *For any $\epsilon > 0$,*

$$\lim_{x \to \infty} \mathbb{P}\left[(e^{-\gamma} - \epsilon)J(x) \le T(x) \le (e^{-\gamma} + \epsilon)J(x)\right] = 1.$$

The proof of Theorem 2.29 goes via an analysis of a random process which removes numbers from the set $\{a_1, a_2, \dots, a_T\}$ without violating the property of square dependence. Specifically, if there exists a prime p which divides some number a_i an odd number of times, and all other numbers a_j an even number of times, then a_i can be discarded. This procedure is then repeated. The authors were able to derive estimates for the number of elements of $\{a_1, a_2, \dots, a_T\}$ which are not discarded by this procedure, and show that when very few numbers are remaining, the first moment calculation as in the proof of Theorem 2.28 becomes applicable.

The Erdős Discrepancy Problem

Around 1932, Erdős [427] conjectured that for any infinite sequence $f : \mathbb{N} \to \{-1, +1\}$ of $+1$s and -1s and any integer C, there exist positive integers n and d such that

$$\left|\sum_{j=1}^{n} f(jd)\right| > C.$$

The problem asking to prove or disprove this conjecture became known as the *Erdős discrepancy problem*. It is one of Erdős' most famous problems, attracted a lot of attention, and was the topic of a polymath project[5] in 2010, but, before 2015, was open even for $C = 2$. In 2015, Konev and Lisitsa [734] proved the $C = 2$ case of the conjecture with an enormous computer-assisted proof whose output takes up 13 gigabytes of data. In 2016, Tao [1181] proved the conjecture for all C.

---

[5] A *polymath project* involves a large group of mathematicians trying to solve a problem collectively by discussing it online. It was mentioned earlier in this book in connection with Theorems 1.4 and 2.19.

Theorem 2.30 *For any sequence $f : \mathbb{N} \to \{-1, +1\}$, the discrepancy*

$$\sup_{n,d \in \mathbb{N}} \left| \sum_{j=1}^{n} f(jd) \right|$$

of f is infinite.

The proof of Theorem 2.30 uses Fourier-analytic methods developed in the polymath project to reduce the problem to the case when f is a completely multiplicative function. After this, Tao used a powerful theory developed for such functions, most importantly Theorem 1.42 and a variant of Theorem 1.44, to finish the proof.

2.6 Colouring of Graphs

The Chromatic Number of the Plane

In 1950, Nelson asked what is the minimum number of colours that are needed to colour the plane so that no two points at distance exactly 1 from each other have the same colour. This minimum number of colours is called the *chromatic number* of the plane. Because the vertices of any equilateral triangle with side length 1 should receive different colours, at least 3 colours are needed, and it is not much harder to see[6] that in fact at least 4 colours are required. As observed by Isbell, also in 1950, 7 colours suffices, see Fig. 2.1. These trivial bounds were not improved for almost 70 years. In 2018, de Grey [528] proved that 4 colours are not enough and at least 5 are needed.

Theorem 2.31 *The chromatic number of the plane is at least* 5.

Theorem 2.31 can be formulated in the language of graph theory. The *chromatic number* of a graph G, denoted $\chi(G)$, is the smallest number of colours needed to colour the vertices of G so that any pair of vertices connected by an edge have different colours. Such a colouring of vertices is called a *proper colouring*. Theorem 2.31 states that $\chi(H) \geq 5$ for an (infinite) graph H whose vertices are points on the plane, and edges connect points at distance 1. Together with Isbell's upper bound, we now know that $5 \leq \chi(H) \leq 7$.

Induced subgraphs of H are called *unit-distance graphs*. De Grey proved Theorem 2.31 by constructing an explicit finite unit-distance graph which is not 4 colourable. The construction starts with the 7-vertex, 12-edge unit-distance graph G consisting of the centre and vertices of a regular hexagon of side-length 1. Then

---

[6] If 3 colours would suffice, then in any rhombus $ABCD$ with $\angle BAD = 60^o$ and side lengths 1, vertices A and C would receive the same colour. Hence, any pair of points at distance $\sqrt{3}$ must have the same colour. Then the triangle with side lengths $\sqrt{3}$, $\sqrt{3}$ and 1 leads to a contradiction.

Fig. 2.1 Colouring of the
plane with 7 colours such that
no 2 points at distance 1 have
the same colour

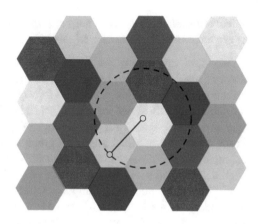

de Grey constructed a unit-distance graph L that contains 52 copies of G and
proved that in any 4-colourings of L, at least one copy of G contains three vertices
coloured in the same way (a monochromatic triple). Finally, he constructed a unit-
distance graph M that contains a copy of G such that there are no 4-colourings
of M in which G contains a monochromatic triple. This leads to a contradiction,
proves Theorem 2.31, and also constructs an explicit finite unit-distance graph N
(consisting of 52 copies of M) which is not 4 colourable.

Perfect Graphs

Theorem 2.31 makes progress in computing the chromatic number of an infinite
graph. In fact, even the study of chromatic numbers of *simple* (that is, with no loops
or multiple edges) finite graphs is a highly non-trivial research area. An obvious
lower bound for the chromatic number $\chi(G)$ of a graph G is the size $w(G)$ of the
largest *clique* (complete subgraph) in G. This bound, however, is not always tight.
For example, if C is a cycle of odd length[7] $n = 2k + 1 \geq 5$ (called an *odd hole*),
then $w(C) = 2$ but $\chi(C) = 3$. Also, if $\bar{C}$ is the complement[8] of such a cycle (called
an *odd antihole*), then $w(\bar{C}) = k$ but $\chi(\bar{C}) = k + 1$.

A graph G is called *perfect* if for every induced subgraph H of G, we have
$\chi(H) = w(H)$. A central open question in the area was to develop a simple
and easy-to-check criteria to determine which graphs are perfect. By the argument
in the previous paragraph, if a graph G is perfect, it cannot contain an odd hole
or an odd antihole as an induced subgraph. In 1961, Berge conjectured that this
obvious necessary condition is also sufficient for G to be perfect. This conjecture
became known as the *strong perfect graph conjecture*, attracted a lot of attention,
but remained open for over 40 years.

In 2006, Chudnovsky et al. [280] proved this conjecture.

---

[7] A *cycle of length n* is the graph with vertices $A_1, \ldots, A_n$ and edges $A_1 A_2, \ldots, A_{n-1} A_n, A_n A_1$.

[8] The *complement* $\bar{G}$ of a graph G is the graph with the same vertex set as G, such that vertices i
and j are connected by an edge in $\bar{G}$ if and only if they are not connected by an edge in G.

Theorem 2.32 *A graph G is perfect if and only if it does not contain an odd hole or an odd antihole as an induced subgraph.*

We next say a few words about the proof of Theorem 2.32. One of the ways of making progress towards the strong perfect graph conjecture has been to investigate the properties of *minimal imperfect graphs*, that is, counterexamples to the conjecture with as few vertices as possible. The aim was to prove that no minimal imperfect graphs exists. As partial progress, reseachers proved a number of results in the form "if a minimal imperfect graph exists, it cannot have property P". For example, we say that two non-adjacent vertices x, y in a graph G form an *even pair* if all induced paths joining x to y have an even number of edges. In 1987, Meyniel [885] proved that no minimal imperfect graph contains an even pair. By 2006, there were many results in this form, so we had an explicit list of properties $P_1, P_2, \ldots$ such that if a minimal imperfect graph exists, it cannot have any of the properties P_i. In 2004, Conforti et al. [294] selected some explicit properties $P_{i_1}, P_{i_2}, P_{i_3}$ from this list, and conjectured that if G is any graph with no odd hole or odd antihole as an induced subgraph then either (i) G is perfect, or (ii) G satisfies one of the properties $P_{i_1}, P_{i_2}, P_{i_3}$, or (iii) the complement of G satisfies one of the properties $P_{i_1}, P_{i_2}, P_{i_3}$. Chudnovsky, Robertson, Seymour, and Thomas [280] proved this conjecture and deduced Theorem 2.32 from it. Indeed, if H is any minimal imperfect graph, it cannot satisfy (i) or (ii). Because the complement of every perfect graph is perfect by a 1972 theorem of Lovász [813], the complement of H is a minimal imperfect graph, and therefore H also cannot satisfy (iii). Hence, H does not exist and Theorem 2.32 follows.

Based on Theorem 2.32, Chudnovsky, Cornuéjols, Liu, Seymour and Vušković [279] developed an efficient method to determine whether any given graph G is perfect.

Theorem 2.33 *There exists a polynomial P and an algorithm which takes a graph G with n vertices as an input, performs at most $P(n)$ operations, and determines whether the graph G is perfect or not.*

In 1988, Grötschel et al. [538] developed an efficient method to determine the chromatic number $\chi(G)$ of every perfect graph G. With this algorithm and Theorem 2.33 in hand, we can, for any graph G, determine in polynomial time whether it is perfect, and if so, find $\chi(G)$.

The Chromatic Number of a Tensor Product

Theorem 2.32 confirms that the strong perfect graph conjecture is true. We next mention another famous conjecture about the chromatic number which turned out to be false.

Let $V(G)$ and $E(G)$ denote the set of vertices and edges of a graph G, respectively. The *tensor product* $G \times H$ of finite simple graphs G and H is the graph with vertex set $V(G) \times V(H)$, such that pairs (g, h) and (g', h') are adjacent if and only if $\{g, g'\} \in E(G)$ and $\{h, h'\} \in E(H)$. It is easy to see that $\chi(G \times H) \leq \min\{\chi(G), \chi(H)\}$ for all graphs G, H. A classical conjecture of Hedetniemi [588], made in 1966, predicted that equality holds for all G and H.

The conjecture attracted a lot of attention, was well-believed, and has been proved in a number of special cases. However, in 2019 Shitov [1116] proved that in general this conjecture is false.

Theorem 2.34 *There exist finite simple graphs G and H such that $\chi(G \times H) < \min\{\chi(G), \chi(H)\}$.*

The counterexample to Hedetniemi's conjecture constructed in [1116] is based on the concept of exponential graphs. For a graph G and a positive integer c, the *exponential graph $E_c(G)$* is the graph whose vertices are all possible maps $V(G) \to \{1, \ldots, c\}$, and two distinct mappings ψ, ϕ are connected by an edge if and only if $\phi(x) \neq \psi(y)$ whenever x and y are connected by an edge in G. Then the tensor product $G \times E_c(G)$ has a proper colouring with c colours in which a vertex (v, ϕ) has colour $\phi(v)$. Shitov [1116] constructed a graph G such that neither G nor $H = E_c(G)$ are c-colourable. Then $\chi(G \times H) \leq c < \min\{\chi(G), \chi(H)\}$.

Inspired by Theorem 2.34, in 2020 He and Wigderson [582] proved the even stronger result that there exists constants $\delta > 0$ and C_0 such that for all $c > C_0$, there exist graphs G and H with $\min\{\chi(G), \chi(H)\} \geq (1+\delta)c$ but $\chi(G \times H) \leq c$.

The Chromatic Number of H-Free Graphs

As mentioned above (see the discussion before Theorem 2.32), the size $w(G)$ of the largest clique in G is a lower bound for its chromatic number $\chi(G)$, but this bound is not tight because, for odd cycles C, $w(C) = 2$ but $\chi(C) = 3$. In the 1940s, Zykov [1295] and Tutte[9] [347] constructed a sequence G_n of graphs with $w(G_n) = 2$ but $\lim_{n \to \infty} \chi(G_n) = \infty$. Graphs with $w(G_n) = 2$ are called *triangle-free* because they do not contain the triangle K_3 as a subgraph. More generally, for any graph H, a graph G is called *H-free* if it does not have a subgraph isomorphic to H. Generalizing Zykov's and Tutte's constructions, in 1959 Erdős [401] constructed, for any graph H containing a cycle, a sequence of H-free graphs with arbitrarily large chromatic numbers.

In 1973, Erdős and Simonovits [419] suggested to investigate whether there exist H-free graphs with arbitrary large chromatic numbers and also large minimum degree.[10] More formally, the *chromatic threshold $\delta_\chi(H)$* of a graph H is the infimum of $d > 0$ such that there exists a $C = C(H, d)$ for which every H-free graph G with minimum degree at least $d|G|$ satisfies $\chi(G) \leq C$. In other words, the chromatic number of H-free n-vertex graphs with minimum degree dn may be arbitrarily large if $d < \delta_\chi(H)$ but it is bounded if $d > \delta_\chi(H)$.

It follows from the 1954 Theorem of Kővári et al. [743] that $\delta_\chi(H) = 0$ for bipartite graphs H. In 2002, Thomassen [1196], confirming a conjecture of Erdős and Simonovits, proved that $\delta_\chi(K_3) = \frac{1}{3}$ for the triangle K_3. More generally, Goddard and Lyle [500] proved in 2011 that $\delta_\chi(K_r) = \frac{2r-5}{2r-3}$, $r \geq 3$, for a clique

---

[9] Here Tutte is writing under the collaborative pseudonym "Blanche Descartes".

[10] We say that a graph G has *minimum degree* at least M if the degree of each vertex of G is at least M.

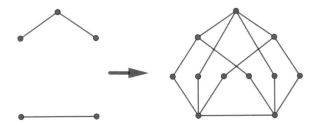

Fig. 2.2 An example of a Zykov graph

K_r with r vertices. This line of research culminated in 2013 when Allen et al. [28] determined the chromatic thresholds of all graphs H.

Theorem 2.35 *The chromatic threshold of every graph H with $\chi(H) = r \geq 3$ must be either $\frac{r-3}{r-2}$, or $\frac{2r-5}{2r-3}$, or $\frac{r-2}{r-1}$.*

Theorem 2.35 lists three possible values for $\delta_\chi(H)$ for every graph H with $\chi(H) \geq 3$. The authors also determined exactly which graphs H have each of these values of $\delta_\chi(H)$. Together with the known fact that $\delta_\chi(H) = 0$ if $\chi(H) \leq 2$, this determines the chromatic thresholds $\delta_\chi(H)$ of all graphs H.

A central step in the proof of Theorem 2.35 was the confirmation of the 2010 conjecture of Łuczak and Thomassé [820]. A graph H is called *near-acyclic* if $\chi(H) = 3$ and H admits a partition into a forest F and an independent set[11] S such that every odd cycle of H meets S in at least two vertices. Łuczak and Thomassé conjectured that a graph H with $\chi(H) = 3$ has $\delta_\chi(H) = 0$ if and only if it is near-acyclic. In fact, only the "if" part of this conjecture remained open and was proved in [28]. An important role in the proof is played by *Zykov graphs*. We start with a graph G' with connected components $C_1, \ldots, C_m$ and then construct a graph G by adding to G', for each m-tuple $u = (u_1, \ldots, u_m) \in C_1 \times \cdots \times C_m$, a vertex v_u adjacent to each u_j, see Fig. 2.2. If G' is triangle-free, then so is G, and Zykov [1295] used this construction to obtain a sequence of triangle-free graphs with high chromatic numbers. The authors of [28] modified this construction in a suitable way, and proved that every graph with linear minimum degree and sufficiently large chromatic number contains a well-structured collection of (modified) Zykov graphs. On the other hand, they proved that for every near-acyclic graph H, if a graph G contains such a collection of Zykov graphs, then G contains H. This implies that $\delta_\chi(H) = 0$ and proves the Łuczak–Thomassé conjecture, which was known to imply Theorem 2.35 for $r = 3$. The authors then generalized this argument to $r > 3$.

The Chromatic Number of Random Graphs

What is the chromatic number of a "typical" large graph? Let $G(n, p)$ denote a random graph with n vertices, such that every possible edge is included independently

---

[11] A *forest* is a graph which contains no cycles as subgraphs. A connected forest is called a *tree*. An *independent set* in a graph is a set of vertices no two of which are connected by an edge.

with probability p. This is the classical model of random graphs introduced by Erdős and Rényi [415] in 1960. In 1988, Bollobás [162] proved that if $0 < p < 1$ is fixed and $n \to \infty$, then almost every random graph $G(n, p)$ has chromatic number

$$\chi(G(n, p)) = \left(\frac{1}{2} + o(1)\right) \log\left(\frac{1}{1-p}\right) \frac{n}{\log n}.$$

In 1991, Łuczak [818] determined the asymptotic growth of $\chi(G(n, p))$ if $p = p(n) \to 0$ but $np \to \infty$.

In the case of random graphs of bounded average degree, that is, $p = \frac{d}{n}$ for constant d, much sharper results are known. In 1997, Alon and Krivelevich [35] proved[12] that for every $d > 0$ there exists a constant k_d, such that, with probability tending to 1 as $n \to \infty$, the chromatic number of $G\left(n, \frac{d}{n}\right)$ is either k_d or $k_d + 1$. However, the exact value of k_d as a function of d remained unknown. In 2005, this value was determined by Achlioptas and Naor [3].

Theorem 2.36 *With probability tending to 1 as $n \to \infty$, the chromatic number of $G\left(n, \frac{d}{n}\right)$ is either k_d or $k_d + 1$, where k_d is the smallest integer k such that $d < 2k \log k$.*

The authors further proved that if $d \in (2k \log k - \log k, 2k \log k)$ then the chromatic number of $G\left(n, \frac{d}{n}\right)$ is almost surely equal to $k + 1$. This determines the exact value of $\chi(G(n, d/n))$ for roughly half of all $d > 0$.

The proof of Theorem 2.36 is based on the 1999 result of Achlioptas and Friedgut [2], who proved that the probability that $G\left(n, \frac{d}{n}\right)$ is k-colourable changes from almost 1 to almost 0 when d crosses an interval whose length tends to 0 with n. This implies that in order to prove that $G\left(n, \frac{d}{n}\right)$ is almost surely k-colourable, it is sufficient to prove that, for some for some $d' > d$ and $\epsilon > 0$, $G\left(n, \frac{d'}{n}\right)$ is k-colourable with probability at least ϵ. This is equivalent to $\mathbb{P}[X > 0] \geq \epsilon$, where X is the number of k-colourings of $G\left(n, \frac{d'}{n}\right)$. The authors then estimated the first and second moments $\mathrm{E}[X]$ and $\mathrm{E}[X^2]$ of X, proved that $\frac{\mathrm{E}[X]^2}{\mathrm{E}[X^2]} \geq \epsilon$ for some $\epsilon > 0$ independent from n, and finished the proof by applying the inequality $\mathbb{P}[X > 0] \geq \frac{\mathrm{E}[X]^2}{\mathrm{E}[X^2]}$, valid for any non-negative random variable X.

Theorem 2.36 covers the important but special case $p = \frac{d}{n}$. More generally, Coja-Oghlan et al. [292] explicitly determined the three possible values of $\chi(G(n, p))$ whenever $p < n^{-\alpha}$ for any $\alpha > 3/4$. In fact, a 1987 theorem of Shamir and Spencer [1109] states that for *any* sequence $p = p(n)$, the chromatic number $\chi(G(n, p))$ is almost surely contained in a (non-explicit) sequence of intervals of length about $\sqrt{n}$. In 1992, Erdős considered the case when $p = \frac{1}{2}$ is constant and asked how accurately $\chi(G(n, 1/2))$ can be estimated. In particular, is it concentrated

---

[12] The theorem of Alon and Krivelevich remains correct, more generally, for $p = n^{-\alpha}$ for any $\alpha > 1/2$.

inside a series of intervals of constant length, like in Theorem 2.36? Erdős predicted that it is not, and this prediction was confirmed by Heckel in a paper [587] posted online in 2019.

Theorem 2.37 *For any constant $c < \frac{1}{4}$, there is no sequence of intervals of length n^c which contain $\chi(G(n, 1/2))$ with high probability.*

The proof of Theorem 2.37 follows from the connection between $\chi(G(n, 1/2))$ and the size $\alpha(G(n, 1/2))$ of the largest independent set in $G(n, 1/2)$. It is known that for most n, $\alpha(G(n, 1/2)) = a(n)$ with high probability, where $a(n)$ is the largest integer not exceeding $2 \log_2 n - 2 \log_2 \log_2 n + 2 \log_2(e/2) + 1$. If X_a denotes the number of independent sets of size a in $G(n, 1/2)$, then the distribution of X_a is known to be approximately Poisson, and X_a is not concentrated in any sequence of intervals of short length. Heckel [587] proved that $\chi(G(n, 1/2))$ varies almost as much as X_a, at least for some values of n, and deduced Theorem 2.37 from this fact.

The List Chromatic Number

A graph G is called *k-choosable* if, whenever for each vertex v of G we assign a list L_v of k colours to v, then it is possible to choose a colour for v from the list L_v, so that no two adjacent vertices receive the same colour. The *list chromatic number* $\chi_l(G)$ is the smallest k such that G is k-choosable.

The list chromatic number $\chi_l(G)$ is a natural generalization of the ordinary chromatic number $\chi(G)$ of G, which corresponds to the case when all L_v are the same. We have $\chi_l(G) \geq \chi(G)$ by definition, but in general $\chi_l(G)$ may be much larger than $\chi(G)$. For example, as observed in [417], $\chi_l(K_{d,d}) = (1 + o(1)) \log_2 d$ for the complete bipartite graph $K_{d,d}$, in sharp contrast with $\chi(K_{d,d}) = 2$.

In 1993, Alon [34] proved that $\chi_l(G) \geq (1/2 + o(1)) \log_2 d$ for all graphs G of minimum degree d. In 2015, Saxton and Thomason [1083] improved this inequality by a factor of 2.

Theorem 2.38 *If G is a simple graph with average degree d, then $\chi_l(G) \geq (1 + o(1)) \log_2 d$ as $d \to \infty$.*

Example $G = K_{d,d}$ shows that the lower bound in Theorem 2.38 is the best possible one.

Theorem 2.38 is just one of many corollaries of the deep theory of hypergraph containers developed by Saxton and Thomason. Recall that an *r-uniform hypergraph* (or *r-graph* for short) is a generalization of a graph in which an edge joins exactly r vertices. An *independent set* in an r-graph is a set of vertices containing no edge. Saxton and Thomason [1083] showed that, for every r-graph G, there exists a relatively small collection C of not very large vertex subsets of G such that every independent set of G is contained within a member of C, see [1083] for details and the exact formulation. This fundamental result has many corollaries, and Theorem 2.38, together with its generalization to hypergraphs, is one of them.

Counting Proper Colourings of Graphs

In addition to determining the minimal number of colours in a proper colouring of a graph G, an interesting problem is to determine the number $\chi_G(q)$ of proper colourings of vertices of G with q colours. As observed by Birkhoff [152] in 1912 for planar[13] graphs, and by Whitney [1255] in 1932 for general graphs, $\chi_G(q)$ is in fact a polynomial

$$\chi_G(q) = \sum_{k=0}^{n} (-1)^{n-k} a_k q^k, \tag{2.14}$$

where $a_0, a_1 \ldots, a_n$ are non-negative integers. The *chromatic polynomial* $\chi_G(q)$ is one of the most beautiful objects of study in graph theory. In 1968, Read [1034] asked for which coefficients $a_0, a_1, \ldots, a_n$ the right-hand side of (2.14) is a chromatic polynomial for some graph. In particular, he conjectured that, for any chromatic polynomial, the coefficient sequence is *unimodal* (that is, $a_0 \leq a_1 \leq \cdots \leq a_{i-1} \leq a_i \geq a_{i+1} \geq \cdots \geq a_n$ for some $0 \leq i \leq n$). In 1974, Hoggar [615] made the stronger[14] conjecture that this sequence is in fact *log-concave*: $a_{i-1} a_{i+1} \leq a_i^2$ for all $0 < i < n$. In 2012, Huh [627] confirmed Hoggar's conjecture and hence Read's conjecture.

Theorem 2.39 *The chromatic polynomial* $\chi_G(q)$ *of any graph* G *is given by* (2.14) *where the sequence* $a_0, a_1 \ldots, a_n$ *is log-concave.*

The proof of Theorem 2.39 uses the theory of matroids. A *(finite) matroid* M is a pair (X, I), where X is a finite set and $I \subset 2^X$ is a family of its subsets, such that

(i) $\emptyset \in I$;
(ii) if $A \in I$ and $A' \subseteq A$ then $A' \in I$; and
(iii) if $A \in I$, $B \in I$, and $|A| > |B|$, then there exists an $x \in A \setminus B$ such that $B \cup \{x\} \in I$.

Sets belonging to the family I are called *independent sets*. For example, let X be a set of vectors, and I a collection of sets of vectors which are linearly independent. Then properties (i)–(iii) hold and (X, I) is a matroid. Matroids representable in this form are called *representable*, or *linear*. In the special case when the vectors have complex entries, we say that the matroid is *representable over* $\mathbb{C}$. As another example, let X be the set of edges of a graph G, and I a collection of edges which form a forest, that is, does not contain a cycle. Then (X, I) is a matroid, which is called a *graphic matroid*. These examples demonstrate the connection of matroid theory to linear algebra and graph theory.

---

[13] A graph G is *planar* if it can be drawn on the plane in such a way that its edges intersect only at their endpoints.

[14] We say that a sequence has no *internal zeros* if the indices of the non-zero elements are consecutive integers. It is easy to see that every nonnegative log-concave sequence with no internal zeros is unimodal.

The *rank* of a matroid $M = (X, I)$ is $r(M) = \max_{B \in I} |B|$. Similarly, the rank of any subset $A \subset X$ is $r(A) = \max_{B \in I, B \subseteq A} |B|$. The polynomial

$$\chi_M(\lambda) = \sum_{A \subseteq X} (-1)^{|A|} \lambda^{r(M)-r(A)}$$

is called the *characteristic polynomial* of the matroid M. In the special case when M is a graphic matroid, the polynomial $\chi_M(\lambda)$ coincides with the chromatic polynomial of the graph (2.14), and counts the number of its proper colourings.

For $k = 0, 1, \ldots, r(M)$, let $w_k(M)$ be the absolute value of the coefficient of $\lambda^{r(M)-k}$ in $\chi_M(\lambda)$. In 1970, Rota [1060] conjectured that the sequence $w_k(M)$ is log-concave. Huh [627] proved a special case of this conjecture for matroids representable over $\mathbb{C}$. Since graphic matroids are representable over $\mathbb{C}$, this implies Rota's conjecture for graphic matroids, or, equivalently, Theorem 2.39.

Shortly after Theorem 2.39 was proved, Huh and Katz [628] extended the argument to prove Rota's conjecture for all representable matroids (not necessary over $\mathbb{C}$). The proof translates the theory developed by Hodge in the 1950s to study algebraic varieties (the solution sets to systems of polynomial equations) to the context of representable matroids. In 2018, Adiprasito, Huh, and Katz [10] developed an analogue of Hodge theory in the context of arbitrary matroids, which allowed them to confirm Rota's conjecture in full generality.

Theorem 2.40 *For any matroid M, the sequence $w_0(M), \ldots, w_{r(M)}(M)$ is log-concave.*

Theorem 2.39 is a beautiful result, but provides no clue how to determine how large $\chi_G(q)$ actually is. In this context, it is convenient to study the normalization

$$f(q, G) = \chi_G(q)^{1/|V(G)|},$$

where $|V(G)|$ is the number of vertices in G, because replacing G by a disjoint union of copies of itself does not change $f(q, G)$. In 2004, Galvin and Tetali [489] initiated the study of the following question: given positive integers d and q, what d-regular[15] graph G maximizes $f(q, G)$? In 2018, Davies, Jenssen, Perkins and Roberts [328] proved that for $d = 3$ the maximizer is the complete bipartite graph $K_{d,d}$. Later in 2018, Davies [327] proved the same result for $d = 4$. In 2020, Sah et al. [1074] proved that $K_{d,d}$ maximizes $f(q, G)$ among the d-regular graphs for all d.

Theorem 2.41 *For every d-regular graph G and every positive integer q, one has $f(q, G) \leq f(q, K_{d,d})$.*

---

[15] A graph G is *d-regular* if all its vertices have the same degree d.

Theorem 2.41 is a corollary of a more general theory of counting graph homomorphisms. A *graph homomorphism* from a graph G to a graph H is a map of vertices $\phi : V(G) \rightarrow V(H)$ such that $\{\phi(u), \phi(v)\}$ is an edge of H whenever $\{u, v\}$ is an edge of G. In general, we allow H to have loops. Let $\hom(G, H)$ be the number of graph homomorphisms of G to H. Galvin and Tetali [489] asked to determine, for each fixed d and H, the d-regular graph G which maximizes $\hom(G, H)^{1/|V(G)|}$. They proved that the answer is $K_{d,d}$ under the additional hypothesis that G is bipartite. Without the bipartite hypothesis, the question seems to be difficult. In 2013, Galvin [488] conjectured that the maximizer is always either $K_{d,d}$ or the complete $d + 1$-vertex graph K_{d+1}, but this was disproved by Sernau [1103] in 2018. Given the difficulties of the general question, it is natural to look at its special cases. Sah et al. [1074] proved that the maximizer is $K_{d,d}$ is we assume that either (i) G is triangle-free, or (ii) $H = K_q$ is a complete q-vertex graph with no loops. Because each proper colouring of G corresponds to its homomorphism to K_q, we have $\hom(G, K_q)^{1/|V(G)|} = \chi_G(q)^{1/|V(G)|} = f(q, G)$, and Theorem 2.41 follows from part (ii).

2.7 The Ramsey Theory of Graphs

Bounding Ramsey Numbers

Let K_n denote the complete graph with n vertices. In 1930, Ramsey [1022] proved that for any positive integers k, m there exists an n such that any red-blue edge colouring of K_n contains either a red K_k or a blue K_m. The smallest natural number n with this property is denoted $r(k, m)$ and is called the *(two-colour) Ramsey number*. The Ramsey numbers $r(k, k)$ are called *diagonal*.

Since the foundational work of Ramsey, the problem of estimating the Ramsey numbers has become one of the central problems in graph theory. In 1935, Erdős and Szekeres [420] proved that $r(k + 1, m + 1) \leq \binom{k+m}{k}$, where $\binom{n}{k} = \frac{n!}{k!(n-k)!}$ denotes the binomial coefficient. In particular, this implies that

$$r(k + 1, k + 1) \leq \binom{2k}{k}. \tag{2.15}$$

In 1987, Thomason [1195] improved it to

$$r(k + 1, k + 1) \leq k^{-1/2 + A/\sqrt{\log k}} \binom{2k}{k} \tag{2.16}$$

for some constant A. In 2009, Conlon [295] made an improvement to Thomason's bound.

Theorem 2.42 *There exist a constant $C > 0$ such that the inequality*

$$r(k+1, k+1) \leq k^{-C \frac{\log k}{\log \log k}} \binom{2k}{k} \tag{2.17}$$

holds for all integers $k \geq 2$.

Theorem 2.42 is a substantial improvement over the previous results. In particular, (2.15) implies that $r(k+1, k+1) \leq C \frac{4^k}{\sqrt{k}}$ for some constant C, (2.16) implies that $r(k+1, k+1) \leq C' \frac{4^k}{k}$, while (2.17) implies that $r(k+1, k+1) \leq C_m \frac{4^k}{k^m}$ for any m.

The proof of (2.15) is very simple. In any red-blue colouring of the complete graph on $r(k, m+1) + r(m, k+1)$ vertices, every vertex v must either be connected to at least $r(k, m+1)$ vertices by red edges or to at least $r(m, k+1)$ vertices by blue edges by the pigeonhole principle. In the first case, the subgraph induced by these $r(k, m+1)$ vertices contains either a blue K_{m+1} or a red K_k. If it contains a red K_k, we have a red K_{k+1} together with vertex v. Hence, we have either a red K_{k+1} or a blue K_{m+1}. The second case is considered similarly, hence

$$r(k+1, m+1) \leq r(k, m+1) + r(m, k+1),$$

and then (2.15) follows from an easy inductive argument.

This proof is based on the fact that if a red-blue colouring of a graph avoids red K_{k+1} and blue K_{m+1} (let us call it a *Ramsey colouring*), then every vertex is adjacent to at most $r(k, m+1) - 1$ red edges and at most $r(k+1, m) - 1$ blue edges. Thomason [1195] observed the stronger property that in every Ramsey colouring every red clique K_r extends to at most $r(k+1-r, m+1)$ red cliques of size $r+1$, while every blue clique K_r extends to at most $r(k+1, m+1-r)$ blue cliques of size $r+1$. He then deduced (2.16) from this via a careful inductive argument. The proof of Theorem 2.42 is based on the same ideas in combination with more careful estimates for the number of red and blue K_r in any Ramsey colouring. In 2020, Sah [1073] posted a preprint online in which he extended the range of r for which we can control the number of red and blue K_r to the optimal scale. This leads to a slightly improved upper bound

$$r(k+1, k+1) \leq e^{-c(\log k)^2} \binom{2k}{k} \tag{2.18}$$

for some $c > 0$ and all $k \geq 3$, but also implies that any further improvement would require a substantially new idea. Note that the upper bound (2.18) still grows faster that $(4 - \epsilon)^k$ for any fixed $\epsilon > 0$. For comparison, we have the lower bound

$$r(k+1, k+1) \geq (1 + o(1)) \frac{k+1}{e} (\sqrt{2})^k \tag{2.19}$$

proved by Erdős [398] in 1947, but no lower bound in the form $r(k + 1, k + 1) \geq (\sqrt{2} + \epsilon)^k$ for some $\epsilon > 0$. Hence, the actual growth rate of $r(k, k)$ remains a mystery.

The numbers $r(k, k)$ can be generalized in various ways. For a graph H, the *Ramsey number* of H, denoted $r(H)$, is the minimum integer n such that in every edge two-colouring of K_n there exists a monochromatic copy of H. Then $r(k, k) = r(H)$ if $H = K_k$ is the complete k-vertex graph. In 1973, Burr and Erdős [234] initiated the study of the "opposite" case when the graph H is "sparse", and the appropriate measure of sparseness in this context is d-degeneracy. A graph H is called *d-degenerate* if all its subgraphs contain a vertex of degree at most d. Burr and Erdős conjectured that, for any fixed d, the Ramsey numbers of k-vertex d-degenerate graphs is bounded by a linear function of k. This is in striking contrast with the exponential growth of $r(K_k)$ guaranteed by (2.19). In 2017, Lee [775] proved this conjecture.

Theorem 2.43 *For every positive integer d there is a constant $c = c(d)$ such that every d-degenerate graph H on k vertices satisfies $r(H) \leq ck$.*

To prove Theorem 2.43, one needs to show that, for $n = ck$, every edge two-colouring of K_n contains a monochromatic copy of H. Lee proved this by developing a random greedy algorithm based on the methodology of dependent random choice, which succeeds in finding such a copy with non-zero probability.

Another natural generalization is to increase the number of colours. Let $R(k, l)$ be the smallest natural number n such that every colouring of the edges of the complete graph K_n in l colours contains a monochromatic K_k. In this notation, $R(k, 2) = r(k, k)$. A simple modification of the proof of (2.15) implies the upper bound

$$R(k, l) \leq l^{lk + o(k)},$$

which remains the best known up to the improvements in the $o(k)$ term. In particular, $R(k, 3) \leq 27^{k + o(k)}$. For the lower bound, a simple probabilistic construction shows that $R(k, 3) \geq (\sqrt{3})^{k - o(k)}$. In 2020, Conlon and Ferber posted online a preprint [296] with a new construction, which shows that $R(k, q + 1) \geq 2^{k/2} q^{3k/8 - o(k)}$ for every prime q. In particular, with $q = 2$, $R(k, 3) \geq 2^{7k/8 - o(k)}$. Note that $2^{7/8} = 1.834...$, while $\sqrt{3} = 1.732...$, hence this is an exponential improvement. In just a few weeks, Wigderson [1259] modified their construction to show that, more generally,

$$R(k, l) \geq \left(2^{3l/8 - 1/4}\right)^{k - o(k)}$$

for all $l \geq 2$.

Explicit Construction of Ramsey Graphs

Colouring the edges of a complete graph K_n with two colours with no large monochromatic cliques is equivalent to constructing an n-vertex graph with no large clique or independent set. A graph on n vertices is called a *k-Ramsey graph* if it contains no clique or independent set of size k. The bound (2.19) implies the existence of a k-Ramsey graph over n vertices with k as low as $(2 + o(1)) \log_2 n$. In fact, the proof uses a probabilistic method and shows that *most n-vertex graphs* are $(2 + o(1)) \log_2 n$-Ramsey. However, it does not provide any specific example of such a graph. For many applications, especially in computer science, it is important to construct k-Ramsey graphs explicitly. By an "explicit construction" we mean a polynomial time algorithm that, given the labels of two vertices in the graph, determines whether there is an edge between them. For an n-vertex graph, the input of such an algorithm has $2 \log n$ bits, hence the running time should be polynomial in $\log n$.

In 1981, Frankl and Wilson [468] constructed k-Ramsey n-vertex graphs for $k \approx 2^{\sqrt{\log n}}$. Between 1981 and 2012, many other constructions were discovered, but, despite using very different methods, all of them gave the same value of k, up to the lower-order terms. Finally, in 2012, Barak et al. [97] were able to break this barrier and achieve a significantly better construction.

Theorem 2.44 *There exist absolute constants $\epsilon > 0$ and n_0 such that there is an explicit construction of a $2^{2^{(\log \log n)^{1-\epsilon}}}$-Ramsey graph over n vertices for every integer $n > n_0$.*

Because $2^{2^{(\log \log n)^{1-\epsilon}}} = 2^{(\log n)^{o(1)}}$, Theorem 2.44 significantly beats the Frankl–Wilson $2^{\sqrt{\log n}}$ construction. In 2019, Cohen [284], and, independently, Chattopadhyay and Zuckerman [265] made the next significant improvement.

Theorem 2.45 *There exist an absolute constant c such that there is an explicit construction of a $2^{(\log \log n)^c}$-Ramsey graph over n vertices.*

Later in 2019, Li [793] presented an even better explicit construction, which builds a k-Ramsey n-vertex graph with $k = (\log n)^{O\left(\frac{\log \log \log n}{\log \log \log \log n}\right)}$. However, even after these breakthroughs, we still do not have a deterministic construction with $k = (\log n)^C$ for any constant C, while random graphs achieve $k = (2 + o(1)) \log_2 n$.

Hypergraph Ramsey Numbers

The *hypergraph Ramsey number $r_s(k, m)$* is the minimum n such that every red-blue colouring of the unordered s-tuples of an n-element set contains a red set of size k or a blue set of size m, where a set is called red (blue) if all s-tuples from this set are red (blue). The "ordinary" Ramsey numbers $r(k, m)$ discussed above correspond to the case $s = 2$. Ramsey's theorem [1022] applies to the case of general s and implies that the numbers $r_s(k, m)$ are finite. As with "ordinary" Ramsey numbers, the central problem for hypergraph Ramsey numbers $r_s(k, m)$ is to find good upper and lower bounds for them.

Here, we will discuss the case when $s = 3$, k is fixed, and m grows. For the upper bound, Erdős and Rado [414] proved in 1952 that

$$\log r_3(k, m) \leq c(k) \frac{m^{2k-4}}{\log^{2k-6} m}. \tag{2.20}$$

For the lower bound, in 1972 Erdős and Hajnal [410] studied the case $k = 4$, showed that $\log r_3(4, m) > cm$ for some $c > 0$, and conjectured that in fact

$$\lim_{m \to \infty} \frac{\log r_3(4, m)}{m} = \infty. \tag{2.21}$$

In 2010, Conlon, Fox, and Sudakov [297] significantly improved both the upper and lower bounds for $r_3(k, m)$.

Theorem 2.46

(i) *For fixed $k \geq 4$ and sufficiently large m,*

$$\log r_3(k, m) \leq \left(\frac{k-3}{(k-2)!} + o(1) \right) m^{k-2} \log m;$$

(ii) *There is a constant $c > 0$ such that*

$$\log r_3(k, m) \geq c k m \log \left(\frac{m}{k} + 1 \right)$$

for all $4 \leq k \leq m$.

Part (i) of Theorem 2.46 improves (2.20) by a factor m^{k-2}, ignoring the $\log m$ factors. Part (ii) of Theorem 2.46 proves conjecture (2.21).

2.8 Expander Graphs

The Existence of Expanders

Informally, expanders are graphs which are simultaneously sparse and highly connected. These properties make them useful in numerous applications in mathematics, computer science, and in physical contexts.

Let G be a graph with vertex set V. For $S \subset V$, let $|\partial S|$ be the number of edges with exactly one endpoint in S. The *edge expansion ratio* of G, denoted $h(G)$, is

$$h(G) = \min_{1 \leq |S| \leq |V|/2} \frac{|\partial S|}{|S|}. \tag{2.22}$$

A sequence of d-regular graphs $G_1, G_2, \ldots$ of increasing size is a *family of expanders* if there exists an $\epsilon > 0$ such that $h(G_i) \geq \epsilon$ for all i.

This definition can be equivalently stated in terms of eigenvalues of the adjacency matrix. The *adjacency matrix* $A = A(G)$ of an n-vertex graph G with vertices marked as $1, 2, \ldots, n$ is an $n \times n$ matrix with entries $a_{ij} = 1$ if vertices i and j are connected by an edge and $a_{ij} = 0$ otherwise. The matrix A has n real eigenvalues which we can arrange in the non-increasing order $\lambda_1(G) \geq \lambda_2(G) \geq \cdots \geq \lambda_n(G)$. If G is d-regular, then $\lambda_1(G) = d$ and $\lambda_n \geq -d$. In 1984, Dodziuk [357] proved that, for any d-regular graph G,

$$\frac{d - \lambda_2(G)}{2} \leq h(G) \leq \sqrt{2d(d - \lambda_2(G))}.$$

In particular, a sequence of d-regular graphs is a family of expanders if and only if there exists a $\delta > 0$ such that $\lambda_2(G_i) \leq d - \delta$ for all i. Hence, the smaller $\lambda_2(G)$, the "better expander" is G. The *Alon–Boppana inequality* states that, for any d-regular n-vertex graph G,

$$\lambda_2(G) \geq 2\sqrt{d - 1} - o_n(1), \tag{2.23}$$

where $o_n(1)$ is a quantity tending to 0 if d is fixed and $n \to \infty$. In 1986, Alon [33] conjectured that, for any $\epsilon > 0$, "almost all" d-regular n-vertex graphs G satisfy the opposite inequality

$$\lambda_2(G) \leq 2\sqrt{d - 1} + \epsilon. \tag{2.24}$$

In other words, Alon's conjecture states that "almost all" d-regular graphs are as good expanders as they possibly can, up to ϵ.

In 2008, Friedman [477] proved this conjecture. For an even $d \geq 4$, by a *random d-regular graph* on n vertices we mean a graph $G_{d,n}$ formed from $d/2$ uniform, independent permutations on $\{1, \ldots, n\}$.

Theorem 2.47 *For fixed d and fixed real $\epsilon > 0$, let p_n be the probability that $G_{d,n}$ satisfies* (2.24). *Then* $\lim_{n \to \infty} p_n = 1$.

Theorem 2.47 was proved using *Wigner's trace method* [1260], which involves estimating the expected value of the trace[16] of a reasonably high power of the adjacency matrix of a random graph. These estimates are then used to bound the eigenvalues. The crucial new idea is to apply this method to the so-called "selective trace" instead of the trace, see [477] for details.

Theorem 2.47, as stated above, resolves Alon's conjecture for even d. In the same paper [477], Friedman also proved the conjecture for other models of a random d-regular graph, including models for odd d. This resolves the conjecture in full.

---

[16] The *trace* of a square matrix A with entries a_{ij} is $\mathrm{tr}[A] = \sum_{i=1}^{n} a_{ii}$.

We next discuss graphs for which (2.24) holds without the ϵ term, which implies that they are even better expanders. A d-regular graph G is *bipartite* if and only if $\lambda_n(G) = -d$. The eigenvalues $\lambda_1(G) = d$ and $\lambda_n(G) = -d$ are called *trivial*. We say that a d-regular bipartite graph is *Ramanujan* if all of its non-trivial eigenvalues lie between $-2\sqrt{d-1}$ and $2\sqrt{d-1}$. In view of (2.23), Ramanujan graphs are the best possible expanders. Many authors tried to construct infinite families of d-regular Ramanujan graphs, but succeeded only for some values of d, for example, such that $d-1$ is a power of a prime number. In 2010, Lubotzky [816] asked whether d-regular bipartite Ramanujan graphs exist for every d. In 2015, Marcus et al. [841] answered this question affirmatively.

Theorem 2.48 *For every* $d \geq 3$ *there exists an infinite sequence of* d-*regular bipartite Ramanujan graphs.*

To prove Theorem 2.48, the authors developed a new technique for controlling the eigenvalues of random matrices, which is based on interlacing polynomials. A polynomial $g(x) = \prod_{i=1}^{n-1}(x - \alpha_i)$ is said to *interlace* with a polynomial $f(x) = \prod_{i=1}^{n}(x - \beta_i)$ if $\beta_1 \leq \alpha_1 \leq \beta_2 \leq \alpha_2 \leq \cdots \leq \alpha_{n-1} \leq \beta_n$. Polynomials $f_1, \ldots, f_k$ have a *common interlacing* if there exists a polynomial g that interlaces each of the f_i. A crucial property of interlacing polynomials is that the largest root of the sum $\sum_{i=1}^{k} f_i$ cannot be less that the largest root of each f_i. The authors proved that the characteristic polynomials of the signed adjacency matrices of graphs form such a family. This allowed them to bound the roots of such polynomials (which are eigenvalues) in terms of roots of their sums, which are easier to analyse. The method of interlacing polynomials turned out to be very powerful, and has many other applications. For example, this method has been used in a crucial way in the resolution of the Kadison–Singer problem, see Theorem 3.94 below.

In a later work [843], the same authors proved the stronger result that n-vertex d-regular bipartite Ramanujan graphs exist for every $d \geq 3$ and every even $n \geq 2$.

Examples of Expanders

Theorem 2.47 states that if we select a large d-regular graph at random, then we get an expander with high probability. Theorem 2.48 proves the existence of graphs with an even better expansion property. However, these theorems provide no explicit examples of expander graphs, and having such examples is important for many applications. The first explicit example[17] of a d-regular family of expanders was constructed by Margulis [844] in 1973 for $d = 8$.

A standard way to generate examples of expander graphs for other values of d is to use the connection with group theory: the Cayley graphs of various groups turn

---

[17] For any integer $m > 0$, let G_m be a graph with vertex set $\mathbb{Z}_m \times \mathbb{Z}_m$, where $\mathbb{Z}_m = \{0, 1, \ldots, m-1\}$, in which the neighbours of any vertex (x, y) are $(x + y, y)$, $(x - y, y)$, $(x, y + x)$, $(x, y - x)$, $(x + y + 1, y)$, $(x - y + 1, y)$, $(x, y + x + 1)$, and $(x, y - x + 1)$, where all operations are modulo m. Margulis proved that $\{G_m\}_{m=1}^{\infty}$ is an 8-regular family of expanders.

out to be expanders. For a finite group G with set of generators[18] S, the *(undirected) Cayley graph* $\Gamma = \Gamma(G, S)$ is the graph whose vertices are elements of G, and vertices g, h are connected by an edge if $g = hs$ or $h = gs$ for some $s \in S$. Given an infinite family $\{G_n\}$ of finite groups, one aims to make their Cayley graphs expanders, using suitably chosen generating sets. In particular, Lubotzky and Weiss [817] asked in 1993 whether the Cayley graphs of the group[19] $SL_2(\mathbb{F}_p)$ form a family of expanders, for various choices of generating sets. In 2008, Bourgain and Gamburd [187] proved that this is indeed the case if the generating sets are chosen at random.

Theorem 2.49 *Fix $k \geq 2$. For any prime p, let S_p be a k-element subset of $SL_2(\mathbb{F}_p)$ whose elements are chosen independently at random, and let $\Gamma_p = \Gamma(SL_2(\mathbb{F}_p), S_p)$ be the corresponding Cayley graph. Then the sequence $\Gamma_2, \Gamma_3, \Gamma_5, \Gamma_7, \Gamma_{11}, \ldots$ forms a family of expanders with probability 1.*

The crucial ingredients in the proof of Theorem 2.49 are that (i) nontrivial eigenvalues of the Cayley graph of $SL_2(\mathbb{F}_p)$ appear with high multiplicity, and (ii) there is an upper bound on the number of short cycles in this Cayley graph. See Theorems 4.18 and 4.19 discussed in the group theory section for another example of generating a family of expanders as Cayley graphs of groups.

A completely different, combinatorial way to construct expander graphs was developed by Reingold et al. [1041] in 2002. For any graph G, a 2-edge path is a set of vertices i, k, j, connected by edges $\{i, k\}$ and $\{k, j\}$. Now, the graph G^2 is the graph with the same vertices as G such that for every 2-edge path i, k, j in G, we put an edge between i and j in G^2.

Denote by $[n]$ the set of integers $1, 2, \ldots, n$. We say that a d-regular graph G is *d-coloured* if its edges are marked with labels from $[d]$ such that no 2 edges with the same mark share a vertex. For a label $i \in [d]$ and a vertex v let $v[i]$ be the neighbour of v along the edge marked i. Let G be a d_1-regular d_1-coloured graph on $[n_1]$, and H be a d_2-regular d_2-coloured graph on $[d_1]$. Then the *zig-zag product* $G \cdot_Z H$ of G and H is a d_2-regular graph on $[n_1] \times [d_1]$ defined as follows: For all $v \in [n_1], k \in [d_1], i, j \in [d_2]$, the edge (i, j) connects the vertex (v, k) to the vertex $(v[k[i]], k[i][j])$.

Theorem 2.50 *The sequence given by $G_1 = H^2$, $G_{i+1} = G_i^2 \cdot_Z H, i \geq 1$ (where H is a d-regular graph of size d^4 and sufficiently high edge expansion) is a d^2-regular family of expanders.*

Explicit examples of expander graphs play a crucial role in, for example, developing deterministic algorithms for combinatorial problems. Many problems have time or memory efficient algorithms based on random graphs, and such random

---

[18] A *set of generators* (or *generating set*) of a group G is a subset $S \subset G$ such that every $g \in G$ can be written as a finite product of elements of S and their inverses.

[19] Here, $\mathbb{F}_p$ is the field $\{0, 1, \ldots, p-1\}$ with operations modulo prime p, and $SL_2(\mathbb{F}_p)$ is the group of 2×2 matrices over $\mathbb{F}_p$ with determinant 1.

graphs can be replaced by explicitly constructed expander graphs. For example, Theorem 7.44 discussed later in this book was proved in this way by using the expander graphs constructed in Theorem 2.50.

2.9 Other Topics in Graph Theory

The Theory of Graph Minors

Given a graph G and edge $\{u, v\}$ in it, one may form a new graph G' by removing this edge and merging the vertices u and v into a new vertex which is connected by edges to all neighbours of u and all neighbours of v. This operation is called *edge contraction* and preserves many important graph properties. For example, if G is a planar graph then so is G'.

A graph H is a *minor* of graph G if H can be obtained from G by a sequence of contractions of edges of G, and deletions of edges and vertices of G. A family $\mathcal{F}$ of graphs is called *minor-closed* if $G \in \mathcal{F}$ implies that $H \in \mathcal{F}$ for all minors H of G. For example, the family of planar graphs is minor closed. Because the complete 5 vertex graph K_5 and complete bipartite graph $K_{3,3}$ are not planar, no planar graph can contain either of these graphs as a minor. In 1937, Wagner [1242] proved that this obvious necessary condition is in fact sufficient for planarity: any finite graph G is planar if and only if it does not have K_5 or $K_{3,3}$ as a minor. In 1970, Wagner [1243] conjectured that, more generally, for any minor-closed family $\mathcal{F}$ of graphs there exist a finite set $H_1, H_2, \ldots, H_n$ of graphs such that $G \in \mathcal{F}$ if and only if G does not have any of the graphs H_i as a minor. This conjecture is equivalent to the statement that in any infinite sequence $G_1, G_2, \ldots$ of graphs, some graph G_i is a minor of some graph G_j. In 1983, Robertson and Seymour [1050] published the first paper in a long series of papers aiming to develop a deep theory of graph minors and prove Wagner's conjecture. In 2004, they finished the proof of the conjecture in the 20th paper in their series [1051]. The total length of all these papers is over 500 pages.

Theorem 2.51 *In any infinite sequence $G_1, G_2, \ldots$ of finite graphs, one can select two graphs, G_i and G_j, such that one of them is isomorphic to a minor of another one.*

Graph Decomposition

We say that graph G is *decomposed* into its subgraphs H_1 and H_2, and write $G = H_1 \oplus H_2$, if G is the edge disjoint union of H_1 and H_2. We say that G is H-*decomposable* if $G = H_1 \oplus H_2 \oplus \cdots \oplus H_k$, where each H_i is isomorphic to H. The theory of graph decompositions studies under what conditions a graph G is H-decomposable. One of the most studied special cases of this problem is the case when $G = K_n$ is the complete graph and $H = C_m$ is the cycle of fixed length m. Obviously, C_m-decomposition of K_n can be possible only if $3 \leq m \leq n$ and the number of edges in K_n is divisible by m. Also, the degrees of all vertices of K_n should be even, hence n should be odd. In 2002, Šajna [1075] proved that

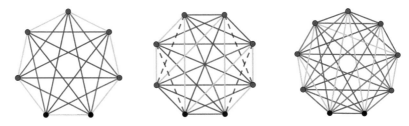

Fig. 2.3 Cycle decompositions of some complete n-vertex graphs

these obvious necessary conditions are in fact sufficient, culminating a long series of papers with partial results. She also established a similar result for even n, in which case a perfect matching is removed from K_n to make the vertices' degrees even.

Theorem 2.52 *(i) For every odd $n \geq 3$ and every m such that $3 \leq m \leq n$ and m divides the number of edges in K_n, the graph K_n is C_m-decomposable; (ii) for every even $n \geq 4$ and every m such that $3 \leq m \leq n$ and m divides the number of edges in K'_n, the graph K'_n is C_m-decomposable, where K'_n denotes the complete graph on n vertices with a perfect matching removed.*

Figure 2.3 illustrates the decomposition in Theorem 2.52 for $n = 7, 8$, and 9, and $m = n$.

Flows of Graphs

Recall that a *digraph* (directed graph) is a set of vertices some of which are connected by directed edges (arcs). An *orientation* of an (undirected) graph $G = (V, E)$ is a digraph obtained from G by replacing each edge $e \in E$ by one of the two possible arcs with the same ends. For each $v \in G$, let $E^+(v)$ and $E^-(v)$ be the set of edges going from v and to v, respectively. For an integer $k > 1$, a *nowhere-zero k-flow* on G is an orientation of G together with a function $f : E \to \{1, \ldots, k-1\}$, such that

$$\sum_{e \in E^+(v)} f(e) = \sum_{e \in E^-(v)} f(e)$$

for every vertex $v \in V$. This concept was introduced by Tutte [1219] in 1949.

A connected graph G is called *h-edge-connected* if it remains connected whenever fewer than h edges are removed. A graph G is called *bridgeless* if each connected component of G is 2-edge-connected. At the heart of the theory of k-flows are the following three conjectures of Tutte.

- *The 3-flow Conjecture.* Every 4-edge-connected graph has a nowhere-zero 3-flow.

Fig. 2.4 The Petersen graph

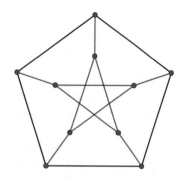

- *The 4-flow Conjecture*. Every bridgeless graph that does not have the Petersen graph (see Fig. 2.4) as a minor has a nowhere-zero 4-flow.
- *The 5-flow Conjecture*. Every bridgeless graph has a nowhere-zero 5-flow.

These conjectures are the best possible ones in several ways. The 3-flow conjecture considers "best" (that is, with lowest k) non-trivial k-flows, because 2-flows have $f(e) = 1$ for all $e \in E$ and are therefore trivial to analyse. On the other hand, the 5-flow conjecture considers the most general family of graphs, because it is easy to see that if G is not bridgeless then it cannot have a nowhere-zero k-flow for any k. The conclusion of the 5-flow conjecture is also the best possible because there are bridgeless graphs (for example, the Petersen graph) which do not have a nowhere-zero 4-flow. The same example demonstrates that (a) the 3-flow conjecture fails for 3-edge-connected graphs, and (b) in the 4-flow conjecture, the minor condition cannot be omitted.

All three conjectures remain open. Partial results establish the existence of a nowhere-zero k-flow (i) for a given family of graphs but with higher k or (ii) with given k but for a less general family of graphs. In 1979, Jaeger [647] proved that 4-edge-connected graphs have nowhere-zero 4-flows, while Seymour [1104] proved in 1981 that every bridgeless graph has a nowhere-zero 6-flow. These results give the "best possible approximation" of the 3-flow and 5-flow conjectures in direction (i). In the direction (ii) for the 3-flow conjecture, the aim is to prove the existence of a nowhere-zero 3-flow for all h-edge-connected graphs for some $h > 4$. In 2012, Thomassen [1197] proved this with $h = 8$. In 2013, Lovász, Thomassen, Wu and Zhang [815] improved this to $h = 6$.

Theorem 2.53 *Every 6-edge-connected graph admits a nowhere-zero 3-flow.*

In 2001, Kochol [727] proved that the 3-flow conjecture for 5-edge-connected graphs implies the conjecture in full. Hence, in terms of the parameter h, Theorem 2.53 came as close to the 3-flow conjecture as possible without resolving it.

Towards the 1-2-3-Conjecture

A *weighted graph* is a graph $G = (V, E)$ together with a function $w : E \to \mathbb{R}$. Let us define the *weight* of each vertex $v \in V$ as the sum of weights of all edges adjacent to it. We say that a weight w induces a *proper colouring* on G if every pair of adjacent vertices have different weights. In 2004, Karoński, Łuczak and Thomason [684] conjectured that for every graph G without isolated edges there exists a weight $w : E \to \{1, 2, 3\}$ which induces a proper colouring on G. This statement became known as the 1-2-3-*conjecture* and has attracted a lot of attention. The conjecture is the best possible in the sense that a weight $w : E \to \{1, 2\}$ with the same property does not exists if G is, for example, a triangle.

A weaker conjecture was the existence of a constant k, independent from G, such that for every graph as above there exists a weight $w : E \to \{1, 2, \ldots, k\}$ with this property. This weaker conjecture was proved by Addario-Berry, Dalal, McDiarmid, Reed and Thomason [6] in 2007, who also showed that it is true with $k = 30$. In 2008, it was improved to $k = 16$ by Addario-Berry et al. [7], and to $k = 13$ by Wang and Yu [1246]. In 2010, Kalkowski et al. [675] proved the same statement with $k = 5$.

Theorem 2.54 *For every graph* $G = (V, E)$ *without components isomorphic to* K_2, *there is a weight* $w : E \to \{1, 2, 3, 4, 5\}$ *such that the induced vertex weights properly colour* V.

Games on Graphs

An active research area in graph theory is to determine the winner in various games on graphs. For example, the *Maker-Breaker game* on a graph consist of a graph G with edge set E, a family $\mathcal{F}$ of subsets of E called the *winning sets*, and positive integers a, b. The Breaker claims b edges in his turn, the Maker answers by claiming a unoccupied edges, and then they continue claiming edges in turns. The Maker wins if she has occupied one of the winning sets in $\mathcal{F}$, the breaker wins otherwise.

A popular special case is the *Hamiltonicity game*, which is played on the complete n-vertex graph K_n, and the Maker wins if she can construct a Hamilton cycle from the edges she has chosen. The research on this game has a long history. In 1978, Chvátal and Erdős [283] showed that in the Hamiltonicity game with $a = b = 1$ (such a game is called *unbiased*) the Maker wins for all sufficiently large n. Moreover, Hefetz and Stich [589] proved in 2009 that the Maker can win this game in just $n + 1$ rounds, which is optimal.

In view of these results, it is natural to consider a version of this game with $a = 1$ but $b = b(n)$ being arbitrary, which is called the "$(1 : b)$ Hamiltonicity game on K_n". Chvátal and Erdős [283] proved that the Breaker wins this game if

$$b(n) \geq (1 + \epsilon) \frac{n}{\log n}, \tag{2.25}$$

where $\epsilon > 0$ is an arbitrarily small constant. In 2011, Krivelevich [750] proved that the bound (2.25) is the best possible, up to the lower order terms.

Theorem 2.55 *The Maker has a strategy to win the* $(1 : b)$ *Hamiltonicity game on* K_n *in at most* $14n$ *moves, for every*

$$b(n) \le \left(1 - \frac{30}{\log^{1/4} n}\right) \frac{n}{\log n},$$

for all large enough n.

The *critical bias* $b^*(n)$ of the $(1 : b)$ Hamiltonicity game on K_n is the maximum possible value of b for which the Maker still wins. The result of Chvátal and Erdős implies that $b^*(n) \le (1 + o(1)) \frac{n}{\log n}$. The theorem provides a matching lower bound and establishes that $b^*(n) = (1 + o(1)) \frac{n}{\log n}$.

Theorem 2.55 was proved by presenting an explicit strategy for the Maker to win the game. The high-level outline of the strategy is to first create an expander, then make it connected, and, finally, transform it into a Hamiltonian cycle.

The Graph Removal Lemma

A graph G is called H-*free* if it does not have a subgraph isomorphic to H. In 1986, Erdős et al. [408] proved that for each $\epsilon > 0$ and every graph H on h vertices there is a $\delta = \delta(\epsilon, H) > 0$ such that every graph on n vertices with at most δn^h copies of H can be made H-free by removing at most ϵn^2 edges. This statement became known as the *graph removal lemma*, and has many applications in graph theory and the theory of algorithms. However, for some applications, it is important to understand how exactly δ depends on ϵ and h.

This dependence can be expressed in terms of the tower function. The *tower function* $T : \mathbb{R} \to \mathbb{R}$ is the function defined by the rules (i) $T(1) = 2$, (ii) $T(n) = 2^{T(n-1)}$ for $n = 2, 3, \ldots$, and (iii), for all real x, $T(x) = T(\bar{x})$, where $\bar{x}$ is the least positive integer such that $x \le \bar{x}$.

The original proof of the graph removal lemma in [408] made a crucial use of famous Szemerédi regularity lemma, which, informally, states that every large graph can be partitioned into a small number of parts such that the edges between different parts behave almost randomly. This was the key lemma in Szemerédi's famous proof [1174] of (2.2), and has been used in numerous applications since then. However, every proof which uses the Szemerédi regularity lemma ends up with a very bad dependence of parameters. In particular, the proof in [408] only implies that the graph removal lemma is true if $\delta \approx T(P(\epsilon^{-1}))^{-1}$ for some polynomial P. In 2011, Fox [463] developed a new proof of the lemma, which avoids the use of the Szemerédi regularity lemma, and therefore leads to better dependence of δ on ϵ.

Theorem 2.56 *For each* $\epsilon > 0$ *and graph* H *on* h *vertices there is a* $\delta = \delta(\epsilon, H) > 0$ *such that every graph on* n *vertices with at most* δn^h *copies of* H *can be made* H-*free by removing at most* ϵn^2 *edges. Moreover, one can choose* $\delta = T(5h^4 \log \epsilon^{-1})^{-1}$.

The Sensitivity Conjecture

Graph theory has numerous applications to other areas of mathematics and computer science. Here we will discuss one particularly nice example.

The *n-dimensional hypercube graph* is the graph Q^n whose vertex set consists of vectors $(x_1, x_2, \ldots, x_n)$ such that each x_i is either 0 or 1, and two vectors are adjacent if and only if they differ in exactly one coordinate. Let $\Delta(H)$ denote the maximum degree of an induced subgraph H of Q^n. It is easy to see that one may select a subgraph H with 2^{n-1} vertices and no edges at all, but $\Delta(H) > 0$ if H has at least $2^{n-1} + 1$ vertices. How small can $\Delta(H)$ be in this case? In 1988, Chung et al. [281] proved that $\Delta(H) \geq (1/2 - o(1)) \log_2 n$. They also remarked that the best lower bound one may hope for is $\Delta(H) \geq \sqrt{n}$, because if $n = k^2$ is a perfect square, then there exist a $(2^{n-1} + 1)$-vertex subgraph H of Q^n with $\Delta(H) = k = \sqrt{n}$. In 2019, Huang [623] proved that the best possible lower bound of $\Delta(H)$ is indeed $\sqrt{n}$.

Theorem 2.57 *For every integer $n \geq 1$, and every $(2^{n-1} + 1)$-vertex induced subgraph H of Q^n, one has $\Delta(H) \geq \sqrt{n}$.*

Theorem 2.57 is a graph-theoretic result, but it has important consequences far beyond graph theory: it implies the truth of the sensitivity conjecture, which was one of the most important open problems in theoretical computer science. Let $\{0, 1\}^n$ be the set of strings of 0s and 1s of length n. For $x \in \{0, 1\}^n$ and $i \in [n] = \{1, \ldots, n\}$, let x^i denote the string which differs from x on exactly the bit i. Similarly, for any subset $B \subseteq [n]$, let x^B denote the string which differs from x on exactly the bits of B. The *sensitivity* $s(f)$ of a Boolean function $f : \{0, 1\}^n \to \{0, 1\}$ is the largest integer k for which there exist $x \in \{0, 1\}^n$ such that there are at least k values of $i = 1, \ldots, n$ with $f(x^i) \neq f(x)$. Similarly, the *block sensitivity* $bs(f)$ is the maximum number t of disjoint subsets $B_1, \ldots, B_t$ of $[n]$ such that $f(x) \neq f(x^{B_j})$ for all $1 \leq j \leq t$ and some $x \in \{0, 1\}^n$. By definition, $s(f) \leq bs(f)$ for any Boolean function f. The *sensitivity conjecture*, posed by Nisan and Szegedy [940] in 1994, predicted the existence of a constant C such that $bs(f) \leq s(f)^C$ for any f. Theorem 2.57, in combination with an earlier result of Gotsman and Linial [506], implies that the sensitivity conjecture is true with $C = 4$.

Menger's Theorem for Infinite Digraphs

Let A and B be two sets of vertices in a finite or infinite directed graph (digraph) G. A path from some $a \in A$ to some $b \in B$ will be called an $A - B$-*path*. We say that a set S of vertices of G *separates* A from B if every $A - B$-path contains at least one vertex $s \in S$. If G is finite, let $\sigma(A, B)$ be the minimal size of a set S which separates A from B, and let $\nu(A, B)$ be the maximal size of a family of vertex-disjoint paths from A to B. Obviously, $\sigma(A, B) \geq \nu(A, B)$. A combination of a 1931 theorem of Kőnig [735] and a 1927 theorem of Menger [880] implies that in fact $\sigma(A, B) = \nu(A, B)$ in every finite digraph G. This implies the existence of a family of disjoint paths and a separating set which chooses exactly one vertex from each path. In 1936, Erdős conjectured that the last statement remains true for infinite digraphs. The conjecture attracted considerable attention and has been proved in a number of special cases. In 1976, Podewski and Steffens [987] proved it for digraphs which are at the same time countable and bipartite. In 1984, Aharoni [13] proved it for all (possibly uncountable) bipartite digraphs. In 1987, Aharoni [14] proved the conjecture for general (not necessary bipartite) countable digraphs. Despite all these efforts, the general case of the conjecture remained open for over 70 years. In 2009, Aharoni and Berger [15] proved the conjecture in full.

Theorem 2.58 *Given two sets of vertices, A and B, in a (possibly infinite) digraph, there exists a family P of disjoint A − B-paths, and a set S separating A from B, such that S consists of a choice of precisely one vertex from each path in P.*

2.10 Counting Problems

Counting Permutations

A central theme in combinatorics is counting objects with various properties. A classical example of such a result is the *Erdős–Ko–Rado theorem*, which counts intersecting subsets of a set. Let $[n] = \{1, 2, \ldots, n\}$ be an n-element set, $r < n/2$, and $\mathcal{F}$ be a family of r-element subsets of $[n]$, such that the intersection of any $A, B \in \mathcal{F}$ is non-empty. How many sets can $\mathcal{F}$ contain? If $\mathcal{F}_i$ is the collection of all r-element subsets containing a fixed element $i \in [n]$, then $|\mathcal{F}_i| = \binom{n-1}{r-1}$. In 1961, Erdős, Ko and Rado [412] proved that in general $|\mathcal{F}| \le \binom{n-1}{r-1}$, and equality holds only if $\mathcal{F} = \mathcal{F}_i$ for some $i \in [n]$. More generally, if $\mathcal{F}$ is a family of r-element subsets of $[n]$, such that $|A \cap B| \ge k$ for all $A, B \in \mathcal{F}$, and n is sufficiently large depending on k and r, then $|\mathcal{F}| \le \binom{n-k}{r-k}$, and equality holds if and only if $\mathcal{F}$ consists of all r-element subsets of $[n]$ containing k fixed elements.

In 1977, Frankl and Deza [467] conjectured that a similar result may be true for permutations. A *permutation* of $[n]$ is a function $\sigma : [n] \to [n]$ such that $\sigma(i) \ne \sigma(j)$ if $i \ne j$. Let S_n be the set of all permutations of $[n]$. We say that two permutations $\sigma, \tau \in S_n$ *k-intersect* if there exist distinct $i_1, i_2, \ldots i_k \in [n]$ such that $\sigma(i_t) = \tau(i_t)$ for $t = 1, 2, \ldots, k$. A subset $I \subset S_n$ is said to be *k-intersecting* if any two permutations in I k-intersect.

Let us fix $\sigma \in S_n$, fix $i_1, i_2, \ldots i_k \in [n]$, and let $I = I(\sigma, i_1, \ldots, i_k)$ be the set of all $\tau \in S_n$ such that $\sigma(i_t) = \tau(i_t)$ for $t = 1, 2, \ldots, k$. Then I is k-intersecting and $|I| = (n - k)!$. Frankl and Deza [467] conjectured that if n is sufficiently large depending on k, then this construction gives the largest possible k-intersecting set of permutations. In 2011, Ellis, Friedgut and Pilpel [388] proved this conjecture. The proof uses eigenvalue techniques and representation theory (see the discussion before Theorem 4.8 for the basic concepts of representation theory).

Theorem 2.59 *For any positive integer k, and any sufficiently large n depending on k, if $I \subset S_n$ is k-intersecting, then $|I| \le (n - k)!$. Equality holds if and only if $I = I(\sigma, i_1, \ldots, i_k)$ for some $\sigma \in S_n$ and $i_1, i_2, \ldots i_k \in [n]$.*

Another interesting research direction is counting permutations avoiding certain patterns. A permutation of $[n] = \{1, 2, \ldots, n\}$ is called an *n-permutation*. We say that an n-permutation σ *contains* a k-permutation π if there exist integers $1 \le x_1 < \cdots < x_k \le n$ such that for all $1 \le i, j \le k$ we have $\sigma(x_i) < \sigma(x_j)$ if and only if $\pi(i) < \pi(j)$. Otherwise, we say that σ *avoids* π. For a fixed permutation π, let $S_n(\pi)$ be the number of n-permutations avoiding π. In the late 1980s, Stanley and Wilf [1263] conjectured that for all π there exists a constant c_π such that $S_n(\pi) \le c_\pi^n$ for all n.

In 2000, Klazar [722] proved that this conjecture would follows from a conjecture of Füredi and Hajnal about permutation matrices. Let A and P be 0-1 *matrices* (that is, matrices with all entries 0 or 1). We say that A *contains* the $k \times l$ matrix P with entries p_{ij} if there exists a $k \times l$ submatrix D of A with entries d_{ij} such that $d_{ij} = 1$ whenever $p_{ij} = 1$. Otherwise we say that A *avoids* P. Let $f(n, P)$ be the maximum number of 1-entries in an $n \times n$ 0-1 matrix avoiding P. A 0-1 matrix P is called a *permutation matrix* if it has exactly one entry of 1 in each row and each column and 0s elsewhere. In 1992, Füredi and Hajnal [480] conjectured that $f(n, P)$ grows at most linearly in n. In 2004, Marcus and Tardos [840] proved this conjecture.

Theorem 2.60 *For all permutation matrices P we have $f(n, P) = O(n)$.*

The proof of Theorem 2.60 is surprisingly simple. The authors used a partitioning of the larger matrix into blocks to prove a linear recursion for $f(n, P)$, from which the theorem follows easily. Together with the result of Klazar [722], Theorem 2.60 implies the truth of the Stanley–Wilf conjecture.

Counting Partitions of Integers

We next discuss counting partitions of integers. A *partition* λ of an integer $m > 0$ is a sequence $0 < \lambda_1 \le \lambda_2 \le \cdots \le \lambda_k$ of integers such that $\sum_{i=1}^k \lambda_i = m$. The number $p(m)$ of such partitions is known as the *partition function*. By convention, we also put $p(0) = 1$ and $p(m) = 0$ for $m < 0$. In addition to estimating $p(m)$, an interesting topic is the study of its divisibility properties. In 1921, Ramanujan [1021] proved that $p(5n+4)$ is always divisible by 5, $p(7n+5)$ by 7, $p(11n+6)$ by 11, but conjectured that "there are no equally simple properties for any moduli involving primes other than these three". In 2003, Ahlgren and Boylan [16] confirmed (a natural formalization of) this conjecture.

Theorem 2.61 *There is no prime $q \ne 5, 7, 11$ and integer b such that $p(qn + b)$ is divisible by q for all integers n.*

The situation changes if we consider divisibility by q of sequences $p(an + b)$, $n = 0, 1, 2, \ldots$ with possibly $a \ne q$. In 1967, Atkin and O'Brien [58] proved that $p(17303n + 237)$ is always divisible by 13. More generally, Ono [951] proved in 2000 that for any prime $q \ge 5$ there exist positive integers a and b such that $p(an + b)$ is divisible by q for all $n \ge 0$.

In 1944, Dyson [378] attempted to explain this type of congruence. He introduced the *rank* of any partition λ as $r(\lambda) = \lambda_k - k$. He conjectured that $N(0, 5; 5n+4) = N(1, 5; 5n+4) = \cdots = N(4, 5; 5n+4)$, where $N(r, q; m)$ denotes the number of partitions λ of m such that $r(\lambda) \equiv r \pmod{q}$. This conjecture was later confirmed by Atkin and Swinnerton-Dyer [59]. Because $\sum_{i=0}^4 N(i, 5; 5n+4) = p(5n+4)$, this explains why $p(5n + 4)$ must be divisible by 5. Many other congruences involving the partition function p can be explained in a similar way.

In 2010, Bringmann and Ono [213] observed that the function $N(r, q; m)$ (known as *Dyson's rank partition function*) itself satisfies congruences of Ramanujan type.

Theorem 2.62 *Let t be a positive odd integer, and let q be a prime such that $6t$ is not divisible by q. If j is a positive integer, then there are infinitely many non-nested arithmetic progressions $An + B$ such that for every $0 \leq r < t$ we have $N(r, t; An + B) \equiv 0 \pmod{q^j}$ $n = 0, 1, 2, \ldots$.*

A standard technique to prove results about the partition function $p(n)$ is by studying properties of its generation function, that is, infinite series in the form

$$f(q) = \sum_{n=0}^{\infty} p(n)q^n.$$

For example, it is known that $f(q)$ can be equivalently written in the infinite product form $f(q) = \prod_{n=1}^{\infty} \frac{1}{1-q^n}$, and this fact is useful for the analysis of $p(n)$. Similarly, the proof of Theorem 2.62 goes through the deep analysis of the properties of the function

$$R(w, q) = 1 + \sum_{m=1}^{\infty} \sum_{r=-\infty}^{\infty} N(m, r)w^r q^m,$$

where $N(m, r)$ denotes the number of partitions of m with rank r.

Counting Graphs and Subgraphs
Other interesting objects to count are graphs with special properties. For example, how many planar graphs are there? For a positive integer n, let $g(n)$ be the number of labelled planar graphs on n vertices. It is not difficult to show the existence of the limit

$$\gamma = \lim_{n \to \infty} \left(\frac{g(n)}{n!} \right)^{1/n},$$

but much more difficult to compute this limit numerically. Many researchers provided better and better lower and upper bounds for γ. In particular, Bender et al. [116] proved in 2002 that $\gamma > 26.18$, while Bonichon et al. [168] proved in 2006 that $\gamma < 30.06$. In 2009, Giménez and Noy [498] derived an exact analytical expression for γ, which allows numerical computation up to arbitrary precision. Moreover, they derived the exact asymptotic formula for $g(n)$. Recall that $g(n) \sim f(n)$ means that $\lim_{n \to \infty} \frac{g(n)}{f(n)} = 1$.

Theorem 2.63 *One has $g(n) \sim Cn^{-7/2}\gamma^n n!$, where $\gamma = 27.22688\ldots$ and $C = 0.0000042609\ldots$ are explicit constants.*

The proof of Theorem 2.63 is based on a singularity analysis of the generating function

$$G(x) = \sum_{n=1}^{\infty} g(n) \frac{x^n}{n!}.$$

In addition to counting graphs, one may also be interested in counting certain subgraphs of a given graph G with special properties, e.g. one might need to count the cliques in G. The famous *Turán's theorem* [1216], proved in 1941, states that for any integer $r \geq 3$, every graph on n vertices with more than $\frac{r-2}{2(r-1)} n^2$ edges contains a clique of size r. A natural question arises to investigate how many cliques a graph with given number of vertices and edges must contain. In 1983, Lovász and Simonovits [814] conjectured a lower bound for this number, which was known as the *clique density conjecture*. This conjecture was proved by Razborov [1030] in 2008 for $r = 3$ and by Nikiforov [939] in 2011 for $r = 4$. In the proof, Razborov used his famous flag algebras technique [1031], while Nikiforov showed how to translate the problem into a question about polynomial forms, and then applied differential techniques to these forms. By combining these ideas, Reiher [1039] proved the conjecture in full in 2016.

We next formulate this result. For $\gamma \in [0, 1/2)$, let s_γ be the unique positive integer such that $\gamma \in \left[\frac{s_\gamma - 1}{2s_\gamma}, \frac{s_\gamma}{2(s_\gamma + 1)} \right)$, and let $\alpha_\gamma \in (0, 1/\gamma]$ be the real number such that $\gamma = \frac{s_\gamma}{2(s_\gamma + 1)} (1 - \alpha_\gamma^2)$.

Theorem 2.64 *For every integer $r \geq 3$ and real $\gamma \in [0, 1/2)$, every graph on n vertices with at least γn^2 edges contains at least*

$$\frac{1}{(s_\gamma + 1)^r} \binom{s_\gamma + 1}{r} (1 + \alpha_\gamma)^{r-1} (1 - (r-1)\alpha_\gamma) \cdot n^r$$

cliques of size r.

For any integers $n > 0$ and $0 < e \leq \frac{n(n-1)}{2}$, denote by $g_r(n, e)$ the minimum number of r-cliques a graph with n vertices and e edges can have. Theorem 2.64 provides a lower bound for $g_r(n, e)$ which, together with the known upper bound, allows us to estimate $g_r(n, e)$ asymptotically in the limit $n \to \infty$ for any fixed r and $\gamma = \frac{e}{n^2}$.

In 2020, Liu et al. [807] derived the *exact* value of $g_r(n, e)$ in the case $r = 3$, n large, and $e \leq \binom{n}{2} - \epsilon n^2$. To formulate this result, we need some notation.[20] For any

---

[20] You may skip this paragraph and just assume that $h^*(n, e)$ is some explicitly given function of n and e.

integer $q > 0$, we can write $n = d_q q + b_q$ for integers d_q and $0 \leq b_q < q$. Define

$$t_q(n) = \frac{n(n-1)}{2} - b_q \frac{d_q(d_q+1)}{2} - (r - b_q)\frac{d_q(d_q-1)}{2}.$$

Let $k = k(n, e)$ be the unique positive integer with $t_{k-1}(n) < e \leq t_k(n)$. Let $a^* = a^*(n, e)$ be the unique integer vector $a^* = (a_1^*, \ldots, a_k^*)$ such that (i) $a_k^* = \min\{a \in \mathbb{N} : a(n-a) + t_{k-1}(n-a) \leq e\}$ and (ii) $a_1^* + \cdots + a_{k-1}^* = n - a_k^*$ and $a_1^* \geq \cdots \geq a_{k-1}^* \geq a_1^* - 1$. Then let $m^* = \sum_{1 \leq i < j \leq k} a_i^* a_j^* - e$, and finally let

$$h^*(n, e) = \sum_{1 \leq h < i < j \leq k} a_h^* a_i^* a_j^* - m^* \sum_{i=1}^{k-2} a_i^*.$$

Theorem 2.65 *For any $\epsilon > 0$, there exists an $n_0 = n_0(\epsilon) > 0$ such that for all positive integers $n \geq n_0$ and any $e \leq \frac{n(n-1)}{2} - \epsilon n^2$, we have that $g_3(n, e) = h^*(n, e)$.*

Liu et al. [807] also described the exact family $H^*(n, e)$ of extremal graphs which has the minimal number of triangles. The proof of Theorem 2.65 is based on the stability result of Pikhurko and Razborov [983], which states that every extremal (and, more generally, every "almost extremal") graph G is within small edit distance to some graph in $H^*(n, e)$. Then the proof proceeds by analyzing G and showing that it must in fact lie in $H^*(n, e)$.

In fact, for any $n > 0$ and $0 < e \leq \frac{n(n-1)}{2}$, there exists a graph with n vertices, e edges, and exactly $h^*(n, e)$ triangles. This implies the upper bound $g_3(n, e) \leq h^*(n, e)$. Liu et al. [807] proved the matching lower bound $g_3(n, e) \geq h^*(n, e)$ under the conditions listed in Theorem 2.65. It is possible that these conditions can be removed and the equality $g_3(n, e) = h^*(n, e)$ holds in general. At least, we currently do not know any values of n and e for which $g_3(n, e) < h^*(n, e)$.

Counting Walks on Lattices

The *honeycomb lattice* (or the *hexagonal lattice*) is the partition of the plane into same-size regular hexagons, see Fig. 2.5. Denote by c_n the number of n-step self-avoiding walks (that is, visiting every vertex at most once) on the honeycomb lattice H starting from some fixed vertex. It is known that there exists a $\mu \in (0, +\infty)$ such that

$$\mu = \lim_{n \to \infty} \sqrt[n]{c_n}.$$

This constant μ is called the *connective constant* of the honeycomb lattice.

Estimating the number of self-avoiding walks on lattices is an interesting mathematical problem, which has applications in chemistry and physics. In 1953, the famous chemist Flory [455] explained the connection of this problem to modelling polymer chains. In 1982, Nienhuis [938] used non-rigorous methods

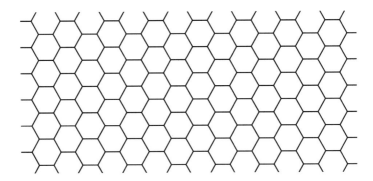

Fig. 2.5 Regular hexagonal grid

from theoretical physics to predict that $\mu = \sqrt{2 + \sqrt{2}}$. In 2012, Duminil-Copin and Smirnov [372] rigorously confirmed this prediction.

Theorem 2.66 *The connective constant of the honeycomb lattice is* $\mu = \sqrt{2 + \sqrt{2}}$.

There are two other ways to cover the plane by regular polygons: triangular and square lattices. The connective constants for these lattices remain unknown. For example, for the square lattice $\mathbb{Z}^2$, numerical experiments suggest that $\mu(\mathbb{Z}^2) = 2.6381...$, see Jensen and Guttmann [653], but we only have rigorous proofs that

$$2.6256 \le \mu(\mathbb{Z}^2) \le 2.6792,$$

where the lower bound was proved by Jensen [652] in 2004, and the upper bound by Pönitz and Tittmann [1000] in 2000.

Counting Matrices

Another research direction is counting matrices with some given properties. Here we discuss the problem of counting alternating-sign matrices, which are important both in pure mathematics (determinant computation) and applications (the so-called "six-vertex model" for ice modelling in statistical mechanics). An $n \times n$ matrix with elements a_{ij} is called an *alternating-sign matrix* (ASM) if (i) all a_{ij} are equal to $-1, 0$, or 1, (ii) the sum of elements in each row and column is 1, and (iii) the non-zero elements in each row and column alternate in sign. In 1983, Mills et al. [894] conjectured that the total number $A(n)$ of $n \times n$ ASMs is

$$A(n) = \prod_{k=1}^{n-1} \frac{(3k+1)!}{(n+k)!}.$$

This conjecture was proved by Zeilberger [1289] in 1996.

In 1991, Robbins [1049] formulated a series of conjectures about the number of ASMs with special symmetries. In particular, an ASM is called

(a) *vertically symmetric* (VSASM) if $a_{ij} = a_{i,n+1-j}$ for all i, j;
(b) *vertically and horizontally symmetric* (VHSASM) if $a_{ij} = a_{i,n+1-j} = a_{n+1-i,j}$ for all i, j;
(c) *half-turn symmetric* (HTSASM) if $a_{ij} = a_{n+1-i,n+1-j}$ for all i, j;
(d) *quarter-turn symmetric* (QTSASM) if $a_{ij} = a_{j,n+1-i}$ for all i, j; and
(e) *diagonally and antidiagonally symmetric* (DASASM) if $a_{ij} = a_{ji} = a_{n+1-j,n+1-i}$ for all i, j.

Robbins conjectured the exact formulas for the number of ASMs with all these symmetries. For many years, these conjectures served as a roadmap, stimulating progress in the area.

In 2002, Kuperberg [752] made the first significant breakthrough in this direction and resolved the Robbins conjectures for VSASM for all n, for HTSASM for even n, and for QTSASM for n divisible by 4.

Theorem 2.67 *Let $A(n)$, $A_V(n)$, $A_{HT}(n)$ and $A_{QT}(n)$ denote the number of $n \times n$ ASMs, VSASMs, HTSASMs, and QTSASMs, respectively. Then $A_V(2n) = 0$,*

$$A_V(2n+1) = (-3)^{n^2} \prod_{i=1}^{2n+1} \prod_{j=1}^{n} \frac{3(2j-i)+1}{2j-i+2n+1},$$

$$A_{HT}(2n) = A(n) \cdot (-3)^{\frac{n(n-1)}{2}} \prod_{i=1}^{n} \prod_{j=1}^{n} \frac{3(j-i)+2}{j-i+n},$$

and

$$A_{QT}(4n) = A_{HT}(2n) \cdot A(n)^2.$$

Theorem 2.67 proved a large portion of the Robbins conjectures and opened the road to the resolution of the remaining ones in subsequent works. In 2006, Razumov and Stroganov [1032, 1033] counted the half-turn and quarter-turn symmetric ASMs for the values of n not covered by Theorem 2.67. Also in 2006, Okada [947] counted vertically and horizontally symmetric ASMs. This leaves open only the last case of those listed by Robbins, the ASMs symmetric in both diagonals. At this point progress stopped for over 10 years, but in 2017 this remaining case was resolved by Behrend, Fischer and Konvalinka [111].

Theorem 2.68 *For any positive integer n, the number of $(2n + 1) \times (2n + 1)$ DASASMs is given by*

$$\prod_{i=0}^{n} \frac{(3i)!}{(n+i)!}.$$

Fig. 2.6 A meander with 8 crossings and 5 minimal arcs

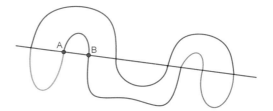

The proof of Theorems 2.67 and 2.68 uses the connection of alternating-sign matrices with the statistical mechanical six-vertex model, see [111, 752] for details.

Counting Meanders

A *meander* is a topological configuration of a line and a simple closed curve in the plane intersecting transversally. We call the points at which the curve and the line intersect *crossings*. There is always an even number of crossings, which we denote by $2N$. Meanders had already been studied by Poincaré over 100 years ago, and appear in various areas of mathematics, and in applications, for example, in physics. A long-standing open problem is to derive the precise asymptotic of how the number of meanders with $2N$ crossings grows with N. It is conjectured that this number is asymptotic to $C R^{2N} N^\alpha$, where C is a constant, $R^2 \approx 12.26$, and $\alpha = -\frac{29+\sqrt{145}}{12} \approx -3.42$, but this remains open.

In 2020, Delecroix et al. [338] solved a version of this problem when the number of minimal arcs is fixed. An *arc* of a meander is the part AB of the curve between crossings A and B which does not contain any other crossings. If the interval AB on the line also does not contain any other crossings, the arc AB is called *minimal*, see Fig. 2.6. Let $M_p(N)$ be the number of meanders with exactly p minimal arcs and with at most $2N$ crossings.

Theorem 2.69 *Let* $p \geq 3$ *be fixed. Then*

$$M_p(N) = C_p N^{2p-4} + o(N^{2p-4}) \quad \text{as} \quad N \to \infty,$$

where $C_p = \frac{2^{p-3}}{\pi^{2p-4} p!(p-2)!} \binom{2p-2}{p-1}^2.$

An important new idea in the proof of Theorem 2.69 is, in the authors' words, the "interpretation of meanders as square-tiled surfaces with one horizontal and one vertical cylinder". See [338] for details.

Counting Independent Sets of a Matroid

Let $M = (X, I)$ be a finite matroid.[21] Let $I_k(M)$ be the number of independent sets $A \in I$ of size $|A| = k$. In 1972, Mason [850] made three conjectures about the sequence $I_k(M)$, in increasing order of strength:

(i) $I_k(M)^2 \geq I_{k-1}(M)I_{k+1}(M)$;

(ii) $I_k(M)^2 \geq \frac{k+1}{k}I_{k-1}(M)I_{k+1}(M)$; and

(iii) $I_k(M)^2 \geq \frac{k+1}{k}\frac{n-k+1}{n-k}I_{k-1}(M)I_{k+1}(M)$.

A sequence $a_0, \ldots, a_n$ of non-negative real numbers is called *log-concave* if $a_{k-1}a_{k+1} \leq a_k^2$ for $0 < k < n$. Hence, conjecture (i) asserts that $I_k(M), k = 0, 1, 2, \ldots$, is a log-concave sequence. A sequence satisfying (iii) is called *ultra log-concave*.

In 2018, Adiprasito et al. [10] proved conjecture (i). Later, Huh et al. [629] proved (ii). In 2020, Brändén and Huh [198], and, independently, Anari et al. [41] proved the strongest conjecture (iii).

Theorem 2.70 *For any n-element matroid M, and any positive integer k,*

$$I_k(M)^2 \geq \frac{k+1}{k}\frac{n-k+1}{n-k}I_{k-1}(M)I_{k+1}(M).$$

The proof of Theorem 2.70 in [198] follows from the deep theory of so-called *Lorentzian polynomials* developed by the authors. This is a family of multivariate polynomials with many useful properties. For example, a bivariate homogeneous polynomial with non-negative coefficients is Lorentzian if and only if the sequence of coefficients form an ultra log-concave sequence with no internal zeros. The authors proved that a certain bivariate polynomial with coefficients $I_k(M)$ is Lorentzian and deduced Theorem 2.70 from there.

Counting Rainbow Bases of a Matroid

Now that we have started a discussion about matroids, we cannot omit one more famous conjecture in this area: *Rota's basis conjecture*. In 1989, Rota (see Huang and Rota [624]) conjectured that in any family $B_1, \ldots, B_n$ of n bases in an n-dimensional vector space V, it is possible to find n disjoint *rainbow bases*, that is, bases consisting of precisely one vector from each of $B_1, \ldots, B_n$. Rota also observed that this conjecture may even be true in the more general setting of matroids: If $B_1, \ldots, B_n$ are disjoint bases (that is, maximal independent sets) in a rank n matroid M, then it might be possible to decompose $B_1 \cup \cdots \cup B_n$ into n disjoint rainbow bases. This conjecture has attracted significant attention and has been proved for many families of matroids but remains open in general. In 2020, Pokrovskiy posted online a preprint [990] in which he proved the asymptotic version of this conjecture.

---

[21] See the discussion before Theorem 2.40 for the definition of matroid and its independent sets.

Theorem 2.71 *Let $B_1, \ldots, B_n$ be disjoint bases in a rank n matroid M. Then there are $n - o(n)$ disjoint rainbow independent sets in $B_1 \cup \cdots \cup B_n$ of size $n - o(n)$.*

2.11 Topological Combinatorics

In 1921, Radon [1017] proved that any set of $d + 2$ points in $\mathbb{R}^d$ can be partitioned into two sets whose convex hulls have a non-empty intersection. In 1966, Tverberg [1220] generalized this theorem and proved that, for any integers $r, d > 1$, any set of $(d+1)(r-1)+1$ points in $\mathbb{R}^d$ can be partitioned into r sets such that all r convex hulls of these sets have a point in common, see Fig. 2.7.

An *m-simplex*, denoted Δ_m, is an m-dimensional polytope which is the convex hull of its $m + 1$ vertices. Radon's theorem is equivalent to the statement that for any *affine* map $f : \Delta_{d+1} \to \mathbb{R}^d$ there are two disjoint faces of Δ_{d+1} whose images have a non-empty intersection. In 1979, Bajmóczy and Bárány [79] proved that this statement, more generally, remains true for any *continuous* map $f : \Delta_{d+1} \to \mathbb{R}^d$. This theorem became known as the *topological Radon theorem*. The authors also conjectured that a similar generalization should be possible for Tverberg's theorem. Specifically, they conjectured that for given integers $r \geq 2, d \geq 1, m \geq (r-1)(d+1)$, and for any continuous map $f : \Delta_m \to \mathbb{R}^d$, there are r pairwise disjoint faces of Δ_m whose images have a point in common. This statement became known as the "topological Tverberg conjecture".

This conjecture was considered by many experts as a major open problem, a holy grail of topological combinatorics. The case when f is an affine map is equivalent to Tveberg's original theorem [1220]. For any continuous f and $r = 2$ it reduces to the topological Radon theorem [79]. In 1981, Bárány et al. [98] proved the topological Tverberg conjecture for all primes r. In 1987, Ozaydin [959] proved it for the case when r is a power of prime. Based on these results, the conjecture was commonly believed to be true in general.

Fig. 2.7 Partition in Tverberg's theorem for $d = 2$ and $r = 3$

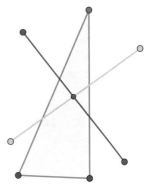

In 2015, Frick posted online a preprint [475] in which he proved, by constructing some counterexamples, that in general the conjecture is false.

Theorem 2.72 *The topological Tverberg conjecture is false for any r that is not a power of a prime and any dimension $d \geq 3r + 1$.*

Some other theorems in topological combinatorics (see Theorems 5.58 and 5.59) are included in the "Geometry and Topology" chapter of this book.

2.12 Design Theory

Let X be a set with n elements, and let $k \geq 2$. A *Steiner system* $S(n, k, r)$ is a collection of k-element subsets of X, called *blocks*, such that each r-element subset of X is contained in exactly one block. Steiner systems are among the most beautiful structures in combinatorics, and have been studied since the work of Plücker (1835), Kirkman (1846) and Steiner (1853), but, before 2014, it was an open question whether any Steiner system with $n > k > r \geq 6$ exists.

More generally, we say that a family S of k-elements subsets of X is a *(combinatorial) design* with parameters (n, k, r, λ) if every r-element subset of X belongs to exactly λ sets in S. A Steiner system is a combinatorial design with $\lambda = 1$. Obvious necessary conditions for the existence of a combinatorial design are $r \leq k \leq n$ and $\binom{k-i}{r-i}$ divides $\lambda \binom{n-i}{r-i}$ for every $0 \leq i \leq r - 1$. If this is true, we say that positive integers (n, k, r, λ) satisfy the *divisibility conditions*.

The famous "existence conjecture" for combinatorial designs predicted that the divisibility conditions are in fact sufficient for the existence of combinatorial designs, provided that n is sufficiently large. This conjecture was the central open question in design theory. In 1975, Wilson [1268] proved the conjecture for $r = 2$, which was already a major result. For larger r, the conjecture remained wide open, even for $\lambda = 1$ (Steiner systems). In 2014, Keevash posted online a paper [699] in which he proved the existence conjecture in general.

Theorem 2.73 *For any fixed positive integers k, r, and λ, there exists an $n_0 = n_0(k, r, \lambda)$ such that if $n \geq n_0$ and (n, k, r, λ) satisfy the divisibility conditions then a combinatorial design with parameters (n, k, r, λ) exists.*

The proof of Theorem 2.73 uses the work of Rödl [1055], who, answering the 1963 question of Erdős and Hanani [411], proved in 1985 the existence of an "approximate design" via a greedy randomized construction. Somewhat surprisingly, Keevash managed to adapt the construction in a clever way and showed that the resulting randomized combinatorial process leads to the exact design with positive probability.

In 1973, Cameron [241] suggested to study a natural analogue of a Steiner system over finite fields. Let $\mathbb{F}_q^n$ be a vector space of dimension n over the finite field $\mathbb{F}_q$. A *q-Steiner system*, denoted $S_q(n, k, r)$, is a set S of k-dimensional subspaces of $\mathbb{F}_q^n$

such that each r-dimensional subspace of $\mathbb{F}_q^n$ is contained in exactly one element of S. A q-Steiner system is called *trivial* if $r = k$ or $k = n$ and *non-trivial* otherwise. q-Steiner systems attracted considerable attention in the literature, however, despite the efforts of many researchers, such systems were known only for $r = 1$, and in the trivial cases $r = k$ and $k = n$. In 2016, Braun, Etzion, Östergård, Vardy and Wassermann [201] provided us with first examples of non-trivial q-Steiner systems with $r \geq 2$.

Theorem 2.74 *There exist non-trivial q-Steiner systems with $r \geq 2$. In particular, there exist over 500 different $S_2(13, 3, 2)$ 2-Steiner systems.*

Chapter 3
Analysis

3.1 Discrete Time Dynamical Systems

Integer-Valued Maps: The Collatz Conjecture

Let X be a set and $f : X \to X$ be a function. For integer $n \geq 0$, the n-th iterate of f, denoted f^n, is defined inductively as $f^0(x) = x$ and $f^n(x) = f(f^{n-1}(x))$ for all $n \geq 1$ and all $x \in X$. Given an initial point $x_0 \in X$, consider the infinite sequence

$$x_n = f^n(x_0), \quad n = 0, 1, 2, \ldots, \tag{3.1}$$

which can also be defined inductively as $x_n = f(x_{n-1})$, $n = 1, 2, \ldots$.

It turns out that the sequence (3.1) can be extremely difficult to analyse, even for very simple functions f. For example, let $X = \mathbb{N}$ be the set of positive integers, and let

$$f(k) = \begin{cases} 3k + 1, & \text{if } k \text{ is odd,} \\ k/2, & \text{if } k \text{ is even.} \end{cases} \tag{3.2}$$

In the 1930s, Collatz conjectured that, for any $k \in \mathbb{N}$, there exists an n such that $f^n(k) = 1$.

This easy-to-state problem turned out to be extremely difficult to solve. Let $f_{\min}(k) = \inf_{n \in \mathbb{N}} f^n(k)$ denote the minimal element of the sequence (3.1) starting with $x_0 = k$. Then the *Collatz conjecture* predicts that $f_{\min}(k) = 1$ for all $k \in \mathbb{N}$. As partial progress, Terras [1194] proved in 1976 that $f_{\min}(k) < k$ for almost all k. In 1979, Allouche [29] improved this to $f_{\min}(k) < k^\theta$ for almost all k, for any fixed constant $\theta > 0.869$. In 1994, Korec [737] proved the same result for any $\theta > \frac{\log 3}{\log 4} = 0.792\ldots$. In 2019, Tao posted online a preprint [1183] in which he proved that one can take any $\theta > 0$, and, more generally, $f_{\min}(k) < g(k)$ for

© The Author(s), under exclusive license to Springer Nature Switzerland AG 2021
B. Grechuk, *Landscape of 21st Century Mathematics*,
https://doi.org/10.1007/978-3-030-80627-9_3

almost all k and any function g going to infinity with k. For example, one can take $g(k) = \log \log \log \log k$.

The definition of "almost all" in Tao's theorem is in terms of "logarithmic density". We say that subset $A \subseteq \mathbb{N}$ has *logarithmic density* 1 if $\lim\limits_{x \to \infty} \left(\frac{1}{\log x} \sum_{k \in A, k \leq x} \frac{1}{k} \right) = 1$. We say that a property P holds for *almost all* $k \in \mathbb{N}$ (in the sense of logarithmic density) if the subset of $\mathbb{N}$ for which P holds has logarithmic density 1.

Theorem 3.1 *Let* $f : \mathbb{N} \to \mathbb{N}$ *be the function defined in* (3.2). *For any function* $g : \mathbb{N} \to \mathbb{R}$ *such that* $\lim\limits_{k \to \infty} g(k) = +\infty$, *one has* $f_{\min}(k) < g(k)$ *for almost all* $k \in \mathbb{N}$ *(in the sense of logarithmic density).*

The proof of Theorem 3.1 uses statistical methods. A key difficulty is that if we have some probability distribution over $\mathbb{N}$, then an application of the map (3.2) can greatly distort it. However, Tao was able to construct an "approximate invariant measure"—a distribution which, after the iterate of (3.2), resembles a smaller version of itself. This method was inspired by a 1994 result of Bourgain [182], who constructed an invariant measure with similar properties for the nonlinear Schrödinger equation.

Tao also provided an informal argument that proving the existence of a constant C_0 such that $f_{\min}(k) < C_0$ for almost all $k \in \mathbb{N}$ "is likely to be almost as hard to settle as the full Collatz conjecture", and concluded that Theorem 3.1 is "about as close as one can get to the Collatz conjecture without actually solving it".

Quadratic Maps

The sequence $\{x_n\}$ defined in (3.1) is called "an orbit of a dynamical system". The central goal in the study of dynamical systems is to understand the behaviour of almost all orbits for almost all functions f in some family. In 2002, Lyubich [824] achieved this goal for the family of quadratic polynomials, which is the simplest non-trivial case.

Let $f : \mathbb{R} \to \mathbb{R}$ be a real-valued quadratic polynomial. Let I be the longest interval which f maps into itself. We say that the sequence (3.1) is a *cycle* if $x_n = x_{n+k}$ for some $k > 0$. If, for (Lebesgue) almost every $x_0 \in I$, the sequence (3.1) converges to a cycle, f is called *regular*. If there exists a function g such that, for almost every $x_0 \in I$, the proportion of terms in the sequence in any interval (a, b) converges to $\int_a^b g(x)\mathrm{d}x$ as $n \to \infty$, f is called *stochastic*. In 2002, Lyubich [824] proved the following dichotomy.

Theorem 3.2 *For almost every* $c \in [-2, 1/4]$, *the polynomial* $f(x) = x^2 + c$ *is either regular or stochastic.*

If f is regular, then, for almost all x_0, the sequence (3.1) converges to a cycle and is therefore "almost periodic". Even if f is not regular, the sequence (3.1) occasionally "nearly repeats itself". For example, if x_n happens to be close to x_0, then $x_{n+1} = f(x_n)$ is close to $x_1 = f(x_0)$, x_{n+2} is close to x_2, and this pattern stays for some time. The closer x_n to x_0, the longer it stays. We may assume that

$x_0 = 0$ and ask how close $x_n = f^n(0)$ can be to 0. In 2005, Avila and Moreira [65] answered this question, confirming a long-standing conjecture of Sinai.

Theorem 3.3 *For almost every $c \in [-1/4, 2]$ such that the quadratic polynomial $f(x) = c - x^2$ is not regular, we have*

$$\limsup_{n \to \infty} \frac{\log(|f^n(0)|)}{\log(n)} = 1.$$

In other words, the set of n such that $|f^n(0)| < n^{-\gamma}$ is finite if $\gamma > 1$ and infinite if $\gamma < 1$.

It is known that any quadratic polynomial can be transformed to the form $f(x) = x^2 + c$, $c \in [-2, 1/4]$, by a change of variables which preserves the properties of the sequence (3.1). Hence, with Theorems 3.2 and 3.3, we may say that sequence (3.1) is now reasonably well understood when f is a quadratic polynomial.

Polynomial Maps

In this section we study the iterates f^n in the case when f is a polynomial of an arbitrary degree. One way to measure how the "complexity" of f^n grows with n is the notion of topological entropy.

Let $I = [a, b]$ be an interval in $\mathbb{R}$. Given $f : I \to I$, assume that I can be decomposed into finitely many subintervals $I_0, \ldots, I_m$ on which f is monotone. The smallest number $m + 1$ of such intervals is called the *lap number* $l(f)$ of f. The *topological entropy* $h_{\text{top}}(f)$ of f is

$$h_{\text{top}}(f) = \lim_{n \to \infty} \frac{\log l(f^n)}{n}.$$

In 1992, Milnor [896] formulated the so-called "Monotonicity Conjecture" stating that, for certain families of polynomials, the isentropes (the sets of parameters for which $h_{\text{top}}(f)$ is constant) are connected. To formulate the conjecture rigorously, we need some notation. Let $\partial I = \{a, b\}$ be the set of endpoints of $I = [a, b]$. Let P^d be the set of real polynomials of degree $d + 1$ mapping I (and ∂I) into itself, with precisely d non-degenerate critical points,[1] each of which is real and is contained in the interior of I. For $\epsilon = \pm 1$, let $P^d_\epsilon \subset P^d$ be the set of polynomials which are increasing (respectively decreasing) at the left endpoint of I when $\epsilon = 1$ (respectively $\epsilon = -1$). The set $\{f \in P^d_\epsilon : h_{\text{top}}(f) = h_0\}$ of polynomials in P^d_ϵ with fixed topological entropy is called the *isentrope*.

For each polynomial $P(x) = \sum_{i=0}^{d+1} a_i x^i$ one can associate a point $a(P) = (a_0, \ldots, a_{d+1}) \in \mathbb{R}^{d+2}$. A subset $S \subseteq P^d_\epsilon$ is called *connected* if $\{a(P), P \in S\}$ is a connected subset of $\mathbb{R}^{d+2}$. Milnor's conjecture states that the isentropes are connected. This fact was known for quadratic polynomials even before the

---

[1] A complex number $z \in \mathbb{C}$ is a *critical point* of a polynomial P if $P'(z) = 0$. It is *non-degenerate* if $P''(z) \neq 0$.

conjecture was formulated. In 2000, Milnor and Tresser [897] proved it for the cubic polynomials. In 2015, Bruin and van Strien [216] proved the conjecture in general.

Theorem 3.4 *For $\epsilon \in \{-1, +1\}$, each positive integer d, and each $h_0 \geq 0$, the isentrope $\{f \in P_\epsilon^d : h_{\text{top}}(f) = h_0\}$ is connected.*

One important ingredient in the proof of Theorem 3.4 is the fact that the set of hyperbolic real polynomials is dense in the set of all real polynomials, see Theorem 3.90 below.

The m-Fold Maps of the Unit Interval

Let $f : X \to X$ be arbitrary. A set $S \subset X$ is called *invariant under* f if $f(x) \in S$ for every $x \in S$. In particular, if $x_0 \in S$, then the sequence (3.1) is fully contained in S.

In this section, we discuss the case when $X = [0, 1)$ is the unit interval and $f = T_m$ is the *m-fold map* of $[0, 1)$. Specifically, for a real number x, let $\lfloor x \rfloor$ be the largest integer not exceeding x, and let $\text{frac}(x) = x - \lfloor x \rfloor$. For an integer m, let $T_m : [0, 1) \to [0, 1)$ be the function given by

$$T_m(x) = \text{frac}(mx).$$

In 1967, Furstenberg [482] proved that if $p \geq 2$ and $q \geq 2$ are integers such that $\frac{\log p}{\log q}$ is irrational, then no infinite proper closed subset of $[0, 1]$ can be simultaneously invariant under T_p and T_q. The simplest example of multiplicatively independent integers p and q is $p = 2, q = 3$, and in this case the result above is known as *Furstenberg's $\times 2, \times 3$ theorem*. Moreover, Furstenberg proposed a series of conjectures which state, in various ways, that sets X and Y invariant under T_p and T_q must be "very different". In particular, he conjectured that

$$\dim(X + Y) = \min\{1, \dim X + \dim Y\}, \tag{3.3}$$

where $X + Y = \{x + y : x \in X; y \in Y\}$, and "dim" refers to the Hausdorff dimension defined in (1.23). Equality (3.3) is a formalization of the intuition that X and Y are "very different", because the inequality $\leq$ is always true, and this inequality is strict if X and Y have some "shared structure". In 2012, Hochman and Shmerkin [610] confirmed this conjecture.

Theorem 3.5 *Let $p, q \geq 2$ be positive integers such that $\frac{\log p}{\log q} \notin \mathbb{Q}$. Let $X, Y \subset [0, 1)$ be closed sets which are invariant under T_p and T_q, respectively. Then for any $s \neq 0$ one has*

$$\dim(X + s \cdot Y) = \min\{1, \dim X + \dim Y\}.$$

An even more natural way to formalize the intuition that sets X and Y are "very different" is to prove that their intersection is "small". In 1970, Furstenberg [483] conjectured that if X and Y are as in Theorem 3.5, then

$$\dim(X \cap Y) \leq \max\{0, \dim(X) + \dim(Y) - 1\}. \tag{3.4}$$

In 2019, Shmerkin [1119] and independently Wu [1276] proved a strong form of this conjecture.

Theorem 3.6 *Let $p, q \geq 2$ be positive integers such that $\frac{\log p}{\log q} \notin \mathbb{Q}$. Let $X, Y \subset [0, 1)$ be closed sets which are invariant under T_p and T_q, respectively. Then for all real numbers $u \neq 0$ and v, $\dim((u \cdot X + v) \cap Y) \leq \max\{0, \dim(X) + \dim(Y) - 1\}$.*

Furstenberg's conjecture (3.4) corresponds to the case $u = 1$ and $v = 0$ in Theorem 3.6.

The Structure of Cantor Sets
A basic question one may ask about the sequence (3.1) is whether it stays bounded or diverges to infinity. For the function

$$f(x) = \begin{cases} 3x, & \text{if } x \leq 0.5, \\ 3(1 - x), & \text{if } x \geq 0.5, \end{cases}$$

(3.1) stays bounded if and only if $x_0 \in [0, 1]$ and its expansion in base 3 does not contain the digit 1. The set of all such x_0 is known as the *ternary Cantor set*. More generally, a *Cantor set* is a closed set consisting entirely of boundary points. A Cantor set arising as a set of x_0 for which (3.1) stays bounded for some f is called *regular*.

In 1987, Palis [963] made a conjecture about the arithmetic difference[2] of typical regular Cantor sets. In 2001, Moreira and Yoccoz [908] confirmed this conjecture.

Theorem 3.7 *The arithmetic difference of a generic pair of regular Cantor sets of the real line either has measure zero or contains an interval.*

Since 2001, Theorem 3.7 and methods developed for its proof have been used to prove many other interesting results, including, in particular, Theorem 1.56 discussed earlier in this book.

The Julia Set of Polynomials
In this section we discuss the behaviour of the sequence of iterates (3.1) if $f : \mathbb{C} \to \mathbb{C}$ is a function of a complex variable. One of the most basic questions one can ask is for which x_0 does the sequence (3.1) remain bounded, that is, $|x_n| \leq B$, $\forall n$ for

---

[2] For any sets $C_1 \subset \mathbb{R}$ and $C_2 \subset \mathbb{R}$, their *arithmetic difference* is the set $C_1 - C_2 = \{x - y \mid x \in C_1, y \in C_2\}$.

some $B \in \mathbb{R}$. Let $K_f \subset \mathbb{C}$ be the set of all x_0 with this property. The boundary J_f of K_f is called the *Julia set* of f.

The Julia set is a fundamental concept in the theory of complex dynamics, because it consists of values x_0 such that an arbitrarily small perturbation can cause significant changes in the sequence of iterated function values. However, describing the geometry of Julia sets of even simple functions f is a difficult problem. In 2004, Petersen and Zakeri [977] solved this problem for almost all quadratic polynomials in the form $f(z) = z^2 + cz$, which is one of the simplest non-trivial families of functions.

Theorem 3.8 *For almost every complex number c with $|c| = 1$, the Julia set of the quadratic polynomial $f(z) = z^2 + cz$ is locally connected*[3] *and has Lebesgue measure zero.*

In fact, Petersen and Zakeri gave a precise arithmetic sufficient condition on c for the conclusion of Theorem 3.8 to hold. However, Theorem 3.8 covers almost all but not all quadratic polynomials. In 2012, Buff and Chéritat [222] proved that there are "exceptional" polynomials whose Julia set looks quite different.

Theorem 3.9 *There exist quadratic polynomials that have a Julia set of positive Lebesgue measure.*

Given a *specific* quadratic polynomial, it may be quite difficult to decide if its Julia set has zero or positive Lebesgue measure. For example, this problem was open for a long time for the Feigenbaum polynomial, a polynomial of central importance arising from renormalization theory in dynamics, which we discuss next.

Renormalization Theory in Dynamics
In the mid 1970s, the physicist Feigenbaum [446] studied the behaviour of the sequence of iterates (3.1) for quadratic polynomials of the form $f(x) = ax(1 - x)$ on $x \in [0, 1]$ for various parameters a. He noticed that when $a > 0$ is small, then, for almost all initial points x_0, the sequence (3.1) converges to a fixed point. When we increase a, and it passes some value a_1, the sequence (3.1) starts to converge to an attracting cycle of length 2, again for almost all x_0. Then, when a passes some other value a_2, almost all sequences start to converge to attracting cycles of length 4. This effect continues, and there is a sequence $\{a_n\}_{n=1}^{\infty}$ such that when a passes each a_i, the length (or period) of the limiting attracting cycles doubles. This effect is known as a *period-doubling bifurcation*. Feigenbaum also noticed that the sequence $\{a_n\}$ converges to some limit, and, moreover, it converges exponentially fast, in the sense that the limit

$$\lambda = \lim_{n \to \infty} \frac{a_{n-1} - a_{n-2}}{a_n - a_{n-1}}$$

---

[3] A set S is called *connected* if it cannot be partitioned into two disjoint open sets. A set S is called *locally connected* if for every $z \in S$ and every open set U containing z there is an open set V containing z such that the intersection $V \cap S$ is connected and contained in U.

exists and is equal to $\lambda = 4.669....$ Feigenbaum then studied the sequence (3.1) for a completely different family of maps $f(x) = b\sin(x)$, and observed a similar effect: there is a sequence of parameters $\{b_n\}_{n=1}^{\infty}$ at which the length of the limiting attracting cycles doubles, and, amazingly, the limit $\lim_{n\to\infty} \frac{b_{n-1}-b_{n-2}}{b_n-b_{n-1}}$ exists and is equal to the same value $\lambda = 4.669....$

Feigenbaum realized that there should be a deep reason for this "coincidence", and revealed the underlying mechanism for this phenomenon. This mechanism is the deep and beautiful theory of renormalization, which relates properties of a family of dynamical systems at various scales. For example, let I be some interval, and let $f : I \to I$ be a function. Let $I_1 \subset I$ be a small subinterval of I. For each $x_0 \in I_1$, perform iterates (3.1) of f until we find the smallest positive integer $n = n(x_0)$ such that $f^n(x_0) \in I_1$. The map $f_1 : I_1 \to I_1$ sending x_0 to $f^n(x_0)$ is called the *first return map*, and describes the dynamics of f "at scale" I_1. Next, we can rescale the interval I_1 back to the original size by a suitable (for example, linear) change of variables, after which f_1 transforms into a map $Rf : I \to I$ which is called the "renormalization" of f. Then we can find the renormalization $R^2 f$ of Rf, and, more generally, this process can then be iterated to build a sequence of maps $R^k f$. In some cases, this sequence converges to some special map $f^* : I \to I$, called the *renormalization fixed point*, with the property $Rf^* = f^*$, which implies that f^* is self-similar at all scales. An example of such a "special map" is the Feigenbaum polynomial defined below.

We say that a point z_0 is *periodic* if $f^n(z_0) = f(z_0)$ for some positive integer n, and the smallest n such that this holds is called the *period* of z_0. For each n, let r_n be the (unique) real number such that 0 is a periodic point with period 2^n of the polynomial $f(z) = z^2 + r_n$. It is known that the limit $r_\infty = \lim_{n\to\infty} r_n$ exists and is numerically equal to $-1.401155....$ The polynomial $f(z) = z^2 + r_\infty$ is called the *Feigenbaum polynomial*. As explained above, this polynomial arises in renormalization theory, and the question of whether its Julia set has zero area was a long-standing open question. In 2020, Dudko and Sutherland [367] resolved this question affirmatively.

Theorem 3.10 *The Julia set of the Feigenbaum polynomial has zero Lebesgue measure.*

In fact, Dudko and Sutherland proved the stronger statement that the Julia set of the Feigenbaum polynomial has Hausdorff dimension less than 2. The proof is based on a new sufficient condition for the Julia set of a polynomial to have Hausdorff dimension less than 2. This is an easier-to-verify version of the sufficient conditions developed by Avila and Lyubich [64] in 2008. Using computer-assisted but rigorous computation with explicit bounds on errors, the authors were able to verify that their sufficient condition is satisfied for the Feigenbaum polynomial, and Theorem 3.10 follows.

Intersections of Orbits

In some applications, it is important to understand the joint behaviour of the sequences of iterates (3.1) of two different functions f and g. For example, under what conditions do these two sequences have infinite intersections? This is obviously the case if the functions f and g have a common iterate, that is, if $f^n = g^n$ for some n. In 2008, Ghioca et al. [496] proved that, if f and g are non-linear polynomials of the same degree, then this obvious sufficient condition is in fact necessary.

Let $\mathbb{C}[X]$ be the set of polynomials in one complex variable. For $f \in \mathbb{C}[X]$, and initial point $x_0 \in \mathbb{C}$, the *orbit* $O_f(x_0)$ is the set

$$O_f(x_0) = \{x_0, f(x_0), f(f(x_0)), \ldots, f^n(x_0), \ldots\}.$$

Theorem 3.11 *Let $x_0, y_0 \in \mathbb{C}$ and $f, g \in \mathbb{C}[X]$ with $\deg(f) = \deg(g) > 1$. If $O_f(x_0) \cap O_g(y_0)$ is infinite, then f and g have a common iterate.*

Iterates of Transcendental Functions

The sequences of iterates (3.1) can be studied not only in the case when the function f is a polynomial. A function $f : \mathbb{C} \to \mathbb{C}$ is called *entire* if $f'(z)$ exists for all $z \in \mathbb{C}$. An entire function f is called *transcendental* if it is not a polynomial. The study of the limiting behaviour of the sequence (3.1) for a transcendental entire function f goes back to a 1926 paper of Fatou [439]. As in the case when f is a polynomial, the central question is when this sequence remains bounded and when it diverges to infinity. In 1989, Eremenko [429] analysed properties of the *escaping set* $I(f) = \{z \in \mathbb{C} : \lim_{n \to \infty} |f^n(z)| = \infty\}$ of an arbitrary transcendental entire function f. A *path-component* of $I(f)$ is the set of all $y \in I(f)$ which can be connected by a path $P \subset I(f)$ to any fixed $x \in I(f)$. Eremenko asked whether the path-components of $I(f)$ are unbounded, that is, whether every point $z \in I(f)$ can be joined to ∞ by a curve in $I(f)$. This question became one of the most famous open problems in the field of transcendental dynamics. In 2011, Rottenfusser et al. [1064] answered it negatively.

Theorem 3.12 *There exists a transcendental entire function f whose escaping set $I(f)$ has only bounded path-components.*

Linear Recurrences

The sequence (3.1) can be defined inductively starting from x_0 and applying a function of one variable: $x_n = f(x_{n-1})$, $n = 1, 2, \ldots$. More generally, given any function f in k variables, and k initial points $x_0, \ldots, x_{k-1}$, we may define the infinite sequence as

$$x_{n+k} = f(x_n, \ldots, x_{n+k-1}), \quad n = 0, 1, 2, \ldots \tag{3.5}$$

A simple special case of (3.5) is when the function f is linear, that is

$$x_{n+k} = \sum_{i=0}^{k-1} c_i x_{n+i}, \quad n = 0, 1, 2, \ldots, \tag{3.6}$$

for some $k \geq 1$ and complex numbers $c_0, c_1, \ldots, c_{k-1}$. The sequence (3.6) is called a *linear recurrence*.

In 1988, van der Poorten [1004], confirming a conjecture of Pisot, proved that if the ratio x_n/y_n of linear recurrences $\{x_n\}_{n=0}^{\infty}$ and $\{y_n\}_{n=0}^{\infty}$ is an integer for all large n, then the sequence $\{x_n/y_n\}_{n=0}^{\infty}$ is itself a linear recurrence. In 1989, van der Poorten [1005] asked whether a similar result can be proved under the much weaker assumption that x_n/y_n is an integer infinitely often. In 2002, Corvaja and Zannier [305] characterized linear recurrences with this property.

Theorem 3.13 *For any linear recurrences $\{x_n\}_{n=0}^{\infty}$ and $\{y_n\}_{n=0}^{\infty}$ such that x_n/y_n is an integer for infinitely many values of n, there exists a nonzero polynomial $P(X) \in \mathbb{C}[X]$ and positive integers q, r such that both sequences $\{P(n)x_{qn+r}/y_{qn+r}\}_{n=0}^{\infty}$ and $\{y_{qn+r}/P(n)\}_{n=0}^{\infty}$ are linear recurrences.*

Interval Exchange Transformations

In the previous subsections we have discussed theorems related to iterates f^n of a function $f : X \to X$. In this section we consider the case when X is a probability space[4] with measure μ, and f is such that $\mu(f^{-1}(A)) = \mu(A)$ for all measurable $A \subset X$. Then f is called a *measure-preserving transformation*, and the pair (X, f) is a *measure-preserving dynamical system*.

A measure-preserving transformation $f : X \to X$ is called

- *ergodic* if for every measurable $A \subset X$ with $f^{-1}(A) = A$, either $\mu(A) = 0$ or $\mu(A) = 1$;
- *uniquely ergodic* if there is no other Borel probability measure ν on X such that $\nu(f^{-1}(A)) = \nu(A)$ for all measurable $A \subset X$;
- *weakly mixing* if for every pair of measurable sets $A, B \subset X$,

$$\lim_{n \to \infty} \frac{1}{n} \sum_{k=1}^{n-1} \left| \mu(f^{-k}(A) \cap B) - \mu(A)\mu(B) \right| = 0;$$

- *mixing* if $\lim_{n \to \infty} \mu(f^{-n}(A) \cap B) = \mu(A)\mu(B)$ for all measurable $A, B \subset X$.

---

[4] A *probability space* is a triple $(X, \mathcal{B}, \mu)$, where X is a set, $\mathcal{B}$ is a σ-algebra over X (sets in $\mathcal{B}$ are called *measurable*), and $\mu : \mathcal{B} \to [0, 1]$ is a *probability measure*, that is, a measure such that $\mu(X) = 1$ and $\mu(\emptyset) = 0$.

Here, $f^{-n}(A)$ denotes the set $\{x \in X : f^n(x) \in A\}$. It is known that every mixing f is weakly mixing, and every weakly mixing f is ergodic. Also, every uniquely ergodic f is ergodic.

Basic examples of measure-preserving transformations are interval exchange transformations of the interval $I = [0, 1)$ with Lebesgue measure. Let $d \geq 2$ be a fixed integer. A permutation π of $\{1, 2, \ldots, d\}$ is called *irreducible* if $\pi(\{1, \ldots, k\}) \neq \{1, \ldots, k\}$ for all $1 \leq k < d$. Let Λ be the set of vectors $\lambda = (\lambda_1, \ldots, \lambda_d)$ such that $\lambda_i \geq 0$ for all i and $\sum_{i=1}^{d} \lambda_i = 1$. Given an irreducible π and $\lambda \in \Lambda$, an *interval exchange transformation* (IET) $f = f(\lambda, \pi)$ is a map $f : I \rightarrow I$ which divides $I = [0, 1)$ into sub-intervals $I_i = \left[\sum_{j<i} \lambda_j, \sum_{j \leq i} \lambda_j \right)$, $i = 1, 2 \ldots, d$ and rearranges the I_i according to π (it maps every $x \in I_i$ to $x + \sum_{\pi(j) < \pi(i)} \lambda_j - \sum_{j<i} \lambda_j$).

In 1982, Masur [851] and Veech [1230] proved that for every irreducible π and almost every $\lambda \in \Lambda$ the IET $f = f(\lambda, \pi)$ is uniquely ergodic and therefore ergodic. On the other hand, Katok [692] proved in 1980 that a typical IET is not mixing. This left open the question whether a typical IET is weakly mixing.

This is not the case if π is a rotation. A permutation π of $\{1, 2, \ldots, d\}$ is called a *rotation* if $\pi(i + 1) \equiv \pi(i) + 1 \pmod{d}$, for all $i \in \{1, 2, \ldots, d\}$. In 2007, Avila and Forni [60] proved that a typical IET which is not a rotation is indeed weakly mixing.

Theorem 3.14 *For every integer $d \geq 2$, every irreducible permutation π of $\{1, 2, \ldots, d\}$ which is not a rotation, and for Lebesgue almost every $\lambda \in \Lambda$, $f(\lambda, \pi)$ is weakly mixing.*

Let π be an irreducible permutation of $\{1, 2, \ldots, d\}$ which is not a rotation, and let $S \subset \Lambda$ be the set of all λ such that $f(\lambda, \pi)$ is not weakly mixing. Theorem 3.14 states that the set S has Lebesgue measure 0 in Λ. In 2016, Avila and Leguil [63] proved the stronger result that the set S is not even full-dimensional, or, in other words, has codim$(S) > 0$, where dim(S) and codim$(S) = d - 1 - \dim(S)$ denote the Hausdorff dimension and Hausdorff codimension of a set $S \subset \Lambda$, respectively. This raises the problem of determining codim(S). In 2020, Chaika and Masur [257] solved this problem for the permutation $\pi_d^* = (1, \ldots, d) \rightarrow (d, \ldots, 1)$ with odd d.

Theorem 3.15 *For every odd $d \geq 3$, the set S of all λ such that $f(\lambda, \pi_d^*)$ is not weakly mixing has codim$(S) = 1/2$.*

In fact, Theorem 3.15 is a corollary of a more general result establishing the inequality codim$(S) \leq 1/2$ for a broader class of permutations π, see [257] for details.

As mentioned above, Masur [851] and Veech [1230] proved that almost every IET is uniquely ergodic. In 2012, Chaika [256] extended this result to the product of almost all IETs. The product of maps $f : I \rightarrow I$ and $g : I \rightarrow I$ is the map $f \times g : I \times I \rightarrow I \times I$ given by $(f \times g)(x, y) = (f(x), g(y))$.

Theorem 3.16 *For almost every pair f, g of IETs their product $f \times g$ is uniquely ergodic.*

The proof of Theorem 3.16 is based on the properties of rigidity sequences. Given an IET f, a sequence $n_1, n_2, \ldots$ of positive integers is called a *rigidity sequence* for f if $\int_I |f^{n_i}(x) - x| dx$ converges to 0 as $i \to \infty$. In 1984, Veech [1231] proved that almost every IET has a rigidity sequence. Chaika strengthened this result and proved that if A is any sequence of natural numbers with density 1, then almost every IET has a rigidity sequence contained in A. He then deduced Theorem 3.16 from this fact.

Ergodic Dynamical Systems

In the previous subsection we introduced ergodic dynamical systems and discussed one example of such a system: (almost all) interval exchange transformations. Here we discuss such systems in general. One of the central results in this area is Birkhoff's famous *ergodic theorem* [153], stating that, for any ergodic dynamical system (X, f) and any[5] $g \in L^1(X)$ the limit

$$\lim_{N \to \infty} \frac{1}{N} \sum_{k=1}^{N} g(f^k(x))$$

exists for almost all $x \in X$ (in fact, this limit is equal to $\int_X g(x) d\mu$). An important research direction is to understand for which sequences $\{n_k\}_{k=1}^{\infty}$ the corresponding limit

$$\lim_{N \to \infty} \frac{1}{N} \sum_{k=1}^{N} g(f^{n_k}(x)) \tag{3.7}$$

exists for almost all $x \in X$. In 1989, Bourgain [179] proved this for the sequence $\{k^2\}_{k=1}^{\infty}$ of perfect squares, provided that $g \in L^p(X)$ for some $p > 1$, and asked if the same is true for all $g \in L^1(X)$.

In 2010, Buczolich and Mauldin [219] answered this question negatively. A sequence $\{n_k\}_{k=1}^{\infty}$ is called L^1-*universally bad* if for all ergodic dynamical systems (X, f) there is some $g \in L^1(X)$ and a measurable set $A \subset X$ with $\mu(A) > 0$, such that the limit (3.7) fails to exist for all $x \in A$.

Theorem 3.17 *The sequence $\{k^2\}_{k=1}^{\infty}$ is L^1-universally bad.*

---

[5] $L^p(X)$ denotes the set of functions $g : X \to \mathbb{R}$ such that $\int_X |g(x)|^p d\mu < \infty$.

3.2   Continuous Time Dynamical Systems

Billiards in Rational Polygons
One of the simplest examples of a continuous time dynamical system is a *billiard*.
Let Q be a *rational polygon*, that is, a polygon whose angles measured in radians are
rational multiples of π. A "billiard" in Q is a point moving inside Q with constant
speed, with the usual rule that the angle of incidence equals the angle of reflection.
Its trajectory is called *equidistributed* if, for every $M \subset Q$ of area $S(M)$,

$$\lim_{T \to \infty} \frac{f_M(T)}{T} = \frac{S(M)}{S(Q)},$$

where $S(Q)$ is the area of Q, and $f_M(T)$ is the time the point spent inside M during
the time interval $[0, T]$. A direction, parametrized by an angle $\alpha \in [0, 2\pi)$, leading
to an equidistributed trajectory is called *ergodic*, and all other directions are called
non-ergodic, see Fig. 3.1.

A fundamental theorem proved by Kerckhoff et al. [703] in 1986 states that,
in any rational polygon Q, almost all directions are ergodic. In other words, if
$NE(Q) \subset [0, 2\pi)$ denotes the set of non-ergodic directions in Q, then the Lebesgue
measure of $NE(Q)$ is 0. In 1992, Masur [852] proved the much stronger result that
in fact $\dim(NE(Q)) \le 1/2$, where dim is the Hausdorff dimension defined in (1.23).
In 2003, Cheung [273] proved that the upper bound $1/2$ in Masur's theorem is the
best possible.

Theorem 3.18 *There exists a rational polygon Q with* $\dim(NE(Q)) = 1/2$.

Theorem 3.18 was proved by constructing an explicit example of such a polygon
Q. The example is a polygon $BCDEFEA$ consisting of a rectangle $ABCD$ with
a line interval EF inside it, see Fig. 3.1. Using the theory of Diophantine numbers
and continued fractions, the authors proved the existence of lengths EF such that
$\dim(NE(Q))$ for this polygon is $1/2$.

An interesting problem related to billiards in rational polygons is the *illumination
problem*. Fix any point x in a rational polygon Q. A point $y \in Q$ is called
illuminated if there is a billiard starting from x which passes through y. In 1969,
Klee [724] asked whether any point x in any (possibly non-convex) rational polygon

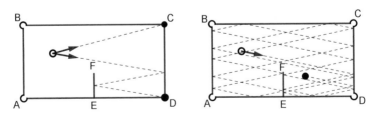

Fig. 3.1 Non-ergodic directions (left) and an ergodic direction (right) in a rational polygon

Q illuminates the whole Q. In 1995, Tokarsky [1205] answered this question negatively by constructing an example in which one point y is not illuminated. In 2016, Lelievre et al. [782] proved that the answer to Klee's question is positive in any rational polygon if we allow for finitely many exceptions.

Theorem 3.19 *In any rational rational polygon Q, and for any point $x \in Q$, there are at most finitely many points $y \in Q$ which are not illuminated from x.*

In fact, Theorem 3.19 is just one of many corollaries of the deep theorem of Eskin and Mirzakhani [431], which has been named the "Magic Wand Theorem" because it serves as a "Magic Wand" for numerous applications.

Billiards in Arbitrary Triangles

The study of billiards in arbitrary (not necessarily rational) polygons is a much more difficult research direction. Even the existence of periodic billiard paths in triangles is not well-understood! A billiard path is called *periodic* if, at some moment in time, the point trajectory starts to repeat itself. The more that two-hundred-year-old *triangular billiards conjecture* predicts that every triangle has a periodic billiard path. The conjecture is relatively easy for rational triangles (those whose angles are all rational multiples of π), and also for right triangles. In 1775, Fagnano proved it for acute triangles. However, there was essentially no progress for general (possibly not rational) obtuse triangles. In 2009, Schwartz [1101] made the first substantial progress over the centuries and proved the conjecture for all triangles with largest angle at most 100 degrees.

Theorem 3.20 *Every triangle with all angles at most one hundred degrees has a periodic billiard path.*

Theorem 3.20 was first discovered using McBilliards, a graphical user interface developed by Hooper and Schwartz to study billiard trajectories in triangles. The theorem was then rigorously proved using a combination of traditional mathematics and exact integer computation. As the reader may guess, 100 degrees is not the absolute limit of the techniques, but a nice round number at which the author decided to stop the computation. With sufficient efforts, the proof can go a bit further, but the author argues that a substantially new idea is needed to go beyond $\frac{5\pi}{8}$ radians (112.5 degrees).

Outer Billiards

An interesting version of billiards is "outer billiards". Given a bounded convex set $S \subset \mathbb{R}^2$, and a point $x_0 \in \mathbb{R}^2$ outside S, let $x_1 \in \mathbb{R}^2$ be the point such that the segment $x_0 x_1$ is tangent to S at its midpoint and a person walking from x_0 to x_1 would see S on the right. Such x_1 exists and is unique for almost every $x_0 \in \mathbb{R}^2$, and the map $x_0 \to x_1$ is called the *outer billiards map*. Iterating this map starting from x_0, we get an infinite sequence $x_0 \to x_1 \to x_2 \to \ldots$, which is called an *outer billiards orbit* of x_0.

The notion of outer billiards was introduced and popularized by Neumann [931] in the 1950s. In particular, Neumann asked the simple-looking question of whether an outer billiards orbit can be unbounded. For decades, this question remained open

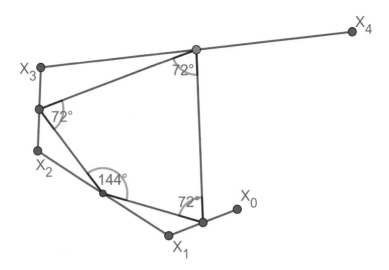

Fig. 3.2 Outer billiards with the Penrose kite

and was a guiding question in the field. Vivaldi and Shaidenko [1238] proved that if S is either a regular polygon or a polygon whose vertices has rational coordinates, then all outer billiards orbits must be bounded. Similar results are known under other restrictions on the set S, for example, if S has a sufficiently smooth boundary. With these results, one might expect that the answer to the Neumann question is negative.

In 2007, Schwartz [1100] answered this question affirmatively by constructing an example in which the outer billiards orbit is unbounded.

Theorem 3.21 *There exists a set $S \subset \mathbb{R}^2$ and $x_0 \in \mathbb{R}^2$ such that the outer billiards orbit of x_0 is well-defined and unbounded.*

In fact, Schwartz provided an explicit example of the set S in Theorem 3.21: it is the convex quadrilateral with interior angles 72, 72, 72, and 144 degrees, known as the *Penrose kite*, see Fig. 3.2. The proof of the theorem is computer-assisted, and combines methods from arithmetic dynamics and self-similar tilings.

Dynamics of Mobile Autonomous Agents
There are many examples of dynamical systems other than billiards, important for theory and/or applications. Here, we discuss a system arising from an application, namely from the theory of coordination of groups of mobile autonomous agents.

Consider n autonomous agents (points) which are moving in the plane with the same speed but with different directions. Fix a constant $r > 0$. The agents are called *neighbours* at time t if the distance between them is at most r. At any moment in time t, let $G(t)$ be a graph with vertices being agents, in which neighbours are connected by edges. At discrete times $t = 0, 1, 2, \ldots$ each agent's direction is updated and becomes the average of its own direction and the directions of all its neighbours.

This simple but important model was suggested by Vicsek et al. [1234] in 1995. The authors performed a number of numerical simulations demonstrating that, after some time, the agents start moving in the same direction, despite the absence of central coordination. In 2003, Jadbabaie et al. [646] provided a theoretical explanation for this observation.

Theorem 3.22 *Assume that, in the model described above, there exists an infinite sequence of continuous, non-empty, bounded, non-intersecting time-intervals $[t_i, t_i')$, $i = 0, 1, 2 \ldots$, starting at $t_0 = 0$, in which the graph $G(t)$ is connected. Then in the limit all agents will eventually move in the same direction.*

The proof of Theorem 3.22 uses elementary methods, and the result is not very deep from the perspective of pure mathematics. However, this work is invaluable for applications in the areas of multiagent systems, robotics, and cooperative control, and, at the time of writing, the paper [646] has over 8,000 citations in google scholar.

3.3 Functions of Complex Variables and Generalizations

Nevanlinna Theory

Let $\mathbb{C}$ be the set of complex numbers. A function $f : \mathbb{C} \to \mathbb{C}$ is called *entire* if $f'(z)$ exists for all $z \in \mathbb{C}$. A function f is called a *meromorphic function* if $f(z) = \frac{g(z)}{h(z)}$ for some entire functions $g(z)$ and $h(z)$ with $h(z)$ not identically 0. Such a function f is defined everywhere except for a set of isolated points z in which $\frac{1}{f}(z) = \frac{h(z)}{g(z)} = 0$. These points are called *poles* of f. The *multiplicity* of a pole z is the multiplicity of the zero of $1/f$ at z.

For a meromorphic function f and real $r \geq 0$, let $n(r, f)$ be the number of poles z_i of f, counting with multiplicity, such that $|z_i| \leq r$. The *Nevanlinna characteristic* $T(r, f)$ of f is $T(r, f) = m(r, f) + N(r, f)$, where

$$m(r, f) = \frac{1}{2\pi} \int_0^{2\pi} \log^+ |f(re^{i\theta})| d\theta,$$

where $\log^+ x = \max(\log x, 0)$, and

$$N(r, f) = \int_0^r (n(t, f) - n(0, f)) \frac{dt}{t} + n(0, f) \log r.$$

The Nevanlinna characteristic $T(r, f)$ measures the rate of growth of a meromorphic function f, and can be used to describe the asymptotic distribution of solutions of the equation $f(z) = a$ as a varies.

The *order* of a meromorphic function f is

$$\sigma(f) = \limsup_{r \to \infty} \frac{\log^+ T(r, f)}{\log r}.$$

In 2008, Chiang and Feng [275] proved that $T(r, f(z + \eta)) \sim T(r, f)$ for finite-order meromorphic functions.

Theorem 3.23 *For every meromorphic function f of order $\sigma(f) < \infty$, any fixed complex number $\eta \neq 0$, and any $\epsilon > 0$, we have*

$$T(r, f(z + \eta)) = T(r, f) + O(r^{\sigma(f)-1+\epsilon}) + O(\log r).$$

The authors demonstrate various applications of Theorem 3.23 to, for example, difference equations, see [275] for details.

Quasiregular Maps

The famous *Little Picard Theorem* states that the range of any non-constant entire function $f : \mathbb{C} \to \mathbb{C}$ is either the whole complex plane or the plane minus a single point. In 1980, Rickman [1044] proved a higher-dimensional generalization of this result for quasiregular maps, which are natural higher-dimensional analogues of differentiable functions of complex variables.

A function $f : \mathbb{R}^n \to \mathbb{R}^n$ is called *differentiable* at $x \in \mathbb{R}^n$ if there exists a linear map $D_{f,x} : \mathbb{R}^n \to \mathbb{R}^n$ such that

$$\lim_{h \to 0} \frac{||f(x + h) - f(x) - D_{f,x}(h)||}{||h||} = 0.$$

Let $||D_{f,x}|| = \max_{y \in \mathbb{R}^n} \frac{||D_{f,x}(y)||}{||y||}$ be the norm of $D_{f,x}$. The function f can be written as $f = (f_1, \ldots, f_n)$ with $f_i : \mathbb{R}^n \to \mathbb{R}$, $1 \leq i \leq n$. If f is differentiable at $x = (x_1, \ldots, x_n)$ then all partial derivatives $\frac{\partial f_i}{\partial x_j}$ exist for $1 \leq i, j \leq n$. Let $J_{f,x}$ be the $n \times n$ matrix with entries $\frac{\partial f_i}{\partial x_j}$. A function $f : \mathbb{R}^n \to \mathbb{R}^n$, differentiable at every $x \in \mathbb{R}^n$, is called *K-quasiregular* if

$$||D_{f,x}||^n \leq K|\det(J_{f,x})|$$

for all $x \in \mathbb{R}^n$. A function f is called *quasiregular* if it is K-quasiregular for some $K \geq 1$.

Rickman's theorem states that given any $K > 1$ and $n \geq 2$ there exists a q depending only on K and n so that a non-constant K-quasiregular mapping $f : \mathbb{R}^n \to \mathbb{R}^n$ omits at most q points. In 1985, Rickman [1045] showed that this result is sharp in dimension $n = 3$, in the sense that, given any finite set P in $\mathbb{R}^3$, there exists a quasiregular mapping $f : \mathbb{R}^3 \to \mathbb{R}^3$ omitting exactly P. In 2015, Drasin and Pankka [361] proved a similar sharpness result in all dimensions.

Theorem 3.24 *Given any integer $n \geq 3$, and any finite set $S \subset \mathbb{R}^n$, there exists a quasiregular map from $\mathbb{R}^n$ onto $\mathbb{R}^n \setminus S$.*

3.4 Properties of Polynomials

Littlewood Polynomials

Many elementary-looking questions about polynomials turned out to be surprisingly difficult to answer. One such question is the existence of flat Littlewood polynomials.

A polynomial $P(z)$ of degree n in a complex variable z is called a *Littlewood polynomial* if all its coefficients are -1 or 1, that is,

$$P(z) = \sum_{k=0}^{n} \epsilon_k z^k,$$

where $\epsilon_k \in \{-1, 1\}$ for all $0 \leq k \leq n$. The study of such polynomials goes back to the 1916 work of Hardy and Littlewood [565]. In 1957, Erdős [427] posed the problem of existence of Littlewood polynomials of every degree $n \geq 2$ satisfying

$$\delta \sqrt{n} \leq |P(z)| \leq \Delta \sqrt{n} \tag{3.8}$$

for all $z \in \mathbb{C}$ with $|z| = 1$, where δ and Δ are positive constants independent of n. Polynomials satisfying (3.8) are called *flat*. In 1966, Littlewood [805] conjectured that flat Littlewood polynomials exist. In 1959 Shapiro [1115] and Rudin [1071] constructed Littlewood polynomials satisfying the upper bound in (3.8), but not the lower bound. In 1977, Carroll et al. [253] proved the existence of Littlewood polynomials of large degree n satisfying the weaker lower bound $|P(z)| \geq n^{0.431}$ for $|z| = 1$. In a different direction, Kahane [671] proved the existence of polynomials satisfying the even stronger bounds

$$|P(z)| = (1 + o(1))\sqrt{n} \qquad \text{whenever} \qquad |z| = 1$$

(such polynomials are called *ultra-flat*), if we allow the coefficients to be any complex numbers with absolute value 1.

In 2020, Balister, Bollobás, Morris, Sahasrabudhe and Tiba [87] proved that flat Littlewood polynomials exist.

Theorem 3.25 *There exist constants $\Delta > \delta > 0$ such that, for all $n \geq 2$, there exists a Littlewood polynomial $P(z)$ of degree n satisfying (3.8) for all $z \in \mathbb{C}$ with $|z| = 1$.*

The proof of Theorem 3.25 constructs, for every $n \geq 2$, a Littlewood polynomial $P(z)$ of degree n satisfying the conditions of the theorem. The polynomial $P(z)$

consists of a real cosine polynomial similar to the one in the Rudin–Shapiro construction, and an imaginary sine polynomial. To ensure that $P(z)$ satisfies the lower bound in (3.8), the sine polynomial is made large when the cosine polynomial is small.

The existence of ultra-flat Littlewood polynomials is still an open question.

Roots of Polynomials

Other elementary-looking but difficult to solve questions ask about properties of roots of polynomials. Let $\mathbb{Z}[x]$ be the set of polynomials in one variable x with integer coefficients. An irreducible[6] polynomial $P \in \mathbb{Z}[x]$ is called *cyclotomic* if P is a divisor of $x^n - 1$ for some integer $n \geq 1$. It is immediate from the definition that all roots of cyclotomic polynomials have absolute value 1. In 1965, Schinzel and Zassenhaus [1088] conjectured that any other monic[7] irreducible polynomial of degree n must have a (possibly complex) root α satisfying

$$|\alpha| \geq 1 + \frac{c}{n},$$

where $c > 0$ is a universal constant. In 2007, Borwein et al. [176] proved this conjecture (and in fact a more general conjecture, see Theorem 3.27 below) for certain families of polynomials, including polynomials with odd coefficients, but the general case remained open. In 2019, Dimitrov [350] posted online a preprint in which he proved the conjecture in full.

Theorem 3.26 *Every non-cyclotomic monic irreducible polynomial $P(x) \in \mathbb{Z}[x]$ of degree $n > 1$ has at least one root $\alpha \in \mathbb{C}$ satisfying $|\alpha| \geq 2^{1/4n}$.*

Because $2^{1/4n} \geq 1 + \frac{\log 2}{4n}$, Theorem 3.26 confirms the Schinzel–Zassenhaus conjecture in full generality, with $c = \frac{\log 2}{4}$.

An even more general conjecture about the distribution of roots of polynomials was made by Lehmer. Every polynomial $P \in \mathbb{Z}[x]$ of degree n can be written as $P(x) = a \prod_{i=1}^{n} (x - \alpha_i)$, where $a \in \mathbb{Z}$ and α_i are (possibly complex) roots of P. *Mahler's measure* of P is

$$M(P) = |a| \prod_{i=1}^{n} \max\{1, |\alpha_i|\}.$$

In 1933, Lehmer [780] asked if for every $\epsilon > 0$ there exists a polynomial $P \in \mathbb{Z}[x]$ satisfying $1 < M(P) < 1 + \epsilon$. It is conjectured that the answer to this question is negative, and this is known as *Lehmer's conjecture*. The Schinzel–Zassenhaus conjecture (now Theorem 3.26) can be easily derived from it.

---

[6] A polynomial $Q(x) \in \mathbb{Z}[x]$ is called a *divisor* of $P(x) \in \mathbb{Z}[x]$ if $P(x) = Q(x)R(x)$ for some $R(x) \in \mathbb{Z}[x]$. If $P(x) \in \mathbb{Z}[x]$ cannot be written as $P(x) = Q(x)R(x)$ for non-constant $Q(x), R(x) \in \mathbb{Z}[x]$, we say that $P(x)$ is *irreducible.*.

[7] A *monic* polynomial is a polynomial with leading coefficient 1.

For integer $m \geq 2$, let $D_m := \{\sum_{i=0}^{n} a_i x^i \in \mathbb{Z}[x] \mid a_i \equiv 1 \pmod{m}, \ 0 \leq i \leq n\}$. In 2007, Borwein et al. [176] proved Lehmer's conjecture for polynomials in D_m for some m.

Theorem 3.27 *The inequality*

$$\log M(P) \geq c_m \left(1 - \frac{1}{n+1}\right)$$

holds for every $P \in D_m$ *of degree n and no cyclotomic divisors, where* $c_2 = (\log 5)/4$ *and* $c_m = \log(\sqrt{m^2 + 1}/2)$ *for* $m > 2$.

In particular, the case $m = 2$ of Theorem 3.27 implies that Lehmer's conjecture is true for polynomials with odd coefficients.

Another basic question about polynomial roots is whether they are real or complex. In particular, an old question, mentioned already in a work of Waring in 1782, is whether a "typical" (or random) polynomial of even degree has any real roots. More formally, let $a_0, a_1, a_2, \ldots$ be an infinite sequence of zero-mean, unit-variance, independent identically distributed (i.i.d.) random variables possessing finite moments of all orders.[8] Let

$$f_n(x) = \sum_{i=0}^{n} a_i x^i,$$

and let P_n be the probability that $f_n(x) \neq 0$, $\forall x \in \mathbb{R}$. In 1999, Poonen and Stoll [1002], motivated by applications in arithmetic geometry, conjectured that $P_{2n} = (2n)^{-b+o(1)}$ for some $b > 0$. In 2002, Dembo et al. [344] confirmed this prediction.

Theorem 3.28 *For an integer* $k \geq 0$, *let* $P_{n,k}$ *be the probability that a random polynomial of degree n has exactly k real roots, all of which are simple. Then*

$$\lim_{n \to \infty} \frac{\log P_{2n+k,k}}{\log(2n)} = -b,$$

where $b > 0$ *is a constant.*

The proof of Theorem 3.28 first treats the special case where all the coefficients have Gaussian distribution, and then extends the result to general distributions (with finite moments) using the technique for approximating partial sums of independent random variables developed by Komlós et al. [733] in 1975.

Dembo et al. also proved that the constant b in Theorem 3.28 satisfies $0.4 \leq b \leq 2$ and predicted that $b \approx 0.76 \pm 0.03$. With $k = 0$, Theorem 3.28 implies

---

[8] We say that a random variable X possesses *finite moments* of all orders if the expectation of $|X|^k$ is finite for all $k > 0$.

that $P_{2n} = (2n)^{-b+o(1)}$. More generally, the authors proved that a random degree n polynomial has $o\left(\frac{\log n}{\log \log n}\right)$ real roots with the same probability $n^{-b+o(1)}$.

Stable Polynomials

A polynomial P of degree n is called *stable* (or *hyperbolic*) if it has n real roots (counting multiplicity). It is easy to see that if a polynomial P is stable, then so is its derivative. Also, the product $P \cdot Q$ of two stable polynomials is stable. In other words, the property of stability is preserved by differentiation, and by multiplication by a stable polynomial. A long-standing open problem was to describe *all* linear operators T acting on the set of polynomials which preserve stability. In 1914, Pólya and Schur [994] did this for operators such that $T(x^k) = \lambda_k x^k, k \geq 0$, where $\{\lambda_k\}_{k=0}^{\infty}$ is a given sequence of numbers. In 2009, Borcea and Brändén [169] provided a characterization of all stability-preserving operators.

To formulate this result, we need some notation. Let $\mathcal{P}_k, k = 1, 2$, be the set of all polynomials in k variables with real coefficients. We say that a linear operator $T : \mathcal{P}_1 \to \mathcal{P}_1$ *preserves stability* if $T(P)$ is a stable polynomial whenever P is stable. We say that stable polynomials $P, Q \in \mathcal{P}_1$ are *interlacing* if either $\alpha_1 \leq \beta_1 \leq \alpha_2 \leq \beta_2 \leq \ldots$ or $\beta_1 \leq \alpha_1 \leq \beta_2 \leq \alpha_2 \leq \ldots$, where $\alpha_1 \leq \alpha_2 \leq \cdots \leq \alpha_n$ and $\beta_1 \leq \beta_2 \leq \cdots \leq \beta_m$ are the roots of $P(x)$ and $Q(x)$, respectively. A polynomial $P \in \mathcal{P}_2$ is called *stable* if $Q(t) = P(a + bt, c + dt) \in \mathcal{P}_1$ is stable for any real a, b, c, d such that $b > 0$ and $d > 0$. For a linear operator $T : \mathcal{P}_1 \to \mathcal{P}_1$, let $S_T : \mathcal{P}_2 \to \mathcal{P}_2$ be a linear operator such that $S_T(x^k y^l) = T(x^k)y^l$, for all $k \geq 0$ and $l \geq 0$.

Theorem 3.29 *A linear operator $T : \mathcal{P}_1 \to \mathcal{P}_1$ preserves stability if and only if*

(a) $T(x^k) = a_k P(x) + b_k Q(x)$, *where* $a_k, b_k, k = 0, 1, 2, \ldots$ *are real numbers, and $P(x)$ and $Q(x)$ are some fixed (independent of k) stable interlacing polynomials; or*

(b) $S_T[(x + y)^k] \in \mathcal{P}_2$ *is a stable polynomial for all $k = 0, 1, 2, \ldots$; or*

(c) $S_T[(x - y)^k] \in \mathcal{P}_2$ *is a stable polynomial for all $k = 0, 1, 2, \ldots$.*

The B. and M. Shapiro Conjecture

Let $\mathcal{P} = \{P_1(z), P_2(z), \ldots, P_k(z)\}$ be a finite set of polynomials in a complex variable z with complex coefficients. The complex span S of $\mathcal{P}$ is the set of polynomials $P(z)$ of the form

$$P(z) = \sum_{i=1}^{k} \lambda_i P_i(z)$$

for some $\lambda_i \in \mathbb{C}, i = 1, \ldots, k$. The *Wronskian* $W(z)$ of $\mathcal{P}$ is the determinant of the $k \times k$ matrix with entries $P_i^{(j-1)}(z), i = 1, \ldots, k, j = 1, \ldots, k$, where $P^{(j)}(z)$ denotes the j-th derivative of the polynomial $P(z)$. In 1993, B. and M. Shapiro conjectured that if $W(z)$ has only real roots, then S has a basis consisting of polynomials with real coefficients. This conjecture has important consequences in real algebraic geometry, and was studied in many works. In 2002, Eremenko and Gabrielov [428] resolved the $k = 2$ case of the conjecture.

Theorem 3.30 *If for polynomials $P(z)$, $Q(z)$ all the solutions of the equation $P(z)Q'(z) - P'(z)Q(z) = 0$ are real, then there exist complex numbers a, b, c, d such that $ad - bc \neq 0$ and $aP(z) + bQ(z)$ and $cP(z) + dQ(z)$ are polynomials with real coefficients.*

In 2009, Mukhin et al. [916] proved the B. and M. Shapiro conjecture in full.

Theorem 3.31 *If all roots of the Wronskian of a set $\mathcal{P}$ of polynomials are real, then the complex span of $\mathcal{P}$ has a basis consisting of polynomials with real coefficients.*

The proof of Theorem 3.31 is based on the *Bethe ansatz method*, a method for finding the wavefunctions of the one-dimensional quantum many-body model called the *Gaudin model*. The proof also crucially uses the fact that a symmetric linear operator on Euclidean space has a real spectrum.

The Waring Problem for Polynomials

A famous theorem of Lagrange states that every natural number is the sum of at most 4 perfect squares. Subsequent authors proved that any natural number can also be written as a sum of 9 cubes, or 19 fourth powers, etc. *Waring's problem* asks whether for each positive integer k there exists a positive integer $g(k)$ such that every natural number is the sum of at most $g(k)$ k-th powers. The *Hilbert–Waring theorem*, proved by Hilbert [606] in 1909, gives a positive answer, and an exact formula for $g(k)$ has been derived in later works.

A similar question can be asked for polynomials. The *Waring rank* $\mathrm{rk}(P)$ of a homogeneous polynomial (or form) P of degree d is the least value of s for which there exist linear forms[9] $L_1, \ldots, L_s$ such that $P = \sum_{i=1}^{s} L_i^d$. In 1995, Alexander and Hirschowitz [27] described the Waring rank of a *generic* form of degree d, but determining the $\mathrm{rk}(P)$ for a given specific P remains an open question. In 2011, Ranestad and Schreyer [1024] proved that $\mathrm{rk}(P) = (m + 1)^{n-1}$ if P is a monomial of the form $P = (x_1 \cdots x_n)^m$. In 2012, Carlini et al. [251] determined the Waring rank for all monomials.

Theorem 3.32 *Let $M = x_1^{a_1} x_2^{a_2} \ldots x_n^{a_n}$ be a monomial in $n \geq 2$ variables, and let $1 \leq a_1 \leq \cdots \leq a_n$. Then*

$$\mathrm{rk}(M) = \prod_{i=2}^{n} (a_i + 1).$$

Polynomials Over Fields

Many interesting research questions can be asked about polynomials over arbitrary fields. One such question concerns counting zero-patterns.

---

[9] A *linear form* in n variables in an expression of the form $L = c_1 x_1 + \cdots + c_n x_n$, where $c_1, c_2, \ldots, c_n$ are some complex coefficients.

Let $f = (f_1, \ldots, f_m)$ be a sequence of polynomials in n variables of degree at most d over a field F, where $m \geq n$. The *zero-pattern* of this sequence at $u \in F^n$ is the set of indexes i ($1 \leq i \leq m$) for which $f_i(u) = 0$. Let $Z_F(f)$ denote the number of zero-patterns of f as u ranges over F^n. Upper bounds for $Z_F(f)$ are useful, for example, in the complexity analysis of quantifier elimination algorithms. Motivated by this application, Heintz [590] proved in 1983 that $Z_F(f)$ is at most $(1 + md)^n$. In 2001, Rónyai et al. [1058] improved this bound by a factor of about $(n/e)^n$.

Theorem 3.33 *Let F be an arbitrary field, $f = (f_1, \ldots, f_m)$ be a sequence of polynomials in n variables over F, and let d_i be the degree of f_i. Then the number of zero-patterns of f is*

$$Z_F(f) \leq \binom{n + \sum_{i=1}^{m} d_i}{n}.$$

If $d = \max_i d_i$, Theorem 3.33 implies that $Z_F(f) \leq \binom{md+n}{n}$. In fact, the authors proved an even stronger upper bound

$$Z_F(f) \leq \binom{md - (d-2)n}{n}.$$

The authors also note that if $m \geq nd$, then these bounds are optimal up to a factor of $(7.25)^n$.

While the proofs of most results about zero-patterns use deep techniques from real algebraic geometry, the proof of Theorem 3.33 is ingeniously simple, and based on elementary linear algebra. If the sequence $f = (f_1, \ldots, f_m)$ has M zero patterns, let $u_1, \ldots, u_M \in F^n$ be vectors such that each u_i corresponds to a different zero-pattern, let S_i be the support of the zero-pattern corresponding to u_i, and let $g_i = \prod_{k \in S_i} f_k$. The authors use simple linear algebra to prove that the polynomials $g_1, \ldots, g_M$ are linearly independent over F, and deduce Theorem 3.33 from this fact.

Generalized Polynomials

The notion of polynomial can be generalized in various useful ways, and here we mention one of them. A *generalized polynomial* is a real-valued function which is obtained from a conventional polynomial in one or several variables and applying in arbitrary order the operations of taking the integer part,[10] addition, and multiplication. The simplest non-trivial examples of generalized polynomials are $\lfloor P \rfloor$ and $\text{frac}(P) = P - \lfloor P \rfloor$, where P is an arbitrary conventional polynomial. A classical 1916 theorem of Weyl [1254] states that if at least one coefficient of P is irrational, then the sequence $\{\text{frac}(P(n))\}_{n \in \mathbb{Z}}$ is uniformly distributed on $[0, 1]$. More generally, one may study the sequence $\{u(n)\}_{n \in \mathbb{Z}}$ for an arbitrary bounded generalized polynomial u, and ask whether it has some regular behaviour.

---

[10] The *integer part* $\lfloor z \rfloor$ of a real number z is the largest integer not exceeding z.

In particular, Bergelson and Håland [133] posed in 1996 the question of existence of the limit

$$\lim_{N \to \infty} \frac{1}{N} \sum_{n=1}^{N} e^{2\pi i u(n)}. \tag{3.9}$$

In 2007, Bergelson and Leibman [135] answered this question affirmatively.

Theorem 3.34 *The limit* (3.9) *exists for any bounded generalized polynomial u.*

In fact, Theorem 3.34 is just one of the many corollaries of the deep theory of the distribution of values of bounded generalized polynomials developed by the authors, see [135] for details.

3.5 The Riemann Zeta Function and Other Special Functions

The Zeta Functions
In 1650, Mengoli asked what the value of the infinite sum $\sum_{n=1}^{\infty} \frac{1}{n^2}$ is. This problem became known as the "Basel problem" and was solved by Euler in 1734, who proved that the answer is $\frac{\pi^2}{6}$. More generally, Euler noted that series $\sum_{n=1}^{\infty} \frac{1}{n^s}$ converges for any $s > 1$ and introduced a function

$$\zeta(s) = \sum_{n=1}^{\infty} \frac{1}{n^s}, \quad s > 1. \tag{3.10}$$

In this notation, Euler's theorem states that $\zeta(2) = \frac{\pi^2}{6}$. It is also known that $\zeta(4) = \frac{\pi^4}{90}$, and, more generally, $\frac{\zeta(k)}{\pi^k}$ is a rational number for all even $k \geq 2$.
   A natural multivariate generalization of (3.10) is the function

$$\zeta(k_1, k_2, \ldots k_n) = \sum_{0 < m_1 < \cdots < m_n} \frac{1}{m_1^{k_1} m_2^{k_2} \ldots m_n^{k_n}},$$

known as the *multiple zeta function*. This function was also studied by Euler, who, in particular, derived expressions for $\zeta(1, m)$. For example, $\zeta(1, 2) = \zeta(3)$, $\zeta(1, 3) = \frac{\zeta(4)}{4} = \frac{\pi^4}{360}$, etc. In 1994, Zagier [1286] conjectured a nice $2n$-variable generalization of the last formula:

$$\zeta(\{1, 3\}^n) = \frac{2\pi^{4n}}{(4n + 2)!}, \tag{3.11}$$

where $\zeta(\{a, b\}^n)$ means $\zeta(a, b, a, b, \ldots, a, b)$, where the arguments a and b are repeated n times. In 2001, Borwein, Bradley, Broadhurst and Lisoněk [175] proved this conjecture.

Theorem 3.35 *Equation* (3.11) *is true for all integers* $n \geq 1$.

Theorem 3.35 is a rare example when one is able to derive an exact analytical expression for an infinite family of multiple zeta functions. Another such example is a 2012 theorem of Zagier [1287], which derives an expression for $\zeta(2, \ldots, 2, 3, 2, \ldots, 2)$.

Theorem 3.36 *Let*

$$H(n) = \zeta(\underbrace{2, \ldots, 2}_{n}) = \frac{\pi^{2n}}{(2n + 1)!} \ and$$

$$H(a, b) = \zeta(\underbrace{2, \ldots, 2}_{a}, 3, \underbrace{2, \ldots, 2}_{b}).$$

Then, for all integers $a, b \geq 0$, *we have*

$$H(a, b) = \sum_{r=1}^{a+b+1} (-1)^r \left[\binom{2r}{2a + 2} - (1 - 2^{-2r}) \binom{2r}{2b + 1} \right] H(a + b - r + 1) \zeta(2r + 1).$$

$$(3.12)$$

Theorem 3.36 was proved by computing the generating functions F and G of both sides of (3.12) in closed form. The author then proved the equality of the functions F and G at some specific points, and also proved that both functions are entire functions of exponential growth. From this, he deduced that the generating functions F and G in fact coincide, and Theorem 3.36 follows.

The Riemann Hypothesis

Euler established the connection of the function (3.10) to the theory of prime numbers by noticing that

$$\zeta(s) = \prod_{k=1}^{\infty} \frac{1}{1 - p_k^{-s}}, \quad s > 1,$$

where p_k denotes the k-th prime. In 1859, Riemann [1046] observed that this connection becomes even more useful if we extend the function ζ to a meromorphic function on the whole complex plane by analytic continuation. This extended function, also denoted $\zeta(s)$, is called the *Riemann zeta function*. Riemann conjectured that if $\zeta(z) = 0$ for a complex number z, then either $z = -2k$ for some integer $k > 0$ or $\text{Re}(z) = \frac{1}{2}$, and observed that some useful information about the distribution of primes can be deduced from this. The conjecture, now known as the *Riemann Hypothesis* (RH), is still open and has become one of the most important open problems in the whole of mathematics. The RH was included as problem 8 in Hilbert's list [605] of problems for the twentieth century, and then in the list [252] of the Millennium Prize Problems for which a million dollars is offered for a solution. Many dozens of other important conjectures in number theory and beyond have been proved conditionally on RH.

The RH has a number of equivalent formulations, one of which we present below. Let $\Gamma(z)$ be the analytic continuation of the integral

$$\Gamma(z) = \int_0^\infty x^{z-1} e^{-x} dx,$$

defined for $\text{Re}(z) > 0$. Let $\Lambda(z) = \pi^{-z/2} \Gamma(z/2) \zeta(z)$, and let the sequence $\{\gamma(n)\}_{n=0}^\infty$ be defined by

$$(-1 + 4z^2)\Lambda\left(\frac{1}{2} + z\right) = \sum_{n=0}^\infty \frac{\gamma(n)}{n!} z^{2n}.$$

The *Jensen polynomial* for the Riemann zeta function of degree d and shift n is the polynomial

$$J_\gamma^{d,n}(x) = \sum_{j=0}^n \binom{d}{j} \gamma(n+j) x^j.$$

We say that a polynomial with real coefficients is *hyperbolic* if all of its zeros are real. In 1927, Pólya [993] proved that the RH is equivalent to the statement that the polynomials $J_\gamma^{d,n}(x)$ are hyperbolic for all integers $d \geq 0$ and $n \geq 0$. Before 2019, this statement was known to hold only for $d \leq 3$. In 2019, Griffin et al. [529] proved the hyperbolicity of $J_\gamma^{d,n}(x)$ for all d, assuming that n is sufficiently large (depending on d).

Theorem 3.37 *For any $d \geq 1$ there is a constant $N = N(d)$ such that the polynomial $J_\gamma^{d,n}(x)$ is hyperbolic for all $n \geq N$.*

As a corollary of Theorem 3.37, the authors also proved the hyperbolicity of $J_\gamma^{d,n}(x)$ for all n if $d \leq 8$. Overall, Theorem 3.37 is considered by many mathematicians as the biggest progress towards the Riemann Hypothesis we currently have.

Another interesting reformulation of the Riemann Hypothesis involves the *De Bruijn–Newman constant*. In 1950, De Bruijn [330] introduced the family of functions $H_t : \mathbb{C} \to \mathbb{C}$ with real parameter $t \in \mathbb{R}$ given by

$$H_t(z) = \int_0^\infty e^{tu^2} \Phi(u) \cos(zu) du,$$

where $\Phi : \mathbb{R} \to \mathbb{R}$ is a function defined as the sum of the infinite series

$$\Phi(u) = \sum_{n=1}^{\infty} (2\pi^2 n^4 e^{9u} - 3\pi n^2 e^{5u}) \exp(-\pi n^2 e^{4u}).$$

In 1976, Newman [936] showed that there exists a finite constant Λ (which acquired the name "De Bruijn–Newman constant") such that the zeros of H_t are all real if and only if $t \geq \Lambda$. Newman noted that the Riemann Hypothesis is equivalent to the upper bound $\Lambda \leq 0$, and conjectured the complementary lower bound $\Lambda \geq 0$. In Newman's words, this conjecture states that if the Riemann Hypothesis is true, it is only "barely so". In 2020, Rodgers and Tao [1054] proved Newman's conjecture.

Theorem 3.38 *The De Bruijn–Newman constant is non-negative:* $\Lambda \geq 0$.

The proof of Theorem 3.38 assumes that $\Lambda < 0$ and then analyses the dynamics of zeros of H_t in the range $t \in (\Lambda, 0]$ to derive information about the distributions of zeros of H_0 which contradicts the known facts about the distribution of zeros of the Riemann zeta function established by Montgomery [903] in 1973.

With Theorem 3.38, the Riemann Hypothesis is equivalent to $\Lambda = 0$. One may speculate that the better upper bound on Λ we have, the "closer" we are to proving the RH. In 2009, Ki et al. [717] proved that $\Lambda < 1/2$. In 2019, a large group of mathematicians, collaborating online, improved this to $\Lambda \leq 0.22$, see [998].

Other Properties of the Riemann Zeta Function
While the Riemann Hypothesis is the central open question, there are other important questions about the properties of the Riemann zeta function ζ. One of them is how to efficiently compute $\zeta(z)$ approximately. This question is especially important if z belongs to the "critical line", that is, $z = \frac{1}{2} + it$ for $t \in \mathbb{R}$. In 1990, Schönhage [1095] presented a method which, for any fixed λ, computes $\zeta\left(\frac{1}{2} + it\right)$ up to an error of $\pm t^{-\lambda}$ in $t^{3/8 + o_\lambda(1)}$ operations. Later, Heath-Brown improved the exponent $\frac{3}{8}$ in the running time to $\frac{1}{3}$. In 2011, Hiary [603] improved the exponent further to $\frac{4}{13}$.

Theorem 3.39 *Given any constant λ, there is an effectively computable constant $C = C(\lambda)$ and an absolute constant k such that for any $t > 1$ the value of $\zeta\left(\frac{1}{2} + it\right)$ can be computed to within $\pm t^{-\lambda}$ using at most $C(\log t)^k t^{4/13}$ operations.*

It has been known since the work of Schönhage [1095] in 1990 that the problem of efficient computation of values of the Riemann zeta function is connected to the problem of efficient numerical evaluation of exponential sums of the form

$$\frac{1}{K^j} \sum_{k=1}^{K} k^j \exp(2\pi i f(k)),$$

where f is some polynomial with real coefficients. At the heart of the proof of Theorem 3.39 is a novel algorithm for faster computation of such sums in the case when f is a cubic polynomial with a small cubic coefficient.

While the Riemann Hypothesis is a question about zeros of the Riemann zeta function ζ, in other applications it is important to quantify "how large" $\zeta(z)$ is, especially on the critical line $z = \frac{1}{2} + it$. The classical upper bound due to Weyl and Hardy–Littlewood states that, for any $\epsilon > 0$,

$$\left| \zeta \left(\frac{1}{2} + it \right) \right| \le C_\epsilon (1 + |t|)^{1/6+\epsilon}, \tag{3.13}$$

where C_ϵ is a constant depending on ϵ. One may also be interested in how large $\left| \zeta \left(\frac{1}{2} + it \right) \right|$ is "on average". To quantify this, we study how the $2k$-th moment

$$M_k(T) = \int_0^T \left| \zeta \left(\frac{1}{2} + it \right) \right|^{2k} dt, \quad k > 0,$$

grows with T. In 1979, Heath-Brown [583] derived a very accurate approximation for the 4-th moment

$$\frac{M_2(T)}{T} = \frac{1}{T} \int_0^T \left| \zeta \left(\frac{1}{2} + it \right) \right|^4 dt = \sum_{i=0}^4 a_i (\log T)^i + O(T^{-\frac{1}{8}+\epsilon}), \tag{3.14}$$

where the a_i are some explicitly computable constants. Note that the error term in (3.14) decays as a power of T.

For general moments, only weaker upper and lower bounds are known. In 1978, Ramachandra [1020], assuming the RH, proved that the lower bound

$$c_k T (\log T)^{k^2} \le M_k(T)$$

holds for some $c_k > 0$. In 2009, Soundararajan [1146], also assuming the RH, proved an almost (up to ϵ) matching upper bound.

Theorem 3.40 *Assume the Riemann Hypothesis. Then for every $k > 0$ and $\epsilon > 0$ there is a constant $C = C(k, \epsilon)$ such that inequality*

$$M_k(T) \le CT (\log T)^{k^2+\epsilon}$$

holds for all $T > 2$.

The proof of Theorem 3.40 proceeds by establishing that the dominant contribution to $M_k(T)$ comes from values of t such that $\left| \zeta \left(\frac{1}{2} + it \right) \right| \approx (\log T)^k$, and the set of such t has measure about $\frac{T}{(\log T)^{k^2}}$. In 2013, Harper [567] removed the ϵ from the exponent in Theorem 3.40, and proved the upper bound

$$M_k(T) \le C_k T (\log T)^{k^2},$$

matching Ramachandra's lower bound up to a constant factor.

Dirichlet Characters and L-Functions

The Riemann zeta function belongs to a more general family of functions which are called Dirichlet L-functions. To introduce them, we need some preliminary definitions. A function $\chi_q : \mathbb{Z} \to \mathbb{C}$ is called a *Dirichlet character* modulo an integer $q > 0$ if

(i) $\chi_q(n) = \chi_q(n + q)$ for all n,
(ii) if $\gcd(n, q) > 1$ then $\chi_q(n) = 0$; if $\gcd(n, q) = 1$ then $\chi_q(n) \ne 0$, and
(iii) $\chi_q(mn) = \chi_q(m)\chi_q(n)$ for all integers m and n.

An example is the function χ_q^* such that $\chi_q^*(n) = 1$ whenever $\gcd(n, q) = 1$ and $\chi_q^*(n) = 0$ otherwise. This is called a *principal character* and all other characters *non-principal*. The *order* of a Dirichlet character χ_q is the least positive integer m such that $\chi_q(n)^m = \chi_q^*(n)$ for all n. A character χ_q is called *primitive* if there is no integer $0 < d < q$ such that $\chi_q(a) = \chi_q(b)$ whenever $\gcd(a, q) = \gcd(b, q) = 1$ and $a - b$ is a multiple of d.

For any primitive Dirichlet character χ_q, and complex number z with $\mathrm{Re}(z) > 1$, let

$$L(z, \chi_q) = \sum_{n=1}^{\infty} \frac{\chi_q(n)}{n^z}.$$

By analytic continuation, the function $L(z, \chi_q)$ can be extended to a meromorphic function on the whole complex plane, and it is called a *Dirichlet L-function*. The Riemann zeta function ζ is an example of a Dirichlet L-function: $\zeta(z) = L(z, \chi_1^*)$.

The generalized Riemann Hypothesis (GRH) states that for every primitive Dirichlet character χ_q and every complex number s with $L(s, \chi_q) = 0$, either s is a negative real number, or $\mathrm{Re}(s) = \frac{1}{2}$. The GRH has important consequences, especially in number theory, many of which are not known to follow from the RH.

By analogy with (3.13), one may ask for upper bounds for the values of a general Dirichlet L-function on the critical line $z = \frac{1}{2} + it$. Such bounds have important applications in number theory. In 1963, Burgess [231] proved that

$$L\left(\frac{1}{2}, \chi_q\right) \le C_\epsilon q^{\frac{3}{16}+\epsilon}.$$

It is conjectured that the exponent $\frac{3}{16}$ in this bound can be improved to $\frac{1}{6}$. In 2020, Petrow and Young [979] achieved this improvement in the case when q is *cube-free*, that is, not divisible by m^3 for any integer $m \ge 2$.

Theorem 3.41 *For any $\epsilon > 0$ there exist a constant C_ϵ such that the inequality*

$$\left| L\left(\frac{1}{2} + it, \chi_q\right)\right| \leq C_\epsilon (q(1 + |t|))^{1/6+\epsilon} \tag{3.15}$$

holds for every primitive Dirichlet character χ_q with a cube-free q, and for all real t.

Building on the methodology developed to prove Theorem 3.41, the same authors managed to remove the cube-free hypothesis, and prove in a subsequent work [978] that the bound (3.15) remains correct for every primitive Dirichlet character χ_q.

For a prime q, denote by $\mathcal{P}(q)$ the set of all primitive Dirichlet characters χ_q modulo q, and let $\phi^*(q)$ be the number of such characters. The expression

$$\frac{1}{\phi^*(q)} \sum_{\chi_q \in \mathcal{P}(q)} \left| L\left(z, \chi_q\right)\right|^{2k}$$

is called the *2k-th moment of Dirichlet L-functions at z*. In 2011, Young [1284] computed the fourth moment at $z = \frac{1}{2}$ for prime moduli q, with the error term decreasing as a power of q. This result is in direct analogue with the Heath-Brown formula (3.14) for the fourth moment of the Riemann zeta function, which also has a power saving error term. The estimates of this accuracy are especially useful but difficult to prove.

Theorem 3.42 *For any $\epsilon > 0$ there exist a constant C_ϵ such that for any prime $q \neq 2$*

$$\left| \frac{1}{\phi^*(q)} \sum_{\chi \in \mathcal{P}(q)} \left| L\left(\frac{1}{2}, \chi_q\right)\right|^4 - \sum_{i=0}^4 c_i (\log q)^i \right| \leq C_\epsilon q^{-\frac{5}{512}+\epsilon},$$

where the c_i are some explicitly computable absolute constants.

Dirichlet characters χ_q are interesting in their own right, not necessarily in connection with the L-functions. Of special interest are upper bounds for the maximal size of character sums

$$M(\chi_q) = \max_{1 \leq m \leq q} \left| \sum_{n=1}^m \chi_q(n)\right|.$$

In 1918, Pólya and Vinogradov independently proved that

$$M(\chi_q) = O(\sqrt{q} \log q)$$

for any non-principal Dirichlet character χ_q. This is known as the Pólya–Vinogradov inequality. It has numerous applications, but it is an open question whether a better bound for $M(\chi_q)$ in terms of q is possible. In 2007, Granville and

Soundararajan [516] established an improved upper bound on $M(\chi_q)$ for characters of odd, bounded order.

Theorem 3.43 *If χ_q is a primitive character modulo q of odd order g, then*

$$M(\chi_q) \leq C_g \sqrt{q} (\log q)^{1 - \frac{\delta_g}{2} + o(1)}$$

where $\delta_g = 1 - \frac{g}{\pi} \sin \frac{\pi}{g}$, and C_g is a constant depending only on g.

The proof of Theorem 3.43 is based on a novel characterization discovered by the authors of characters χ for which $M(\chi)$ is large. This characterization leads to the discovery of some hidden structure among such characters. For example, if k is a not very large integer, and $M(\chi)$ is large, then $M(\chi^k)$ is small if k is even and large if k is odd. The authors also derived many other (more technical) properties of χ with large $M(\chi)$, and deduced Theorem 3.43 using these properties.

3.6 Inequalities in Analysis

The Grothendieck Inequality

In 1953, Grothendieck [537] proved the existence of a universal constant $K < \infty$ such that, given any $m \times n$ matrix with real entries a_{ij} and arbitrary unit vectors $x_1, \ldots, x_m, y_1, \ldots, y_n$ in $\mathbb{R}^{m+n}$, there are signs $\epsilon_1, \ldots, \epsilon_m, \delta_1, \ldots, \delta_n \in \{-1, +1\}$ such that the inequality

$$\sum_{i=1}^{m} \sum_{j=1}^{n} a_{ij} \langle x_i, y_j \rangle \leq K \sum_{i=1}^{m} \sum_{j=1}^{n} a_{ij} \epsilon_i \delta_j \tag{3.16}$$

holds.[11] This is called the *Grothendieck inequality*. The minimal constant K for which it holds is called the *Grothendieck constant*, and is denoted by K_G. The Grothendieck inequality, despite being innocent-looking at first glance, turned out to be important in many areas of mathematics and applications, such as functional analysis, harmonic analysis, operator theory, quantum mechanics, and computer science. For many applications, it is important to have it with the best possible constant. However, despite substantial efforts, even the second digit of K_G remains unknown. The best lower bound is $K_G \geq 1.67696...$ proved by Davie [325] in 1985. For the upper bound, Krivine [751] proved in 1979 that

$$K_G \leq \frac{\pi}{2 \log(1 + \sqrt{2})} = 1.7822...,$$

---

[11] Here, $\langle x, y \rangle = \sum_{i=1}^{n} x_i y_i$ denotes the inner product in $\mathbb{R}^n$.

and conjectured that K_G is actually equal to this value. This conjecture was well-believed, and many researchers focused on proving the matching lower bound. However, Braverman et al. [202] proved in 2013 that this conjecture is false, and the Grothendieck constant is strictly smaller than Krivine's bound.

Theorem 3.44 *There exist a constant $\epsilon_0 > 0$ such that $K_G < \dfrac{\pi}{2\log(1+\sqrt{2})} - \epsilon_0$.*

The constant ϵ_0 in Theorem 3.44 is small, hence the improvement in the upper bound is negligible for practical applications. The importance of Theorem 3.44, however, is the proof that Krivine's bound is non-optimal, and hence it makes sense to try to improve it further.

The idea of the proof is to project (suitably transformed) input vectors $x_1, \ldots, x_m, y_1, \ldots, y_n$ onto a random two-dimensional plane. Then the authors suggested a partition of this plane into two regions, S^+ and S^-, and put each ϵ_i (respectively, δ_j) equal to $+1$ or -1 depending on whether the vector x_i (respectively, y_j) is projected into S^+ or S^-. The authors then demonstrated that with this choice of ϵ_i and δ_j, inequality (3.16) holds with the improved constant K.

In 1999, Nemirovski et al. [929] proved a quadratic variant of the Grothendieck inequality. For a positive integer n, let K_n be the minimal constant such that for every $n \times n$ matrix with entries a_{ij} and all unit vectors $x_1, \ldots, x_n$ in $\mathbb{R}^n$ there are signs $e_1, \ldots, e_n \in \{-1, +1\}$ for which

$$\sum_{i \neq j} a_{ij}(x_i, x_j) \leq K_n \sum_{i \neq j} a_{ij} e_i e_j.$$

Nemirovski, Roos and Terlaky [929] proved that $K_n \leq C \log n$ for some constant C. In 2001, Megretski [871] asked if this can be improved to $K_n \leq C$. In 2004, Charikar and Wirth [264] derived from the inequality $K_n \leq C \log n$ various algorithms for combinatorial problems, and noticed that a better upper bound on K_n could lead to algorithms with better accuracy. However, Alon et al. [36] proved in 2006 that the inequality $K_n \leq C \log n$ is optimal up to a constant factor.

Theorem 3.45 *There exist a constant $c > 0$ such that $K_n \geq c \log n$ for all n.*

The authors of [36] also developed a theory of Grothendieck inequalities on arbitrary graphs, which nicely unifies all the previous results. In this theory, the original Grothendieck constant K_G corresponds to the bipartite graphs, while the constant K_n in Theorem 3.45 is the Grothendieck constant of the complete n-vertex graph.

The Hardy–Littlewood Maximal Inequality

Let $\mathcal{L}^1(\mathbb{R})$ be the set of all functions $f : \mathbb{R} \to \mathbb{R}$ such that the integral $\int_{-\infty}^{+\infty} |f(x)| dx$ exists and is finite. We will denote the value of this integral by $\|f\|_1$. The centered *Hardy–Littlewood maximal operator* is the map which assigns to any function $f \in \mathcal{L}^1(\mathbb{R})$ a function

$$M_f(t) = \sup_{\epsilon > 0} \frac{1}{2\epsilon} \int_{t-\epsilon}^{t+\epsilon} |f(x)| dx.$$

For any $\delta > 0$, denote by $|\{M_f > \delta\}|$ the Lebesgue measure of the set of real numbers t such that $M_f(t) > \delta$. The *centered Hardy–Littlewood maximal inequality* asserts that

$$|\{M_f > \delta\}| \leq C\frac{\|f\|_1}{\delta} \tag{3.17}$$

for some constant $C > 0$. It has various applications in the theories of differentiation and integration. Let us denote by C^* the best possible (smallest) constant C such that (3.17) holds. Menarguez and Soria [875] suggested that $C^* = 1.5$, but this was disproved by Aldaz [23] in 1998. In 2002, Melas [873] proved that $\frac{11+\sqrt{61}}{12} \leq C^* \leq \frac{5}{3}$ and conjectured that the lower bound is the actual value of C^*. In 2003, Melas [874] confirmed this conjecture.

Theorem 3.46 *Inequality* (3.17) *holds for every* $f \in \mathcal{L}^1(\mathbb{R})$ *and for every* $\delta > 0$ *with* $C = \frac{11+\sqrt{61}}{12} \approx 1.5675$, *and this constant is the best possible one.*

The proof of Theorem 3.46 uses a discretization technique introduced by De Guzmán [333] in 1981, which reduces the problem to the search for the best possible constant in a similar inequality involving finite sums instead of integration. The latter problem was solved using some arguments of a combinatorial nature.

The centered Hardy–Littlewood maximal inequality (3.17) has a higher-dimensional version. For $x = (x_1, \ldots, x_d) \in \mathbb{R}^d$, define $\|x\|_\infty = \max\limits_{1 \leq i \leq d} |x_i|$. For $x \in \mathbb{R}^d$ and $r > 0$, let $Q(x, r) = \{y \in \mathbb{R}^d : \|y - x\|_\infty \leq r\}$. Let $\mathcal{L}^1(\mathbb{R}^d)$ be the set of all functions $f : \mathbb{R}^d \to \mathbb{R}$ such that $\|f\|_1 = \int_{\mathbb{R}^d} |f(x)| dx < \infty$. For any $f \in \mathcal{L}^1(\mathbb{R}^d)$, let

$$M_f(x) = \sup_{r>0} \frac{1}{|Q(x,r)|} \int_{Q(x,r)} |f(y)| dy,$$

where $|Q|$ denotes the Lebesgue measure of a subset Q of $\mathbb{R}^d$. Let c_d be the lowest constant c such that inequality

$$|\{x \in \mathbb{R}^d : M_f(x) > \delta\}| \leq c\frac{\|f\|_1}{\delta} \tag{3.18}$$

holds for all $\mathcal{L}^1(\mathbb{R}^d)$ and all $\delta > 0$. Theorem 3.46 states that $c_1 = \frac{11+\sqrt{61}}{12}$. We know that $c_1 \leq c_2 \leq c_3 \leq \ldots$ [25], but no exact value of c_d for $d > 1$ is currently known.

In 1983, Stein and Strömberg [1155] asked whether inequality (3.18) holds with some constant c independent of the dimension. Equivalently, the question is whether the sequence c_d is bounded. Stein and Strömberg proved the upper bound $c_d \leq Cd \log d$ for some constant C. In 1992, Menarguez and Soria [875] proved the lower bound $c_d \geq \left(\frac{1+2^{1/d}}{2}\right)^d$ for all d, but this does not answer the question of Stein and Strömberg, because $\left(\frac{1+2^{1/d}}{2}\right)^d < 2$ for all d.

In 2011, Aldaz [24] answered this question by proving that $\lim_{d \to \infty} c_d = \infty$.

Theorem 3.47 *For any fixed $T > 0$, there exists a $D = D(T)$ such that for every dimension $d \geq D$, one has $c_d \geq T$.*

The Poincaré Inequalities

Let $\Omega \subset \mathbb{R}^n$ be a domain bounded in at least one direction. Let $H^1(\Omega)$ be the set of functions $u : \Omega \to \mathbb{R}$ for which the gradient ∇u exists on Ω and both

$$||u||_{L^2(\Omega)} = \sqrt{\int_\Omega u(x)^2 dx} \quad \text{and} \quad ||\nabla u||_{L^2(\Omega)} = \sqrt{\int_\Omega ||\nabla u(x)||^2 dx}$$

are finite. In 1980, Poincaré [988] proved the existence of a constant c_Ω, depending on Ω, such that the inequality

$$||u||_{L^2(\Omega)} \leq c_\Omega ||\nabla u||_{L^2(\Omega)} \tag{3.19}$$

holds for all $u \in H^1(\Omega)$ such that $\int_\Omega u(x) dx = 0$. This result, known as the *Poincaré inequality*, is useful because it allows us to obtain bounds on a function using bounds on its derivatives. However, for general domains, it is unclear how to explicitly express the constant c_Ω in terms of parameters of the set Ω. In 1960, Payne and Weinberger [969] published a theorem stating that if $\Omega \subset \mathbb{R}^n$ is a convex set with diameter d, then the Poincaré inequality (3.19) holds with $c_\Omega = \frac{d}{\pi}$. However, their proof is correct only for $n = 2$. In 2003, Bebendorf [108] proved this result for all n.

Theorem 3.48 *For any convex domain $\Omega \subset \mathbb{R}^n$ with diameter d, inequality (3.19) holds with $c_\Omega = \frac{d}{\pi}$ for all $u \in H^1(\Omega)$ such that $\int_\Omega u(x) dx = 0$.*

The proof of Theorem 3.48 is a modification of the argument in [969], which fixes the gap in their argument.

Another inequality named after Poincaré controls the size of (weighted) deviations of Lipschitz continuous functions. A function $f : \mathbb{R}^n \to \mathbb{R}$ is called *Lipschitz* if $|f(x) - f(y)| \leq L||x - y||$ for all $x, y \in \mathbb{R}^n$ and some constant L. Let

$$\operatorname{Lip} f(x) = \limsup_{y \to x} \frac{|f(x) - f(y)|}{||x - y||}.$$

For a function $w : \mathbb{R}^n \to \mathbb{R}$ (which we call a *weight function*) denote by $|S|_w = \int_S w(x) dx$ the "weighted volume" of a set $S \subset \mathbb{R}^n$. For a function $f : \mathbb{R}^n \to \mathbb{R}$ denote by

$$f_{S,w} = \frac{1}{|S|_w} \int_S f(x) w(x) dx$$

its weighed average on S, and

$$\Delta(f, S, w) = \frac{1}{|S|_w} \int_S |f(x) - f_{S,w}| w(x) dx$$

its weighted deviation from the average. We say that a weight function $w : \mathbb{R}^n \to \mathbb{R}$ is *p-admissible*, $p \geq 1$, if:

(i) there is a constant $C \geq 1$ such that $|B(x, 2r)|_w \leq C|B(x, r)|_w$ holds for all
     $x \in \mathbb{R}^n$ and $r > 0$, where $B(x, r) = \{y \in \mathbb{R}^n : ||y - x|| < r\}$ is the ball with
     centre x and radius $r > 0$,
(ii) $0 < |B|_w < \infty$ for every ball B, and
(iii) there are constants $C' \geq 1$ and $0 < t \leq 1$ such that the inequality

$$\Delta(f, B(x_0, tr), w) \leq 2rC' \left(\frac{1}{|B|_w} \int_B (\operatorname{Lip} f(x))^p w(x) dx \right)^{1/p} \qquad (3.20)$$

holds for all balls $B = B(x_0, r)$, and for every Lipschitz continuous function
$f : \mathbb{R}^n \to \mathbb{R}$.

Inequality (3.20) is called the $(1, p)$-*Poincaré inequality*. It controls the size of weighted deviations of Lipschitz continuous functions f in terms of $\operatorname{Lip} f(x)$, and the smaller p the better control, because the right-hand side of (3.20) is non-decreasing in p. This motivates the question to find, for a weight function w, the smallest p such that w is p-admissible. However, Keith and Zhong [700] proved in 2008 that if w is not 1-admissible then such smallest p does not exist.

Theorem 3.49 *If $w : \mathbb{R}^n \to \mathbb{R}$ is a p-admissible weight for some $p > 1$, then there exists an $\epsilon > 0$ such that w is q-admissible for every $q > p - \epsilon$.*

The Trudinger–Moser Inequality
Let $\Omega \subset \mathbb{R}^2$ be bounded, and let

$$||u||_D := \left(\int_\Omega \left(|\nabla u(x)|^2 \right) dx \right)^{1/2}$$

denote the Dirichlet norm of a differentiable function $u : \Omega \to \mathbb{R}$. In analysis, it is important to understand what is the fastest-growing function g such that

$$\sup_{||u||_D \leq 1} \int_\Omega g(u(x)) dx < \infty.$$

The answer is given by the classical *Trudinger–Moser inequality* [912, 1214], stating that

$$\sup_{||u||_D \leq 1} \int_\Omega \left(e^{\alpha u^2} - 1 \right) dx = c(\Omega) < \infty$$

for $\alpha = 4\pi$, but the supremum becomes infinite for $\alpha > 4\pi$. However, the constant $c(\Omega)$ depends on the domain, grows to infinity if the area of Ω grows, and becomes infinite for unbounded domains.

In 2005, Ruf [1072] proved that with the Sobolev norm

$$||u||_S := \left(\int_\Omega \left(|\nabla u(x)|^2 + |u(x)|^2 \right) dx \right)^{1/2}$$

in place of the Dirichlet norm, the Trudinger–Moser inequality holds with constant independent of Ω, and therefore is valid for all domains, including the unbounded ones.

Theorem 3.50 *There exists a constant $d > 0$ such that for any $\Omega \subset \mathbb{R}^2$, one has*

$$\sup_{||u||_S \leq 1} \int_\Omega \left(e^{4\pi u^2} - 1 \right) dx \leq d.$$

3.7 Differentiability

Differentiability on the Real Line

At the beginning of the nineteenth century, some scientists believed that every continuous function f should be differentiable everywhere except on a set of isolated points. Moreover, it was believed that the real line can be partitioned into countably many intervals on which the function f is monotone. In 1872, Weierstrass [1251] constructed an example of a function $f : \mathbb{R} \to \mathbb{R}$ which is continuous at every $x \in \mathbb{R}$ but nowhere monotone[12] and nowhere differentiable. In 1915, Denjoy [345] constructed an example of a function $f : \mathbb{R} \to \mathbb{R}$ that is differentiable at every $x \in \mathbb{R}$ but still nowhere monotone.

These examples were initially considered as "exotic". However, Gurariy [544] proved in 1966 that the Weierstrass example is quite "typical" in a well-defined sense. Let $C(\mathbb{R})$ be the set of all functions $f : \mathbb{R} \to \mathbb{R}$ that are continuous at every $x \in \mathbb{R}$. A subset $S \subset C(\mathbb{R})$ is called *lineable* in $C(\mathbb{R})$ if $S \cup \{0\}$ contains an infinite-dimensional vector space. Gurariy proved that the set of nowhere differentiable functions is lineable in $C(\mathbb{R})$. In 2005, Aron et al. [48] proved the same result for differentiable but nowhere monotone functions, thus showing that Denjoy's example is also "typical".

Theorem 3.51 *There exists an infinite-dimensional vector space of differentiable functions on $\mathbb{R}$, every non-zero element of which is nowhere monotone. In other*

---

[12] A function $f : \mathbb{R} \to \mathbb{R}$ is called *nowhere monotone* if there are no real numbers $a < b$ such that f is monotone on (a, b).

words, the set of differentiable functions $f : \mathbb{R} \to \mathbb{R}$ *that are nowhere monotone is lineable in* $C(\mathbb{R})$.

To prove Theorem 3.51, the authors developed point-wise analogues of classical results about differentiability of uniformly convergent series, see [48] for details.

Differentiability in $\mathbb{R}^n$

Let n and m be positive integers. A function of several real variables $f : \mathbb{R}^n \to \mathbb{R}^m$ is called *differentiable* at a point $x_0 \in \mathbb{R}^n$ if there exists a linear map $D : \mathbb{R}^n \to \mathbb{R}^m$ such that

$$\lim_{h \to 0} \frac{||f(x_0 + h) - f(x_0) - D(h)||_m}{||h||_n} = 0,$$

where $|| \cdot ||_m$ and $|| \cdot ||_n$ are the norms in $\mathbb{R}^m$ and $\mathbb{R}^n$, respectively. A function $f : \mathbb{R}^n \to \mathbb{R}^m$ is called *Lipschitz* if there exists a constant $K \geq 0$ such that $||f(x) - f(y)||_m \leq K||x - y||_n$ for all $x, y \in \mathbb{R}^n$. The classical *Rademacher theorem* states that Lipschitz functions are differentiable almost everywhere. In other words, if a Lipschitz function $f : \mathbb{R}^n \to \mathbb{R}^m$ is differentiable at no point of a set $A \subset \mathbb{R}^n$, then A must be Lebesgue null (have Lebesgue measure 0). The natural converse of this statement asks: given a Lebesgue null set $A \subset \mathbb{R}^n$, does there exist a Lipschitz function $f : \mathbb{R}^n \to \mathbb{R}^m$ which is differentiable at no point of A? In 2010, Alberti, Csörnyei and Preiss [22] proved that the answer is "Yes" for $m \geq n = 2$. In 2011, Csörnyei and Jones [313] proved that the "Yes" answer remains valid for any $m \geq n$. In 2015, Preiss and Speight [1007] proved that in the remaining case $m < n$ the answer is "No".

Theorem 3.52 *For all integers* $n > m \geq 1$, *there exists a Lebesgue null set* $N \subseteq \mathbb{R}^n$ *such that every Lipschitz function* $f : \mathbb{R}^n \to \mathbb{R}^m$ *is differentiable at some point* $x_0 \in N$.

Differentiability in Banach Spaces

The differentiability of Lipschitz functions can be studied in a more general setting. Let X, Y be Banach spaces with norms $|| \cdot ||_X$ and $|| \cdot ||_Y$, respectively. A function $f : X \to Y$ is called *Lipschitz continuous* if there is a constant $K \geq 0$ such that $||f(x) - f(y)||_Y \leq K||x - y||_X$ for all $x, y \in X$. A function $f : X \to Y$ is called *Fréchet differentiable* at $x_0 \in X$ if there is a bounded linear function $T : X \to Y$ such that

$$f(x_0 + u) = f(x_0) + T(u) + o(||u||_X) \quad \text{as} \quad u \to 0.$$

Rademacher's theorem states that if X and Y are finite-dimensional, then Lipschitz functions are differentiable Lebesgue almost everywhere. There is no analogue of Lebesgue measure for infinite-dimensional spaces, but there are other ways to formalize the intuition that Lipschitz functions are differentiable at "many" points. For example, Rademacher's theorem implies that if X is finite-dimensional then every countable collection of Lipschitz functions on X has a common point of differentiability. It is an intriguing open question whether this result can be extended to infinite-dimensional spaces. However, before 2003, there was not even a single example of an infinite-dimensional Banach space X in which this property was known to hold. In 2003, Lindenstrauss and Preiss [799] established the existence of such examples.

Theorem 3.53 *There exists infinite-dimensional Banach spaces X and Y such that every sequence $f_1, f_2, \ldots$ of Lipschitz continuous functions from X to Y has a point $x_0 \in X$ such that all f_i are Fréchet differentiable at x_0.*

To prove Theorem 3.53, Lindenstrauss and Preiss [799] established explicit sufficient conditions for X and Y which guarantee that they satisfy the conditions of Theorem 3.53. This allowed them to provide *specific examples* of such X and Y. For example, their conditions are true if X is the space of all infinite sequences of real numbers converging to 0 with the supremum norm and $Y = \mathbb{R}$. The authors then proved that their conditions are also true for some examples when X and Y are both infinite-dimensional, and Theorem 3.53 followed.

A key tool in the proof is the introduction of a family of subsets of X called the Γ-*null sets*, and proving that (i) every Lipschitz continuous function $f : X \to Y$ is Fréchet differentiable outside a Γ-null set, and (ii) any countable union of Γ-null sets is again a Γ-null set, and, in particular, cannot cover the whole of X. Theorem 3.53 then follows immediately from the properties (i) and (ii).

3.8 Differential and Integral Equations

The Newtonian n-Body Problem

Many important problems in physics can be reduced to solving systems of differential equations. One of the most famous examples is the problem of describing the motion of n bodies under gravitation (the *n-body problem*) in space or the plane. Let n point particles with masses $m_i > 0$ and positions $x_i \in \mathbb{R}^d$ move according to Newton's laws of motion:

$$m_j \frac{d^2 x_j}{dt^2} = \sum_{i \neq j} \frac{m_i m_j (x_i - x_j)}{r_{ij}^3}, \qquad 1 \leq j \leq n, \qquad (3.21)$$

where r_{ij} is the distance between x_i and x_j. In general, the solution to (3.21) can be very complicated even for the plane ($d = 2$). An interesting research direction is to at least classify "nice" solutions, known as relative equilibria. A *relative equilibrium* motion in $\mathbb{R}^2$ is a solution to (3.21) of the form $x_i(t) = R(t)x_i(0)$ where $R(t)$ is a uniform rotation with constant angular velocity $v \neq 0$ around some point $c \in \mathbb{R}^2$. Two relative equilibria are *equivalent* if they are related by rotations, translations, and dilations in the plane. The question of whether, for a fixed $n \geq 3$, the number of relative equilibria is finite is a major open problem asked by Chazy [267] in 1916, which was included as problem 6 in Smale's list of problems for the twenty-first century [1137]. For $n = 3$, the solution has been known for centuries: in this case, there are 5 relative equilibria, two equilateral triangles described by Lagrange [757] in 1772, and three collinear configurations discovered by Euler [435] in 1767. In 2006, Hampton and Moeckel [560] resolved this problem for $n = 4$.

Theorem 3.54 *For $n = 4$, there is only a finite number of equivalence classes of relative equilibria, for any positive masses m_1, m_2, m_3, m_4.*

To prove Theorem 3.54, the authors found a way to replace the original system of equations to many "reduced systems" which are easier to analyse. The number of reduced systems is unmanageable for human analysis, but small enough for computer calculations. All the computations are either symbolic or integer-based, and therefore completely rigorous.

Another fundamental open problem about the system of Newton equations (3.21) is: under what initial conditions is the solution to (3.21) well-defined for all $t \geq 0$? The solution may not be well-defined for two reasons: (a) a collision happened, or (b) there are no collisions but a point escapes to infinity in a finite time. Case (b) is known as *non-collision singularity*. In 1897, Painlevé [960] proved that there are no such singularities for $n = 3$, but conjectured their existence for every fixed $n > 3$. In 1992, Xia [1277] proved this conjecture for motions in $\mathbb{R}^3$ for $n \geq 5$. In 2020, Xue [1278] proved that non-collision singularities exist even for the motion of $n = 4$ bodies on the plane. This finishes the proof of the Painlevé conjecture.

Theorem 3.55 *For $d = 2$ and $n = 4$, there is a non-empty set of initial conditions such that all four points moving according to Newtonian law (3.21) escape to infinity in a finite time, avoiding collisions.*

Another research direction in the study of the n-body problem (3.21) is the analysis of hyperbolic motions. The motion of particles is determined by the initial conditions: their masses and their positions and velocities at time $t = 0$. Let Ω be the set of initial configurations such that the motion has no collisions ($r_{ij}(t) > 0$ for all i, j and t) and is well-defined for all $t \geq 0$. A motion is called *hyperbolic* if each particle has a different limit velocity vector, that is,

$$\lim_{t \to \infty} \frac{dx_j}{dt} = a_j \in \mathbb{R}^d$$

and $a_i \neq a_j$ whenever $i \neq j$. The only explicitly known hyperbolic motions are such that the shape of the configuration does not change with time, but it is conjectured that there are only finitely many such motions for any fixed n, see Theorem 3.54 and the discussion before it. The set of all hyperbolic motions is much larger. In 2020, Maderna and Venturelli [825] proved that hyperbolic motions exist for all initial configurations and all choices of the limited velocities.

Theorem 3.56 *For the Newtonian n-body problem (3.21) in a space E of dimension $d \geq 2$, there are hyperbolic motions $x : [0; +\infty) \to E^n$ such that $x(t) = \sqrt{2h}ta +$*

o(t) as t → ∞ *for any choice of* $x_0 = x(0) \in E^n$, *for any* $a = (a_1, \ldots, a_n) \in \Omega$ *normalized by* $\|a\| = 1$, *and for any constant* $h > 0$.

De Giorgi's Conjecture

In addition to the system (3.21), there are many other differential equations and system of equations which arise from physics. A notable example is the *Allen–Cahn equation*

$$\Delta u = u^3 - u, \tag{3.22}$$

where $u : \mathbb{R}^n \to \mathbb{R}$ and $\Delta u = \sum_{i=1}^{n} \frac{\partial^2 u}{\partial x_i^2}$ is its Laplacian. Equation (3.22) arises in the Ginzburg–Landau model of phase transitions, and is important and well-studied. In 1978, De Giorgi [332] conjectured that if $u : \mathbb{R}^n \to \mathbb{R}$ is a solution to this equation satisfying

(i) $|u| < 1$, and
(ii) $\frac{\partial u}{\partial x_n} > 0$ for every $x = (x_1, \ldots, x_n) \in \mathbb{R}^n$,

then all level sets[13] of u are hyperplanes, at least if $n \leq 8$. This conjecture, if true, allows us to write down a formula for u easily. In 1998, Ghoussoub and Gui [497] proved the conjecture for $n = 2$, while in 2000 Ambrosio and Cabré [39] proved it for $n = 3$. In 2009, Savin [1082] proved De Giorgi's conjecture in all dimensions $n \leq 8$, under the additional assumption

(iii) $\lim_{x_n \to \pm\infty} u(x_1, \ldots, x_{n-1}, x_n) = \pm 1$ for every fixed $x_1, \ldots, x_{n-1}$.

Theorem 3.57 *Suppose that* $u : \mathbb{R}^n \to \mathbb{R}$ *is a solution to the differential equation* (3.22) *satisfying* (i), (ii), *and* (iii). *If* $n \leq 8$, *then all level sets of* u *are hyperplanes.*

Complementing Theorem 3.57, Del Pino, Kowalczyk and Wei [337] proved in 2011 that De Giorgi's conjecture fails in dimensions $n \geq 9$.

Theorem 3.58 *For any* $n \geq 9$, *there exists a solution* $u : \mathbb{R}^n \to \mathbb{R}$ *to the differential equation* (3.22) *satisfying* (i) *and* (ii) *but such that the level sets of* u *are not hyperplanes.*

The proofs of Theorems 3.57 and 3.58 explore the well-known connection between De Giorgi's conjecture and the theory of minimal surfaces, surfaces that locally minimizes their area, which we will discuss in detail in Sect. 5.7 below. A minimal graph in $\mathbb{R}^n$ is a minimal hypersurface which is also a graph of a function of $n - 1$ variables. In 1916, Bernstein proved that any minimal graph in $\mathbb{R}^3$ must be a plane. In 1968, Simons [1131] proved that every minimal graph in $\mathbb{R}^n$ is a

---

[13] A *level set* of a function $u : \mathbb{R}^n \to \mathbb{R}$ is a set of the form

$$L_c(f) = \{(x_1, \ldots, x_n) \mid u(x_1, \ldots, x_n) = c\}$$

for some constant $c \in \mathbb{R}$.

hyperplane for $n \leq 8$, and this fact was used in the proof of Theorem 3.57. However, Bombieri et al. [164] proved in 1969 that in any dimension $n \geq 9$ there exists a minimal graph Γ which is not a hyperplane. This graph Γ played a crucial role in the construction of the counterexample to De Giorgi's conjecture in Theorem 3.58.

Onsager's Conjecture

Let $\mathbb{T}$ be the interval $[0, 1)$ with arithmetic operations defined modulo 1 (the standard notation is $\mathbb{T} = \mathbb{R}/\mathbb{Z}$). Let $I \subset \mathbb{R}$ be a non-empty open interval. A pair of continuous functions $v : \mathbb{R} \times \mathbb{T}^3 \to \mathbb{R}^3$ and $p : \mathbb{R} \times \mathbb{T}^3 \to \mathbb{R}$ is called a *weak solution to the 3D incompressible Euler equations* if for all $t \in I$ and all smooth subregions $\Omega \subseteq \mathbb{T}^3$ with boundary $\partial\Omega$, one has

$$\frac{\mathrm{d}}{\mathrm{d}t} \int_{\Omega} v(t, x) \, \mathrm{d}x = \int_{\partial\Omega} v(t, x)(v \cdot n) \, \mathrm{d}S + \int_{\partial\Omega} p(t, x) n \, \mathrm{d}S, \qquad (3.23)$$

and

$$\int_{\partial\Omega} v(t, x) \cdot n(x) \, \mathrm{d}S = 0, \qquad (3.24)$$

where $\cdot$ denotes the scalar product in $\mathbb{R}^3$, $n = n(x)$ is the inward unit normal vector field on the boundary $\partial\Omega$, and $\mathrm{d}S = \mathrm{d}S(x)$ is the surface measure on the boundary. In fact, v has the physical meaning of velocity, p is the pressure, t is time, and the equations describe the movement of an incompressible fluid occupying the region Ω.

One of the central questions in this area is to understand under what conditions the value of the integral

$$\int_{\mathbb{T}^3} \frac{1}{2} |v|^2 (t, x) \, \mathrm{d}x \qquad (3.25)$$

is constant in time. This has the physical meaning of the conservation of total kinetic energy.

In 1997, Shnirelman [1121] proved that there are solutions to (3.23)–(3.24) which are compactly supported and in particular have non-constant energy (3.25). In 2000, Shnirelman [1122] proved that there are solutions where energy is a monotone decreasing function of time. More generally, De Lellis and Székelyhidi [334] proved in 2013 that there are continuous solutions to (3.23)–(3.24) whose energy (3.25) is any prescribed function of time.

Theorem 3.59 *Let $e : [0, 1] \to \mathbb{R}$ be a positive smooth function. Then there is a continuous vector field $v : \mathbb{R} \times \mathbb{T}^3 \to \mathbb{R}^3$ and a continuous scalar field $p : \mathbb{R} \times \mathbb{T}^3 \to \mathbb{R}$ satisfying (3.23)–(3.24) and such that*

$$e(t) = \int_{\mathbb{T}^3} \frac{1}{2} |v|^2 (t, x) \, \mathrm{d}x.$$

To prove Theorem 3.59, the authors developed an iteration procedure such that, during the iterations, the resulting maps are "almost solutions" of the Euler equations, which in the limit converge to the true solution with the prescribed energy.

If the energy (3.25) is not constant in general, what additional conditions on v should be imposed to guarantee energy conservation? A standard "regularity condition" is the *Hölder condition* of order α, which states that

$$|v(t, x + y) - v(t, x)| \leq C|y|^\alpha, \quad \text{for all} \quad t \in I, \; x, y \in \mathbb{R}^3, \tag{3.26}$$

for some $C \geq 0$. An old conjecture, which originates from a 1949 paper by Onsager [952], states that

(i) If $\alpha > 1/3$, then every solution to (3.23)–(3.24) satisfying (3.26) must satisfy conservation of energy: the value of the integral (3.25) does not depend on time;
(ii) conversely, for every $\alpha < 1/3$, there exist solutions to (3.23)–(3.24) satisfying (3.26) such that (3.25) fails to be constant in time.

Onsager's conjecture has been a central open question in turbulence theory. In 1994, Eyink [436] proved a slightly weaker version of part (i) of the conjecture. In the same year, Constantin et al. [299] proved part (i) in full.

Part (ii) turned out to be even more challenging. Building upon the technique introduced in [334] to prove Theorem 3.59, in 2014 De Lellis and Székelyhidi [335] made the first step by proving part (ii) for $\alpha < 1/10$. In his Ph.D. thesis [641], Isett improved this to the range $\alpha < 1/5$. Finally, in 2018, Isett [640] proved the same result for all $\alpha < 1/3$ and thus established Onsager's conjecture in full.

Theorem 3.60 *For any $\alpha < 1/3$, there is a non-zero solution to (3.23)–(3.24) satisfying (3.26) such that v is identically 0 outside a finite time interval. In particular, the solution v fails to conserve energy.*

The proof of Theorem 3.60 is based on a novel technique developed by the authors which they call "Gluing Approximation", that exploits a special structure in the linearization of Eqs. (3.23)–(3.24). See [640] for details.

Equations with Fractional Laplacian
Many important and well-studied equations are formulated using the concept of a fractional Laplacian. Let $L^1(\mathbb{R}^d)$ be the space of measurable functions $u : \mathbb{R}^d \to \mathbb{R}$ with $\int_{\mathbb{R}^d} |u(x)|dx < \infty$. For $0 < s < 1$, the *fractional Laplacian* $(-\Delta)^s$ is an operator which transforms any $u \in L^1(\mathbb{R}^d)$ into a function $(-\Delta)^s u : \mathbb{R}^d \to \mathbb{R}$ defined by

$$(-\Delta)^s u(x) = C_{d,s} \lim_{r \to 0+} \int_{\mathbb{R}^d \setminus B(x,r)} \frac{u(x) - u(y)}{|x - y|^{d+2s}} dy,$$

where $|\cdot|$ is the distance in $\mathbb{R}^d$, $B(x, r) = \{y \in \mathbb{R}^d : |x - y| < r\}$, $C_{d,s} = \frac{4^s \Gamma(d/2+s)}{\pi^{d/2} |\Gamma(-s)|}$ is the normalization factor, and $\Gamma(z) = \int_0^\infty x^{z-1} e^{-x} dx$ is the gamma function. One of the easiest and most-studied equations involving $(-\Delta)^s$ is the linear equation

$$(-\Delta)^s u + V u = 0 \quad \text{in } \mathbb{R}^d, \tag{3.27}$$

where $V : \mathbb{R}^d \to \mathbb{R}$ is a function called the *potential*. A central research direction is to establish the conditions under which Eq. (3.27) has solutions of certain types. The main difficulty is that, in contrast to the classical Laplacian Δ, the fractional Laplacian is non-local, in the sense that the value of $(-\Delta)^s u(x)$ depends not only on the values of u in a small neighbourhood of x. In particular, the well-known and very useful fact that (under suitable conditions on V) the equation $-\Delta u + V u = 0$ has at most one radial solution had no counterpart for Eq. (3.27) until recently. In 2016, such a result was proved by Frank et al. [466].

To formulate the theorem, we need some notation. A function $u : \mathbb{R}^d \to \mathbb{R}$ is called *radial* if $u(x) = u(y)$ whenever $|x| = |y|$. We say that u "vanishes at infinity" if the Lebesgue measure of $\{x \in \mathbb{R}^d : |u(x)| > \alpha\}$ is finite for every $\alpha > 0$. Let $C^{0,\gamma}(\mathbb{R}^d)$ be the set of measurable functions $u : \mathbb{R}^d \to \mathbb{R}$ such that

$$\sup_{x \in \mathbb{R}^d} |u(x)| + \sup_{x \neq y} \frac{|u(x) - u(y)|}{|x - y|^\gamma} < \infty.$$

Theorem 3.61 *Let $d \geq 1$ be an integer and let $s \in (0, 1)$. Let $V : \mathbb{R}^d \to \mathbb{R}$ be a function which is radial, non-decreasing in $|x|$, and belongs to $C^{0,\gamma}(\mathbb{R}^d)$ for some $\gamma > \max\{0, 1 - 2s\}$. Suppose that $u \in L^1(\mathbb{R}^d)$ is a radial and bounded solution of Eq. (3.27) and that u vanishes at infinity. Then $u(0) = 0$ implies that $u(x) = 0$ for all $x \in \mathbb{R}^d$.*

Together with the linearity of (3.27), Theorem 3.61 implies that, with any given initial condition $u(0) = u_0$, Eq. (3.27) has at most one bounded and radial solution that vanishes at infinity.

The Lieb Integral Equation

For a positive integer n and real number $\alpha \in (0, n)$, consider the integral equation

$$u(x) = \int_{\mathbb{R}^n} \frac{1}{|x - y|^{n-\alpha}} u(y)^{\frac{n+\alpha}{n-\alpha}} dy. \tag{3.28}$$

Equation (3.28) arose in the 1983 paper of Lieb [795] on the best possible constant in the so-called Hardy–Littlewood–Sobolev inequality. It also has a connection with a well-known family of semilinear partial differential equations. The solution u of (3.28) is called *regular* if $u \in L^{\frac{2n}{n-\alpha}}(\mathbb{R}^n)$. Lieb posed the classification of all positive regular solutions of (3.28) as an open problem. This problem was open for over 20 years, until it was fully solved by Chen et al. [269] in 2006.

Theorem 3.62 *Every positive regular solution of Eq. (3.28) has the form*

$$u(x) = c \left(\frac{t}{t^2 + |x - x_0|^2} \right)^{\frac{n-\alpha}{2}},$$

for some constant $c = c(n, \alpha)$, *some* $t > 0$ *and some* $x_0 \in \mathbb{R}^n$.

3.9 Transformations of Functions, Operators

The Fourier Transform

The term "operator" is used to name a function $h : X \to Y$ if the domain X is itself a set of functions or a set with some structure like a vector space. If X and Y are sets of functions, then h transforms functions into functions and may also be called a "transform". One of the most prominent examples with countless applications is the *Fourier transform*

$$\hat{f}(t) = \int_{-\infty}^{\infty} f(x) e^{-2\pi i t x} dx, \quad t \in \mathbb{R} \tag{3.29}$$

of an integrable function $f : \mathbb{R} \to \mathbb{C}$.

The classical *Whittaker–Shannon interpolation formula* [1111, 1257] states that if the Fourier transform $\hat{f}$ of a function $f : \mathbb{R} \to \mathbb{R}$ is supported in $[-w/2, w/2]$, then

$$f(x) = \sum_{n=-\infty}^{\infty} f\left(\frac{n}{w}\right) \frac{\sin(\pi(wx - n))}{\pi(wx - n)}.$$

This formula has numerous applications, in particular it allows us to construct a "nice" continuous function which approximates a given sequence of real numbers. However, it does not work for functions whose Fourier transform has unbounded support. In 2019, Radchenko and Viazovska [1014] derived a similar interpolation formula which can be applied to an arbitrary Schwartz function[14] on the real line.

---

[14] A function $f : \mathbb{R} \to \mathbb{R}$ is called a *Schwartz function* if there exist all derivatives $f^{(k)}(x)$ for all $k = 1, 2, 3, \ldots$ and for all $x \in \mathbb{R}$, and, for every k and $\gamma \in \mathbb{R}$, there is a constant $C(k, \gamma)$ such that $|x^\gamma f^{(k)}(x)| \leq C(k, \gamma), \forall x \in \mathbb{R}$.

Theorem 3.63 *There exists a collection of even*[15] *Schwartz functions $a_n : \mathbb{R} \to \mathbb{R}$ with the property that for any even Schwartz function $f : \mathbb{R} \to \mathbb{R}$ and any $x \in \mathbb{R}$ we have*

$$f(x) = \sum_{n=0}^{\infty} a_n(x) f(\sqrt{n}) + \sum_{n=0}^{\infty} \hat{a}_n(x) \hat{f}(\sqrt{n}),$$

where the right-hand side converges absolutely.

Theorem 3.63 has many corollaries. For example, it implies that if $f : \mathbb{R} \to \mathbb{R}$ is an even Schwartz function such that $f(\sqrt{n}) = \hat{f}(\sqrt{n}) = 0$ for $n = 0, 1, 2, \ldots$, then $f(x) = 0$ for all $x \in \mathbb{R}$. The theorem was proved by providing an explicit construction of the interpolating basis $\{a_n(x)\}_{n \geq 0}$. The authors also proved a similar result for odd Schwartz functions, and remarked that it is possible to combine the two results into a general interpolation theorem.

The Legendre Transform

Another function transformation of fundamental importance is the Legendre transform. Let $C(\mathbb{R}^n)$ denote the set of all lower-semicontinuous[16] convex functions $\phi : \mathbb{R}^n \to \overline{\mathbb{R}}$, where $\overline{\mathbb{R}} = \mathbb{R} \cup \{\pm\infty\}$. The *Legendre transform* is the map $\mathcal{L} : C(\mathbb{R}^n) \to C(\mathbb{R}^n)$ given by

$$(\mathcal{L}\phi)(x) = \sup_{y \in \mathbb{R}^n} (\langle x, y \rangle - \phi(y)),$$

where $\langle x, y \rangle = \sum_{i=1}^{n} x_i y_i$ is the scalar product in $\mathbb{R}^n$. The Legendre transform is a central tool in convex analysis and optimization, with numerous applications.

A map $T : C(\mathbb{R}^n) \to C(\mathbb{R}^n)$ is called an *order-reversing involution* if it satisfies the following two properties:

(P1) $TTf = f$ for all $f \in C(\mathbb{R}^n)$ and
(P2) $f \leq g$ implies $Tf \geq Tg$ for all $f, g \in C(\mathbb{R}^n)$.

Is it easy to check that the Legendre transform satisfies these properties. In 2009, Artstein-Avidan and Milman [56] proved a somewhat surprising converse result: any order-reversing involution must in fact be the Legendre transform, up to linear terms.

---

[15] A function $f : \mathbb{R} \to \mathbb{R}$ is called *even* if $f(-x) = f(x)$ for all $x \in \mathbb{R}$, and *odd* if $f(-x) = -f(x)$ for all $x \in \mathbb{R}$.
[16] A function $\phi : \mathbb{R}^n \to \overline{\mathbb{R}}$ is *lower-semicontinuous* if the set $\{x \in \mathbb{R}^n \mid f(x) \leq c\}$ is closed for all $c \in \mathbb{R}$.

Theorem 3.64 *For every map $T : C(\mathbb{R}^n) \to C(\mathbb{R}^n)$ (defined on the whole domain $C(\mathbb{R}^n)$) satisfying (P1) and (P2), there exists a constant $C_0 \in \mathbb{R}$, a vector $v_0 \in \mathbb{R}^n$, and an invertible symmetric linear map[17] B such that $(Tf)(x) = (\mathcal{L}f)(Bx + v_0) + \langle x, v_0 \rangle + C_0$ for all $x \in \mathbb{R}^n$.*

The Hilbert Transform
Another useful function transformation is the Hilbert transform, introduced by Hilbert in 1905. Given a function $f : \mathbb{R} \to \mathbb{R}$, its *Hilbert transform* is the function

$$Hf(x) = \lim_{\epsilon \to 0} \int_{|t| > \epsilon} \frac{f(x - t)}{t} \, dt, \quad x \in \mathbb{R}.$$

The Hilbert transform is important in signal processing and in other applications. It has various generalizations, some of which remain active areas of modern research. In particular, the *bilinear Hilbert transform*, introduced by Calderón in the 1960s, takes two Schwartz functions $f : \mathbb{R} \to \mathbb{R}$ and $g : \mathbb{R} \to \mathbb{R}$ and outputs the function

$$z_{\alpha,\beta}(x) = \lim_{\epsilon \to 0} \int_{|t| > \epsilon} f(x - \alpha t) \, g(x - \beta t) \, \frac{dt}{t}, \quad x \in \mathbb{R},$$

where $\alpha, \beta \in \mathbb{R}$ are real parameters. A long-standing open question was whether the L^p norm of $z_{\alpha,\beta}(x)$ can be bounded in terms of norms of f and g, uniformly in α and β. In 2004, such a bound was derived by Grafakos and Li [513].

Theorem 3.65 *For every $2 < p_1 < \infty$, $2 < p_2 < \infty$ such that $1 < p = \frac{p_1 p_2}{p_1 + p_2} < 2$, there is a constant $C = C(p_1, p_2)$ such that*

$$\sup_{\alpha, \beta} \|z_{\alpha,\beta}\|_p \leq C \|f\|_{p_1} \|g\|_{p_2}$$

for all Schwartz functions $f, g : \mathbb{R} \to \mathbb{R}$.

The Almost Mathieu Operator
In this section we discuss an operator which applies to sequences rather than functions. Let $l^2(\mathbb{Z})$ be the set of all infinite sequences $x = (\ldots, x_{-1}, x_0, x_1, \ldots)$ such that $\sum_{n \in \mathbb{Z}} x_n^2 < \infty$. The *almost Mathieu operator* is the map $H : l^2(\mathbb{Z}) \to l^2(\mathbb{Z})$ mapping each $x \in l^2(\mathbb{Z})$ to

$$(Hx)_n = x_{n+1} + x_{n-1} + 2\lambda \cos 2\pi(\theta + n\alpha) x_n, \quad n \in \mathbb{Z},$$

where $\lambda \neq 0$, α, and θ are real parameters, called *coupling*, *frequency*, and *phase*, respectively. The almost Mathieu operator arises from applications in physics, and a

---

[17] A map $B : \mathbb{R}^n \to \mathbb{R}^n$ is called *linear* if $B(x + y) = B(x) + B(y)$ and $B(\alpha x) = \alpha B(x)$ for all $x, y \in \mathbb{R}^n$ and $\alpha \in \mathbb{R}$, *symmetric* if $\langle Bx, y \rangle = \langle x, By \rangle$ for all $x, y \in \mathbb{R}^n$, and *invertible* if $B(x) \neq 0$ whenever $x \neq 0$.

particularly important research direction is to understand its spectrum. The *spectrum* of H is the set of all $t \in \mathbb{R}$ for which the map $H_t : l^2(\mathbb{Z}) \to l^2(\mathbb{Z})$ given by $(H_t x)_n = t x_n - (Hx)_n$, $n \in \mathbb{Z}$, is not invertible.[18] In 1964, Azbel [66] conjectured that if α is irrational then the spectrum of H is a Cantor set.[19] In 1981, Kac offered ten martinis for anyone who could prove or disprove this conjecture, and since then the problem has been known as "The Ten Martini Problem". In 2009, Avila and Jitomirskaya [61] proved the conjecture.

Theorem 3.66 *The spectrum of the almost Mathieu operator is a Cantor set for all irrational α, all real θ and all $\lambda \neq 0$.*

The Hot Spots Conjecture and Beyond

The famous *hot spots conjecture* predicts, in non-technical English, that if a flat piece of metal is given some initial heat distribution which then flows throughout the metal, then, after some time, the hottest point on the metal will lie on its boundary. To formulate the conjecture rigorously, we need some definitions. Let $\Omega \subset \mathbb{R}^2$ be a *Lipschitz domain*, that is, a bounded domain with boundary $\partial\Omega$ such that every $x \in \partial\Omega$ has a neighbourhood U such that $\partial\Omega \cap U$ is the graph of a Lipschitz function in some orthonormal coordinate system. The *second Neumann eigenvalue* $\mu_2 = \mu_2(\Omega)$ is the smallest positive real number such that there exists a not identically zero, smooth function $u : \Omega \to \mathbb{R}$ that satisfies the equation

$$\Delta u = -\mu_2 \cdot u$$

on Ω and the boundary condition $\frac{\partial u}{\partial n}\big|_{\partial\Omega} \equiv 0$ at the smooth points of $\partial\Omega$, where n denotes the outward pointing normal vector. See Eq. (3.30) below for an equivalent definition of $\mu_2(\Omega)$. A function u that satisfies these conditions is called the *second Neumann eigenfunction* for Ω. The hot spot conjecture, proposed by Rauch [1027] in 1975, predicted that the second Neumann eigenfunction attains its extrema at the boundary of Ω. In 1999, Burdzy and Werner [230] constructed a domain (non-convex, with holes) for which the conjecture fails, but it is still believed to be true for simply connected domains, and in particular for convex domains.

In 2004, Atar and Burdzy [57] made significant progress by proving the conjecture for domains bounded by the graphs of two Lipschitz functions with Lipschitz constant 1.

Theorem 3.67 *Let $\Omega \subset \mathbb{R}^2$ be a bounded, open, connected Lipschitz domain, given by*

$$\Omega = \{(x_1, x_2) : f_1(x_1) < x_2 < f_2(x_1)\},$$

---

[18] A map $T : l^2(\mathbb{Z}) \to l^2(\mathbb{Z})$ is called *invertible* if for every $y \in l^2(\mathbb{Z})$ there exists a unique $x \in l^2(\mathbb{Z})$ such that $T(x) = y$.

[19] See the discussion before Theorem 3.7 for the definition of Cantor sets.

where f_1, f_2 are Lipschitz functions with constant 1. Then every Neumann eigenfunction corresponding to $\mu_2(\Omega)$ attains its maximum and minimum at boundary points only.

Theorem 3.67 is significant progress, but it leaves uncovered many very simple domains. One of the simplest non-trivial examples of a bounded domain is a triangle, and even in this case the hot spot conjecture turned out to be highly non-trivial. In 1999, Banuelos and Burdzy [95] proved the conjecture for obtuse triangles, while Theorem 3.67 implies it for right triangles. In 2012, the conjecture for acute triangles was the topic of a polymath project with a large number of participants. In 2015, Siudeja [1132] proved the conjecture for acute triangles with at least one angle less than $\pi/6$. Culminating this line of research, Judge and Mondal [666] in 2020 proved the hot spots conjecture for all triangles.

Theorem 3.68 *For any triangle Ω, the second Neumann eigenfunction attains its extrema at the boundary of Ω.*

Neumann eigenvalues can be defined for more general domains in any dimension. Let $n \geq 2$, and let $\Omega \subset \mathbb{R}^n$ be a bounded open set. Let $L^2(\Omega)$ be the set of functions $u : \Omega \to \mathbb{R}$ with norm $\|u\|_2 := \left(\int_\Omega |u(x)|^2 dx\right)^{1/2} < \infty$. Let $H^1(\Omega)$ be the set of differentiable functions $u : \Omega \to \mathbb{R}$ with norm $\|u\|_H := \left(\int_\Omega \left(|\nabla u(x)|^2 + |u(x)|^2\right) dx\right)^{1/2} < \infty$. We say that a domain Ω is *regular* if $H^1(\Omega)$ is compactly embedded in $L^2(\Omega)$. Every Lipschitz domain is regular, but the class of regular domains is larger.

For every integer $k \geq 1$, let $\mathcal{S}_k$ be the family of all subspaces of dimension k in $\{u \in H^1(\Omega) : \int_\Omega u(x)dx = 0\}$, and let

$$\mu_k(\Omega) = \min_{S \in \mathcal{S}_k} \max_{u \in S} \frac{\int_\Omega |\nabla u(x)|^2 dx}{\int_\Omega |u(x)|^2 dx}. \tag{3.30}$$

The sequence $\mu_k(\Omega)$ is known as "the spectrum of the Laplace operator with Neumann boundary conditions", and is well-studied in mathematics with applications in physics. In the 1950s, Szegő [1172] and Weinberger [1252] proved that $|\Omega|^{2/n}\mu_1(\Omega)$ is maximized when Ω is a ball in $\mathbb{R}^n$. Since then, researchers have worked on the maximization problem for $|\Omega|^{2/n}\mu_k(\Omega)$ for $k > 1$. In 2019, Bucur and Henrot [218] solved this problem for $k = 2$.

If $B \subset \mathbb{R}^n$ is a ball, let $\mu_2^* = 2^{2/n}|B|^{2/n}\mu_1(B)$. This is a constant which does not depend on B.

Theorem 3.69 *Inequality $|\Omega|^{2/n}\mu_2(\Omega) \leq \mu_2^*$ holds for every regular set $\Omega \subset \mathbb{R}^n$, with equality if Ω is the union of two disjoint, equal balls.*

To prove Theorem 3.69, Bucur and Henrot constructed a set of n test functions which are orthogonal to the constant function and also to the first Neumann eigenfunction on Ω, see [218] for details.

As one of the applications, Theorem 3.69 immediately implies the $k = 2$ case of an important 1954 conjecture of Pólya [992] (known as the Pólya conjecture for the Neumann eigenvalues), which states that

$$\mu_k(\Omega) \leq 4\pi^2 \left(\frac{k}{w_n |\Omega|} \right)^{2/n},$$

where w_n is the volume of the unit ball in $\mathbb{R}^n$.

3.10 Harmonic Analysis

Harmonic Functions and Their Zero Sets
Let V be an open subset of $\mathbb{R}^n$. A function $f : V \to \mathbb{R}$ is called *harmonic* if

$$\frac{\partial^2 f}{\partial x_1^2} + \cdots + \frac{\partial^2 f}{\partial x_n^2} = 0$$

for every $x \in V$. Harmonic functions are studied in many areas of mathematics, such as mathematical physics and the theory of stochastic processes. A central research direction is to understand the zero set $\{x \in V : f(x) = 0\}$ of a harmonic function f. In 1997, Nadirashvili [919] conjectured that every harmonic function f in the unit ball $B^n = \{x \in \mathbb{R}^n : ||x|| \leq 1\}$ such that $f(0) = 0$ has a zero set whose $n - 1$ dimensional Hausdorff measure[20] H^{n-1} is bounded away from 0 by a constant independent of f. In 2018, Logunov [810] proved this conjecture.

Theorem 3.70 *For every dimension $n \geq 3$, there exists a constant $c = c(n) > 0$, depending only on n, such that the inequality*

$$H^{n-1}(\{f = 0\} \cap B^n) \geq c$$

holds for every harmonic function $f : B^n \to \mathbb{R}$ such that $f(0) = 0$.

In particular, Theorem 3.70 implies that the zero set of any non-constant harmonic function $h : \mathbb{R}^3 \to \mathbb{R}$ has an infinite area. Also, Theorem 3.70 is an important step towards proving a more general conjecture of Yau [1282] made in 1982, which predicts a similar result on n-dimensional curved spaces (called "C^∞-smooth Riemannian manifolds") in place of $\mathbb{R}^n$.

---

[20] See (1.22) for the definition of the Hausdorff measure. In fact, for integer n, it coincides with the Lebesgue measure, up to a constant factor.

Harmonic Functions on Graphs

Harmonic functions can also be defined in the discrete setting, on graphs. Let G be a connected graph with infinite vertex set V. For each pair $x, y \in V$ we define a *conductance* $C_{xy} \geq 0$ such that $C_{xy} = C_{yx}$ and also $C_{xy} = 0$ unless x and y are connected by an edge. The graph G together with the conductances C_{xy} is denoted by (G, C) and called a *weighted graph*. For $x \in V$, let $\mu_G(x) = \sum_{y \in V} C_{xy}$. For $A \subset V$, let $\mu_G(A) = \sum_{x \in A} \mu_G(x)$. A function $h : A \to \mathbb{R}$ is called *harmonic* on A if

$$h(x) = \sum_{y \in V} h(y) C_{xy}$$

for all $x \in A$.

For $x \in V$ and $r > 0$, let $B_G(x, r)$ be the set of $y \in V$ with $d_G(x, y) < r$, where $d_G(x, y)$ is the length of the shortest path connecting vertices x and y in graph G. We say that the *elliptic Harnack inequality* holds for (G, C) if there exists a c_1 such that whenever $x_0 \in V$, $r \geq 1$, and h is non-negative and harmonic in $B_G(x_0, 2r)$, then $h(x) \leq c_1 h(y)$ for all $x, y \in B_G(x_0, r)$. The elliptic Harnack inequality was proved by Moser [911] in 1961 in the context of partial differential equations, but since then it has turned out to be useful in many other applications, including weighted graphs. In 2018, Barlow and Murugan [99] proved that this inequality is stable under rough isometries, resolving a long-standing open question.

We say that weighed graphs (G, C) and (H, C') with vertex sets V_G and V_H are *roughly isometric* if there is a function $\phi : V_G \to V_H$ and constants $C_1 > 0$ and $C_2, C_3 > 1$ such that

(i) for every $y \in V_H$ there exists a $x \in V_G$, such that $d_H(y, \phi(x)) \leq C_1$,
(ii) $C_2^{-1}(d_G(x, y) - C_1) \leq d_H(\phi(x), \phi(y)) \leq C_2(d_G(x, y) + C_1)$ for all $x, y \in V_G$, and
(iii) $C_3^{-1} \mu_G(B_G(x, r)) \leq \mu_H(B_H(\phi(x), r)) \leq C_3 \mu_G(B_G(x, r))$ for all $x \in V_G$ and $r > 0$.

Theorem 3.71 *Let (G, C) and (H, C') be two connected bounded degree weighed graphs that are roughly isometric. Then the elliptic Harnack inequality holds for (G, C) if and only if it holds for (H, C').*

While we state Theorem 3.71 only for weighed graphs here, Barlow and Murugan actually proved it in much more general setting, see [99] for details.

The Restriction Conjecture for Paraboloids

For a function $f : \Omega \to \mathbb{C}$ defined on $\Omega \subseteq \mathbb{R}^n$, let $||f||_{L^p(\Omega)} = \left(\int_\Omega |f(\omega)|^p d\omega \right)^{1/p}$, where $p \geq 1$. Let B^n denote the unit ball in $\mathbb{R}^n$. The *extension operator for the paraboloid* is the operator E which puts to every function $f : B^{n-1} \to \mathbb{C}$ with $||f||_{L^p(B^{n-1})} < \infty$ into correspondence with a function $E_f : \mathbb{R}^n \to \mathbb{C}$ given for every $x = (x_1, \ldots, x_n)$ by

$$E_f(x) = \int_{B^{n-1}} e^{i(x_1 w_1 + \cdots + x_{n-1} w_{n-1} + x_n |w|^2)} f(w) dw.$$

In 1979, Stein [1154] conjectured that the inequality

$$||E_f||_{L^p(\mathbb{R}^n)} \leq C_p ||f||_{L^p(B^{n-1})}$$

holds whenever $p > \frac{2n}{n-1}$. This conjecture became known as "the restriction conjecture for paraboloids", and attracted significant attention. In 2018, Guth [547] proved this conjecture under slightly more restrictive assumptions on p.

Theorem 3.72 *Let $n \geq 2$ and let $p > 2\frac{3n+1}{3n-3}$ if n odd and $p > 2\frac{3n+2}{3n-2}$ if n is even. Then there exists a constant $C = C(p)$ such that $||E_f||_{L^p(\mathbb{R}^n)} \leq C||f||_{L^p(B^{n-1})}$ for every f.*

The proof of Theorem 3.72 proceeds by showing that if $||E_f||_{L^p(\mathbb{R}^n)}$ is large, then the region where $|E_f|$ is large has an algebraic structure: it consists of thin neighbourhoods of low-degree algebraic varieties. This structure was found by using the polynomial partitioning method, which was introduced by Katz and Guth [549] in 2015 to prove Theorem 5.18 discussed later in this book.

3.11 Banach Spaces

The Diameter of a Banach Space

A *Banach space* is a complete vector space B with a norm[21] $||\cdot||_B$. Banach spaces B and B' are called *isomorphic* if there exist a bijection $f : B \to B'$, which preserves addition and multiplication by constants, and such that

$$m||x||_B \leq ||f(x)||_{B'} \leq M||x||_B, \quad \forall x \in B$$

for some constants $m > 0$ and $M > 0$. Assuming that the constants m and M are chosen to be best possible, we denote by $d_f(B, B')$ the ratio M/m. Let $d(B, B')$ be the infimum of $d_f(B, B')$ over all f satisfying the conditions above. The diameter $D(B)$ of a Banach space B is the supremum of $d(B', B'')$ over all Banach spaces B' and B'' isomorphic to B.

In 1981, Gluskin [499] proved the existence of a constant $c > 0$ such that the inequality $cN \leq D(B)$ holds for every Banach space B of finite dimension N (it is also known that $D(B) \leq N$). From this, it is natural to conjecture that if the dimension N is infinite, then $D(B)$ should be infinite as well. In fact, this problem was raised by Schäffer [1085] in 1976, even before Gluskin's result. In

---

[21] Recall that a *norm* is a function $B \to \mathbb{R}$ such that (i) $||x||_B \geq 0$ for all $x \in B$, (ii) $||x||_B = 0$ if and only if $x = 0$, (iii) $||\alpha x||_B = |\alpha|||x||_B$ for every scalar α, and (iv) $||x + y||_B \leq ||x||_B + ||y||_B$ for all $x, y \in B$. A normed space B is called *complete* if every Cauchy sequence converges to some limit in B.

2005, Johnson and Odell [662] resolved this problem for all separable[22] Banach spaces.

Theorem 3.73 *If B is a separable infinite-dimensional Banach space, then* $D(B) = \infty$.

An important step in the proof of Theorem 3.73 is the proof that if a separable infinite-dimensional Banach space B with $D(B) < \infty$ existed, it would contain an isomorphic copy of c_0 (the space of sequences converging to 0 with the supremum norm). This is where the separability assumption is used in a crucial way.

The l^2-Space

The simplest example of an infinite-dimensional Banach space is the l^2-*space*: the space of infinite sequences $x = (x_1, x_2, \ldots, x_n, \ldots)$ equipped with coordinate-wise addition and scalar multiplication and norm $|x| := \sqrt{\sum_{i=1}^{\infty} x_i^2} < \infty$. The l^2 space is a Hilbert space with inner product $\langle x, y \rangle = \sum_{i=1}^{\infty} x_i y_i$. In fact, it is (up to isomorphism) the only separable infinite-dimensional Hilbert space.

It is easy to see that the l^2-space is isomorphic to every infinite-dimensional closed subspace of itself. In his famous 1932 book [92], Banach asked whether this is the only such example. In 2002, Gowers [509] gave a positive answer to this old question.

Theorem 3.74 *The l^2-space is (up to isomorphism) the only infinite-dimensional Banach space which is isomorphic to every infinite-dimensional closed subspace of itself.*

An important ingredient in the proof of Theorem 3.74 is the development of a theory resembling the Ramsey theory discussed in Sect. 2.1 and 2.7 in the setting of infinite-dimensional Banach spaces.

Another remarkable property of l^2 is that it is a HAPpy space. A Banach space B is said to have the *approximation property* (AP) if for every compact set K in B and for every $\epsilon > 0$, there is a bounded linear operator $T : B \to B$, whose range is finite-dimensional, and such that $\|Tx - x\|_B \le \epsilon$ for all $x \in K$. We say that a Banach space B has the *hereditary AP* (or is a *HAPpy space*) if all of its subspaces have the AP.

For some time, the l^2-space was the only known example of a HAPpy space and there was a conjecture that no other HAPpy spaces exist. The first examples of HAPpy Banach spaces not isomorphic to a Hilbert space were constructed by Johnson [661] in 1980. However, unlike l^2, these examples do not have a symmetric basis. A sequence $\{x_n\}$ in a Banach space B is called a *basis* of B if every $x \in B$ has a unique representation of the form $x = \sum_{n=1}^{\infty} a_n x_n$ for some real numbers a_n. Two bases $\{x_n\}$ and $\{y_n\}$ of B are called *equivalent* if the series $\sum_{n=1}^{\infty} a_n x_n$ converges in B if and only if $\sum_{n=1}^{\infty} a_n y_n$ converges. A basis $\{x_n\}$ of B is called *symmetric* if

---

[22] A Banach space B is called *separable* if it contains a sequence $x_1, x_2, \ldots, x_n, \ldots$ such that for any $x \in B$ and any $\epsilon > 0$ there exists an n such that $\|x - x_n\|_B < \epsilon$.

every permutation of $\{x_n\}$ is a basis of B equivalent to $\{x_n\}$. Johnson concluded his 1980 paper [661] with the question of whether there are HAPpy spaces other than l^2 which have a symmetric basis.

In 1988, Pisier [986] constructed a large family of examples of HAPpy spaces not isomorphic to l^2, but none of them has a symmetric basis, so Johnson's question remained open. In 2012, Johnson and Szankowski [663] resolved this question affirmatively.

Theorem 3.75 *There exists a HAPpy Banach space, not isomorphic to the Hilbert space l_2, which has a symmetric basis.*

Types in Banach Spaces
Let $\epsilon_1, \epsilon_2, \ldots$ be a sequence of independent random variables, each equal to $+1$ or -1 with equal probabilities. We say that a Banach space X has *Rademacher type* $p \in [1, 2]$ if there exists a constant $C \in (0, \infty)$ such that the inequality

$$E \left\| \sum_{i=1}^{n} \epsilon_j x_j \right\|^p \leq C^p \sum_{j=1}^{n} \|x_j\|^p$$

holds for all $n \geq 1$ and all $x_1, \ldots, x_n \in X$, where E denotes the expectation. The notion of Rademacher type is one of the central and most useful concepts in the theory of Banach spaces, and researchers have long tried to extend it to general metric spaces. However, the definition of Rademacher type uses addition in X substantially, and is difficult to extend. In 1978, Enflo [391] introduced a definition of type which is straightforward to extend to metric spaces, and conjectured that, for Banach spaces, the two notions of type coincide. For $\epsilon = (\epsilon_1, \ldots, \epsilon_n)$, and index j, denote by ϵ^{j-} the vector $(\epsilon_1, \ldots, -\epsilon_j, \ldots, \epsilon_n)$. We say that a Banach space X has *Enflo type p* if there exists a constant $C \in (0, \infty)$ such that the inequality

$$E\|f(\epsilon) - f(-\epsilon)\|^p \leq C^p \sum_{j=1}^{n} E\|f(\epsilon) - f(\epsilon^{j-})\|^p$$

holds for all $n \geq 1$ and every function $f : \{-1, 1\}^n \to X$. If a Banach space X has Enflo type p, then, applying the definition to the function $f(\epsilon) = \sum_{j=1}^{n} \epsilon_j x_j$, it is easy to see that X has Rademacher type p as well. However, the converse direction of Enflo's conjecture remained open for decades, despite intensive study and a number of partial results. In 2020, Ivanisvili et al. [642] resolved the conjecture in full.

Theorem 3.76 *A Banach space X has Rademacher type $p \in [1, 2]$ if and only if X has Enflo type p.*

The proof of Theorem 3.76 is based on a novel version of *Pisier's inequality*. The original version, discovered by Pisier [985] in 1986, states that, for every $n \geq 1$, there is a constant $C = C(n)$, such that for every function $f : \{-1, 1\}^n \to X$,

$$E\|f(\epsilon) - Ef(\epsilon)\|^p \leq C^p E \left\| \sum_{i=1}^n \delta_j \frac{f(\epsilon) - f(\epsilon^{j-})}{2} \right\|^p, \qquad (3.31)$$

where ϵ, δ are independent random vectors in $\mathbb{R}^n$ whose coordinates are independent and equal to $+1$ or -1 with equal probabilities. If (3.31) held with constant C independent from n, Theorem 3.76 would follow trivially, but Talagrand [1178] proved in 1993 that this is not the case. Ivanisvili, van Handel and Volberg established a version of Pisier's inequality which does hold with constant independent of n and deduced Theorem 3.76 from it.

Fixed Point Theorems

One of the cornerstone theorems in the whole of mathematics, with countless applications, is *Banach's fixed point theorem* proved by Banach [91] in 1922. It states that if (X, d) is a non-empty complete[23] metric space and $f : X \to X$ is such that

(*) $d(f(x), f(y)) \leq c \cdot d(x, y)$, $\forall x, y$ for some $c \in (0, 1)$,

then f has a unique fixed point x^*, and

$$\lim_{n \to \infty} f^n(x) = x^* \quad \text{for every} \quad x \in X,$$

where $f^n(x)$ denotes $f(f(\ldots f(x)\ldots))$ with f repeated n times. After this theorem was proved, mathematicians developed fixed point theorems in various contexts, useful in different applications. Many of these theorems became central tools in the corresponding areas of mathematical research. Here we present several such examples.

In some applications, the condition (*) in Banach's original fixed point theorem does not hold. In 2012, Samet et al. [1076] developed a version of the fixed point theorem with a more flexible condition depending on some functions α and ψ. This condition reduces to (*) in the special case $\alpha(x, y) = 1$ and $\psi(t) = ct$, but, by selecting different α and ψ, one can make the new theorem applicable in many cases when Banach's original theorem does not work. As an example, the authors provide such applications to the theory of differential equations.

Formally, let Ψ be the set of all non-decreasing functions $\psi : [0, +\infty) \to [0, +\infty)$ such that $\sum_{n=1}^{\infty} \psi^n(t) < +\infty$ for all $t > 0$, where ψ^n is the n-th iterate of ψ. Let (X, d) be a complete metric space. A mapping $T : X \to X$ is called α-ψ-*contractive* if there exist two functions $\alpha : X \times X \to [0, +\infty)$ and $\psi \in \Psi$ such that

---

[23] Recall that a metric space X is called *complete* if every Cauchy sequence in X converges to some limit in X.

$\alpha(x, y)d(Tx, Ty) \leq \psi(d(x, y))$ for all $x, y \in X$. We say that T is α-admissible if for $x, y \in X$, $\alpha(x, y) \geq 1$ implies that $\alpha(Tx, Ty) \geq 1$.

Theorem 3.77 *Let (X, d) be a complete metric space and $T : X \to X$ be an α-ψ-contractive mapping satisfying the following conditions: (i) T is α-admissible; (ii) there exists an $x_0 \in X$ such that $\alpha(x_0, Tx_0) \geq 1$; and (iii) T is continuous. Then, T has a fixed point, that is, there exists an $x^* \in X$ such that $Tx^* = x^*$.*

In 2004, Ran and Reurings [1023] proved a version of Banach's fixed point theorem for partially ordered sets, in which condition (*) is required to hold only for ordered pairs x, y. Ran and Reurings demonstrated its applications to matrix equations, and subsequent authors found many more applications.

Formally, a *partially ordered set* is a set T together with a binary relation $\leq$ on T such that (i) $a \leq a$ for all $a \in T$ (reflexivity), (ii) if $a \leq b$ and $b \leq a$ then $a = b$ (antisymmetry), and (iii) if $a \leq b$ and $b \leq c$ then $a \leq c$ (transitivity). Note that for some elements $a, b \in T$ we may have neither $a \leq b$ nor $b \leq a$. A map $f : T \to T$ is called *monotone* if it is either *order-preserving* ($x \leq y$ implies $f(x) \leq f(y)$) or *order-reversing* ($x \leq y$ implies $f(y) \leq f(x)$). A point $x \in T$ is called a *fixed point* of f if $f(x) = x$.

Theorem 3.78 *Let T be a partially ordered set such that for every pair $x, y \in T$ there exist $u, v \in T$ such that $u \leq x \leq v$ and $u \leq y \leq v$. Furthermore, let d be a metric on T such that (T, d) is a complete metric space. If $f : T \to T$ is a continuous, monotone map such that*

(i) $\exists c \in (0, 1) : d(f(x), f(y)) \leq c \cdot d(x, y), \forall x \geq y$, and
(ii) $\exists x_0 \in T : x_0 \leq f(x_0)$ or $x_0 \geq f(x_0)$,

then f has a unique fixed point x^*. Moreover, $\lim_{n \to \infty} f^n(x) = x^*$ for every $x \in T$.

In 2012, Bader et al. [77] developed a version of a fixed point theorem for L^1 spaces. The authors demonstrated that it has many applications. For example, it can be used to derive the optimal solution to the so-called "derivation problem", see [77] for details and for more applications.

Let B be a Banach space with norm $|| \cdot ||$. A linear map $f : B \to B$ is called an *isometry* if $||f(x)|| = ||x||$ for all $x \in B$. For a subset $A \subset B$, we say that f preserves A if $f(x) \in A$ for all $x \in A$.

Theorem 3.79 *Let A be a non-empty bounded subset of an L^1 space B. Then there is a point in B fixed by every isometry of B preserving A. Moreover, one can choose a fixed point which minimizes $\sup_{a \in A} ||v - a||$ over all $v \in B$.*

3.12   The Embedding Theory of Metric Spaces

Embedding with Distortion

We say that a metric space (X, ρ_X) admits a *bi-Lipschitz embedding* (or just "embeds" for short) into a metric space (Y, ρ_Y) if there exist constants $m, M > 0$

and a function $f : X \to Y$ such that

$$m\rho_X(x, y) \leq \rho_Y(f(x), f(y)) \leq M\rho_X(x, y), \qquad \forall x, y \in X. \tag{3.32}$$

If (3.32) holds for some constants $m, M > 0$ such that $\frac{M}{m} \leq \alpha$, we say that (X, ρ_X) embeds into (Y, ρ_Y) with *distortion* at most α. Studying embeddings of metric spaces into each other with minimal distortion is an old topic, which has applications to algorithm development for combinatorial problems. In 1985, Bourgain [178] proved that every n-point metric space X can be embedded into $\mathbb{R}^d$ (with the usual norm) with distortion at most $C \log n$ (where the dimension d depends on n but the constant C does not). For general metric spaces, the bound $C \log n$ for the distortion is essentially sharp. To achieve better distortion, we may:

(i) embed some restricted classes of metric spaces rather than all finite metric spaces, or
(ii) embed metric spaces into a different metric space instead of $\mathbb{R}^d$, or
(iii) embed not the whole metric space but a subset of it.

As an example of the approach (i), a better distortion is possible for metric spaces of negative type. We say that a finite metric space (X, ρ) with elements $x_1, x_2, \ldots, x_n$ is of *negative type* if for any real numbers $c_1, \ldots, c_n$ with $\sum_{i=1}^{n} c_i = 0$ we have

$$\sum_{i=1}^{n} \sum_{j=1}^{n} c_i c_j \rho(x_i, x_j) \leq 0.$$

In 2008, Chawla et al. [266] proved that every n-point metric space of negative type can be embedded into a Euclidean space with distortion $O((\log n)^{3/4})$. In the same year, Arora, Lee and Naor [49] improved the distortion to $O(\sqrt{\log n} \log \log n)$, which matches the lower bound $O(\sqrt{\log n})$, established by Enflo [390] in 1970, up to an insignificant $\log \log n$ factor.

Theorem 3.80 *Every n-point metric space of negative type can be embedded into a Euclidean space with distortion $O(\sqrt{\log n} \log \log n)$.*

In 2009, Theorem 3.80 was used to develop a more accurate approximation algorithm for the sparsest cut problem, see Theorem 7.13 discussed later in this book.

In Theorem 3.80, by a Euclidean space we mean $\mathbb{R}^d$ (for some d) with the usual metric $\rho(x, y) = \sqrt{\sum_{i=1}^{n} (x_i - y_i)^2}$. Let $l_1(\mathbb{R}^d)$ be $\mathbb{R}^d$ with the metric

$$\rho(x, y) = \sum_{i=1}^{n} |x_i - y_i|.$$

We say that a metric space (X, ρ) *embeds into* l_1 with distortion α if it can be embedded into $l_1(\mathbb{R}^d)$ for some d with distortion α. It is known that every n-

point subset of Euclidean space embeds into l_1 isometrically (that is, with distortion 1), hence any embedding into Euclidean space automatically implies embedding into l_1 with the same distortion. However, the opposite direction is not known and one may hope to find embeddings into l_1 with better distortion. In fact, Goemans [502] in 1997 and Linial [800] in 2002 conjectured that every finite metric space of negative type embeds into l_1 with constant distortion. This became known as the $(l_2^2, l_1, O(1))$-*conjecture*, and, if true, would imply efficient constant-factor approximation algorithms for some important combinatorial problems. However, Khot and Vishnoi [715] proved in 2015 that the $(l_2^2, l_1, O(1))$-conjecture is false.

Theorem 3.81 *For any* $\delta > 0$ *and all sufficiently large n, there is an n-point metric space of negative type which cannot be embedded into* l_1 *with distortion less than* $(\log \log n)^{1/6 - \delta}$.

The proof of Theorem 3.81 uses the fact that the Goemans–Linial conjecture would imply an algorithm for the sparsest cut problem with constant integrality gap. The authors then proved that such algorithms do not exist using the techniques surrounding the unique game conjecture (see the discussion before Theorem 7.21 later in this book). Note that the proof is unconditional and does not rely on the unique game conjecture.

Theorem 3.81 implies that embedding with constant distortion is not possible into l_1 (and hence also not possible into Euclidean space), even if we consider only metric spaces of negative type. However, Bartal, Linial, Mendel and Naor [102] proved in 2005 that constant distortion is achievable if we are looking for embeddings of large subsets of finite metric spaces rather than the whole spaces.

Theorem 3.82 *There exists a constant* $C > 0$ *such that for every* $\alpha > 1$, *every n-point metric space has a subset of size* $n^{1 - C \frac{\log(2\alpha)}{\alpha}}$ *which can be embedded into* $\mathbb{R}^d$ *with distortion at most* α.

In 2007, Mendel and Naor [877] showed that we can take an even larger subset of size $n^{1 - \frac{C}{\alpha}}$ such that the conclusion of Theorem 3.82 remains correct, and that this result is optimal up to the value of the constant C.

Quantifying Non-embeddability

A classical 1936 Theorem of Paley [961] states that if $2 < q < p < \infty$, then the space[24] L^q is not isomorphic to any subspace of L^p. In 1972, Mankiewicz [835] used this theorem to deduce that, more generally, L^q fails to admit a bi-Lipschitz embedding into L^p. To make the last result quantitative, one may use the notion of a θ-snowflake. For $\theta \in (0, 1]$, the θ-*snowflake* of a metric space (X, ρ_X) is the metric space on the same set X with distance $\rho(x, y) = (\rho_X(x, y))^\theta$, $\forall x, y \in X$.

Because L^q does not embed into L^p, a θ-snowflake of L^q may embed into L^p only if $\theta < 1$. Quantifying by "how much" θ is bounded away from 1 gives

---

[24] By the L^p-*space* we mean, for concreteness, the space of functions $f : [0, 1] \to \mathbb{R}$ with norm $\|f\|_p = \left(\int_0^1 |f(x)|^p dx \right)^{1/p} < \infty$.

an important quantitative refinement of non-embeddability L^q into L^p. However, before 2016, no estimate in the form $\theta \leq 1 - \delta(p, q)$ for any explicit function $\delta(p, q)$ was known. In 2016, Naor and Schechtman [922] provided the first estimate in this form.

Theorem 3.83 *For every* $2 < q < p < \infty$, *if* $\theta \in (0, 1]$ *is such that the θ-snowflake of L^q admits a bi-Lipschitz embedding into L^p, then necessarily* $\theta \leq 1 - \frac{(p-q)(q-2)}{2p^3}$.

At the heart of the proof of Theorem 3.83 is constructing, for every real number $p > 0$, a function g_p which maps the metric space (X, ρ_X) to a real number $g_p(X) \in [0, \infty]$ in such a way that that if $g_p(X) < \infty$, and the metric space (Y, ρ_Y) admits a bi-Lipschitz embedding into X, then $g_p(Y) < \infty$. The authors first proved that if $2 < q < p < \infty$ then $g_p(L^p) \approx \frac{p}{\log p} < \infty$ but $g_p(L^q) = \infty$, which gives an alternative proof that L^q does not embed into L^p. Then Theorem 3.83 is proved in a similar way, by computing g_p of θ-snowflakes.

Let $\theta^* = \theta^*(p, q)$ denote the maximal $\theta \in (0, 1]$ for which the θ-snowflake of L^q embeds into L^p. In 2004, Mendel and Naor [876] proved that $\frac{q}{p} \leq \theta^*$. Theorem 3.82 states that $\theta^* \leq 1 - \frac{(p-q)(q-2)}{2p^3}$. In the same work [922] in which they proved Theorem 3.83, Naor and Schechtman formulated a conjecture about properties of their function g_p, and proved that this conjecture would imply that $\theta^* = \frac{q}{p}$. In 2016, Naor [921] confirmed this conjecture.

Theorem 3.84 *For every* $2 < q < p < \infty$, *the maximal* $\theta \in (0, 1]$ *for which the θ-snowflake of L^q admits a bi-Lipschitz embedding into L^p is equal to* $\frac{q}{p}$.

Uniform and Coarse Embedding

A slightly different type of embedding is uniform embedding. Let (M, d_M) and (N, d_N) be metric spaces. A function $f : M \to N$ is called *uniformly continuous* if for every $\epsilon > 0$ there exists a $\delta > 0$ such that $d_N(f(x), f(y)) < \epsilon$ for every $x, y \in M$ with $d_M(x, y) < \delta$. We say that M embeds uniformly into N if there is an injective function $f : M \to N$ such that both f and f^{-1} are uniformly continuous. There is a large body of literature studying the uniform embedding of metric spaces. However, several fundamental questions remain open. In particular, before 2008 it was not known for which positive real numbers p and q the space L^p embeds uniformly into L^q. In 2008, Mendel and Naor [878] gave a complete answer to this question.

Theorem 3.85 *For real numbers* $p, q > 0$, L^p *embeds uniformly into L^q if and only if* $p \leq q$ *or* $q \leq p \leq 2$.

In fact, Mendel and Naor [878] developed a deep and general theory of "metric cotype", which nicely complements the theory of types, see Theorem 3.76. Theorem 3.85 is just one of many corollaries of this theory.

A seemingly weaker notion of embedding was introduced by Gromov. We say that a metric space (X, d_X) *coarsely embeds* into a metric space (Y, d_Y) if there is a function $f : X \to Y$ and non-decreasing functions $\rho_1, \rho_2 : [0, \infty) \to [0, \infty)$ such that $\lim_{t \to \infty} \rho_1(t) = +\infty$ and

$$\rho_1(d_X(x, y)) \le d_Y(f(x), f(y)) \le \rho_2(d_X(x, y)) \tag{3.33}$$

for all $x, y \in X$. A Banach space is called *coarsely minimal* if it coarsely embeds into every infinite-dimensional Banach space. For a long time, there was an important open problem asking whether the l^2-space is coarsely minimal. There was some evidence and partial results supporting that the answer may be positive. If it were true, then it would imply the solution of some other important problems in the field. However, Baudier et al. [107] proved in 2018 that the answer is negative in a strong sense: not only is l^2 not coarsely minimal, but, moreover, there are no coarsely minimal infinite-dimensional Banach spaces at all.

Theorem 3.86 *There is no coarsely minimal infinite-dimensional Banach space.*

3.13   The Kakeya Conjecture

A *Besicovitch set*, also known as a *Kakeya set*, is a compact subset of $\mathbb{R}^n$ which contains a unit line segment in each direction. It is named after Besicovitch, who proved in 1920 that there exists such set in $\mathbb{R}^2$ with Lebesgue measure 0. However, Davies [329] proved in 1971 that every Besicovitch set in $\mathbb{R}^2$ has Hausdorff dimension 2. A conjecture known as the *Kakeya conjecture* predicts that, more generally, every Besicovitch set in $\mathbb{R}^n$ has Hausdorff dimension n. This conjecture has connections to problems in number theory, harmonic analysis, and the analysis of partial differential equations, but remains open for every $n > 2$.

The simplest unsolved case is $n = 3$. Davies' result implies that every Besicovitch set S in $\mathbb{R}^3$ has Hausdorff dimension $\dim(S)$ at least 2. In 1991, Bourgain [180] improved the lower bound to $\dim(S) \ge \frac{7}{3}$. In 1995, Wolff [1272] improved it further to $\dim(S) \ge \frac{5}{2}$. This bound remained the best known for over 20 years, until a small improvement was achieved by Katz and Zahl [693] in 2019.

Theorem 3.87 *There exists a constant $\epsilon_0 > 0$ such that every Besicovitch set in $\mathbb{R}^3$ has Hausdorff dimension at least $\frac{5}{2} + \epsilon_0$.*

While the improvement in the lower bound in Theorem 3.87 is very small, the importance of this theorem lies in the fact that there are sets of Hausdorff dimension 5/2 which closely resemble Besicovitch sets, and Katz and Zahl demonstrated, for the first time, how to deal with this difficulty. After this, there is hope for more rapid further progress.

In 1999, Wolff [1271] suggested to study a finite field analogue of the Kakeya conjecture, in the hope that the resolution of this easier problem may give some insight into how to approach the original conjecture. Let $\mathbb{F}$ be a finite field with $|\mathbb{F}|$ elements, and let $\mathbb{F}^n$ be a vector space over $\mathbb{F}$ of dimension n. A Kakeya set in $\mathbb{F}^n$ is a subset $K \subset \mathbb{F}^n$ such that for every $x \in \mathbb{F}^n$ there exists a point $y \in \mathbb{F}^n$ such that the set $L = \{y + a \cdot x \mid a \in \mathbb{F}\}$ (called a *line*) is contained in K. The Kakeya conjecture in finite fields predicted that every Kakeya set in $\mathbb{F}^n$ has at least $C_n \cdot |\mathbb{F}|^n$ elements. In 2009, Dvir [375] confirmed this conjecture.

Theorem 3.88 *The size of every Kakeya set in $\mathbb{F}^n$ is at least $C_n \cdot |\mathbb{F}|^n$, where C_n is a constant that depends only on n.*

The idea of the proof is the observation that any homogeneous polynomial P of degree $|\mathbb{F}| - 2$ in n variables over $\mathbb{F}$ can be "reconstructed" given its values on points from any Kakeya set in $K \subset \mathbb{F}^n$. This implies that the size of K cannot be less that the dimension of the space of such polynomials. Since 2009, this "polynomial method" has found many other applications, see, for example, Theorems 2.17 and 2.18.

However, Theorems 3.88, 2.17 and 2.18 are about finite fields, while the original Kakeya conjecture is about Euclidean space. In 2010, Guth [546] demonstrated that the polynomial method can be used in the Euclidean setting by resolving, with the help of this method, the multilinear version of the Kakeya conjecture formulated by Bennett et al. [122] in 2006. This conjecture says, informally, that cylinders pointing in different directions cannot overlap too much. The authors of [122] made progress on this conjecture, but could not prove the last case of it, which they call "the endpoint case". In 2010, Guth [546] resolved this remaining case using the polynomial method and thus finished the proof of the conjecture.

To avoid technicalities, here we formulate a weaker but easier to state version of the conjecture, see [546] for the full version. A cylinder of radius R around a line $L \subset \mathbb{R}^n$ is the set of all points $x \in \mathbb{R}^n$ within a distance R from the line L. The line L is called the *core* of the cylinder. Suppose that we have a finite collection of cylinders $T_{j,a} \subset \mathbb{R}^n$, where $1 \leq j \leq n$, and $1 \leq a \leq A$ for some integer A. Each cylinder $T_{j,a}$ has radius 1 and the angle between the core of $T_{j,a}$ and the x_j-axis is at most $(100n)^{-1}$. Let $I = \bigcap_{j=1}^{n} \bigcup_{a=1}^{A} T_{j,a}$ be the set of points that belong to at least one cylinder in each direction.

Theorem 3.89 *For each dimension n there is a constant $C = C(n)$ such that $\mathrm{Vol}(I) \leq C A^{n/(n-1)}$ for any collection of cylinders as above.*

3.14   Approximation of Functions

Approximation by Hyperbolic Functions
In 1998, Fields medalist Stephen Smale formulated a list of problems [267] which he considers as the most important open problems in mathematics for the twenty-first century. Problem 11(b) asks whether every smooth function $g : [0, 1] \to [0, 1]$ can be approximated by a hyperbolic function. Because every "sufficiently smooth" function g can be approximated by polynomials, it is sufficient to resolve this question for polynomials.

To formulate the problem more rigorously, we need some definitions. The *n-th functional power* of a function $f : \mathbb{R} \rightarrow \mathbb{R}$ is the function $f^n(x) = f(f(\ldots f(x)\ldots))$, where f is repeated n times. A point x_0 is called *periodic* for f if $f^k(x_0) = x_0$ for some $k \geq 1$. The smallest k for which this holds is called the *period* of x_0. A periodic point x_0 with period k is called *hyperbolic* if $|(f^k)'(x_0)| \neq 1$, and it is called *hyperbolic attracting* if $|(f^k)'(x_0)| < 1$. For a polynomial $f : \mathbb{R} \rightarrow \mathbb{R}$, and $x_0 \in \mathbb{R}$, either

(i) $\lim\limits_{n \to \infty} |f^n(x_0) - f^n(x^*)| = 0$ for some hyperbolic attracting periodic point x^*, or

(ii) $\lim\limits_{n \to \infty} |f^n(x_0)| = \infty$, or

(iii) neither (i) nor (ii) happens.

Let S_f be the set of all $x_0 \in \mathbb{R}$ for which case (iii) holds. A polynomial f is called *hyperbolic* if there exist constants $C > 0$ and $\lambda > 1$ such that $|(f^n)'(x)| > C\lambda^n$, $\forall n, \forall x \in S_f$. Hyperbolic polynomials are central objects of study in dynamical systems.

In 2007, Kozlovski et al. [744] resolved Smale's problem 11(b) for polynomials and hence in general.

Theorem 3.90 *For every real polynomial $f(x) = \sum_{i=0}^{d} a_i x^i$ and any $\epsilon > 0$, there exists a hyperbolic real polynomial $h(x) = \sum_{i=0}^{d} b_i x^i$ such that $|a_i - b_i| < \epsilon$, $i = 0, 1, \ldots, d$.*

Approximation by Polynomials

In approximation theory, it is common to approximate a function f by a polynomial P at some discrete points $x_1, \ldots, x_N$, and then prove that this approximation works well on some interval I. In other words, a central problem is to understand to what extent the fact that $|f(x_i) - P(x_i)|$ is small for $i = 1, \ldots, N$ implies that $|f(x) - P(x)|$ is small for all $x \in I$.

This problem is non-trivial even for $f = 0$. Let $P_{n,N}$ be the set of polynomials P of degree at most n with $|P(x_k)| \leq 1, k = 1, 2, \ldots, N$, where $x_k = -1 + \frac{2k-1}{N}$, $k = 1, 2, \ldots, N$, is the sequence of N equidistant points on $I = [-1, 1]$. Let

$$K_{n,N}(x) = \max_{P \in P_{n,N}} |P(x)|.$$

The idea is that if a degree n polynomial P approximates $f = 0$ with error at most 1 at points x_k, then it approximates 0 with error at most $K_{n,N}(x)$ at x.

In 1992, Coppersmith and Rivlin [302] proved the existence of some absolute constants $c_1, c_2 > 0$ such that

$$e^{c_1 n^2 / N} \leq \max_{x \in I} K_{n,N}(x) \leq e^{c_2 n^2 / N}. \tag{3.34}$$

In particular, (3.34) implies that $\max_{x \in I} K_{n,N}(x)$ is bounded if and only if n^2/N is bounded, or $n \leq C\sqrt{N}$ for some constant C. However, it is not always possible to find a good approximation of a function at N points by a polynomial of degree $n \leq C\sqrt{N}$. In fact, we may need a polynomial of degree as large as $n = N - 1$ in the worst case. If $n \approx N$, (3.34) implies that $\max_{x \in I} K_{n,N}(x)$ grows exponentially fast with N. However, Rakhmanov [1019] proved in 2007 that for any $n < N$ the function $K_{n,N}(x)$ is uniformly bounded on any compact subinterval of $(-r, r)$, where $r = \sqrt{1 - \frac{n^2}{N^2}}$. Moreover, Rakhmanov also showed $(-r, r)$ is the "maximal" subinterval of $[-1, 1]$ with this property.

Theorem 3.91 *There exists a constant C such that the inequality*

$$K_{N,n}(x) \leq C \log \frac{\pi}{\arctan\left(\frac{N}{n}\sqrt{r^2 - x^2}\right)}$$

holds for all $n < N$ and all $x \in (-r, r)$, where $r = \sqrt{1 - \frac{n^2}{N^2}}$.

Fitting a Smooth Function to Data

For a function $f : \mathbb{R}^n \to \mathbb{R}$ and vector $\beta = (\beta_1, \beta_2 \ldots, \beta_n)$ of non-negative integers, denote by $\partial^\beta f$ the partial derivative $\frac{\partial^{|\beta|} f}{\partial x_1^{\beta_1} \partial x_2^{\beta_2} \ldots \partial x_n^{\beta_n}}$ of order $|\beta| = \beta_1 + \beta_2 + \cdots + \beta_n$. Let $C^m(\mathbb{R}^n)$ denote the space of functions $f : \mathbb{R}^n \to \mathbb{R}$ whose derivatives of order $\leq m$ are continuous and bounded on $\mathbb{R}^n$.

Given a function $f : E \to \mathbb{R}$, where E is a subset of $\mathbb{R}^n$, how can we decide whether f extends to a $C^m(\mathbb{R}^n)$ function F on $\mathbb{R}^n$? This is a classical question which was answered by Whitney [1256] in 1934, and the result is known as the *Whitney extension theorem*. In 2005, Fefferman [443] solved a version of this problem in which F is required to be $C^{m-1,1}(\mathbb{R}^n)$, where $C^{m-1,1}(\mathbb{R}^n)$ denotes the space of functions whose derivatives of order $m - 1$ are Lipschitz with constant 1.

Theorem 3.92 *For any integers $m \geq 1$ and $n \geq 1$ there exists an integer k, depending only on m and n, for which the following holds. Let $f : E \to \mathbb{R}$ be a function defined on an arbitrary subset E of $\mathbb{R}^n$. Suppose that, for any k distinct points $x_1, \ldots, x_k \in E$ there exist polynomials $P_1, \ldots, P_k$ on $\mathbb{R}^n$ of degree $m - 1$ satisfying*

(a) $P_i(x_i) = f(x_i)$ for $i = 1, \ldots, k$;
(b) $|\partial^\beta P_i(x_i)| \leq M$ for $i = 1, \ldots, k$ and $|\beta| \leq m - 1$; and
(c) $|\partial^\beta (P_i - P_j)(x_i)| \leq M|x_i - x_j|^{m-|\beta|}$ for $i, j = 1, \ldots, k$ and $|\beta| \leq m - 1$, where M is a constant independent of $x_1, \ldots, x_k$.

Then f extends to a $C^{m-1,1}$ function on $\mathbb{R}^n$.

Theorem 3.92 is sharp in the sense that the conditions (a), (b) and (c) in it are necessary for the conclusion.

In applications, $E \subset \mathbb{R}^n$ is often a finite set of cardinality N, and the function $f : E \to \mathbb{R}$ represents N data points, to which we want to fit a smooth function F with the smallest possible $C^m(\mathbb{R}^n)$ norm (or $C^{m-1,1}(\mathbb{R}^n)$ norm). We may want an exact fit to the data, or an approximate one, with the approximation quality measured by a parameter σ defined below. In 2009, Fefferman and Klartag [441] proved the existence of an efficient algorithm which approximately computes the smallest possible $C^m(\mathbb{R}^n)$ norm of F we can achieve.

To formulate this result, we need some notation. Let $E \subset \mathbb{R}^n$ be a finite set of cardinality N. Let $f : E \to \mathbb{R}$ and $\sigma : E \to [0, \infty)$ be given functions on E. Let $||f||_{C^m(E,\sigma)}$ denote the infimum of all $M > 0$ for which there exists an $F \in C^m(\mathbb{R}^n)$ such that $||F||_{C^m(\mathbb{R}^n)} \leq M$ and $|F(x) - f(x)| \leq M\sigma(x)$ for all $x \in E$. We say that two numbers $X, Y \geq 0$ determined by E, f, σ, m, n have the same *order of magnitude* if $cX \leq Y \leq CX$, where the constants c and C depend only on m and n. By "compute the order of magnitude of X" we mean compute some Y such that X and Y have the same order of magnitude.

Theorem 3.93 *There is an algorithm which, given E, f, σ, m, n, computes the order of magnitude of $||f||_{C^m(E,\sigma)}$ in time at most $CN \log N$ and using memory at most CN, where the constant C depends only on m and n.*

As mentioned above, the algorithm in Theorem 3.93 approximately computes the smallest possible $C^m(\mathbb{R}^n)$ norm of function F which fits the data. In a subsequent paper [442], the same authors also presented an algorithm which computes F itself.

3.15   Other Topics in Analysis

The Kadison–Singer Problem

In 1959, Kadison and Singer [670] formulated a problem in functional analysis, which became known as the *Kadison–Singer problem*. In 2006, Casazza et al. [254] discovered that numerous open problems in pure mathematics, applied mathematics, engineering and computer science are all equivalent to this problem. Hence, it was sufficient to solve one of these problems to solve them all. In 2015, Marcus et al. [842] achieved exactly this. Below we present the formulation they solved.

Let $\mathbb{C}^d$ be the set of vectors $x = (x_1, \ldots, x_d)$ with complex components x_i. Denote by $\langle x, y \rangle = \sum_{i=1}^d x_i \bar{y}_i$ the inner product in $\mathbb{C}^d$. Let $||x|| = \sqrt{\langle x, x \rangle}$ be the norm in $\mathbb{C}^d$. We say that $u \in \mathbb{C}^d$ is a unit vector if $||u|| = 1$.

Theorem 3.94 *There exist universal constants $\eta \geq 2$ and $\theta > 0$ so that the following holds. Let $w_1, \ldots, w_m \in \mathbb{C}^d$ satisfy $||w_i|| \leq 1$ for all i and suppose $\sum_{i=1}^m |\langle u, w_i \rangle|^2 = \eta$ for every unit vector $u \in \mathbb{C}^d$. Then there exists a partition S_1, S_2 of $\{1, \ldots, m\}$ so that $\sum_{i \in S_j} |\langle u, w_i \rangle|^2 \leq \eta - \theta$, for every unit vector $u \in \mathbb{C}^d$ and each $j \in \{1, 2\}$.*

The proof uses the method of interlacing polynomials, introduced in [841] to prove Theorem 2.48. The authors proved that the characteristic polynomials of certain matrices that arise in Theorem 3.94 forms such a family, used this to extract information about the largest roots of these polynomials, and deduced Theorem 3.94 from there.

The Combinatorial Dimension

The *combinatorial dimension* is a way to measure "how large a set of functions is". Let Ω be a set, $\mathcal{A}$ be the set of all functions $f : \Omega \to \mathbb{R}$, and let A be any subset of $\mathcal{A}$. We say that a subset σ of Ω is *t-shattered* by A if there exists a function h on σ such that, given any decomposition $\sigma = \sigma_1 \cup \sigma_2$ with $\sigma_1 \cap \sigma_2 = \emptyset$, one can find a function $f \in A$ with $f(x) \leq h(x)$ if $x \in \sigma_1$ but $f(x) \geq h(x) + t$ if $x \in \sigma_2$. The *combinatorial* (or *shattering*) *dimension* $v(A, t)$ of A is the maximal cardinality of a set t-shattered by A.

There are other ways to measure "how large a set of functions is", and an important research direction is to establish the connections among different "measures". One of these "measures" is the *covering number* of a set of bounded functions. Let μ be a probability measure on Ω. Let B_1 be the set of all functions $f : \Omega \to [-1, 1]$, and let A be any subset of B_1. Denote by $N(A, t, \mu)$ the covering number of A, that is, the minimal number of functions whose linear combination can approximate any function in A within an error of t in the $L^2(\mu)$ norm. It is well-known that the covering numbers of a set are at most exponential in its linear algebraic dimension. Talagrand's famous *entropy problem* asked if this result can be extended to the combinatorial dimension. This would be especially useful because the combinatorial dimension never exceeds the linear algebraic dimension, and is often much smaller than it. In 2003, Mendelson and Vershynin [879] proved that Talagrand's entropy problem has a positive solution.

Theorem 3.95 *There exist positive absolute constants K and c such that inequality*

$$N(A, t, \mu) \leq \left(\frac{2}{t}\right)^{K \cdot v(A, ct)}$$

holds for all A and all $0 < t < 1$.

Another way to measure "how large a set of functions is" is the metric entropy. For k points $x_1, x_2, \ldots, x_k \in \Omega$ and two functions f and g in $\mathcal{A}$, define

$$d_{x_1, \ldots x_k}(f, g) = \sqrt{\frac{1}{k} \sum_{i=1}^{k} (f(x_i) - g(x_i))^2}.$$

For any $t > 0$, let $N_{x_1, \ldots x_k}(A, t)$ be the maximal n for which there exist functions $f_1, f_2, \ldots f_n \in A$ such that $d_{x_1, \ldots x_k}(f_i, f_j) \geq t$ for all $i \neq j$. The quantity

$$D(A, t) := \log \left(\sup_k \sup_{x_1, \ldots x_k} N_{x_1, \ldots x_k}(A, t) \right)$$

is called the *Koltchinskii–Pollard entropy*, or just *metric entropy* of A. In 2006, Rudelson and Vershynin [1068] proved a theorem which establishes the connection between metric entropy and combinatorial dimension.

Theorem 3.96 *If there exists a constant $b > 1$ such that*

$$v(A, bt) \leq \frac{1}{2}v(A, t), \quad \forall t > 0, \tag{3.35}$$

then the inequalities

$$c \cdot v(A, 2t) \leq D(A, t) \leq C \cdot v(A, ct)$$

hold for all $t > 0$, where $c > 0$ is an absolute constant, and C depends only on b.

The condition (3.35) is known as the minimal regularity condition, and it is known that the conclusion of Theorem 3.96 does not hold without it. Theorem 3.96 shows that, under this condition, two ways of measuring "how large a set of functions is" are equivalent, up to constant factors.

Spherical Designs
Let S^d be a sphere in $\mathbb{R}^{d+1}$ normalized such that $\mu_d(S^d) = 1$, where μ_d is the d-dimensional Lebesgue measure. A set of points $x_1, \ldots, x_N \in S^d$ is called a *spherical t-design* if the equality

$$\int_{S^d} P(x) d\mu_d(x) = \frac{1}{N} \sum_{i=1}^{N} P(x_i)$$

holds for all polynomials P in $d + 1$ variables, of total degree at most t. The concept of a spherical design was introduced by Delsarte et al. [340] in 1977, and, as follows from the definition, it is useful for evaluating integrals. Of course, the smaller N, the easier it is to compute $\frac{1}{N} \sum_{i=1}^{N} P(x_i)$. In 1993, Korevaar and Meyers [738] conjectured the existence of spherical t-designs in S^d with as little as $c_d t^d$ points, which is optimal up to a constant factor. As partial progress, Korevaar and Meyers themselves showed that $N = c_d t^{(d^2+d)/2}$ points suffices, improving earlier constructions of $N = c_d t^{Cd^4}$ and $N = c_d t^{Cd^3}$ points due to Wagner and Volkmann [1241] and Bajnok [80], respectively. In 2010, Bondarenko and Viazovska [167] showed that $N = c_d t^{\frac{2d(d+1)}{d+2}}$ points suffices. Finally, in 2013, Bondarenko et al. [166] improved the exponent from $\frac{2d(d+1)}{d+2}$ to d, thus confirming the Korevaar–Meyers conjecture in full.

Theorem 3.97 *For every $d \in \mathbb{N}$ there exist a constant $c_d > 0$ such that for each $t \in \mathbb{N}$ and each $N \geq c_d t^d$, there exists a spherical t-design in S^d consisting of N points.*

Falconer's Distance Set Problem
For a set $E \subset \mathbb{R}^d$, define the distance set

$$\Delta(E) = \{|x - y| : x \in E, y \in E\}.$$

In 1985, Falconer [438] posed the following problem. What is the smallest constant $c(d)$ such that every compact set E in $\mathbb{R}^d$ with Hausdorff dimension greater than $c(d)$ must have distance set $\Delta(E)$ of positive Lebesgue measure? Falconer proved that $\frac{d}{2} \leq c(d)$ by constructing a lattice-based example of a set $E \subset \mathbb{R}^d$ of dimension $\frac{d}{2}$ such that $\Delta(E)$ has Lebesgue measure 0. He then conjectured that in fact $c(d) = \frac{d}{2}$ for all $d \geq 2$, and this is known as *Falconer's distance conjecture*.

Falconer himself proved the first upper bound $c(d) \leq \frac{d+1}{2}$ for all $d \geq 2$. In 1999, Wolff [1270] proved that $c(2) \leq \frac{4}{3}$. In 2005, Erdogan [393] extended this result to higher dimensions and proved that $c(d) \leq \frac{d}{2} + \frac{1}{3}$ for every $d \geq 2$. In 2018, Du, Guth, Ou, Wang, Wilson and Zhang [363] proved that $c(3) \leq \frac{9}{5}$ and $c(d) \leq \frac{d}{2} + \frac{1}{4} + \frac{d+1}{4(2d+1)(d-1)}$ for $d \geq 4$. In 2019, Du and Zhang [364] proved that $c(d) \leq \frac{d^2}{2d-1}$. This upper bound recovers the above-mentioned bounds in dimensions $d = 2, 3$ and improves them in all dimensions $d \geq 4$.

Theorem 3.98 *For every $d \geq 2$, if $E \subset \mathbb{R}^d$ is a compact set with Hausdorff dimension greater than $\frac{d^2}{2d-1}$, then $\Delta(E)$ has positive Lebesgue measure. Moreover, there exists a fixed point $x \in E$ such that its pinned distance set*

$$\Delta_x(E) = \{|x - y| : y \in E\}$$

has positive Lebesgue measure.

For the plane, Falconer's distance conjecture predicts that $c(2) = 1$. Interestingly, the above-mentioned works [364, 393, 1270], despite using very different proof techniques, all prove the same upper bound $c(2) \leq \frac{4}{3}$, so this bound started to look like some barrier for all known techniques. It was therefore a great achievement when Guth et al. [548] in 2020 improved this bound to $c(2) \leq \frac{5}{4}$.

Theorem 3.99 *If $E \subset \mathbb{R}^2$ is a compact set with Hausdorff dimension greater than $5/4$, then $\Delta(E)$ has positive Lebesgue measure.*

Properties of Jordan Curves

A *Jordan curve* is a simple closed curve in the plane $\mathbb{R}^2$, or, more formally, the image of an injective continuous map of a circle into the plane. The fundamental *Jordan curve theorem* states that the complement $\mathbb{R}^2/\Gamma$ of every such curve Γ consists of exactly two connected components, Ω^+ and Ω^-, exactly one of which is bounded, and $\Gamma = \partial\Omega^+ = \partial\Omega^-$ is the boundary of each component.

Here, we discuss a conjecture of Carleson, which gives a characterization (up to a set of measure 0) of the tangent points of a Jordan curve. The concept of "tangent point" is clear intuitively, but requires some notation to define formally. For $x \in \mathbb{R}^2$ and $r > 0$ let $B(x, r)$ be the ball with centre x and radius r. Also, for any $x \in \mathbb{R}^2$, unit vector u, and real parameter $a \in (0, 1)$, define

$$X_a(x, u) = \{y \in \mathbb{R}^2 : |\langle y - x, u \rangle| > a|y - x|\},$$

where $\langle \cdot, \cdot \rangle$ is the inner product. Then, for an open set $\Omega^+ \in \mathbb{R}^2$ with boundary $\Gamma = \partial \Omega^+$, we say that $x \in \Gamma$ is a *tangent point* of Γ if there exists a unit vector u such that, for all $a \in (0, 1)$, there exists some $r > 0$ such that (i) $\Gamma \cap X_a(x, u) \cap B(x, r) = \emptyset$ and (ii) one component of $X_a(x, u) \cap B(x, r)$ is contained in Ω^+ and the other one is in Ω^-.

Next, for $x \in \mathbb{R}^2$ and $r > 0$, let $I^+(x, r)$ and $I^-(x, r)$ be the longest open arcs of the circle $\partial B(x, r)$ contained in Ω^+ and Ω^-, respectively. Then let

$$\epsilon(x, r) = \frac{1}{r} \max\{|\pi r - \mathcal{H}_1(I^+(x, r))|, |\pi r - \mathcal{H}_1(I^-(x, r))|\},$$

where $\mathcal{H}_1$ is the one-dimensional Hausdorff measure.[25] The function

$$\mathcal{E}_\Gamma(x) = \sqrt{\int_0^1 \epsilon(x, r)^2 \frac{dr}{r}}$$

is called the *Carleson ϵ^2-function*. It follows from a classical argument of Beurling (see [154]) that $\mathcal{E}_\Gamma(x) < \infty$ for almost every tangent point of Γ. A famous conjecture of Carleson stated that the converse to this statement is also true, and remained open for decades. In 2019, Jaye et al. [651] proved this conjecture.

Theorem 3.100 *Let Γ be a Jordan curve in $\mathbb{R}^2$. Except for a set of $\mathcal{H}_1$-measure zero, a point $x \in \Gamma$ is a tangent point of Γ if and only if $\mathcal{E}_\Gamma(x) < \infty$.*

Bases of Exponentials

Let $S \subseteq \mathbb{R}^d$ be a bounded, measurable set of positive measure, and let $L^2(S)$ be the set of functions $f : S \to \mathbb{C}$ for which the integral $\int_S |f(x)|^2 dx$ exists and is finite. For many applications, it is useful to represent every function $f \in L^2(S)$ as a combination of exponentials

$$f(x) = \sum_{\lambda \in \Lambda} c_\lambda e^{2\pi i \langle \lambda, x \rangle}. \tag{3.36}$$

for some countable set $\Lambda \subset \mathbb{R}^d$, where $\langle \lambda, x \rangle$ is the inner product and $c_\lambda \in \mathbb{C}$ are some coefficients. It would be ideal if the functions $\{e^{2\pi i \langle \lambda, x \rangle}\}_{\lambda \in \Lambda}$ form an orthogonal basis of $L^2(S)$. Formally, for $f, g \in L^2(S)$, define $(f, g) = \int_S f(x) \bar{g}(x) dx$ and $\|f\| = \sqrt{(f, f)}$. A sequence $\{f_n\}_{n=1}^\infty$ of functions in $L^2(S)$ is called *orthogonal* if $(f_n, f_m) = 0$ whenever $n \neq m$. Also, a sequence $\{f_n\}_{n=1}^\infty$ is called *complete* if the only $f \in L^2(S)$ satisfying $(f, f_n) = 0$ for all n is $f = 0$. Any orthogonal and complete sequence $\{f_n\}_{n=1}^\infty$ is called an *orthogonal basis* in $L^2(S)$. We say that a set S is *spectral* if $L^2(S)$ contains an orthogonal basis of exponential functions. In this case, the representation (3.36) exists, is unique, and

---

[25] Intuitively, $\mathcal{H}_1$ is just "length" of an arc, see (1.22) with $f(r) = r$ for the formal definition.

the coefficients c_λ are easy to calculate. For these reasons, a central question in this area is to understand which sets are spectral.

Much of the research on this question has been motivated by a famous 1974 conjecture of Fuglede [479], who predicted that a set $S \subseteq \mathbb{R}^d$ is spectral if and only if it can tile $\mathbb{R}^d$ by translations. Formally, this means that there exists a countable set $A \subset \mathbb{R}^d$ such that the collection of translated copies $\{S + a\}_{a \in A}$ of S forms a partition of $\mathbb{R}^d$ up to measure zero. This conjecture, if true, would give a beautiful geometric characterization of spectral sets.

The conjecture remained open for over 30 years, until Tao [1180] constructed a counterexample in all dimensions $d \geq 5$, by showing the existence of spectral sets which cannot tile $\mathbb{R}^d$ by translations. In 2006, Kolountzakis and Matolcsi [731] presented counterexamples to both directions of the conjecture in dimensions $d \geq 3$. The conjecture remains open in dimensions $d = 1$ and $d = 2$.

All the known counterexamples are highly non-convex, which gave the hope that Fuglede's conjecture can still be true for convex bodies. And indeed, Venkov [1232] provided in 1954 a complete characterization of all convex bodies which tile the plane by translations, and this characterization easily implies that all such bodies are spectral. Hence, only the "spectral implies tiling" direction of the convex case of Fuglede's conjecture remained open. This direction was proved in the plane $\mathbb{R}^2$ by Iosevich et al. [637] in 2003, and for convex polytopes in $\mathbb{R}^3$ by Greenfeld and Lev [527] in 2017. Finally, Lev and Matolcsi in 2019 posted online a preprint [785] in which they proved the convex case of Fuglede's conjecture in full.

Theorem 3.101 *A convex body in $\mathbb{R}^d$ is a spectral set if and only if it can tile $\mathbb{R}^d$ by translations.*

The proof of Theorem 3.101 is based on the aforementioned characterization of Venkov [1232] which states that a convex body $S \subset \mathbb{R}^d$ can tile the plane by translations if and only if it satisfies four conditions: (a) S must be a convex polytope, (b) S is centrally symmetric, (c) all the facets of S are centrally symmetric, and one more condition (d) which is a bit too long to be stated here, see [785] if you are interested. Hence, Theorem 3.101 reduces to verifying properties (a)–(d) for every spectral set. In 2000, Kolountzakis [730] proved (b). In 2017, Greenfeld and Lev [527] proved (c) assuming (a). The first main result of [785] proves (a), and the second one proves (d) assuming (a). In combination with the earlier works, these results imply Theorem 3.101.

Theorem 3.101 implies that only very "special" convex bodies are spectral. For non-convex sets, the situation is no better. For example, in 2013 Iosevich and Kolountzakis [638] constructed a set $S \subset \mathbb{R}$ which is as simple as the union of two disjoint intervals, but still not spectral. For such S, representation (3.36) with an orthonormal basis does not exist. So, what should we do in applications?

If an orthogonal basis is not available, the next best choice is a Riesz basis. A complete sequence $\{f_n\}_{n=1}^{\infty}$ is called a *Riesz basis* if there exist positive constants c and C such that

$$c \sum_{n=1}^{\infty} |a_n|^2 \leq \left\| \sum_{n=1}^{\infty} a_n f_n \right\| \leq C \sum_{n=1}^{\infty} |a_n|^2$$

for every sequence $\{a_n\}_{n=1}^{\infty}$ of complex numbers such that $\sum_{n=1}^{\infty} |a_n|^2 < \infty$. With a Riesz basis consisting of exponential functions $\left\{ e^{2\pi i \langle \lambda, x \rangle} \right\}_{\lambda \in \Lambda}$, we still have a unique representation of every function $f \in L^2(S)$ in the form (3.36). However, there are relatively few examples of sets S for which such a basis is known to exist. For example, it was known that it exists if $S \subset \mathbb{R}$ is an interval. In 2015, Kozma and Nitzan [746] proved the existence of a Riesz basis of exponentials in the important case when $S \subset \mathbb{R}$ is a finite union of intervals.

Theorem 3.102 *If $S \subseteq \mathbb{R}$ is a finite union of intervals, then there exists a set $\Lambda \subset \mathbb{R}$ such that the functions $\left\{ e^{i\lambda t} \right\}_{\lambda \in \Lambda}$ form a Riesz basis in $L^2(S)$. Moreover, if $S \subseteq [0, 2\pi]$ then Λ may be chosen to satisfy $\Lambda \subseteq \mathbb{Z}$.*

The main difficulty in the proof of Theorem 3.102 is the well-known fact that the union of two Riesz bases for two intervals is not necessarily a Riesz basis for the union. However, the authors derived some extra conditions under which the union of bases is a Riesz basis for the union. After this, the problem reduces to finding a Riesz basis on an interval satisfying these conditions. The analysis of properties of Riesz bases on an interval is a well-studied research direction, with some deep results (like the Paley–Wiener stability theorem [962]) available. This machinery helped the authors to find bases with the desired properties and finish the proof.

In 2016, Kozma and Nitzan [747] proved a multi-dimensional version of Theorem 3.102, which states that a Riesz basis of exponentials exists for any finite union of axis-parallel rectangles in $\mathbb{R}^d$.

Some Topics in Measure Theory
Let $\mathcal{B}$ be a Boolean algebra of sets.[26] A map $v : \mathcal{B} \to \mathbb{R}$ is called a *submeasure* if the following holds:

(i) $v(\emptyset) = 0$,
(ii) $v(A) \leq v(B)$ for every $A, B \in \mathcal{B}$ such that $A \subset B$, and
(iii) $v(A \cup B) \leq v(A) + v(B)$ for every $A, B \in \mathcal{B}$.

A submeasure v is called a (*finitely additive*) *measure* if

(iv) $v(A \cup B) = v(A) + v(B)$ whenever sets $A, B \in \mathcal{B}$ are disjoint.

---

[26] That is, $\mathcal{B}$ is a family of subsets of some set X such that (i) $X \setminus A \in \mathcal{B}$ for all $A \in \mathcal{B}$, (ii) $\emptyset \in \mathcal{B}$, and (iii) $A \cup B \in \mathcal{B}$ for all $A, B \in \mathcal{B}$.

A submeasure ν is called *exhaustive* if $\lim_{n \to \infty} \nu(E_n) = 0$ for every sequence $\{E_n\}_{n=1}^{\infty}$ of elements of $\mathcal{B}$ such that $E_n \cap E_m = \emptyset$ whenever $n \neq m$. We say that a submeasure ν_1 is *absolutely continuous* with respect to a submeasure ν_2 if for every $\epsilon > 0$ there exists an $\alpha > 0$ such that $\nu_1(A) \leq \epsilon$ for every $A \in \mathcal{B}$ with $\nu_2(A) \leq \alpha$.

Any measure is exhaustive. Moreover, if a submeasure is absolutely continuous with respect to a measure, it is exhaustive. One of the many equivalent formulations of the famous *Maharam's problem*, formulated by Maharam [828] in 1947, is whether the converse is true. In 2008, Talagrand [1179] gave a negative answer to this question.

Theorem 3.103 *There exists a Boolean algebra $\mathcal{B}$ of sets, and a non-zero exhaustive submeasure ν on it, which is not absolutely continuous with respect to any measure.*

Central objects of study in geometric measure theory are rectifiable and purely unrectifiable sets. The set $\{y \in \mathbb{R}^m \mid y = f(x)\}$ for some K-Lipschitz[27] $f : \mathbb{R}^n \to \mathbb{R}^m$ is called the *Lipschitz image* of $\mathbb{R}^n$. A subset $S \subset \mathbb{R}^m$ is called *n-rectifiable* if it can be covered by a set of H^n measure[28] zero and a countable number of Lipschitz images of $\mathbb{R}^n$. A set S is *purely n-unrectifiable* if all of its n-rectifiable subsets have H^n measure zero.

In 2020, Bate [105] proved that for "almost every" bounded 1-Lipschitz function $f : \mathbb{R}^n \to \mathbb{R}^m$ we have $H^n(f(S)) = 0$ if and only if the set S is purely n-unrectifiable. This gives a complete and useful characterization of purely unrectifiable sets.

To formulate the result rigorously, we need to formalize the notion of "almost every" Lipschitz function. A standard formalization of this is the notion of a residual set. Let $\mathrm{Lip}_1(n, m)$ be the set of all bounded 1-Lipschitz functions $f : \mathbb{R}^n \to \mathbb{R}^m$ equipped with the norm $\|f\| = \sup_{x \in \mathbb{R}^n} \|f(x)\|_m$. A subset $\mathcal{F} \subseteq \mathrm{Lip}_1(n, m)$ is called *residual* if it contains a countable intersection of dense open sets.

Theorem 3.104 *For any purely n-unrectifiable $S \subset \mathbb{R}^k$ with $H^n(S) < \infty$, the set of all $f \in \mathrm{Lip}_1(k, m)$ with $H^n(f(S)) = 0$ is residual in $\mathrm{Lip}_1(k, m)$. Conversely, if S is not purely n-unrectifiable, then the set of $f \in \mathrm{Lip}_1(k, m)$ with $H^n(f(S)) > 0$ is residual.*

In fact, Bate also derived a similar characterization of purely unrectifiable sets in general metric spaces, see the original paper [105] for details.

---

[27] Recall that a function $f : \mathbb{R}^n \to \mathbb{R}^m$ is called *K-Lipschitz* if $\|f(x) - f(y)\|_m \leq K\|x - y\|_n$ for all $x, y \in \mathbb{R}^n$, where $\|\cdot\|_n$ is the Euclidean norm in $\mathbb{R}^n$.

[28] As usual, H^n is the n-dimensional Hausdorff measure, see (1.22) for the definition.

Chapter 4
Algebra

4.1 Linear Algebra

Matrices and Their Properties

Matrices are central objects of study in linear algebra, and a large body of literature is devoted to studying their properties. In some cases, the aim is to extend to matrices some results and methods which are known to work with real numbers.

For example, to prove that a multivariate polynomial in several real variables is non-negative, a standard approach is to try to represent it as a sum of squares of other polynomials. This method, however, does not always work because there exist non-negative polynomials, e.g. the polynomial

$$P(x, y, z) = x^6 + y^4 z^2 + y^2 z^4 - 3x^2 y^2 z^2$$

discovered by Motzkin [915] in 1967, which has no sum of squares representations. Hilbert's famous 17th problem asks if every non-negative polynomial P can be represented as a sum of squares of rational functions, that is,

$$P = \sum_{k=1}^{m} \left(\frac{P_k}{Q_k} \right)^2$$

for some integer $m > 0$ and polynomials $P_1, \ldots, P_m, Q_1, \ldots, Q_m$. In 1927, Artin [52] proved that this is indeed the case, while Delzell [341] developed an algorithm to actually find such a sum-of-squares decomposition. This gives a general method for proving that a non-negative polynomial P is indeed non-negative.

In 2002, Helton [598] obtained a similar result in the non-commutative (matrix) setting. To formulate the result, we need a few definitions. Consider a set of $2n$ variables, which we denote by $x_1, x_2, \ldots, x_n, x_1^T, x_2^T, \ldots, x_n^T$. A *non-commutative monomial* in these variables is any expression of the form $a y_1 y_2 \ldots y_m$, where a

© The Author(s), under exclusive license to Springer Nature Switzerland AG 2021
B. Grechuk, *Landscape of 21st Century Mathematics*,
https://doi.org/10.1007/978-3-030-80627-9_4

is a real coefficient and each y_i is equal to either x_j or x_j^T for some j. A *non-commutative polynomial* $Q(\cdot)$ is the sum of any number of such monomials. If we substitute into $Q(\cdot)$ any real matrices $A_1, \ldots, A_n$ of any dimension $r \times r$ instead of the variables $x_1, x_2, \ldots, x_n$, and their transposes $A_1^T, \ldots, A_n^T$ instead of the variables $x_1^T, \ldots, x_n^T$, then $Y = Q(A_1, \ldots, A_n, A_1^T, \ldots, A_n^T)$ is an $r \times r$ matrix. Let $S(Q)$ be the set of all matrices Y which can be obtained in this way. If all matrices in $S(Q)$ are symmetric, we call $Q(\cdot)$ *symmetric*. If all matrices in $S(Q)$ are positive semidefinite,[1] we call $Q(\cdot)$ *matrix-positive*. Finally, we say that $Q(\cdot)$ is a *sum of squares* if $Q(\cdot) = \sum_{i=1}^k h_i(\cdot)^T h_i(\cdot)$, where $h_i(\cdot)$, $i = 1, 2, \ldots, k$, are non-commutative polynomials in the same set of variables.

We now are ready to state Helton's theorem [598].

Theorem 4.1 *A non-commutative symmetric polynomial is matrix positive if and only if it is a sum of squares.*

There are many other methods, concepts, and operations which can be generalized from numbers to matrices. One prominent example is exponentiation. The *exponential* $\exp(A)$ of an $n \times n$ matrix A with complex entries a_{ij} is the $n \times n$ matrix given by

$$\exp(A) = \sum_{k=0}^{\infty} \frac{1}{k!} A^k.$$

The *trace* of A is

$$\operatorname{tr}[A] = \sum_{i=1}^{n} a_{ii}. \tag{4.1}$$

In 1975, Bessis et al. [136] conjectured that the function $f(t) = \operatorname{tr}[\exp(A - tB)]$, $t \geq 0$, can be represented as the Laplace transform of a positive measure μ on $[0, \infty)$. This conjecture became known as the *BMV conjecture*, was extensively studied and had a number of equivalent formulations. In 2013, Stahl [1151] proved the conjecture.

Theorem 4.2 *Let A and B be two $n \times n$ Hermitian matrices and let B be positive semidefinite. For $t \geq 0$, let $f(t) = \operatorname{tr}[\exp(A - tB)]$. Then there exists a positive measure μ on $[0, \infty)$ (which may depend on A and B), such that*

$$f(t) = \int_0^{\infty} e^{-ts} d\mu(s) \tag{4.2}$$

for all $t \geq 0$.

---

[1] In general, the *eigenvalues* of an $n \times n$ matrix A with complex entries a_{ij} are the (possibly complex) solutions to the equation $\det(A - \lambda I) = 0$, where det denotes the determinant. A matrix A is called *Hermitian* if $a_{ij} = \bar{a}_{ji}$ for all i, j. A Hermitian matrix is called *positive semidefinite* if all its eigenvalues λ_i are non-negative, and *positive definite* if all λ_i are positive.

Our next example of generalization from numbers to matrices is the concept of least square mean. Given n real numbers $x_1, \ldots, x_k$, their *least square mean* is the real number x which minimizes $\sum_{i=1}^{k} (x_i - x)^2$. An appropriate generalization of this concept to positive definite matrices is the Karcher mean. Let P_n be the set of all positive definite $n \times n$ matrices. For $A, B \in P_n$, we write $A \geq B$ if $A - B$ is positive semidefinite. The *trace metric distance* between $A, B \in P_n$ is

$$\delta(A, B) = \sqrt{\sum_{i=1}^{n} \log^2 \lambda_i (A^{-1} B)},$$

where $\lambda_i(X), i = 1, \ldots, n$, denote the eigenvalues of $n \times n$ matrix X. The *least square mean* (also known as the *Karcher mean*) of k matrices $A_1, \ldots, A_k \in P_n$ is

$$\sigma(A_1, \ldots, A_k) = \arg \min_{X \in P_n} \sum_{i=1}^{k} \delta^2(X, A_i).$$

The concept of the Karcher mean was developed by Grove and Karcher [539] in 1973 for general metric spaces, and applied to positive definite matrices by Moakher [900] in 2005. In 2006, Bhatia and Holbrook [145, 146] derived some properties of the Karcher mean (consistency with scalars, joint homogeneity, permutation invariance, etc.) and conjectured that it satisfies another important property which they called "monotonicity". In 2011, Lawson and Lim [773] proved this conjecture.

Theorem 4.3 *If $B_i \leq A_i, i = 1, \ldots, k$, then $\sigma(B_1, \ldots, B_k) \leq \sigma(A_1, \ldots, A_k)$.*

The next concept we will discuss is the spectral radius. The *spectral radius* $\rho(A)$ of a square matrix A is the largest absolute value of an eigenvalue of A. The spectral radius is useful in a wide range of contexts, for example, in the study of dynamical systems defined by the repeated action of the linear transformation defined by the matrix A. To study dynamical systems defined by several transformations, the concept of generalized spectral radius is useful. Let Σ be a finite set of $n \times n$ matrices. Let

$$\rho_k(\Sigma) = \max\{\rho(A_1 A_2 \ldots A_k)^{1/k}, A_i \in \Sigma, i = 1, 2, \ldots, k\}.$$

The *generalized spectral radius* of Σ is the quantity $\rho(\Sigma) = \limsup_{k \to \infty} \rho_k(\Sigma)$. It is known that $\rho(\Sigma) \geq \rho_k(\Sigma)$ for all $k \geq 0$. In 1995, Lagarias and Wang [756] conjectured that, for every Σ, there is a k such that $\rho(\Sigma) = \rho_k(\Sigma)$. This statement became known as the *finiteness conjecture*. In 2002, Bousch and Mairesse [194] disproved this conjecture.

Theorem 4.4 *There exists a finite set Σ of matrices such that $\rho(\Sigma) > \rho_k(\Sigma)$ for all $k \geq 0$.*

Theorem 4.4 was proved by constructing an example of such a set Σ, which in fact consists of two 2×2 matrices.

Tensors

Tensors are natural generalizations of matrices to higher dimensions and are central in many applications. In particular, 3-dimensional tensors serve as a natural representation of bilinear systems of equations in the same way as matrices are used to represent linear systems. In 1973, Strassen [1166] conjectured that the *complexity* (that is, the number of multiplicative operations needed for solution) of the union of two bilinear systems depending on different variables is equal to the sum of the complexities of both systems. This conjecture can be conveniently formulated in the language of tensors.

Let $n_1, n_2, \ldots, n_d$ be positive integers. Let $\mathcal{I}$ be the set of vectors $i = (i_1, i_2, \ldots, i_d)$ such that all i_j are integers and $1 \leq i_j \leq n_j$ for $j = 1, 2, \ldots, d$. A function $T : \mathcal{I} \to \mathbb{R}$ is called a *d-dimensional $n_1 \times n_2 \times \cdots \times n_d$ tensor*. For example, a 3-dimensional tensor with real entries is a function $T : I \times J \times K \to \mathbb{R}$, where I, J, K are finite sets. Such a tensor T is called *decomposable* if there exist real numbers a_i, $i \in I$, and b_j, $j \in J$, and c_k, $k \in K$, such that $T(i, j, k) = a_i b_j c_k$ for all $(i, j, k) \in I \times J \times K$. The *rank* rank$(T)$ of a tensor T is the smallest r for which T can be written as a sum of r decomposable tensors. The *direct sum* $T = T_1 \oplus T_2$ of tensors $T_1 : I_1 \times J_1 \times K_1 \to \mathbb{R}$ and $T_2 : I_2 \times J_2 \times K_2 \to \mathbb{R}$ (with disjoint indexing sets $I_1, I_2, J_1, J_2, K_1, K_2$) is the tensor $T : I \times J \times K \to \mathbb{R}$, where $I = I_1 \cup I_2$, $J = J_1 \cup J_2$, $K = K_1 \cup K_2$, such that

(i) $T(i, j, k) = T_1(i, j, k)$ if $i \in I_1, j \in J_1, k \in K_1$,
(ii) $T(i, j, k) = T_2(i, j, k)$ if $i \in I_2, j \in J_2, k \in K_2$, and
(iii) $T(i, j, k) = 0$ for all other $(i, j, k) \in I \times J \times K$.

Strassen's 1973 conjecture stated that rank$(T_1 \oplus T_2) = $ rank$(T_1) + $ rank(T_2) for any tensors T_1, T_2. A similar statement is true for matrices, and was widely believed to be true for tensors. This belief was supported by some partial positive results. For example, Buczynski et al. [220] proved the conjecture in the case when the rank of one of the tensors is at most 6. However, Shitov [1117] proved in 2019 that in general the conjecture is false.

Theorem 4.5 *There exist 3-dimensional tensors T_1 and T_2 with real entries such that*

$$\text{rank}(T_1 \oplus T_2) < \text{rank}(T_1) + \text{rank}(T_2).$$

In fact, Shitov [1117] also disproved the same conjecture for tensors whose entries lie in an arbitrary infinite field.

Tensors are extensively used in computation. However, storing a d-dimensional tensor requires memory which grows exponentially in d, and performing any operations requires exponential time. In 2011, Oseledets [956] introduced the idea of a tensor-train decomposition, a way to store tensors and do computations on them with significantly less time and memory resources.

Let A be a d-dimensional $n_1 \times n_2 \times \cdots \times n_d$ tensor. Let $r_k, k = 0, 1, 2, \ldots, d$ be positive integers such that $r_0 = r_d = 1$. We say that A has a *tensor-train decomposition* (*TT decomposition* for short) with *TT-ranks* r_k if $A(i_1, i_2, \ldots, i_d) = G_1(i_1) G_2(i_2) \ldots G_d(i_d)$, where $G_k(i_k)$ is an $r_{k-1} \times r_k$ matrix. For $k = 0, 1, \ldots, d$, the unfolding matrix A_k of A is a $(\prod_{s=1}^{k} n_s) \times (\prod_{s=k+1}^{d} n_s)$ matrix with rows indexed by $(i_1, \ldots, i_k)$ and columns indexed by $(i_{k+1}, \ldots, i_d)$ such that

$$A_k((i_1, \ldots, i_k), (i_{k+1}, \ldots, i_d)) = A(i_1, \ldots, i_d).$$

Theorem 4.6 *For each d-dimensional tensor A there exists a TT decomposition with TT-ranks $r_k \leq \mathrm{rank}(A_k)$, $k = 0, 1, \ldots, d$.*

Theorem 4.6 guarantees the existence of a TT decomposition and provides upper bounds on the TT-ranks. Oseledets [956] also explained how to use the TT decomposition for performing basic linear algebra operations with high-dimensional tensors more efficiently.

Linear Groups
The set of all invertible $n \times n$ matrices with entries from a field F form a group with matrix multiplication as the group operation. This group is denoted $\mathrm{GL}_n(F)$ and is called the *general linear group* of degree n over field F. A group is said to be *linear* if it is isomorphic to a subgroup of $\mathrm{GL}_n(F)$ for some natural number n and some field F. Linear groups have some useful properties and establishing that a given group G is linear gives us these properties "for free". However, proving the linearity of a group may be a highly non-trivial task. For example, this question has long been open for the braid groups.

Let X_1 be a set of m points in the coordinate plane with coordinates $(x, 1)$, $(x, 2)$, $\ldots$, (x, m), respectively, and X_2 be a parallel set of points with coordinates $(y, 1)$, $(y, 2), \ldots, (y, m)$ for some $y > x$. Assume that we connect points from X_1 to some points from X_2 using m (not necessarily straight) strands, going from left to right, such that no strands have a common end. If two strands intersect, one should be put above the other, and this then cannot be changed without destroying the strands. This whole configuration is called a *braid*. The exact form of the strands, as well as the exact values of x and y, are not important: if one configuration can be transformed to another without destroying the strands, it is considered to be the same braid. What is important, is how the strands are intertwined. Let X be a braid as above, and let Y be another braid, going from points in X_2 to a set of points X_3 with coordinates $(z, 1), (z, 2), \ldots, (z, m)$ for some $z > y > x$. Then X and Y form another braid Z, consisting of m strands going from points in X_1 to points in X_3, see Fig. 4.1. Then we call Z the product of X and Y, and write $Z = X \times Y$. The set B_m of all braids with m strands form a group with the $\times$ operation, which is called the *braid group*.

In 1935, Burau [229] suggested a way to represent elements of any braid group B_m as matrices over some field. However, he did not prove whether this representation is faithful (that is, whether different braids corresponds to different matrices). In 1991, Moody [904] showed that this not always the case. In 2002, Krammer [749]

Fig. 4.1 Braids and their products

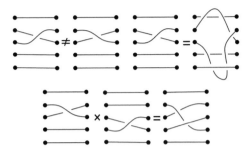

and indepedently Bigelow [147] proved that a different representation, suggested by Krammer [748] in 2000, is faithful, thus establishing that all braid groups are linear.

Theorem 4.7 *For every m, there exist a natural number n, a field K, and a map ρ assigning to any braid X with m strands an $n \times n$ matrix $\rho(X)$ over the field K, such that*

(a) $\rho(X \times Y) = \rho(X) \cdot \rho(Y)$ *for any two braids X,Y, and*
(b) $\rho(X) \neq \rho(Y)$ *whenever $X \neq Y$ (different matrices correspond to different braids).*

In other words, all braid groups are linear.

Representation Theory
Theorem 4.7 states that the braid groups are linear, that is, their elements can be represented as matrices in such a way that (a) multiplication is preserved and (b) different group elements correspond to different matrices. In this section, we discuss representations of group elements as matrices without requiring (b). For simplicity, we only consider matrices with complex entries.

Let G be a finite group. A *representation* of G is a map which puts into correspondence to every $g \in G$ an invertible $k \times k$ matrix $\rho(g)$ with complex entries such that

$$\rho(g \cdot h) = \rho(g) \cdot \rho(h), \quad \text{for all} \quad g, h \in G,$$

where $\cdot$ on the left-hand side is the operation in G and $\cdot$ on the right-hand side is matrix multiplication. A subspace $W \subset \mathbb{C}^k$ is called an *invariant subspace* of ρ if $\rho(g)v \in W$ for every $v \in W$ and every $g \in G$. Subspaces $W = \{0\}$ and $W = \mathbb{C}^k$ are always invariant subspaces, which are called *trivial*. A representation ρ is called *irreducible* if it has no non-trivial invariant subspaces.

The map $\chi_\rho : G \to \mathbb{C}$ given by $\chi_\rho(g) = \text{tr}(\rho(g))$, where tr is the trace defined in (4.1), is called the *character* of the representation ρ. If ρ is an irreducible representation, χ_ρ is called an *irreducible character*. The size k of the matrix $\rho(g)$ is called the *degree* of the character χ_ρ. Equivalently, the degree of χ_ρ is $\chi_\rho(1)$, where 1 is the identity element of G.

It is known that every finite group has a finite number of irreducible characters. Thus, it makes sense to count characters with various properties, for example, irreducible characters of odd and even degrees. In 1972, McKay [865] made a fundamental conjecture about the number of irreducible characters of odd degree in a group G. Let us denote this number by $m_2(G)$. To formulate the conjecture, we need a few more definitions. The order n of a finite group G is the number of elements in it. For a prime p, let m be the largest integer such that n is divisible by p^m. Any subgroup H of G of order p^m is called a *Sylow p-subgroup* of G. The *normalizer* $N_G(H)$ in G of a subgroup H is the set of elements $g \in G$ such that $g^{-1}Hg = H$. It is known that $N_G(H)$ is a subgroup of G containing H. McKay conjectured that if H_2 is a Sylow 2-subgroup of G, then $m_2(G) = m_2(N_G(H_2))$.

The conjecture was made by just looking at the number of odd degree irreducible characters of small groups. However, it turned out to be very deep, and had a significant influence on the development of representation theory. Researchers have made a series of stronger and far-reaching conjectures, but, for a long time, could not prove even the original one. Finally, culminating a long chain of partial results, Malle and Späth [833] proved the conjecture in 2016.

Theorem 4.8 *Let G be a finite group and let H_2 be a Sylow 2-subgroup of G. Then the groups G and $N_G(H_2)$ have the same number of irreducible characters of odd degree.*

In addition to counting the number of irreducible characters of odd degree, one may also study their values. If χ is a character of a finite group G, the field of values $\mathbb{Q}(\chi)$ of χ is the smallest field containing all values $\chi(g)$ for all $g \in G$. A deep question in representation theory is to understand, given a prime p, what fields can be represented as $\mathbb{Q}(\chi)$ for some irreducible character χ of a finite group G, such that its degree $\chi(1)$ is not divisible by p. This question is highly non-trivial even for the special case of $p = 2$ (corresponding to the odd degree characters) and fields $\mathbb{Q}(\sqrt{d})$, where d is a square-free integer and $\mathbb{Q}(\sqrt{d})$ is the field of numbers of the form $a + b\sqrt{d}$ for rational a, b. In 2019, Isaacs et al. [639] proved that if χ is an odd degree irreducible character of a finite group G, and $\mathbb{Q}(\chi) = \mathbb{Q}(\sqrt{d})$ for a square-free $d \neq -1$, then d must be congruent to 1 modulo 4. In 2020, Navarro and Tiep [926] proved the highly non-trivial converse statement: for every square-free d congruent to 1 modulo 4, one can find a finite group G which has an odd degree irreducible character χ whose field of values $\mathbb{Q}(\chi)$ is exactly $\mathbb{Q}(\sqrt{d})$. Together, the main results of [639] and [926] imply the following characterization.

Theorem 4.9 *If $d \neq -1$ is a square-free integer, then $\mathbb{Q}(\chi) = \mathbb{Q}(\sqrt{d})$ for some odd-degree irreducible character χ of some finite group G precisely when $d \equiv 1 \pmod 4$.*

In fact, Theorem 4.9 is just one of many corollaries of the main theorems of [926] that give a complete characterization of all fields which can be represented as $\mathbb{Q}(\chi)$ for some odd-degree irreducible character χ of a finite group G, thus resolving the $p = 2$ case of the deep question described above. The authors also formulated a plausible conjecture and show how the truth of their conjecture would imply the answer to this question for all p.

4.2   Growth in Groups

Groups of Intermediate Growth

A *generating set* of a group G is a subset $S \subset G$ such that every $g \in G$ can be written as a finite product of elements of S and their inverses. A group G is called *finitely generated* if it has a finite generating set S. Let $\gamma_S(n)$ be the number of elements of G which are the product of at most n elements in $S \cup S^{-1}$. We say that a group G

(i) has a *polynomial growth rate* if $\gamma_S(n) \le C(n^k + 1)$ for some constants $C, k < \infty$;

(ii) has an *exponential growth rate* if $\gamma_S(n) \ge a^n$ for some $a > 1$, and

(iii) is of *intermediate growth* if neither (i) nor (ii) is true.

Properties (i)–(iii) do not depend on the choice of the generating set S but are properties of the underlying group G. The rate of growth of $\gamma_S(n)$ can tell us a lot about the structure of G, and studying this rate is one of the central research directions in group theory. The existence of groups of intermediate growth was an open question (asked by Milnor in 1968), until the first examples were constructed by Grigorchuk [531] in 1980. The simplest such example is the first Grigorchuk group, defined below.

Let T be the set of all finite strings of 0s and 1s together with the empty string $\emptyset$. For $x, y \in T$, we write $x < y$ if x is an initial segment of y. Let Aut(T) be the set of all length-preserving permutations $\sigma : T \to T$ such that $\sigma(x) < \sigma(y)$ whenever $x < y$. With the composition operation $(\sigma_1 \cdot \sigma_2)(x) = \sigma_1(\sigma_2(x))$, Aut($T$) forms a group. Let a, b, c, d be specific elements of Aut(T) defined recursively by the following relations, which hold for all strings x: $a(0) = 1, a(1) = 0, b(0) = c(0) = d(0) = 0, b(1) = c(1) = d(1) = 1, a(0x) = 1x, a(1x) = 0x, b(0x) = 0a(x),$ $b(1x) = 1c(x), c(0x) = 0a(x), c(1x) = 1d(x), d(0x) = 0x, d(1x) = 1b(x)$. The *first Grigorchuk group* G^* is defined as the subgroup of Aut(T) generated by a, b, c, d.

For any generating set S of G^*, let $\gamma_S^*(n)$ be the number of elements of G^* which are the product of at most n elements in $S \cup S^{-1}$. In 1984, Grigorchuk [532] proved that

$$\exp(cn^{1/2}) \le \gamma_S^*(n) \le \exp(Cn^\beta),$$

where c, C are positive constants, and $\beta = \log_{32} 31 < 1$. This implies that the group G^* is of intermediate growth. From Grigorchuk's bound it is clear that the limit $\lim_{n \to \infty} \frac{\log \log \gamma_S^*(n)}{\log n}$ (which is called the *volume exponent* of G^*), if it exists, is between $1/2$ and $\log_{32} 31$. In 2020, Erschler and Zheng [430] proved that this limit exists and computed it exactly.

Theorem 4.10 *For any generating set S of the first Grigorchuk group G^*, we have*

$$\lim_{n \to \infty} \frac{\log \log \gamma_S^*(n)}{\log n} = \alpha_0,$$

where $\alpha_0 = 0.7674...$ is the constant defined by $\alpha_0 = \frac{\log 2}{\log \lambda_0}$, where λ_0 is the positive solution to the equation $x^3 - x^2 - 2x - 4 = 0$.

The proof of Theorem 4.10 proceeds by constructing a random walk on G^* with power-law tail decay such that the rate of this decay is related to the volume exponent of G^*, and then computing this decay rate.

While the first Grigorchuk group remains the simplest known example of a group with intermediate growth, later authors constructed many other examples with interesting properties. In particular, Kassabov and Pak [689] constructed in 2013 examples for which the volume exponent does not exist.

Theorem 4.11 *For every increasing function $\mu : \mathbb{N} \to \mathbb{R}_+$ such that $\lim_{n\to\infty} \frac{\mu(n)}{n} = 0$, there exists a finitely generated group G and its finite generating set S such that (i) $\gamma_S(n) < \exp(n^{4/5})$ for infinitely many $n \in \mathbb{N}$, but (ii) $\gamma_S(n') > \exp(\mu(n'))$ for infinitely many $n' \in \mathbb{N}$.*

In other words, Theorem 4.11 proves the existence of groups whose growth is not only intermediate but *oscillating*: $\gamma_S(n)$ is smaller than $\exp(n^{4/5})$ for some values of n, but is higher than, say, $\exp(n/\log\log\log n)$ for other values of n.

A group G is called *simple* if it does not have any normal subgroups,[2] except for itself and $\{e\}$, where $\{e\}$ is the set consisting only of the identity element. In 1984, Grigorchuk [532] asked whether there exists groups of intermediate growth which are at the same time simple. All the examples of groups of intermediate growth constructed by Grigorchuk [532], Kassabov and Pak [689], and many other authors, are not simple, and this question remained open until 2018. In 2018, Nekrashevych [928] answered it affirmatively.

Theorem 4.12 *There exist finitely generated simple groups of intermediate growth.*

Theorem 4.12 has some important corollaries. In particular, it generalizes (and immediately implies) Theorem 4.35 discussed later in this book.

Uniform Exponential Growth
For a finitely generated group G with generating set S, let

$$e_S(G) = \lim_{n \to \infty} \sqrt[n]{\gamma_S(n)}.$$

---

[2] A subgroup H of a group G is called *normal* if $g \cdot a \cdot g^{-1} \in H$ for any $a \in H$ and $g \in G$.

It is known that if $e_S(G) > 1$ for some S then in fact $e_S(G) > 1$ for all S, and, in this case, we say that G has *exponential growth*. If, moreover,

$$\inf_S e_S(G) > 1,$$

where the infimum is taken over all finite generating sets S of G, we say that G has *uniform exponential growth*. In 1981, Gromov [534] asked whether every group of exponential growth has uniform exponential growth. The question was answered affirmatively in many special cases. In particular, Eskin et al. [432] gave a positive answer for the subgroups of $\mathrm{GL}_n(\mathbb{C})$.

Theorem 4.13 *Let G be a finitely generated subgroup of $\mathrm{GL}_n(\mathbb{C})$. If G has exponential growth, then it has uniform exponential growth.*

However, Wilson [1267] proved in 2004 that in general the answer to Gromov's question is negative.

Theorem 4.14 *There exists a finitely generated group G which has exponential growth but not uniform exponential growth.*

The Dehn Function of a Finitely Presented Group

Let S be a set of symbols which we call *generators*. For each $x \in S$, we also introduce a symbol x^{-1}. A *word* is any finite string of symbols $w_1 \ldots w_n$ where each w_i is either x or x^{-1} for some $x \in S$. Let us call the words xx^{-1} and $x^{-1}x$, $x \in S$ "trivial". Words w_1 and w_2 are called *equivalent* if they can be transformed into each other by insertion of trivial words in any place or their deletion. The set of (equivalence classes of) all words from a group with operation $w_1 \cdot w_2 = w_1 w_2$ and the empty word as the identity element. This group is denoted F_S and is called the *free group* with generating set S.

Let $R \subset F_S$ be a set of words. We say that words $w_1 \in F_S$ and $w_2 \in F_S$ are *R-equivalent* if they can be transformed into each other by insertions or deletions either of trivial words or words from the set R. The set of equivalence classes of words under R-equivalence form a group with the same operation $w_1 \cdot w_2 = w_1 w_2$ which we denote by $\langle S|R \rangle$. If a group G is isomorphic to a group $\langle S|R \rangle$ for some set S and some $R \subset F_S$, we will call $\langle S|R \rangle$ a *presentation* of G, where S is called the set of *generators*, and R is called the set of *relators*. It is easy to see that a group G is finitely generated if and only if it has a presentation $\langle S|R \rangle$ with a finite set S. A group G is called *finitely presented* if it has a presentation $\langle S|R \rangle$ in which both S and R are finite sets.

Let G be a finitely presented group with presentation $\langle S|R \rangle$. Let $E_G \subset F_S$ be the set of words R-equivalent to the empty word. For each $w \in E_G$, there is a sequence of steps that reduces w to the empty word, where each step is an insertion or deletion of a relator (insertions and deletions of trivial words are not counted). We call the number of steps in this sequence its *cost*, and let $T(w)$ be the minimum cost of a sequence that transforms w into an empty word. The *Dehn function* of G

is $\text{Dehn}_G(n) = \max_w T(w)$, where the maximum is taken over words $w \in E_G$ of length at most n.

For functions $f; g : \mathbb{N} \to \mathbb{N}$, we write $g \succeq f$ if there is a constant c such that $f(n) \leq cg(cn + c) + c$ for all n, and

$$f \sim g \quad \text{if} \quad g \succeq f \quad \text{and} \quad f \succeq g. \tag{4.3}$$

The *growth rate* of a function f is the set of all functions g such that $f \sim g$. It is known that the growth rate $\text{Dehn}_G(n)$ of the Dehn function depends only on the group G and not on its finite presentation $\langle S|R \rangle$. The growth rate of the Dehn function is a fundamental invariant of a finitely presented group, which connects group theory, the theory of algorithms, and geometry.

Let $\text{SL}_n(\mathbb{Z})$ be the group of $n \times n$ matrices with integer entries and determinant 1. For $G = \text{SL}_n(\mathbb{Z})$, the Dehn function grows linearly for $n = 2$, and exponentially fast for $n = 3$, see [392, Ch. 10.4]. For $n \geq 4$, Thurston conjectured that it grows quadratically, that is, $\text{Dehn}_G(n) \sim n^2$. In 2013, Young [1285] confirmed this conjecture for $n \geq 5$.

Theorem 4.15 *For any $n \geq 5$, the Dehn function of $\text{SL}_n(\mathbb{Z})$ is quadratic.*

After Theroem 4.15 was proved, Thurston's conjecture remained open only for $n = 4$. In 2017, Leuzinger and Young posted online a preprint [784] in which they confirmed this conjecture in full.

Theorem 4.16 *For any $n \geq 4$, the Dehn function of $\text{SL}_n(\mathbb{Z})$ is quadratic.*

In fact, Leuzinger and Young proved a more general conjecture of Gromov, which predicted the quadratic growth of the Dehn function for a much broader family of groups, see [784] for details.

Growth in Semigroups

The growth rate can also be studied for semigroups. A *semigroup* is a set together with an associative binary operation on it. A *generating set* of a semigroup S is a subset $X \subset S$ such that every $s \in S$ can be written as a finite product of elements of X. Let X be finite and let $\gamma_X(n)$ be the number of elements of S which are the product of at most n elements from X. If X and Y are two finite generating subsets of S then $\gamma_X(n) \sim \gamma_Y(n)$, where $\sim$ is defined in (4.3) and is called "asymptotic equivalence". Hence, all finite generating sets are equivalent and we may speak about the growth function of S.

In 2019, Bell and Zelmanov posted online a preprint [112] in which they completely characterized, up to equivalence, all possible functions that can occur as growth functions of semigroups. If $\mathbb{N}$ is the set of non-negative integers, and $F : \mathbb{N} \to \mathbb{N}$ is a map, denote by $F'(n) = F(n) - F(n - 1)$, $F'(0) = F(0)$ the *discrete derivative* of $F(n)$.

Theorem 4.17 *A function $G : \mathbb{N} \to \mathbb{N}$ is asymptotically equivalent to the growth function of some semigroup if and only if G is asymptotically equivalent to a con-*

stant function, a linear function, or a non-decreasing function $F : \mathbb{N} \to \mathbb{N}$ such that (i) $F'(n) \geq n + 1$ for all n and (ii) $F'(m) \leq F'(n)^2$ for $n \geq 1$ and $m \in \{n, \ldots, 2n\}$.

We remark that, despite all efforts, there is no theorem of this level of generality in the case of groups. For example, an important conjecture of Grigorchuk predicts that there are no groups whose growth is superpolynomial but bounded above by $e^{\alpha n}$ for some $\alpha < 1/2$. Given that this remains open, it is clear that we are very far from a complete characterization of all possible growth functions for groups.

4.3   Simple Groups

Finite Simple Groups: Examples, Classification, and Expansion

Recall that a subset S of group G is called *normal* if $g \cdot a \cdot g^{-1} \in S$ for any $a \in S$ and $g \in G$. A group G is called *simple* if it does not have any normal subgroups, except for itself and $\{e\}$, where $\{e\}$ is the set consisting only of the identity element.

The simplest examples of simple groups are the *cyclic groups* $\mathbb{Z}_p$ for prime p, that is, groups of integers $\{0, 1, \ldots, p-1\}$ where the operation is addition modulo p.

Other examples are the alternating groups. Let $\mathrm{Sym}(n)$ be the set of all bijections $f : [n] \to [n]$, where $[n] = \{1, 2, \ldots, n\}$. The set $\mathrm{Sym}(n)$ forms a group with the composition operation, and this group is called the *symmetric group*. For $f \in \mathrm{Sym}(n)$, let $M(f)$ be the number of pairs $1 \leq i < j \leq n$ such that $f(i) > f(j)$. The set $\mathrm{Alt}(n)$ of all f such that $M(f)$ is even forms a subgroup of $\mathrm{Sym}(n)$ which is called the *alternating group*. The alternating groups $\mathrm{Alt}(n)$ are simple for all $n \geq 5$.

Our next examples are groups of matrices. Let $\mathbb{F}_p$ be the field of integers $\{0, 1, \ldots, p - 1\}$ with arithmetic operations defined modulo p. The set $\mathrm{SL}_n(\mathbb{F}_p)$ of $n \times n$ matrices over $\mathbb{F}_p$ with determinant 1 forms a group with matrix multiplication as the operation. If we call matrices A and $-A$ "equivalent", the set of equivalence classes form a subgroup of $\mathrm{SL}_n(\mathbb{F}_p)$ which is denoted $\mathrm{PSL}_n(\mathbb{F}_p)$. For $n = 2$, the group $\mathrm{PSL}_2(\mathbb{F}_p)$ is simple for every prime $p \geq 5$.

A major achievement of twentieth century mathematics is the classification theorem for finite simple groups. It states that if G is a finite simple group then G is either a cyclic group $\mathbb{Z}_p$ for some prime p, or an alternating group $\mathrm{Alt}(n)$ for some $n \geq 5$, or $\mathrm{PSL}_2(\mathbb{F}_p)$ for some prime $p \geq 5$, or... The statement of the theorem continues in this way and eventually all possible examples (some infinite families and some individual groups) of finite simple groups are explicitly listed! This classification theorem is very useful, because if we want to prove that all finite simple groups satisfy some property, we can "just" verify this property for cyclic groups, alternating groups, groups $\mathrm{PSL}_2(\mathbb{F}_p)$, and so on down the list until the end.

For example, this approach has been used to prove the important theorem that the Cayley graphs of non-abelian[3] finite simple groups are expanders. Recall that

---

[3] Recall a group G is called *abelian* if $a \cdot b = b \cdot a$ for all $a, b \in G$ and *non-abelian* otherwise.

for a finite group G with set of generators S_G, the (undirected) *Cayley graph* $\Gamma = \Gamma(G, S_G)$ is the graph whose vertices are elements of G, and vertices g, h are connected by an edge if $g = hs$ or $h = gs$ for some $s \in S_G$. Given $\epsilon > 0$, we say that $\Gamma(G, S_G)$ is an ϵ-*expander* if

$$|A \cdot S_G| \geq (1 + \epsilon)|A| \qquad \text{for all } A \subset G \text{ with } |A| < |G|/2, \qquad (4.4)$$

where $A \cdot S_G = \{a \cdot s \mid a \in A, \, s \in S_G\}$. Given an infinite family $\mathcal{G}$ of finite groups with generating sets S_G for $G \in \mathcal{G}$, a family $\Gamma(G, S_G)$ of their Cayley graphs form a family of *bounded degree expanders* if $|S_G| \leq L$ for some universal constant L and (4.4) holds for all $G \in \mathcal{G}$ with the same $\epsilon > 0$. See the discussion before Theorem 2.49 for the importance of expander graphs and their explicit generation as Cayley graphs of groups.

Many authors (including, for example, Babai, Hetyei, Kantor, Lubotzky and Seress [71]) asked if this method works if $\mathcal{G}$ is a family of symmetric groups $\mathrm{Sym}(n)$ or alternating groups $\mathrm{Alt}(n)$. In 2007, Kassabov [687] answered this question affirmatively.

Theorem 4.18 *There exist universal constants L and $\epsilon > 0$ such that the following holds. For every n there exists a generating set S_n of the symmetric group $\mathrm{Sym}(n)$ such that $|S_n| \leq L$ and the Cayley graphs $\Gamma(\mathrm{Sym}(n), S_n)$ form a family of ϵ-expanders. Similarly, there exist generating sets S'_n of the alternating groups $\mathrm{Alt}(n)$ with the same property.*

Theorem 4.18 became a cornerstone of the major project of proving the same property for all non-abelian finite simple groups. We can use the classification theorem and verify this property case by case. We do not need to consider the cyclic groups because they are abelian. The alternating groups are covered in Theorem 4.18. We next need to consider $\mathrm{PSL}_2(\mathbb{F}_p)$ groups, and so on down the list. Kassabov et al. [688] proceeded in this way and proved the result for all classes of non-abelian finite simple groups, except for one infinite family of groups called *Suzuki groups*.[4] In 2011, Breuillard et al. [209] successfully treated this remaining case and finished the proof of the following theorem.

Theorem 4.19 *There exist universal constants L and $\epsilon > 0$ such that for every non-abelian finite simple group G there exists a generating set S_G of size at most L such that (4.4) holds. In other words, the Cayley graphs $\Gamma(G, S_G)$ form a family of ϵ-expanders.*

Diameters of Finite Simple Groups

While the classification theorem is an extremely useful tool, it does not reduce all theorems about finite simple groups to easy case-by-case verification, because some properties are very difficult to prove even for specific families of groups like the alternating groups. Here we discuss one example. For a finite group G with

---

[4] See [209, 1169] for the definition of this class of groups.

generating set S, let $d(G, S)$ be the minimal integer m such that every $g \in G$ can be expressed as a product of at most m elements of S and their inverses. The *diameter* $\mathrm{diam}(G)$ of a group G is the maximal value of $d(G, S)$ over all choices of the generating set S. In 1992, Babai [74] conjectured the existence of constants M and c such that

$$\mathrm{diam}(G) \le M (\log |G|)^c \qquad (4.5)$$

for every non-abelian finite simple group G with $|G|$ elements.

With the classification theorem, we can verify the conjecture for classes of non-abelian finite simple groups one by one. In 2008, Helfgott [592] proved the conjecture for $G = \mathrm{PSL}_2(\mathbb{F}_p)$.

Theorem 4.20 *There exist absolute constants M and c such that, for every prime p, (4.5) holds for $G = \mathrm{SL}_2(\mathbb{F}_p)$ and $G = \mathrm{PSL}_2(\mathbb{F}_p)$.*

Theorem 4.20 was proved by showing that every subset of $G = \mathrm{SL}_2(\mathbb{F}_p)$ or $G = \mathrm{PSL}_2(\mathbb{F}_p)$ grows rapidly when it acts on itself by the group operation. Specifically, a key lemma states that for any prime p, and any subset $A \subset G$ which (i) is not contained in a proper subgroup of G and (ii) has $|A| < p^{3-\delta}$ elements for some fixed $\delta > 0$, we have

$$|A \cdot A \cdot A| > c|A|^{1+\epsilon}, \qquad (4.6)$$

where $c > 0$ and $\epsilon > 0$ are constants which depend only on δ. This is an example of a product theorem in a group, see Theorem 4.44 below and the discussion before it for more details on this fascinating topic. On the other hand, the authors proved that if $|A| > p^\delta$ for some fixed $\delta > 0$, then there is an integer $k = k(\delta) > 0$ such that every element $g \in G$ can be expressed as a product of at most k elements of $A \cup A^{-1}$. Theorem 4.20 is then deduced from the combination of these two facts.

In 2016, Pyber and Szabó [1012] proved several far-reaching generalizations of Theorem 4.20. Here, we present the following result, which is not the most general one from [1012] but is the easiest to state.

Theorem 4.21 *Let $n \ge 1$ be an integer, p be a prime, and let G be a simple subgroup of the group $\mathrm{SL}_n(\mathbb{F}_p)$. Then $\mathrm{diam}(G) \le (\log |G|)^{c(n)}$, where the constant $c(n)$ depends only on n but not on p and G.*

Theorem 4.21 was proved by first establishing a version of (4.6) for this family of groups (but with constant c depending on n), and then proceeding similarly to the proof of Theorem 4.20.

Theorem 4.20 proves Babai's conjecture (4.5) for the subgroup $\mathrm{PSL}_2(\mathbb{F}_p)$ of $\mathrm{SL}_2(\mathbb{F}_p)$. Theorem 4.21 proves it, more generally, for all simple subgroups of $\mathrm{SL}_n(\mathbb{F}_p)$ for every fixed n. However, Babai's conjecture predicts that Theorem 4.21 should be true with the universal constant c independent of n. If one could prove such a "uniform" version of Theorem 4.21 (and, more generally, of other theorems in [1012]), this would imply Babai's conjecture for all the families of finite simple groups except for the alternating groups. A key difficulty is that the product theorem (4.6) fails for such groups. If $G = \mathrm{Alt}(n)$ or $G = \mathrm{Sym}(n)$ Babai and Seress [73] proved in 1988 that

$$\mathrm{diam}(G) \leq \exp((1 + o(1))\sqrt{\log |G|}),$$

and this upper bound remained unimproved for over 25 years. In 2014, Helfgott and Seress [595] improved it significantly.

Theorem 4.22 *There exist a constant C such that*

$$\mathrm{diam}(G) \leq \exp(C (\log n)^4 \log \log n)$$

for all n, where G is either $\mathrm{Alt}(n)$ *or* $\mathrm{Sym}(n)$.

In terms of $|G|$, the bound in Theorem 4.22 is $\exp((\log \log |G|)^{O(1)})$. Such a bound is called *quasi-polynomial* in $\log |G|$. Whether the polynomial is $\log |G|$ bound (4.5) remains open.

Bound (4.5) is equivalent to saying that $d(G, S) \leq M (\log |G|)^c$ for every generating set S of G. In 2001, Liebeck and Shalev [798] proved an even stronger upper bound on $d(G, S)$ in the special case when the generating set S is normal.

Theorem 4.23 *There exists a constant c such that if G is a finite simple group and* $S \neq \{e\}$ *is a normal subset of G, then, for any* $m \geq c \log |G| / \log |S|$, *any element of G can be expressed as a product of m elements of S.*

In other words, Theorem 4.23 proves that $d(G, S) \leq c \log |G| / \log |S|$ for normal S. This bound is the best possible up to a constant factor. Theorem 4.23 has many applications, and one of them is discussed in the next section.

Waring's Problem for Finite Simple Groups
Using Theorem 4.23, Liebeck and Shalev [798] derived many interesting results. We mention one of them, which is about a Waring-type problem for finite simple groups. The 1909 *Hilbert–Waring theorem* [606] states that for each positive integer k there exists a positive integer $g(k)$ such that every natural number is the sum of at most $g(k)$ k-th powers. In 1996, Martinez and Zelmanov [849] proved a similar result for groups: for every integer $k > 0$ there exists an integer $c = c(k) > 0$ such that every element g of every non-abelian finite simple group G can be written as

$$g = x_1^k x_2^k \ldots x_c^k \tag{4.7}$$

for some $x_1, \ldots, x_c \in G$.

More generally, let $w = w(x_1, \ldots, x_d)$ be a non-trivial group word, that is, a non-identity element of the free group F_d with generators $x_1, \ldots, x_d$. For any group G, w induces a word map $w : G^d \to G$ such that $w(g_1, \ldots, g_d)$ is obtained by substitution of $g_1, \ldots, g_d$ into w instead of $x_1, \ldots, x_d$, respectively, and performing the group operation. More formally, if $w = \prod_{j=1}^{r} x_{i_j}^{\epsilon_j}$ for some

$i_1, \ldots, i_r \in \{1, \ldots, d\}$ and $\epsilon_j = \pm 1$, $j = 1, \ldots, r$, then

$$w(g_1, \ldots, g_d) = \prod_{j=1}^{r} g_{i_j}^{\epsilon_j}. \tag{4.8}$$

Denote by $w(G) = \{g \in G : g = w(g_1, \ldots, g_d), g_1, \ldots, g_d \in G\}$ the image of the word map w. Then, generalizing (4.7), one may ask whether there is a constant $c = c(w)$ such that[5]

$$w(G)^c = G \tag{4.9}$$

for every non-abelian finite simple group G. By (4.7), the answer is "yes" for the power word $w = x_1^k$. In 1994, Wilson [1266] proved this for the commutator word $w = x_1^{-1} x_2^{-1} x_1 x_2$. In 2001, Liebeck and Shalev [798] deduced from Theorem 4.23 that the answer is "yes" for all non-trivial group words.

This result sounds like the end of the story but in fact it was only the beginning. In (4.9), the constant $c = c(w)$ is unspecified and is allowed to depend on w. In 2009, Shalev [1106] proved that, for sufficiently large groups, the result remains true with a constant c independent of w. Moreover, one can take $c = 3$.

Theorem 4.24 *For any non-trivial group word w there exists a positive integer $N = N(w)$ such that for every finite simple group G with $|G| \geq N(w)$ we have $w(G)^3 = G$.*

The proof of Theorem 4.24 uses some probabilistic ideas, in combination with deep methods from algebraic geometry and representation theory. In particualr, the proof relies on studying the properties of the function $\xi_G(s) = \sum_{\chi \in \mathrm{Irr}(G)} \chi(e)^{-s}$, where e is the identity element of G and $\mathrm{Irr}(G)$ is the set of irreducible characters, see the discussion before Theorem 4.8 for the definition.

After proving Theorem 4.24, Shalev and co-authors started to investigate whether the result remains correct even with $c = 2$. Even more generally, Shalev conjectured that we have $w_1(G)w_2(G) = G$ for any *pair* of non-trivial group words w_1, w_2 and every large finite simple group G. To verify this conjecture, one can use the classification theorem and proceed case by case. In 2009, Larsen and Shalev [763] made the first significant step in this direction by proving the conjecture for alternating groups $\mathrm{Alt}(n)$.

Theorem 4.25 *For each pair of non-trivial words w_1, w_2 there exists an $N = N(w_1, w_2)$ such that for all integers $n \geq N$ we have $w_1(G)w_2(G) = G$, where $G = \mathrm{Alt}(n)$.*

---

[5] For subsets A, B of a group G, let $A \cdot B = \{g \in G : g = a \cdot b, a \in A, b \in B\}$. Denote by A^c the set $A \cdot A \cdots A$ (c times).

It is interesting to note that the proof of Theorem 4.25, in addition to probabilistic arguments and representation theory, uses results from analytic number theory, e.g. a version of Vinogradov's theorem about representation odd integers as sums of three primes, see Theorem 1.15 and the discussion before it.

In 2011, Larsen et al. [764] finished verifying the remaining cases and proved Shalev's conjecture in full.

Theorem 4.26 *For each pair of non-trivial words w_1, w_2 there exists an $N = N(w_1, w_2)$ such that for every finite non-abelian simple group G with $|G| \geq N$ we have $w_1(G) \cdot w_2(G) = G$.*

In the special case $w_1 = w_2 = w$, Theorem 4.26 implies that $w(G)^2 = G$. In other words, (4.9) is actually true with $c = 2$. In general, this result is the best possible, and (4.9) cannot be true with $c = 1$: for example, for the power word $w = x_1^k$ we have $w(G) \neq G$ for every finite simple group G of order non-coprime to k. However, for some *special* words even $c = 1$ is achievable.

A notable example is the commutator word $w = x_1^{-1} x_2^{-1} x_1 x_2$. An element $a \in G$ is called a *commutator* if $a = g^{-1} h^{-1} gh$ for some $g \in G$ and $h \in G$. In any abelian group, $g^{-1} h^{-1} gh = g^{-1}(h^{-1}h)g = g^{-1}eg = e$, hence only the identity element is a commutator. In general, the size of the set of commutators can be viewed as an indicator of "how far" the group is from being abelian. In 1951, Ore [954] conjectured that in every finite non-abelian simple group the set of commutators is the whole group, which makes such groups as "far" from being abelian as they possibly can. In the same paper [954], Ore proved the conjecture for the alternating groups.

As mentioned above, Wilson [1266] proved in 1994 that $w(G)^c = G$ for the commutator word w and every finite non-abelian simple group G. The Ore conjecture states that this is true with $c = 1$. In 2009, Shalev [1106] proved that one can take $c = 2$ for large G. In 2010, Liebeck et al. [797] proved the Ore conjecture in full.

Theorem 4.27 *Every element of every finite non-abelian simple group is a commutator.*

Similar to the proofs of many other theorems in this section, the proof of Theorem 4.27 uses representation theory. In this case, the connection to representation theory is based on a classical theorem of Frobenius, which states that that an element g of a finite group G is a commutator if and only if $\sum_{\chi \in \mathrm{Irr}(G)} \frac{\chi(g)}{\chi(e)} \neq 0$. In some important special cases, the authors managed to prove that $\frac{\chi(g)}{\chi(e)}$ is small for every non-trivial character[6] χ, and, because the trivial character contributes $+1$ to the sum, the whole sum becomes positive and cannot be 0.

There are some other words w such that $w(G) = G$ for every finite non-abelian simple group G. For example, Liebeck et al. [796] proved this for the word $w =$

---

[6] A character χ is called *trivial* if $\chi(g) = 1$ for all $g \in G$ and *non-trivial* otherwise.

x^2y^2 in 2012. In 2018, Guralnick et al. [543] extended this result to an infinite family of words in the form $w = x^N y^N$ for every N which has at most two distinct primes in its prime factorization.

Theorem 4.28 *Let p, q be prime numbers, let a, b be non-negative integers, and let $N = p^a q^b$. Then every element g of every finite non-abelian simple group G can be written as $g = x^N y^N$ for some $x, y \in G$.*

If G is a finite group and $g \in G$, let $N_G^w(g)$ be the number of $(g_1, \ldots, g_d) \in G^d$ such that $w(g_1, \ldots, g_d) = g$, where $w(g_1, \ldots, g_d)$ is the word map defined in (4.8). Two words w_1, w_2 are said to be *disjoint* if they are words in disjoint sets of variables. Let $w_1, w_2 \neq e$ be disjoint words and let $w = w_1 w_2$. In this notation, Theorem 4.26 states that for every sufficiently large finite non-abelian simple group G, we have $N_G^w(g) > 0$ for every $g \in G$. In the language of probability theory, this means that if we select $g_1, \ldots, g_d$ from G at random, and compute $w(g_1, \ldots, g_d)$, we can get any element $g \in G$ in this way with non-zero probability. In 2013, Shalev [1107] further asked whether $w(g_1, \ldots, g_d)$ has almost uniform distribution in G with respect to the L^1 norm.[7] In 2019, Larsen et al. [765] answered this question affirmatively.

Theorem 4.29 *Let $w_1, w_2 \neq e$ be disjoint words and let $w = w_1 w_2$. Then*

$$\lim_{|G| \to \infty} \left\| \frac{N_G^w(g)}{|G|^d} - \frac{1}{|G|} \right\|_{L^1} = 0,$$

where G ranges over all finite non-abelian simple groups.

Similar to the proofs of Theorems 4.24 and 4.26, the proof of Theorem 4.29 relies heavily on the classification of finite simple groups and proceeds using a case-by-case analysis.

Theorem 4.29 states that if w is the union of two disjoint words, then the distribution of $w(g_1, \ldots, g_d)$ in large non-abelian simple groups is approximately uniform. One may also ask for which words this distribution is *exactly* uniform in *all* finite groups. A word w is called *measure-preserving with respect to a finite group* G if $N_G^w(g) = N_G^w(h)$ for every $g, h \in G$. We say that w is *measure-preserving* if it is measure-preserving with respect to every finite group G. A word w is called *primitive* if it belongs to some basis (free generating set)[8] of the free group F_d. It is not difficult to check that every primitive word is measure-preserving, and several authors have conjectured that the converse is also true. In 2014, Puder [1010] proved this conjecture for words in F_2. In 2015, Puder and Parzanchevski [1011] confirmed the conjecture in full.

---

[7] For a function $f : G \to \mathbb{R}$, we write $\|f\|_{L^1} = \sum_{g \in G} |f(g)|$.

[8] A subset $S \subset F_d$ is a *free generating set* if F_d has a presentation $\langle S | R \rangle$ in which the set of relators R is the empty set.

Theorem 4.30 *Every measure-preserving word is primitive.*

Simple Groups and Algebraic Geometry

In some cases, it is a highly non-trivial problem to check if a given (infinite) group is simple. Here we discuss one such example arising from algebraic geometry. This is a field of mathematics which, roughly speaking, studies zeros of multivariate polynomials. It started a few centuries ago with the study of polynomials in two real variables, but has greatly expanded since then. Most modern theorems in this field are too technical to be included in this book, but in this section we will discuss one exception.

Algebraic geometers often work with projective spaces. Let k be a field. Let k^n be the n-dimensional vector space[9] over k. A vector $x \in k^n$ is a *zero vector* if all $x_i = 0$, where 0 is the additive identity of k, and *non-zero* otherwise. Let us call non-zero vectors $x, y \in k^n$ equivalent if $x_i = \lambda y_i$, $i = 1, \ldots, k$ for some $\lambda \in k$. For non-zero $x = (x_0, \ldots, x_n) \in k^{n+1}$, let us denote by $(x_0 : x_1 : \cdots : x_n)$ the equivalence class containing x. The set of all such equivalence classes is denoted $\mathbb{P}_k^n$ and is called the *n-dimensional projective space* over k.

Now interpret $x = (x_0, \ldots, x_n)$ as a vector of variables, and consider a sequence $\phi_0(x), \ldots, \phi_m(x)$ of same-degree homogeneous polynomials over[10] k. We say that this sequence has *no common factor* if there are no homogeneous polynomials $f_i(x)$ and $h(x)$, with degree of h greater than 0, such that $\phi_i(x) = h(x)f_i(x)$, $i = 0, \ldots, m$. Any such sequence defines a map ϕ from $\mathbb{P}_k^n$ to $\mathbb{P}_k^m$ which maps every $(x_0 : x_1 : \cdots : x_n) \in \mathbb{P}_k^n$ into $(\phi_0(x) : \cdots : \phi_m(x)) \in \mathbb{P}_k^m$. This map is not well-defined if $\phi_0(x) = \cdots = \phi_m(x) = 0$ but is well-defined for all other $x \in \mathbb{P}_k^n$. Any such map ϕ is called a *rational map* from $\mathbb{P}_k^n$ to $\mathbb{P}_k^m$. If there exists a rational map ϕ^{-1} from $\mathbb{P}_k^m$ to $\mathbb{P}_k^n$ such that $\phi^{-1} \circ \phi$ and $\phi \circ \phi^{-1}$ are both the identity maps wherever they are well-defined, where $\circ$ is the composition operation, then rational map ϕ is called *birational*. The set $\mathrm{Bir}(\mathbb{P}_k^n)$ of all the birational maps from $\mathbb{P}_k^n$ to itself forms a group with composition operation $\circ$, and this group is called the *Cremona group* of rank n over k.

Birational maps and the Cremona groups are among the central objects of study in a subfield of algebraic geometry called *birational geometry*. However, despite much effort, one of the simplest questions about the structure of such groups remained open for over a century. In 1895, Enriques asked whether the Cremona groups for $n \geq 2$ are simple groups, and this basic question remained open for a long time even if $k = \mathbb{C}$ is the field of complex numbers and $n = 2$. The first

---

[9] That is, the space of vectors $x = (x_1, \ldots, x_n)$ with each $x_i \in k$, and with addition and scalar multiplication defined component-wise.

[10] A *monomial* of degree d in variables $x_0, \ldots, x_n$ is an expression of the form $a \prod_{i=0}^{n} x_i^{m_i}$, where $a \in k$ is a coefficient and m_i are non-negative integers with $\sum_{i=0}^{n} m_i = d$. A *homogeneous polynomial* of degree d is the sum of any finite number of such monomials.

breakthrough happened in 2013, when Cantat et al. [248] resolved this problem for any algebraically closed[11] field k, which includes the case $k = \mathbb{C}$ as a special case.

Theorem 4.31 *If k is an algebraically closed field, the Cremona group $\mathrm{Bir}(\mathbb{P}^2_k)$ is not a simple group.*

The techniques developed to prove Theorem 4.31 opened the road for further progress. In 2016, Lonjou [812] extended Theorem 4.31 to the case when k is an arbitrary field. In 2019, Blanc et al. posted online a preprint [155] in which they also treated the much more difficult case of dimension $n \geq 3$, at least when k is a subfield of $\mathbb{C}$.

Theorem 4.32 *For each subfield $k \subseteq \mathbb{C}$ and each $n \geq 3$, the Cremona group $\mathrm{Bir}(\mathbb{P}^n_k)$ is not a simple group.*

4.4   Existence of Groups with Special Properties

Conjugacy Classes
Two elements a and b of a group G are called *conjugate* if $gag^{-1} = b$ for some $g \in G$. It is easy to see that conjugacy is an equivalence relation and therefore partitions G into equivalence classes, which are called *conjugacy classes*. In an abelian group $gag^{-1} = a$, hence conjugacy classes are one-element sets. An "opposite extreme" is a group G in which all non-identity elements are conjugate, hence G has exactly two conjugacy classes, $\{e\}$ and $G \setminus \{e\}$. A trivial example of such a group is the 2-element cyclic group $\mathbb{Z}_2$. Surprisingly, the existence of any other finitely generated group with exactly two conjugacy classes has long been an open question. In 2010, Osin [957] answered this question affirmatively, and in fact proved a much more general result which we formulate below.

We say that a group G is *2-generated* if it has a 2-element generating set. We say that a group H can be *embedded* into a group G if G has a subgroup isomorphic to H. The *order* of a non-identity element $a \in G$ is the lowest positive integer n such that $a^n = e$, with the convention that the order is infinite if no such n exists. For a group G, let $\pi(G)$ be the set of all (finite) positive integers n such that there exists an element $a \in G$ of order n.

Theorem 4.33 *Any countable group H can be embedded into a 2-generated group G such that any two elements of the same order are conjugate in G and $\pi(H) = \pi(G)$.*

A group is called *torsion-free* if all non-identity elements in it have infinite order. Applying Theorem 4.33 to a torsion-free group H, we obtain that any countable torsion-free group can be embedded into a torsion-free 2-generated

---

[11] Recall that a field k is called *algebraically closed* if every non-constant univariate polynomial with coefficients in k has a root in k.

group with exactly 2 conjugacy classes. In particular, this implies that there exist uncountably many pairwise non-isomorphic groups with these properties. More generally, Theorem 4.33 implies that for any integer $n \geq 2$ there are uncountably many pairwise non-isomorphic finitely generated groups with exactly n conjugacy classes.

Amenable Groups

In 1924, Banach and Tarski [93] proved that one can decompose a unit ball into a finite number of disjoint pieces, move and rotate the pieces, and construct two unit balls! This is known as the *Banach–Tarski paradox*. In 1929, von Neumann [933] introduced the notion of amenable groups, and proved that the Banach–Tarski paradox arises because the group $SO(3)$ of rotations in $\mathbb{R}^3$ is not amenable.

A group G is called *amenable* if it has a left-invariant finitely-additive probability measure, that is, a function μ from subsets of G to $[0, 1]$ such that

(i) $\mu(G) = 1$,
(ii) $\mu(A \cup B) = \mu(A) + \mu(B)$ whenever $A \cap B = \emptyset$, and
(iii) $\mu(gA) = \mu(A)$ for all $g \in G$, where $gA = \{h \in G \mid h = ga, \, a \in A\}$.

Von Neumann [933] proved that the group $SO(3)$ is not amenable. In fact, he proved that any group containing a free non-cyclic[12] subgroup is not amenable, and asked if the converse is true. In 1980, Ol'shanskii [949] proved the existence of a counterexample to von Neumann's problem. However, it remained open whether such counterexamples can be finitely presented. In 2003, Ol'shanskii and Sapir [950] answered this question affirmatively.

Theorem 4.34 *There exists a finitely presented non-amenable group without free non-cyclic subgroups.*

To prove Theorem 4.34, one does not need to prove any general statement about all groups, all that needed is to construct one example of a group satisfying the conditions of the theorem. Given this intuition, it may be surprising for some readers that the paper [950] with the construction and the proof that it works is 127 pages long.

Another long-standing open question about amenable groups was the existence of a finitely generated *simple* group G which is infinite and amenable. In 2013, Juschenko and Monod [667] proved that there are uncountably many such groups.

Theorem 4.35 *There exist finitely generated simple groups that are infinite and amenable. In fact, there are uncountably many non-isomorphic such groups.*

Theorem 4.35 was proved by confirming a 2012 conjecture of Grigorchuk and Medynets [530] that a specific group, called "the topological full group of a minimal Cantor system" is amenable. In an earlier work [858], Matui showed how to deduce Theorem 4.35 from this conjecture.

---

[12] A free group is called *non-cyclic* if it has at least 2 generators.

In 2018, Nekrashevych [928] proved the more general result that there exist finitely generated simple groups of intermediate growth (Theorem 4.12). This theorem generalizes (and implies) Theorem 4.35 because it is known that every group of intermediate growth is amenable.

A weak version of amenability, which has many equivalent formulations and is sufficient for many applications, is the notion of coarse amenability. A metric space $(X; d)$ is called *uniformly discrete* if there exists a constant $r > 0$ such that, for any $x, y \in X$, we have either $x = y$ or $d(x, y) > r$. A uniformly discrete metric space $(X; d)$ is *coarsely amenable* if for every $\epsilon > 0$ and $R > 0$, there exist a constant $S > 0$ and a collection of finite subsets $\{A_x\}_{x \in X}$, $A_x \subseteq X \times \mathbb{N}$ for every $x \in X$, such that (a) $|A_x \Delta A_y|/|A_x \cap A_y| \leq \epsilon$ when $d(x, y) \leq R$, and (b) $A_x \subseteq B(x, S) \times \mathbb{N}$. A finitely generated group G is called *coarsely amenable* if it is so for the word metric[13] with respect to a finite generating set S, and is called *coarsely non-amenable* otherwise. This property does not depend on the choice of S.

The notion of coarse amenability is related to the notion of coarse embeddability, see (3.33). We say that a finitely generated group G is *coarsely embeddable* into a metric space $(Y; d_Y)$ if it is so for the word metric with respect to a finite generating set S. Once again, this property does not depend on the choice of S. In particular, it is known that any finitely generated coarsely amenable group is coarsely embeddable into the Hilbert space l^2. Whether the converse is true (are groups coarsely embeddable into l^2 coarsely amenable?) is a natural question raised by a number of researchers. In 2020, Osajda [955] proved that this question has a negative answer.

Theorem 4.36 *There exist finitely generated coarsely non-amenable groups that are coarsely embeddable into the infinite-dimensional Hilbert space l^2.*

The main tool in the proof of Theorem 4.36 is the theory of small cancellation labellings of graphs. A *labelling* of a directed graph is an assignment of labels to its edges, and such a labelling is said to satisfy a "small cancellation condition" when no two different long paths receive the same labels. The authors reduced Theorem 4.36 to the problem of finding small cancellation labellings of certain graphs, and then the resolution of this problem finished the proof.

Left Orderable Groups
A group G with group operation $\cdot$ is called *left orderable* if there exists a total order $\leq$ on G such that $a \leq b$ implies $c \cdot a \leq c \cdot b$ for all $a, b, c \in G$. An example of such a group is the set of real numbers with operation $+$. This easy-looking property turns out to be surprisingly difficult to verify for some groups. For example, it was an intriguing open question whether the group of boundary fixing disk homeomorphisms is left-orderable.

Let $B = \{x \in \mathbb{R}^2 : |x| \leq 1\}$ and $S^1 = \{x \in \mathbb{R}^2 : |x| = 1\}$ be the unit disk and the unit circle in $\mathbb{R}^2$, respectively. A function $f : B \to B$ is called

---

[13] That is, the distance $d(g, h)$ between $g, h \in G$ is the smallest integer k such that $g^{-1}h$ can be written as a product of k elements of S or their inverses.

a *homeomorphism* if (i) f is a bijection, (ii) f is continuous, and (iii) the inverse function f^{-1} is continuous. We say that a function $f : B \to B$ is boundary fixing if $f(x) = x$ for all $x \in S^1$. We say that a function $h : B \to B$ is the *composition* of functions $f : B \to B$ and $g : B \to B$ if $h(x) = f(g(x))$ for all $x \in B$. It is known that the set of all boundary fixing homeomorphisms $f : B \to B$ with composition operation forms a group, which we denote by $\mathrm{Homeo}(B, S^1)$. The question whether the group $\mathrm{Homeo}(B, S^1)$ is left-orderable has been asked by many researchers and has been included in several collections of open problems in group theory, see, for example, [716]. In 2019, this question was resolved by Hyde [634].

Theorem 4.37 *The group* $\mathrm{Homeo}(B, S^1)$ *of boundary fixing homeomorphisms of B is not left-orderable.*

Theorem 4.37 was proved by constructing an explicit finitely generating subgroup H of $\mathrm{Homeo}(B, S^1)$ which is not left-orderable. Since any left-order of $\mathrm{Homeo}(B, S^1)$ would imply a left order on H, Theorem 4.37 follows. The whole proof is just a few pages long.

Theorem 4.37 is interesting but it deals with the specific group $\mathrm{Homeo}(B, S^1)$. A more fundamental question was posed by Rhemtulla in 1980: does there exist a finitely generated simple left orderable group? This question attracted a lot of attention and has been re-asked many times, see, for example, [716]. In 2019, Hyde and Lodha [635] answered it affirmatively.

Theorem 4.38 *There exists a finitely generated infinite simple left orderable group. In fact, there are continuum many non-isomorphic such groups.*

Profinitely Rigid Groups

The (*left*) *coset* of a subgroup N in a group G with respect to $g \in G$, denoted gN, is the set $\{g \cdot a : a \in N\}$. If N is normal, the cosets form a group with operation $(gN) \cdot (hN) = (g \cdot h)N$, which is called the *quotient group* and denoted G/N. If G/N is a finite group, we say that N has a *finite index*. A group G is called *residually finite* if the intersection of all its normal subgroups of finite index is the identity element. Let $c(G)$ be the set of all finite quotients of G. A finitely generated, residually finite group G is called *profinitely rigid in the absolute sense*, if, for any finitely generated residually finite group H, $c(G) = c(H)$ (up to isomorphism) implies that H is isomorphic to G.

One of the fundamental questions in the theory of finitely generated groups is to what extent such groups are determined by their set of finite quotients. To have a hope to recover a group G from its finite quotients $c(G)$, we must assume that G is residually finite. It is conjectured that a broad class of residually finite groups are recoverable, but, before 2020, this was known only for some special groups like abelian groups. In particular, no examples of full-sized[14] groups with this property were known. In 2020, Bridson et al. [210] provided the first such examples.

---

[14] A group is called *full-sized* if it has a non-abelian free subgroup.

Theorem 4.39 *There exist full-sized groups that are profinitely rigid in the absolute sense.*

4.5   Other Topics in Group Theory

The Rubik's Cube and God's Number

The Rubik's Cube, invented by Ernő Rubik in 1974, is widely considered to be the world's best-selling toy. By a solution of a position of a Rubik's Cube we mean the sequence of rotations of the cube's faces after which each face of the Rubik's Cube has only one colour. The least integer n such that every position can be solved in at most n rotations is known as *God's number*. The problem of determining God's number (let us denote it by g) has been of interest ever since the Rubik's Cube was invented. A simple counting argument proves that $g \geq 18$. The first upper bound $g \leq 52$ was proved by Thistlethwaite in 1981. Then many researchers proved better and better lower and upper bounds. Culminating this line of research, in 2014 Rokicki et al. [1057] determined God's number exactly: $g = 20$.

Theorem 4.40 *Every starting position of Rubik's Cube can be solved in* 20 *moves or less. The constant* 20 *in this statement is the best possible.*

This may not be immediately obvious, but Theorem 4.40 is actually a result in group theory. The set of all possible moves (where a "move" is the effect of any sequence of rotations of the cube's faces) form a group with respect to the composition operation (call it G), which is generated by the set S of face rotations. Theorem 4.40 determines the diameter[15] of G with respect to S.

Because the lower bound $g \geq 20$ had been proved by Reid [1036] in 1995, only the upper bound $g \leq 20$ remained to be proved. The authors of [1057] achieved this by partitioning all the positions into about two billion cosets of a carefully selected subgroup H of G, using symmetry to significantly reduce the number of cosets that need to be treated, and then writing a highly efficient computer program to search for at most 20-move solutions for the positions in each coset.

On the Rank of Subgroups of Free Groups

The *rank* of a group G, denoted rank(G), is the smallest cardinality of a generating set for G. In 1954, Howson [622] showed that if $\mathcal{K}, \mathcal{L}$ are nontrivial, finitely generated subgroups of a free group $\mathcal{F}$, then $\mathcal{K} \cap \mathcal{L}$ is finitely generated, and, moreover,

$$\mathrm{rank}(\mathcal{K} \cap \mathcal{L}) - 1 \leq 2\, \mathrm{rank}\mathcal{K}\, \mathrm{rank}\mathcal{L} - \mathrm{rank}\mathcal{K} - \mathrm{rank}\mathcal{L}.$$

---

[15] See the discussion before Theorem 4.20 for the definition of diameter.

In 1957, Neumann [932] improved this to

$$\text{rank}(\mathcal{K} \cap \mathcal{L}) - 1 \leq 2\,(\text{rank}(\mathcal{K}) - 1)(\text{rank}(\mathcal{L}) - 1),$$

and conjectured that the factor 2 in this bound can be removed. This conjecture became known as the *Hanna Neumann conjecture*, and was open for over 50 years. In 2012, Mineyev [898], and, independently Friedman [478] proved the conjecture.

Theorem 4.41 *If $\mathcal{K}$, $\mathcal{L}$ are nontrivial, finitely generated subgroups of a free group $\mathcal{F}$, then $\mathcal{K} \cap \mathcal{L}$ is finitely generated, and*

$$\text{rank}(\mathcal{K} \cap \mathcal{L}) - 1 \leq (\text{rank}(\mathcal{K}) - 1)(\text{rank}(\mathcal{L}) - 1).$$

In fact, both Mineyev and Friedman proved a stronger version of the conjecture formulated by Neumann [935] in 1990, see [478, 898] for details.

Kazhdan's Property (T)
Kazhdan's property (T), introduced by Kazhdan [696] in 1967, is one of the most important concepts in analytic group theory. It has a number of equivalent definitions and many applications. For example, any group with property (T) can be used to make explicit constructions of expander graphs.[16] There are many equivalent definitions of property (T). We present one suggested by Ozawa [958] in 2016. The *real group algebra* $\mathbb{R}(G)$ of a group G is the set of formal sums

$$\sum_{g \in G} \xi(g)g$$

where $\xi : G \to \mathbb{R}$ is a finitely supported function, with the addition and multiplication defined by

$$\sum_{g \in G} \xi(g)g + \sum_{g \in G} \psi(g)g = \sum_{g \in G} (\xi(g) + \psi(g))g,$$

and

$$\sum_{g \in G} \xi(g)g \cdot \sum_{h \in G} \psi(h)h = \sum_{g \in G} \sum_{h \in G} (\xi(g)\psi(h))gh.$$

Let the *-operation be defined by $\xi^*(g) = \xi(g^{-1})$. For a finite symmetric generating set S of G, the *Laplacian* Δ is the element of $\mathbb{R}(G)$ defined by $\Delta = |S|e - \sum_{g \in S} g$, where e is the identity element of G. We say that the group G satisfies *Kazhdan's*

---

[16] If we fix a finite set S of generators of a group G with property (T), then the Cayley graphs $\Gamma(G/N, S)$ form a family of expanders when N runs over the finite index normal subgroups of G.

property (T) if there exists a $\lambda > 0$ and finitely many elements $\xi_i \in \mathbb{R}(G)$ such that

$$\Delta^2 - \lambda\Delta = \sum_i \xi_i^* \xi_i.$$

Ozawa [958] proved that this property does not depend on the choice of S and is equivalent to the original definition of property (T) given in [696]. The advantage of the Ozawa definition is that it makes it easier to verify that a given infinite group (or family of groups) has this property. In particular, Kaluba et al. [677] used this approach to prove property (T) for the automorphism group[17] $\mathrm{Aut}(F_n)$ of the free group with $n \geq 6$ generators.

Theorem 4.42 *The group* $\mathrm{Aut}(F_n)$ *has Kazhdan's property (T) for all* $n \geq 6$.

The question for which n the group $\mathrm{Aut}(F_n)$ has Kazhdan's property (T) was a long-standing open question. McCool [864] proved in 1989 that $\mathrm{Aut}(F_n)$ does not have property (T) for $n \leq 3$. In 2019, Kaluba, Nowak and Ozawa [678] proved that $\mathrm{Aut}(F_n)$ has property (T) for $n = 5$, while Theorem 4.42 proves this for all $n \geq 6$. In 2020, Nitsche [941] posted online a preprint in which he gave a computer-assisted proof of property (T) for $\mathrm{Aut}(F_4)$, thus resolving the last remaining open case.

The Spread of a Group
Recall that a group G is called 2-*generated* if it has a 2-element generating set $\{g, h\}$, which is also called a *generating pair*. A combination of the classification theorem for finite simple groups and a 1962 theorem of Steinberg [1157] imply that every finite simple group is 2-generated. In 2000, Guralnick and Kantor [542] proved a much stronger property: every non-trivial element of a finite simple group G belongs to some generating pair. Groups with this property are called $\frac{3}{2}$-*generated*. In general, it is known that if G is $\frac{3}{2}$-generated then every proper quotient of G is cyclic. In 2008, Breuer et al. [207] conjectured that the converse statement is true for all finite groups.

More generally, the *spread* $s(G)$ of a group G is the largest integer k such that for any non-trivial elements $g_1, \ldots, g_k$ in G, there exists an $h \in G$ such that $\{g_i, h\}$ is a generating pair for every $i = 1, \ldots, k$. If no such largest integer exists, then we write $s(G) = \infty$. In this terminology, a group G is $\frac{3}{2}$-generated if and only if $s(G) \geq 1$. Interestingly, there are infinitely many examples of finite groups with $s(G) = 2$, but no examples of finite groups with $s(G) = 1$. In 1975, Brenner and Wiegold [206] asked if there are infinitely many such groups.

In 2020, Burness et al. [233] posted online a preprint in which they answered both of these questions.

Theorem 4.43 *If every proper quotient of a finite group G is cyclic, then $s(G) \geq 2$.*

---

[17] An *automorphism* of a group G is a one-to-one function $\phi : G \to G$ such that $\phi(g \cdot h) = \phi(g) \cdot \phi(h)$ for all $g, h \in G$. The set of all automorphisms forms a group with composition operation, which is called the *automorphism group* of G.

Theorem 4.43 implies that for a finite group G, the following statements are equivalent: (i) $s(G) \geq 2$, (ii) $s(G) \geq 1$, and (iii) every proper quotient of G is cyclic. In particular, it implies that the conjecture of Breuer et al. [207] is true, and that there are no finite groups G with $s(G) = 1$.

The starting point in the proof of Theorem 4.43 is the 2008 result of Breuer et al. [207], who proved the statement of the theorem in the case when G is a simple group. Many authors then tried to extend this result to the family of so-called almost simple groups.[18] The classification theorem for simple groups reduces this task to a case-by-case analysis, and most of the cases had been solved by 2020. The authors solved the remaining case and thus finished the proof of the theorem for almost simple groups, and then deduced the general case from there.

Product Theorems in Groups

For sets of integers A, B, let $A + B = \{a + b \mid a \in A, \ b \in B\}$. If A is an arithmetic progression, then $|A + A| < 2|A|$. A deep 1973 theorem of Freiman [473] describes all possible examples of sets A such that $|A + A| \leq k|A|$ for fixed k. An important research direction is to obtain a result similar to Freiman's theorem under the weaker assumption $|A + A| \leq |A|^{1+o(1)}$.

Similar questions can be studied in arbitrary groups. For subsets A and B of a group G let $A \cdot B = \{g \in G \mid g = ab, \ a \in A, \ b \in B\}$ and $A^k = A \cdot A \cdots \cdot A$, where A is repeated k times. If $|A^k|$ is much smaller than $|A|^k$ (for example, $|A^k| \leq |A|^{1+o(1)}$), what can be said about the structure of A? It turns out that it is impossible to deduce the structure of A if $|A^2|$ is small, but sometimes it is possible if $|A^3|$ is small. In particular, Razborov [1029] proved in 2014 that if A is a subset of a free group, then $|A^3|$ may be small only if $ab = ba$ for all $a, b \in A$.

Theorem 4.44 *If A is a finite subset of a free group F_m with at least two non-commuting elements, then*

$$|A^3| \geq \frac{|A|^2}{(\log |A|)^{O(1)}}.$$

The central part of the proof of Theorem 4.44 is estimating the number of collisions, that is, the number of different triples $a, b, c \in A$ and $a', b', c' \in A$ such that $abc = a'b'c'$. The authors proved an upper bound on the number of collisions, which then easily translates into a lower bound for $|A^3|$.

Invariants and Semi-Invariants

An important research direction in finite group theory is the study of invariants, and, more generally, semi-invariants. Let G be a finite subgroup of $GL_n(\mathbb{C})$. Let $H_n(\mathbb{C})$ be the set of homogeneous polynomials in n complex variables with complex coefficients. A polynomial $P \in H_n(\mathbb{C})$ is called a *semi-invariant* of G if for every matrix $A \in G$ we have $P(Az) = B_A P(z)$, $\forall z \in \mathbb{C}^n$, where $B_A \in \mathbb{C}$ is a constant

---

[18] A finite group G is *almost simple* if it has a unique minimal normal subgroup G_0 that is nonabelian and simple.

depending on A. Let $d(G)$ be the minimum integer k such that G has a semi-invariant of degree k. In 1981, Thompson [1198] proved that, if $n > 1$ is any integer and G is any finite subgroup of $\mathrm{GL}_n(\mathbb{C})$, then $d(G) \le 4n^2$. He further conjectured that this quadratic upper bound can be improved to a linear one. In 2016, Tiep [1202] confirmed this conjecture.

Theorem 4.45 *There exists a constant $C < \infty$ such that $d(G) \le Cn$ for every finite subgroup G of $\mathrm{GL}_n(\mathbb{C})$. In fact, $C = 1\,184\,036$ works.*

Approximating Semigroups in $\mathbb{Z}^n$

A *semigroup* is a set together with an associative binary operation on it. Let $S \subset \mathbb{Z}^n$ be a semigroup. Let G be the subgroup of $\mathbb{Z}^n$ generated by S, L be the subspace of $\mathbb{R}^n$ spanned by S, and C be the smallest closed convex cone (with apex at the origin) containing S. The semigroup $S' = C \cap G$ is called the *regularization* of S. From the definition, S' contains S. In 2012, Kaveh and Khovanskii [695] proved that the regularization S' asymptotically approximates the semigroup S.

Theorem 4.46 *Let $C' \subset C$ be a closed strongly convex cone that intersects the boundary (in the topology of the linear space L) of the cone C only at the origin. Then there exists a constant $N > 0$ (depending on C') such that any point in the group G that lies in C' and whose distance from the origin is bigger than N belongs to S.*

The authors call this result "the approximation theorem". It is not very difficult to prove, but the authors showed that it has many applications in convex and algebraic geometry, intersection theory, and other areas of mathematics. See [695] for details.

On Homomorphisms Between Polish Groups

An old classical theorem states that any Lebesgue measurable function $f : \mathbb{R} \to \mathbb{R}$ satisfying the functional equation $f(x + y) = f(x) + f(y)$, $x, y \in \mathbb{R}$ must be continuous. In 1971, Christensen [277] conjectured a far-reaching generalization of this fact.

To formulate it, we need some definitions. A *topological group G* is a topological space that is also a group such that the group operations of product $G \times G \to G : (x, y) \to xy$ and taking inverses $G \to G : x \to x^{-1}$ are continuous. A topological space is called *Polish space* if it is homeomorphic to a complete metric space that has a countable dense subset. A *Polish group* is a topological group G that is also a Polish space. A *Borel probability measure* on a topological space is a probability measure that is defined on all open sets. A subset of a Polish group G is called *universally measurable* if it is measurable with respect to every Borel probability measure on G. A homomorphism $\phi : G \to H$ between Polish groups G and H is called *universally measurable* if $\phi^{-1}(U)$ is a universally measurable set in G for every open set $U \subseteq H$. Christensen [277] asked whether every universally measurable Polish group homomorphism is continuous. In 2019, Rosendal [1059] answered this question affirmatively.

Theorem 4.47 *Every universally measurable homomorphism between Polish groups is automatically continuous.*

4.6   Rings

Parametrizations of Low-Rank Rings
Each commutative ring R is an abelian group with respect to addition. We say that R has *rank n* if this group is isomorphic to $\mathbb{Z}^n$. Rings with rank $n = 2, 3, 4$ and 5 are called quadratic, cubic, quartic, and quintic, respectively. The only ring of rank 1 is $\mathbb{Z}$. For $n = 2$, there is an explicit one-to-one correspondence between quadratic rings and the set of integers $\mathbb{D} = \{D \in \mathbb{Z} : D \equiv 0 \text{ or } 1 \,(\mathrm{mod}\, 4)\}$, and we say that quadratic rings can be *parametrized* by $\mathbb{D}$. In 1964, Delone and Faddeev [339] derived a parametrization for cubic rings. In 2004, Bhargava [137] achieved this for the quartic rings.

To formulate the result, we need some definitions. An *integral ternary quadratic form* is an expression of the form $ax^2 + by^2 + cz^2 + dxy + exz + fyz$ with $a, b, c, d, e, f \in \mathbb{Z}$. We say that two such forms A and B are *linearly independent* over $\mathbb{Q}$ if $uA + vB = 0$ with rational u, v is possible only if $u = v = 0$. Let $GL_n(\mathbb{Z})$ be the set of invertible $n \times n$ matrices with integer entries, and let $GL_2^{\pm 1}(\mathbb{Q})$ be the set of 2×2 matrices with rational entries and determinant ± 1.

Theorem 4.48 *There exists a canonical bijection between isomorphism classes of non-trivial quartic rings and $GL_3(\mathbb{Z}) \times GL_2^{\pm 1}(\mathbb{Q})$-equivalence classes of pairs (A, B) of integral ternary quadratic forms where A and B are linearly independent over $\mathbb{Q}$.*

The statement of Theorem 4.48 means that it is possible to associate to every non-trivial quartic ring a pair (A, B) of integral ternary quadratic forms, such that different pairs (A, B) corresponds to different rings unless they can be transformed into each other using a matrix $\gamma_1 \in GL_3(\mathbb{Z})$ and a matrix $\gamma_2 \in GL_2^{\pm 1}(\mathbb{Q})$, see [137] for details.

In 2008, Bhargava [139] also obtained an explicit parametrization for quintic rings, but the formulas are too difficult to be presented here. An interested reader may find them in the original paper [139].

Integral Group Rings
For any finite group G, an *integral group ring* is the ring whose elements are all possible sums $\sum\limits_{g \in G} a_g g$, where a_g are integers, addition is defined as

$$\sum_{g \in G} a_g g + \sum_{g \in G} b_g g = \sum_{g \in G} (a_g + b_g)g,$$

and multiplication is given by

$$\left(\sum_{g \in G} a_g g\right) \cdot \left(\sum_{h \in G} b_h h\right) = \sum_{g \in G} \sum_{h \in G} a_g b_h gh.$$

Obviously, if two groups G and H are isomorphic, then so are their integral group rings. In 1940, Higman [604] asked if the converse is true, that is, whether different finite groups have different integral group rings. The question was answered positively in many special cases. However, Hertweck [600] proved in 2001 that in general the answer is negative.

Theorem 4.49 *There exist two finite non-isomorphic groups G and H such that their integral group rings are isomorphic.*

Theorem 4.49 was proved by constructing an example of groups G and H satisfying the conditions of the theorem. To give the reader an intuition about the non-triviality of the construction, we only mention that the groups G and H have $2^{21} \cdot 97^{28}$ elements each.

Zero-Divisor Graphs of Rings

Let R be a commutative ring. An element $a \in R$ is called a *zero-divisor* if there exists a non-zero $x \in R$ such that $ax = 0$. Let $Z(R)$ be the set of zero-divisors of R. The *zero-divisor graph* $\Gamma(R)$ of R is the graph with vertices $Z(R) \setminus \{0\}$ such that there is an (undirected) edge between vertices a and b if and only if $a \neq b$ and $ab = 0$. Studying properties of zero-divisor graphs of rings is important for understanding the structure of rings and their zero divisors. In particular, Anderson et al. [43] asked an interesting question: which finite commutative rings have planar zero-divisor graphs? In 2003, Akbari et al. [20] answered this question for local rings. An element $u \in R$ is a *unit* if there is a $v \in R$ such that $uv = vu = 1$. A ring R is called *local* if $1 \neq 0$ and the sum of any two non-units in R is a non-unit.

Theorem 4.50 *If R is a local ring with at least 33 elements, and $Z(R) \neq \emptyset$, then $\Gamma(R)$ is not planar.*

4.7 Fields

Counting Number Fields

A *number field* is a field F that contains $\mathbb{Q}$ and has finite dimension n when considered as a vector space over $\mathbb{Q}$. The number n is called the *degree* of F. A fundamental numerical invariant associated with each number field is its discriminant. A set $e = \{e_1, e_2, \ldots, e_n\}$ of n elements of F is called a *basis* of F if every $x \in F$ can be written as $x = \sum_{i=1}^{n} c_i(x)e_i$ with coefficients $c_i(x) \in \mathbb{Q}$. The sum $\text{Tr}(x) = \sum_{i=1}^{n} c_i(x \cdot e_i)$ does not depend on the choice of basis e and is called the *trace* of x.

An element $x \in F$ is called an *algebraic integer* if it is a root of a monic polynomial with integer coefficients. If, for every algebraic integer $x \in F$, all $c_i(x)$ are integers, e is called an *integral basis* of F. The determinant of the $n \times n$ matrix with entries $\text{Tr}(e_i \cdot e_j)$, $i = 1, \ldots, n$, $j = 1, \ldots, n$, does not depend on the choice of integral basis e and is called the *discriminant* of F.

For $X > 0$, let $N_n(X)$ denote the number of non-isomorphic number fields of degree n with absolute value of the discriminant at most X. It is known that $N_n(X) < \infty$ for any fixed n and X, and there is an old folklore conjecture that

$$\lim_{n \to \infty} \frac{N_n(X)}{X} = c_n > 0 \tag{4.10}$$

for every fixed n. The conjecture is trivial for $n \leq 2$. In 1971, Davenport and Heilbronn [322] proved it for $n = 3$. In 2005, Bhargava [138] proved the conjecture for $n = 4$.

Theorem 4.51 *The limit* $\lim_{X \to \infty} \frac{N_4(X)}{X}$ *exists and is equal to*

$$c_4 = \frac{5}{24} \prod_{p \in \mathcal{P}} \left(1 + p^{-2} - p^{-3} - p^{-4}\right) = 0.253...,$$

where $\mathcal{P}$ is the set of prime numbers.

In 2010, Bhargava [140] made the next step and proved the conjecture for $n = 5$.

Theorem 4.52 *The limit* $\lim_{X \to \infty} \frac{N_5(X)}{X}$ *exists and is equal to*

$$c_5 = \frac{13}{120} \prod_{p \in \mathcal{P}} \left(1 + p^{-2} - p^{-4} - p^{-5}\right) = 0.149...,$$

where $\mathcal{P}$ is the set of prime numbers.

The proof of Theorem 4.51 relies on the parametrization of quartic rings (Theorem 4.48), while the proof of Theorem 4.52 uses the corresponding parametrization of quintic rings [139].

For general n, only weaker upper bounds for $N_n(X)$ in terms of X are known. In 1995, Schmidt [1091] proved that $N_n(X) \leq B_n X^{(n+2)/4}$, where B_n is a constant depending on n. This bound was significantly improved by Ellenberg and Venkatesh [386] in 2006.

Theorem 4.53 *There exists a constant B_n, depending only on n and an absolute constant C, such that the inequality*

$$N_n(X) \leq B_n X^{\exp(C\sqrt{\log n})}$$

holds for all $n > 2$ and all $X > 0$.

In 2020, Couveignes [306] achieved the next dramatic improvement.

Theorem 4.54 *There exists a constant $C > 0$ such that the inequality*

$$N_n(X) \leq n^{Cn \log^3 n} X^{C \log^3 n}$$

holds for all integers $n \geq C$ and $X \geq 1$.

Also in 2020, Oliver and Thorne posted online a preprint [948] in which they improved the bound further to

$$N_n(X) \leq B_n X^{c \log^2 n}$$

for $n \geq 6$, where one can take $c = 1.564$. However, even after all these breakthroughs, we still do not even have a bound of the form $N_n(X) \leq B_n X^C$ with a constant C independent of n. Conjecture (4.10) predicts that this should be true with $C = 1$.

The Hopf Condition Over Arbitrary Fields
It is easy to check that $(a^2 + b^2)(c^2 + d^2) = (ac + bd)^2 + (ad - bc)^2$ for all real numbers a, b, c, d. More generally, for a field F and positive integers r, s, and n, by a *sums-of-squares formula of type* $[r, s, n]$ *over* F we mean an identity of the form

$$\left(\sum_{i=1}^{r} x_i^2 \right) \cdot \left(\sum_{i=1}^{s} y_i^2 \right) = \sum_{i=1}^{n} z_i^2, \tag{4.11}$$

where each z_i is a bilinear form, that is, an expression of the form

$$z_i = \sum_{j=1}^{r} \sum_{k=1}^{s} c_{ijk} x_j y_k,$$

with some coefficients $c_{ijk} \in F$. An important research direction is to understand for which values of r, s, n a sums-of-squares formula (4.11) exists. In 1941, Hopf [619] proved that, over the field $\mathbb{R}$ of real numbers, (4.11) may exist only if the numbers $\frac{n!}{i!(n-i)!}$ are even integers for all i such that $n - r < i < s$. This condition became known as the *Hopf condition*. In 2007, Dugger and Isaksen [369] proved that this condition remains true over an arbitrary field F (except for fields of characteristic[19] 2, in which sums-of-squares formulas of type $[r, s, n]$ trivially exist for all r, s, n).

Theorem 4.55 *If F is a field of characteristic not equal to 2, and a sums-of-squares formula* (4.11) *of type* $[r, s, n]$ *exists over* F, *then the numbers* $\frac{n!}{i!(n-i)!}$ *are even integers for all i such that $n - r < i < s$.*

The u-Invariant of a Field
A positive integer $u(F)$ is called a *u-invariant* of a field F if (a) the equation

$$\sum_{i=1}^{n} \sum_{j=1}^{n} a_{ij} x_i x_j = 0 \tag{4.12}$$

---

[19] We say that a field F with additive identity 0 and multiplicative identity 1 is of *characteristic* 2 if $1 + 1 = 0$.

(where $x_1, x_2, \ldots, x_n$ are variables and a_{ij} are some coefficients from F) in $n = u(F)$ variables has no solutions, except for $x_1 = x_2 = \cdots = x_n = 0$, and (b) the same equation in $n = u(F) + 1$ variables always has a non-zero solution. For example, $u(\mathbb{C}) = 1$ for a field $\mathbb{C}$ of complex numbers, $u(F) = 2$ for any finite field F, and $u(F_n) = 2^n$ for the field F_n of rational functions in n variables over $\mathbb{C}$. In 1953, Kaplansky [680] conjectured that the u-invariant of any field, if it exists, is always a power of 2. In 1991, Merkur'ev [882] disproved this conjecture by showing that the u-invariant can be any even number, but left open whether it can be an odd number greater than 1. It is known that the u-invariant can never be 3, 5, or 7. In 2001, Izhboldin [644] proved that the u-invariant can be 9.

Theorem 4.56 *There exists a field F with u-invariant* 9.

The definition of the u-invariant given above is simple and intuitive, but is not applicable for fields in which Eq. (4.12) may have no non-zero solutions for all n. For example, this is the case for the field $\mathbb{R}$ of real numbers, and, more generally, for all formally real fields. A field F is called *formally real* if there is a total order $<$ on F such that, for all $a, b, c \in F$, (i) if $a < b$ then $a + c < b + c$ and (ii) if $0 < a$ and $0 < b$, then $0 < a \cdot b$. One may define $u(F) = \infty$ for all such fields F, but this is not very informative. In 1973, Elman and Lam [389] suggested a meaningful way to extend the notion of u-invariant to formally real fields.

A subset S of a field K is *algebraically independent* over a subfield F if the elements of S do not satisfy any non-trivial polynomial equation with coefficients in F. The largest cardinality of such S is called the *transcendence degree* of K over F, which we denote by $d_F(K)$. If $K \neq F$ but $d_F(K) = 0$, K is called a *proper algebraic extension* of F. A formally real field is *real-closed* if it admits no proper formally real algebraic extensions. Any formally real field F with given ordering $<$ has a unique real-closed algebraic extension K preserving $<$, which we denote by $\mathrm{cl}(F, <)$.

Every symmetric $n \times n$ matrix A with entries $a_{ij} \in F$ generates a quadratic form

$$f = \sum_{i=1}^{n} \sum_{j=1}^{n} a_{ij} x_i x_j$$

of dimension n over F. This form is called *non-degenerate* if $\det(A) \neq 0$. It is called *anisotropic* if there is no non-zero vector $(x_1, \ldots, x_n)$ on which the form evaluates to zero. Let $K(F)$ be the set of all non-degenerate anisotropic forms over F. Quadratic forms generated by matrices A and B are *equivalent* if $A = MBM^T$ for some $n \times n$ matrix M over F with $\det(M) \neq 0$. If F is real-closed, any $f \in K(F)$ is equivalent to a form

$$\sum_{i=1}^{r} x_i^2 - \sum_{i=r+1}^{r+s} x_i^2$$

for some positive integers r, s. The difference $r - s$ is called the *signature* of f. If F is formally real, the signature of $f \in K(F)$ with respect to the ordering $<$ is the signature of f as a form over $\mathrm{cl}(F, <)$. Let $K_0(F)$ be the set of all $f \in K(F)$ whose signature with respect to any ordering $<$ of F is 0. If F is not formally real, define $K_0(F) = K(F)$. Now, define the u-invariant $u(F)$ of a field F as the maximal dimension of any form $f \in K_0(F)$.

In 1982, Pfister [980] conjectured that if a field F has transcendence degree d over $\mathbb{R}$, then $u(F) \leq 2^d$. In 2019, Benoist [125] verified the $d = 2$ case of this conjecture.

Theorem 4.57 *If F is any field of transcendence degree 2 over $\mathbb{R}$, then $u(F) \leq 4$.*

Chapter 5
Geometry and Topology

5.1 Packing Problems

Sphere Packings in $\mathbb{R}^3$

What is the densest possible way to pack congruent balls in $\mathbb{R}^3$? More formally, by an *open ball* $B(A, R)$ in $\mathbb{R}^d$ with center A and radius R we mean the set of points $X \in \mathbb{R}^d$ such that $|AX| < R$, where $|\cdot|$ denotes the usual Euclidean distance in $\mathbb{R}^d$. By a "packing of congruent balls", or "sphere packing", we mean an (infinite) set E of non-overlapping open balls with the same radii. For a fixed point $O \in \mathbb{R}^d$, define

$$\delta(E) = \limsup_{R \to \infty} \frac{|B(O, R) \cap E|}{|B(O, R)|}, \qquad (5.1)$$

where $|\cdot|$ denotes the volume in $\mathbb{R}^d$. It can be shown that $\delta(E)$ depends only on E but not on O and it is called the *upper density* of E. If the limit (as opposed to limsup) exists in (5.1), $\delta(E)$ is called the *density* of E.

In $\mathbb{R}^2$, there is an obvious way to arrange non-overlapping circles, called the "regular hexagonal packing", see Fig. 5.1a, which has density $\frac{\pi}{\sqrt{12}} = 0.9069....$ In 1773, Lagrange proved that this packing is the densest one among the *lattice packings*, that is, packings E for which the centres of the circles form a lattice.[1] In 1910, Thue [1199] gave a (somewhat non-rigorous) proof that the regular hexagonal packing is optimal among all packings. In 1943, Tóth [448] gave a fully rigorous proof of this fact.

In $\mathbb{R}^3$, two packings of congruent balls (called the "cubic close packing" and "hexagonal close packing") with the same density $\frac{\pi}{3\sqrt{2}}$ have been known for centuries. The construction of one packing is illustrated in Fig. 5.1b: we can start

---

[1] A *lattice* Λ in $\mathbb{R}^d$ is a set of the form $\Lambda = \left\{ \sum_{i=1}^{d} a_i v_i \ \middle|\ a_i \in \mathbb{Z} \right\}$, where $v_1, \ldots, v_d$ is a basis for $\mathbb{R}^d$.

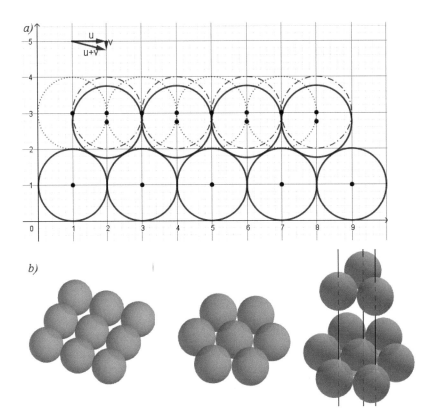

Fig. 5.1 Optimal circle and sphere packings in dimensions 2 and 3

with spheres whose centers form a cubic grid, then rearrange spheres in each layer according to the optimal circle packing on the plane, and then shift the layers and move them closer to each other. In 1611, Kepler [702] conjectured that these packings are the densest possible ones. In 1831, Gauss [493] proved this conjecture for lattice packings, but the general case remained open. In 1900, Hilbert [605] included Kepler's conjecture in his famous list of 23 unsolved problems in mathematics (as part of the 18th problem). In 1953, Tóth [449] reduced the general case of the conjecture to a finite but very large computation. By refining this approach, Hales [557] gave a complete computer-assisted proof of Kepler's conjecture in 2005.

Theorem 5.1 *No packing of congruent balls in* $\mathbb{R}^3$ *has upper density greater than* $\frac{\pi}{3\sqrt{2}} = 0.74048....$

Because the proof of Theorem 5.1 uses a computer and is difficult for humans to verify, Hales initiated a project to give a fully rigorous computer-assisted verification of the proof back to the axioms of mathematics. The project was completed in 2015 [558], giving 100% confidence in the proof's correctness.

Sphere Packings in Higher Dimensions
In this section we discuss the sphere packing problem in higher dimensions $d > 3$. We start with an easier version of the problem in which we restrict ourselves to lattice packings only. The densest possible lattice packing was found in dimensions $4 \le d \le 5$ by Korkine and Zolotareff [739, 740] in the 1870s, and in dimensions $6 \le d \le 8$ by Blichfeldt [156] in 1935. For example, the answer in $\mathbb{R}^8$ is the packing in which the sphere centres form the "E_8 root lattice" defined as

$$\left\{ x = (x_1, \ldots, x_8) \in \mathbb{Z}^8 \cup \left(\mathbb{Z} + \frac{1}{2} \right)^8 : \sum_i x_i \equiv 0 \,(\mathrm{mod}\, 2) \right\}.$$

This packing has density $\frac{\pi^4}{384}$.

After 1935, the problem in other dimensions remained unresolved for over 70 years, and the next solved case was, somewhat surprisingly for non-specialists, $d = 24$. The explanation is that in $\mathbb{R}^{24}$ there is a particularly nice lattice, called the *Leech lattice*, which can be defined as the unique lattice L^* in $\mathbb{R}^{24}$ which is (i) unimodular,[2] (ii) even, that is, $\sum_{i=1}^{24} x_i^2$ is an even integer for every $x = (x_1, \ldots, x_{24}) \in L^*$, and (iii) $|x| \ge 2$ for every non-zero $x \in L^*$. Property (iii) ensures that spheres with centres in L^* form a valid sphere packing, and its density $\frac{\pi^{12}}{12!}$ has been conjectured to be optimal. In 2009, Cohn and Kumar [288] confirmed this conjecture.

Theorem 5.2 *No lattice sphere packing in $\mathbb{R}^{24}$ has upper density greater than* $\frac{\pi^{12}}{12!} = 0.001929....$

The proof of Theorem 5.2 uses human reasoning to reduce the problem to verification of certain properties of a large list of explicit polynomials, and then uses a computer to perform this verification.

The E_8 root lattice in $\mathbb{R}^8$ and the Leech lattice in $\mathbb{R}^{24}$ have long been conjectured to be optimal among *all* sphere packings in these dimensions, not only among the lattices packings. In 2003, Cohn and Elkies [287] came very close to proving these conjectures. To formulate their result, we need one more definition. A function $f : \mathbb{R}^d \to \mathbb{R}$ is called *admissible* if there exist positive constants ϵ, C_1, C_2 such that

$$|f(x)| \le \frac{C_1}{(1 + \|x\|)^{n+\epsilon}} \quad \text{and} \quad |\hat{f}(x)| \le \frac{C_2}{(1 + \|x\|)^{n+\epsilon}} \quad \forall x \in \mathbb{R}^d,$$

where $\|x\|$ is the norm in $\mathbb{R}^d$, and $\hat{f}(x)$ is the Fourier transform defined in (3.29).

Theorem 5.3 *Suppose $f : \mathbb{R}^d \to \mathbb{R}$ is an admissible function, is not identically zero, and such that (i) $f(x) \le 0$ for $\|x\| \ge 1$, and (ii) $\hat{f}(t) \ge 0$ (which implies that*

---

[2] See the discussion before Theorem 1.70 for the definition of a unimodular lattice.

$\hat{f}(t)$ *is real) for all* $t \in \mathbb{R}^d$. *Then the upper density of any sphere packing in* $\mathbb{R}^d$ *is bounded above by* $\frac{f(0)}{2^d \hat{f}(0)} V_d$, *where* V_d *is the volume of the unit ball in* $\mathbb{R}^d$.

Theorem 5.3 provides a general method for proving upper bounds for sphere packing densities, by constructing, in any dimention d, a function $f : \mathbb{R}^d \to \mathbb{R}$ satisfying the conditions of the theorem with $\frac{f(0)}{2^d \hat{f}(0)}$ as small as possible. In the same paper [287], Cohn and Elkies used this method to derive improved upper bounds in dimensions $4 \le d \le 36$. In particular, for $d = 8$, they derive the upper bound $1.000001 \Delta_8$, where $\Delta_8 = \frac{\pi^4}{384}$ is the density of the E_8 packing. Moreover, they conjectured the existence of a function $f^* : \mathbb{R}^8 \to \mathbb{R}$ which, when substituted into Theorem 5.3, gives the upper bound Δ_8 exactly. In 2017, Viazovska [1233] constructed such a function and hence resolved the sphere packing problem in dimension $d = 8$.

Theorem 5.4 *No packing of congruent balls in* $\mathbb{R}^8$ *can have upper density greater than* $\frac{\pi^4}{384}$.

Viazovska constructed the function f^* needed to deduce Theorem 5.4 from Theorem 5.3 as a Laplace transform (4.2) of a modular form of a certain kind, see the discussion before Theorem 1.64 for a brief introduction to this topic, and the original paper [1233] for full details.

After proving Theorem 5.4, Viazovska joined the team of Cohn, Kumar, Miller and Radchenko, and they together found out how to use the same method to resolve the sphere packing problem in dimension $d = 24$ as well. Specifically, they proved [289] that the Leech lattice with density $\frac{\pi^{12}}{12!}$ is indeed optimal among all sphere packings.

Theorem 5.5 *No packing of congruent balls in* $\mathbb{R}^{24}$ *can have upper density greater than* $\frac{\pi^{12}}{12!}$.

In dimensions d other than $1, 2, 3, 8$ and 24, the problem of determining the optimal sphere packing upper density ρ_d^* remains open. For general d, we have bounds in the form

$$2(d - 1)2^{-d} \le \rho_d^* \le 2^{(-0.599\cdots+o(1))d}, \qquad (5.2)$$

where the lower and upper bounds are due to Ball [88] and Kabatyanskiı and Levenshteın [668], respectively. This shows that ρ_d^* decays exponentially in d, but the exact decay rate remains unknown.

Universally Optimal Configurations

Sphere packing theorems find the densest possible configurations of points (sphere centres) in $\mathbb{R}^d$ subject to the constraint that the distance between any two points is at least 2. In physics, this corresponds to the densest particle configuration in the presence of a repulsive force which is infinitely strong for any particles $x, y \in \mathbb{R}^d$ at distance $|x - y| < 2$, but disappear completely if $|x - y| \ge 2$. Real physical forces do not act like this; instead, they decay continuously with distance. In general, maximal density configurations in the presence of such forces are complicated and depend on the forces.

In some dimensions, however, numerical experiments show that in the presence of any force from a broad family of repulsive forces, the optimal configurations stay the same. Such configurations are called *universally optimal*. This is an extremely strong property, with a number of important consequences. In particular, if a configuration C is universally optimal, then spheres with centres in C form the densest possible sphere packing in the corresponding dimension. However, before 2019, no configuration in any dimension $d \geq 2$ had been proved to be universally optimal. For example, it has been long conjectured that the centres of hexagons in the regular hexagonal grid (see Fig. 2.5) form a universally optimal configuration on the plane, but the proof remains elusive.

In 2019, Cohn et al. posted online a preprint [290] in which they proved that the E_8 root lattice and the Leech lattice are universally optimal in dimensions $d = 8$ and $d = 24$, respectively. This is a far-reaching generalization of Theorems 5.4 and 5.5.

We now formulate this result rigorously. A *point configuration C* is a non-empty subset of $\mathbb{R}^d$ such that every ball in $\mathbb{R}^d$ contains only finitely many points of C. We say that C has *density ρ* if

$$\rho = \lim_{r \to \infty} \frac{|C \cap B_r^d(0)|}{\mathrm{Vol}(B_r^d(0))}$$

where $B_r^d(0)$ denotes the closed ball in $\mathbb{R}^d$ of radius r with centre 0. For any function $p : (0, \infty) \to \mathbb{R}$, the *lower p-energy* of a point configuration C in $\mathbb{R}^d$ is

$$E_p(C) := \liminf_{r \to \infty} \frac{1}{|C \cap B_r^d(0)|} \sum_{x, y \in C \cap B_r^d(0),\, x \neq y} p(|x - y|). \tag{5.3}$$

If the limit (as opposed to liminf) exists in (5.3), then $E_p(C)$ is called the *p-energy* of C. Intuitively, $p(r)$ represents the potential energy of a pair of particles at distance r, and $E_p(C)$ is the average energy per particle. We say that C *minimizes energy for p* if its p-energy $E_p(C)$ exists and every point configuration in $\mathbb{R}^d$ of density ρ has lower p-energy at least $E_p(C)$.

A function $p : (0, \infty) \to \mathbb{R}$ is called a *completely monotonic function of squared distance* if $p(r) = g(r^2)$ for some infinitely differentiable function $g : (0, \infty) \to \mathbb{R}$ such that $(-1)^k g^{(k)} \geq 0$ for all $k \geq 0$, where $g^{(k)}$ is the k-th derivative of g. In particular, g is non-negative, weakly decreasing, convex, and so on. Two important examples of such functions are inverse power laws $p(r) = 1/r^s$ for $s > 0$ and Gaussians $p(r) = e^{-\alpha r^2}$ for $\alpha > 0$. We say that a point configuration C in $\mathbb{R}^d$ with density $\rho > 0$ is *universally optimal* if it minimizes p-energy whenever $p : (0, 1) \to \mathbb{R}$ is a completely monotonic function of squared distance. We are now ready to state the main result of [290].

Theorem 5.6 *The E_8 root lattice and the Leech lattice are universally optimal in $\mathbb{R}^8$ and $\mathbb{R}^{24}$, respectively.*

The proof proceeds similarly to the proofs of Theorems 5.4 and 5.5. The authors first established a stronger version of Theorem 5.3 which proves Theorem 5.6 conditional on the existence of an auxiliary function f with suitable properties. To construct such a function f, the authors developed a new interpolation theorem, which allowed them to reconstruct a radial Schwartz function from the values of its radial derivatives and its Fourier transform at specific points. This result can be viewed as a multidimensional generalization of Theorem 3.63, and is of independent interest.

The Dodecahedral Conjecture

So far we have discussed sphere packings which are optimal globally. One may also ask for the densest possible way to pack spheres *locally*, around one fixed sphere. To formulate this problem rigorously, we need the notion of a Voronoi cell. Let Λ be any set of points in $\mathbb{R}^3$ with pairwise distance between any of them at least 2. Equivalently, Λ is a set of centres of non-overlapping unit spheres. The *Voronoi cell* $\Omega(\Lambda, v)$ around $v \in \Lambda$ consists of points of space that are closer to v than to any other point $w \in \Lambda$. The volume of the Voronoi cell $\Omega(\Lambda, v)$ measures "how densely" the spheres are packed around the fixed sphere with center v. In the hexagonal close packing, which is the densest sphere packing in $\mathbb{R}^3$ by Theorem 5.1, the volumes of all Voronoi cells are $4\sqrt{2} = 5.6568....$ However, we can do better locally. Let D^* be a regular dodecahedron whose inscribed sphere has radius 1. If Λ consists of the centre O of D^* and 12 mirror images of O with respect to faces of D^*, then the Voronoi cell $\Omega(\Lambda, O)$ is exactly D^*, see Fig. 5.2, and its volume is $V^* = 10\sqrt{130 - 38\sqrt{5}} = 5.55029... < 4\sqrt{2}$. In 1943, Tóth [448] conjectured that this volume is the smallest possible one. This statement became known as the *dodecahedral conjecture*. In 2010, Hales and McLaughlin [555] confirmed this conjecture.

Theorem 5.7 *Let Λ be any set of points in $\mathbb{R}^3$ with pairwise distance between any of them at least 2, and let $v \in \Lambda$. Then the volume of the Voronoi cell $\Omega(\Lambda, v)$ is at least V^*.*

The proof of Theorem 5.7 follows a similar strategy as the proof of Theorem 5.1. In particular, both proofs involve long computer calculations, and in fact the corresponding computer codes significantly overlap. However, neither of the theorems is a corollary of the other.

Packing of Regular Tetrahedra

The problem of finding the densest possible packings in $\mathbb{R}^3$ can be studied for other shapes, not necessary for spheres. For regular tetrahedra, the problem goes back at least to Aristotle, who mistakenly believed that a perfect packing (that is, one that fully covers the space) is possible. In 2006, Conway and Torquato [300] constructed a packing of regular tetrahedra with density about 0.72. In just a few years after this,

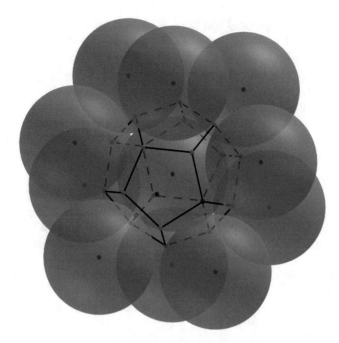

Fig. 5.2 Construction of 12 spheres touching one such that the Voronoi cell is a regular dodecahedron. One sphere is not depicted for better visibility of the dodecahedron

many authors developed denser and denser packings, culminating in the following 2010 theorem of Chen et al. [268].

Theorem 5.8 *There exists a regular tetrahedra packing in* $\mathbb{R}^3$ *with density* $\frac{4000}{4671} = 0.856347\ldots$.

The packing in Theorem 5.8 remains the densest known one but is not guaranteed to be optimal. It is known that Aristotle was wrong and regular tetrahedra do not fully cover $\mathbb{R}^3$. Moreover, Gravel et al. [518] proved in 2011 that no packing of regular tetrahedra in $\mathbb{R}^3$ can have upper density greater than $1 - 2.6 \cdot 10^{-25}$. However, the gap between this upper bound and the lower bound in Theorem 5.8 remains substantial.

The starting point in the proof of Theorem 5.8 is an earlier work of Kallus, Elser and Gravel [676], who achieved the density $\frac{100}{117} = 0.854700\ldots$ by discovering a one-parameter family of tetrahedra packings called "dimer cluster packings". In 2009, Torquato and Jiao [1208] improved the density to $\frac{12250}{14319} = 0.855506\ldots$ by introducing a two-parameter family of such packings. Theorem 5.8 was proved by discovering a three-parameter family of dimer packings, such that packings from [676] and [1208] are both special cases of this family. The authors then found the *optimal* packing in this family to get the density in Theorem 5.8. This optimality is an indication that a substantially new approach is needed to improve beyond this.

In 1972, Ulam conjectured that the optimal density for packing of congruent spheres is smaller than that of copies of any other convex body. With Theorem 5.1, this conjecture states that, for any convex body K in $\mathbb{R}^3$, there is a packing of non-overlapping congruent copies of K with upper density at least $\frac{\pi}{3\sqrt{2}} = 0.74048....$
For a long time, no packing of regular tetrahedra with this density was known, and it was possible that the regular tetrahedron is a counterexample to this conjecture. More recent results like Theorem 5.8 imply that this is not the case, and regular tetrahedra can be packed more densely than spheres. Ulam's conjecture survived and remains a fascinating open question.

Covering Space by Convex Bodies

The packing problem asks to pack copies of a convex body K in $\mathbb{R}^d$ in the densest possible way without intersections. A related problem is the *covering problem*: what is the most economical way to *cover* $\mathbb{R}^d$ by overlapping copies of K? A simpler but still highly non-trivial special case is when the translates of K are arranged in a lattice.

Let us now formulate this question more rigorously. Let $\mathcal{L}_d$ be the set of all unimodular lattices[3] in $\mathbb{R}^d$. For $L \in \mathcal{L}_d$ and a convex body $K \subset \mathbb{R}^d$, the K-*covering density* of L, denoted $\Theta_K(L)$, is the minimal volume of a body $r \cdot K$ such that $\mathbb{R}^d = L + r \cdot K$. Finally, let

$$\Theta_{d,K} = \inf_{L \in \mathcal{L}_d} \Theta_K(L).$$

If $K = B$ is a ball in $\mathbb{R}^d$, then $\Theta_{d,B}$ grows linearly in the dimension, up to logarithmic factors. Specifically, Rogers [1056] proved in 1959 the existence of positive constants c and γ such that $cd \leq \Theta_{d,B} \leq cd(\log d)^\gamma$ for all $d \geq 3$. By comparing this with (5.2), we see that we can cover space by balls much more economically than packing balls in space.

For general convex bodies, the question is much more difficult. Rogers [1056] proved the upper bound $\Theta_{d,K} \leq d^{\log_2 \log d + c}$ for some $c > 0$, and this super-polynomial bound remained unimproved for over 60 years. In particular, it was not known whether $\Theta_{d,K}$ grows polynomially in d. In 2020, Ordentlich, Regev and Weiss posted online a preprint [953] in which they proved that in fact $\Theta_{d,K}$ grows at most quadratically in the dimension!

Theorem 5.9 *There exists a constant $c > 0$ such that for any convex body $K \subset \mathbb{R}^d$ in any dimension $d \geq 1$ we have $\Theta_{d,K} \leq cd^2$.*

Theorem 5.9 was proved using a probabilistic argument. Instead of constructing a specific lattice $L = L(K) \in \mathcal{L}_d$ with small $\Theta_K(L)$, the authors proved that a random lattice works with high probability. Specifically, there exists a (natural) probability measure on $\mathcal{L}_d$ and positive constants c_1, c_2, c_3, c_4 such that for any d, any convex body $K \subset \mathbb{R}^d$, any $M \in [c_1 d^2, c_2 d^3]$, and random $L \in \mathcal{L}_d$,

---

[3] See the discussion before Theorem 1.70 for the definition of a unimodular lattice.

the probability that $\Theta_K(L) > M$ is less than $c_3 e^{-c_4 M/d^2}$. Theorem 5.9 follows immediately from this fact. As a key ingredient in the proof, the authors use (and strengthen) the bounds on the sizes of Kakeya-type sets in finite vector spaces, see Theorem 3.88.

Kissing Numbers

A problem closely related to sphere packing is the problem of determining the *kissing number* $k(d)$, which is defined as the highest number of equal non-overlapping spheres in $\mathbb{R}^d$ that can touch another sphere of the same size. It is easy to see that $k(1) = 2$ and $k(2) = 6$. Determining $k(3)$ (the maximal number of white billiard balls that can touch a black ball) is an old problem which goes back to at least 1694, when it was the subject of a famous discussion between Newton and Gregory. Newton conjectured that the construction with 12 touching spheres, like the one in Fig. 5.2, is optimal, and $k(3) = 12$, while Gregory thought that $k(3)$ may be equal to 13. In 1952, Schütte and van der Waerden [1099] proved that $k(3) = 12$, confirming Newton's prediction. In 1979, Levenshtein [786], and independently Odlyzko and Sloane [944] proved that $k(8) = 240$ and $k(24) = 196560$. The optimal configurations in these dimensions are the E_8 lattice and the Leech lattice, respectively. Before 2008, the problem was open in all other dimensions.

In $\mathbb{R}^4$, there is a known construction[4] in which a unit sphere touches 24 other unit spheres, hence $k(4) \geq 24$. In 1963, Coxeter [308] made a conjecture which implies that $k(4) \leq 26$. In 1978, Böröczky [171] proved this conjecture. In 1979, Odlyzko and Sloane [944] proved that $k(4) \leq 25$. In 2008, Musin [917] completed this line of research by proving that $k(4) = 24$.

Theorem 5.10 *The kissing number in $\mathbb{R}^4$ is 24.*

The proofs of the equalities $k(8) = 240$ and $k(24) = 196560$ are based on the method introduced by Delsarte in the 1970s, which, in every dimension, constructs a linear program whose solution is an upper bound for $k(d)$. In 1997, Arestov and Babenko [47] proved that this method cannot improve the upper bound $k(4) \leq 25$. Despite this, Musin proved Theorem 5.10 by discovering a clever modification of Delsarte's method that produces an improved bound $k(4) < 25$, which implies that $k(4) = 24$.

For all other dimensions, only the upper and lower bounds for $k(d)$ are known. For example, the best bounds known before 2008 for $d = 5$ were $40 \leq k(5) \leq 46$, while the best bounds for $d = 6$ were $72 \leq k(6) \leq 82$. In 2008, Bachoc [76] developed a new method for computing better upper bounds in all dimensions in which the problem is open. The method defines a semidefinite program whose optimal solution gives an improved upper bound for $k(d)$. In particular, this method implies that the kissing numbers in dimensions $5, 6, 7, 9$ and 10 are at most $45, 78, 135, 366$ and 567, respectively.

---

[4] Take one unit sphere with centre $(0, 0, 0, 0)$ surrounded by 24 spheres with centres $(\pm\frac{1}{\sqrt{2}}, \pm\frac{1}{\sqrt{2}}, 0, 0)$, where you can choose the signs and also permute the coordinates.

Theorem 5.11 *Let $k(d)$ be the kissing number in $\mathbb{R}^d$. Then $k(5) \le 45$, $k(6) \le 78$, $k(7) \le 135$, $k(9) \le 366$, and $k(10) \le 567$.*

After 2008, the upper bounds in Theorem 5.11 have been improved only slightly. As of 2020, the best known upper bounds in dimensions $5, 6, 7, 9$ and 10 are $44, 78, 134, 363$ and 553, respectively, see Caluza Machado and de Oliveira Filho [240].

In the limit when the dimension d goes to infinity, the kissing number $k(d)$ is known to grow exponentially fast with d. In the 1950s several authors, including, for example, Shannon [1112], used a random choice procedure to prove that

$$k(d) \ge (1 + o(1)) \sqrt{\frac{3\pi d}{8}} \left(\frac{2}{\sqrt{3}} \right)^d.$$

This lower bound implies that $\frac{\log_2 k(d)}{d} \ge \log_2 \frac{2}{\sqrt{3}} = 0.2075\ldots$ for large d. On the other hand, Kabatyanskiĭ and Levenshteĭn [669] proved in 1978 that $\frac{\log_2 k(d)}{d} \le 0.4041\ldots$.

A long-standing open problem was whether the exponential growth remains if we restrict ourselves to lattice configurations only. Recall that a *lattice L* in $\mathbb{R}^d$ is a set of the form $L = \left\{ \sum_{i=1}^d a_i v_i \mid a_i \in \mathbb{Z} \right\}$, where $v_1, \ldots, v_d$ is a basis for $\mathbb{R}^d$. Let $\lambda_1(L)$ be the length of the shortest non-zero vector in L. The *kissing number* $k(L)$ of L is the number of vectors of length $\lambda_1(L)$ in L. The *lattice kissing number* $k_l(d)$ in dimension d is the maximum value of $k(L)$ over all lattices L in $\mathbb{R}^d$. Obviously $k_l(d) \le k(d)$, but the inequality is believed to be strict for all large d. In particular, it was not known whether $k_l(d)$ grows exponentially with d. In 2019, Vlăduţ [1239] resolved this problem affirmatively.

Theorem 5.12 *There exist a constant $c > 0$ such that $k_l(d) \ge e^{cd}$ for any $d \ge 1$.*

Apollonian Circle Packings

We next discuss a particularly nice way to pack circles on the plane called "Apollonian circle packings". A set of four mutually tangent circles in the plane with distinct points of tangency is called a *Descartes configuration*. Given a Descartes configuration, one can construct four new circles, each of which is tangent to three of the given ones. Continuing to repeatedly fill the interstices between mutually tangent circles with further tangent circles, we arrive at an infinite circle packing. It is called a *bounded* Apollonian circle packing (bACP), see Fig. 5.3. Apollonian circle packing is named after Apollonius of Perga, who lived more than 2000 years ago, and has been studied by many researchers since then.

One of the central research directions in this area is to count circles in the packing with radius bounded from below. Given a bounded Apollonian circle packing $\mathcal{P}$ and number $T > 0$, let $N^{\mathcal{P}}(T)$ be the number of circles in $\mathcal{P}$ of radius at least $\frac{1}{T}$. It is easy to see that $N^{\mathcal{P}}(T) < \infty$ for any fixed $T > 0$, and the goal is to determine the asymptotic growth of $N^{\mathcal{P}}(T)$ as $T \to \infty$. In 1977, Wilker [1264], based on

Fig. 5.3 Apollonian circle
packing

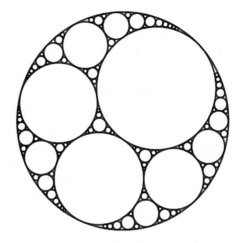

computer experiments, predicted that

$$\lim_{T \to \infty} \frac{\log N^{\mathcal{P}}(T)}{\log T} = \alpha,$$

where $\alpha = 1.30568...$ is a constant independent of $\mathcal{P}$. In 1982, Boyd [195] confirmed this prediction. In 2011, Kontorovich and Oh [736] established the exact asymptotic growth of $N^{\mathcal{P}}(T)$. Recall that we write $f(T) \sim g(T)$ if $\lim_{T \to \infty} \frac{f(T)}{g(T)} = 1$.

Theorem 5.13 *For every bounded Apollonian circle packing $\mathcal{P}$ there exists a constant $c = c(\mathcal{P}) > 0$ such that $N^{\mathcal{P}}(T) \sim c \cdot T^{\alpha}$ as $T \to \infty$, where $\alpha = 1.30568...$ is an absolute constant.*

A nice property of every bACP is that if the original four circles have integer curvatures,[5] all of the circles in the bACP will have integer curvatures as well. In this case, the bACP is called *integer*. For an integer bACP $\mathcal{P}$, let $\kappa(\mathcal{P}, X)$ denote the number of distinct integers up to X occurring as curvatures in the packing.

In 2003, Graham et al., Wilks [514] proved that $\kappa(\mathcal{P}, X) \geq c' \sqrt{X}$ for some constant c' and conjectured that this can be improved to $\kappa(\mathcal{P}, X) \geq cX$. This statement became known as the *positive density conjecture* for integer bACPs. In his 2008 unpublished letter to Lagarias, Sarnak made substantial progress by proving that

$$\kappa(\mathcal{P}, X) \geq c \frac{X}{\sqrt{\log X}} \tag{5.4}$$

---

[5] Recall that the *curvature* of a circle is $1/r$ where r is its radius.

for some $c > 0$. In 2009, Fuchs circulated a preprint with improved bound $\kappa(\mathcal{P}, X) \geq c\frac{X}{(\log X)^{\epsilon}}$ for $\epsilon = 0.150\ldots$. Finally, Bourgain and Fuchs [186] confirmed the positive density conjecture in full in 2011. Let $X_{\mathcal{P}}$ be the curvature of the largest circle in $\mathcal{P}$.

Theorem 5.14 *For any integer bACP $\mathcal{P}$ there exists a constant $c > 0$ depending on $\mathcal{P}$ such that $\kappa(\mathcal{P}, X) \geq cX$ for all $X \geq X_{\mathcal{P}}$.*

Theorem 5.14 was proved by refining the method developed by Sarnak to prove (5.4). This method first considers a fixed circle $C_0 \in \mathcal{P}$ and determines the set $S(C_0)$ of integers which occur as curvatures of circles tangent to C_0. Sarnak observed that the set $S(C_0)$ contains all integers represented by a certain binary quadratic form, which helped him to derive a lower bound on the number of integers in $S(C_0)$. This lower bound was then used to prove (5.4). Bourgain and Fuchs proved Theorem 5.14 using a clever modification of this method.

In fact, the positive density conjecture proved in Theorem 5.14 is just a special case of a much more general conjecture, called the *local-global conjecture*, which describes exactly which large integers occur as curvatures of circles in $\mathcal{P}$.

More formally, let $b(C)$ denote the curvature of a circle C. For an integer bACP $\mathcal{P}$, let $\mathcal{B} = \{b(C) : C \in \mathcal{P}\}$ be the set of all curvatures in $\mathcal{P}$. The integer bACP $\mathcal{P}$ is called *primitive* if $\gcd(\mathcal{B}) = 1$. A positive integer m is called *admissible* for $\mathcal{P}$ if for any integer $q \geq 1$, there exists a $k \in \mathcal{B}$ such that $k - m$ is divisible by q. All integers in $\mathcal{B}$ are admissible by definition. The local-global conjecture predicts that, conversely, every sufficiently large admissible integer belongs to $\mathcal{B}$. Theorem 5.14 implies that a positive percentage of integers satisfy this conjecture. In 2014, Bourgain and Kontorovich [189] proved a much stronger theorem which implies that the percentage of integers up to N satisfying this conjecture approaches 100% as N grows.

Let $f_{\mathcal{P}}(N)$ be the number of admissible integers not exceeding N which do not belong to $\mathcal{B}$.

Theorem 5.15 *For any primitive integer bACP $\mathcal{P}$ there exist constants $\eta > 0$ and $C < \infty$ such that $f_{\mathcal{P}}(N) \leq CN^{1-\eta}$ for all $N > 0$.*

The proof of Theorem 5.15 is based on the Hardy–Littlewood circle method, which we have briefly discussed in the context of the proof of Theorem 1.20.

The theory of Apollonian circle packings has connections to many other areas of mathematics. One example of the connection to harmonic analysis is presented below.

So far we have discussed bounded Apollonian circle packings consisting of circles only. In general, an Apollonian circle packing may also contain lines. By a *general circle* on the plane we mean a circle or a line. Every triple T of pairwise tangent general circles has exactly two general circles (called *Soddy general circles*) tangent to all three. An Apollonian circle packing (ACP) generated by T is a minimal collection of general circles containing T that is closed under the addition of Soddy general circles. For $k \in \mathbb{Z}$, let $\mathcal{P}_k$ be the ACP generated by the lines $x = 2k$, $x = 2k + 2$ and the circle $C((2k + 1, 0), 1)$, where $C((x, y), r)$ denotes the circle with centre $(x, y) \in \mathbb{R}^2$ and radius r. Let $\mathcal{P}^* = \bigcup_{k \in \mathbb{Z}} \mathcal{P}_k$. To each circle $C((x, y), r) \in \mathcal{P}^*$ we associate the matrix

$$A_C = \frac{1}{2} \begin{bmatrix} r + x & y \\ y & r - x \end{bmatrix}.$$

Next we discuss the connection between ACPs and integer superharmonic matrices. For $x, y \in \mathbb{Z}^2$ we write $x \sim y$ if $||x - y|| = 1$. For a function $g : \mathbb{Z}^2 \to \mathbb{Z}$, let

$$\Delta g(x) = \sum_{y \sim x} (g(y) - g(x)).$$

A 2×2 matrix A is called *integer superharmonic* if there exists a function $g : \mathbb{Z}^2 \to \mathbb{Z}$ such that $g(x) = \frac{1}{2} x^T A x + o(||x||^2)$ and $\Delta g(x) \leq 1$ for all $x \in \mathbb{Z}^2$. Integer superharmonic matrices arise from the theory of partial differential equations, and it was an open question to characterize them. In 2017, Levine et al. [788] achieved this via the connection to matrices A_C arising from ACPs.

Theorem 5.16 *Let S_2 be the set of all 2×2 real symmetric matrices. A matrix $A \in S_2$ is integer superharmonic if and only if either[6] $\text{trace}(A) \leq 2$ or there exists a circle $C \in \mathcal{P}^*$ such that the matrix $A_C - A$ is positive semidefinite.*

For non-specialists, Theorem 5.16 may look surprising because it connects areas of mathematics which look totally unrelated. Such theorems are useful because they allow methods and results from one area to be applied to another.

5.2 Combinatorial Geometry

The "Happy Ending" Problem

We say that m points on the plane are in *general position* if no three of them are on a line. For every integer $n \geq 3$, let $ES(n)$ be the minimal integer m such that any set of m points in the plane in general position contains n points which are the vertices of a convex n-gon. In 1933, Esther Klein proved that $ES(4) = 5$, that is, any set of five points in the plane in general position has a subset of four points that form the vertices of a convex quadrilateral. This problem became known as the "Happy Ending" problem, because it led to her marriage with George Szekeres. Shortly after this, Makai proved that $ES(5) = 9$, see Fig. 5.4.

In 1935, Erdős and Szekeres [420] proved that $ES(n)$ is finite for every $n \geq 3$. Moreover, they proved an explicit upper bound for $ES(n)$ growing as $4^{n-o(n)}$. Based on the values $ES(3) = 3$, $ES(4) = 5$, and $ES(5) = 9$ known to them at that time,

---

[6] Recall that the *trace* of a square matrix is the sum of its diagonal entries.

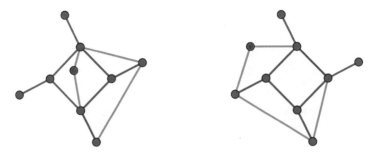

Fig. 5.4 (Left) A configuration of 8 points (blue) in general position, no 5 of which are vertices of a convex pentagon. (Right) For 9 points, this is not possible

Erdős and Szekeres conjectured that

$$ES(n) = 2^{n-2} + 1 \qquad \text{for all} \quad n \geq 3. \tag{5.5}$$

In 2008, Szekeres and Peters [1173] gave a computer-assisted proof that $ES(6) = 17$, confirming the $n = 6$ case of (5.5). The conjecture remains open for every $n > 6$.

In 1960, Erdős and Szekeres [422] proved the lower bound $ES(n) \geq 2^{n-2} + 1$, so it remains to prove a matching upper bound. However, the original upper bound $4^{n-o(n)}$ for $ES(n)$ remained unimproved for over 80 years, except for the improvements in the $o(n)$ term. In 2017, Suk [1168] dramatically improved the upper bound from $4^{n-o(n)}$ to $2^{n+o(n)}$, thus establishing the Erdös–Szekeres conjecture up to the $o(n)$ term in the exponent.

Theorem 5.17 *There exists an absolute constant n_0 such that*

$$ES(n) \leq 2^{n+6n^{2/3} \log_2 n}$$

for all $n \geq n_0$.

Given how long the problem has been open, it is a bit surprising that the proof of Theorem 5.17 is short, combinatorial, and does not rely on any "heavy" machinery. The proof is based on Ramsey theory, which states that any large set contains some "organized" subsets, see Sects. 2.1 and 2.7 for the different aspects of this theory. In 2020, Holmsen et al. [617] used a similar argument to improve the upper bound slightly to $ES(n) \leq 2^{n+O(\sqrt{n \log n})}$.

The Distinct Distances Problem

In 1946, Erdős [397] posed the question: what is the smallest number of distinct distances that can be determined by N points in the plane? Erdős showed that if the points are arranged in a square grid, then the number of distinct distances is proportional to $\frac{N}{\sqrt{\log N}}$, and conjectured that this is optimal up to a constant factor.

More formally, let S be a set of N points on the plane. Let $D(S)$ be the set of distinct real numbers $t > 0$ such that $t = d(A, B)$ for some $A \in S$ and $B \in S$. Let $|D(S)|$ be the number of elements in set $D(S)$. Erdős' conjecture predicts that $|D(S)| \geq c\frac{N}{\sqrt{\log N}}$ for some constant $c > 0$. Many authors proved better and better lower bounds for $|D(S)|$, with the best result before 2015 being the 2004 theorem of Katz and Tardos [694] stating that $|D(S)| \geq cN^{0.8641}$. In 2015, Guth and Katz [549] proved a much better lower bound for $|D(S)|$, which almost matches the Erdős prediction.

Theorem 5.18 *There exists a constant $c > 0$ such that*

$$|D(S)| \geq c\frac{N}{\log N}$$

for every set S of $N \geq 2$ points on the plane.

The starting point of the proof of Theorem 5.18 is the 2011 result of Elekes and Sharir [384], which reduces the distinct distances problem to the problem of estimating the number of point-line incidences in $\mathbb{R}^3$, see Theorem 5.19 below for a higher-dimensional version of this problem. The authors then introduced a novel method based on the polynomial ham sandwich theorem[7] to bound the number of such point-line incidences.

Incidence Theorems

Let P be a finite set of distinct points in the plane, L be a finite set of distinct lines, and let $I(P, L) := \{(p, l) \in P \times L : p \in l\}$ be the set of incidences. The famous *Szemerédi–Trotter theorem* [1175], proved in 1983, states that

$$|I(P, L)| \leq C(|P|^{2/3}|L|^{2/3} + |P| + |L|) \tag{5.6}$$

for some absolute constant C. This bound is the best possible up to a constant factor. In a paper which first appeared online in 2005, Tóth [1209] conjectured that the same bound holds for k-dimensional affine subspaces of $\mathbb{R}^{2k}$ in place of lines on the plane. In 2012, Solymosi and Tao [1144] confirmed this conjecture, up to ϵ loss.

Theorem 5.19 *Let P be a finite set of points in $\mathbb{R}^d$, and let L be a finite set of k-dimensional affine subspaces in $\mathbb{R}^d$, such that any two distinct spaces in L intersect in at most one point. Then for every real $\epsilon > 0$, and all integers $k \geq 1$ and $d \geq 2k$,*

---

[7] The original *ham sandwich theorem* states that if $U_1, \ldots, U_n$ are opens sets in $\mathbb{R}^n$, then there is a hyperplane that divides each U_i into two equal-volume parts. The *polynomial ham sandwich theorem* is the generalization proved by Stone and Tukey [1164] in 1942, stating that if we are allowed to use a curved surface defined by a polynomial of degree $d > 1$ instead of a hyperplane, then we can simultaneously bisect a large number $M = \binom{n+d}{n} - 1$ of open sets in $\mathbb{R}^n$.

there exists a constant $A = A(\epsilon, k) > 0$ such that

$$|I(P, L)| \leq A|P|^{2/3+\epsilon}|L|^{2/3} + \frac{3}{2}|P| + \frac{3}{2}|L|.$$

The proof of Theorem 5.19 uses the polynomial ham sandwich theorem, and proceeds by induction on both the dimension d and the number of points in set P.

In 2016, Basu and Sombra [104] conjectured that an appropriate version of the Szemerédi–Trotter estimate should also hold for hypersurfaces in $\mathbb{R}^d$. In 2020, Walsh [1245] confirmed this conjecture. To formulate it, we need some notation. We say that $V \subset \mathbb{R}^d$ is a *hypersurface* of degree m if there exists a polynomial $P(x_1, \ldots, x_d)$ of degree m such that V is the solution set to the equation $P(x_1, \ldots, x_d) = 0$. If P is a finite set of points in $\mathbb{R}^d$, and $\mathcal{H}$ is a finite set of hypersurfaces, let $I(P, \mathcal{H}) := \{(p, V) \in P \times \mathcal{H} : p \in V\}$ be the set of incidences.

Theorem 5.20 *Let $d \geq 2$, $k, c \geq 1$, and let P and $\mathcal{H}$ be finite sets of points and hypersurfaces in $\mathbb{R}^d$ satisfying the following conditions:*

(a) the degrees of the hypersurfaces in $\mathcal{H}$ are bounded by c;
(b) the intersection of any family of d distinct hypersurfaces in $\mathcal{H}$ is finite, and
(c) for any subset of k distinct points in P, the number of hypersurfaces in $\mathcal{H}$ containing them is bounded by c.

Then

$$|I(P, \mathcal{H})| \leq C_{d,k,c} \left(|P|^{\frac{k(d-1)}{dk-1}} |\mathcal{H}|^{\frac{d(k-1)}{dk-1}} + |P| + |\mathcal{H}| \right),$$

where $C_{d,k,c}$ is a constant depending on d, k, c.

Equiangular Sets of Lines

A set of lines through the origin in $\mathbb{R}^n$ is called *equiangular* if any pair of lines defines the same angle. Let $N(n)$ denote the maximum cardinality of an equiangular set of lines in $\mathbb{R}^n$. Let $N_\theta(n)$ denote the maximum number of equiangular lines in $\mathbb{R}^n$ with common angle θ, where θ is a constant not depending on the dimension.

Equiangular sets of lines appear naturally in many areas of mathematics, and the problem of estimating the maximum size of such sets has been studied starting from at least the work of Haantjes [550] in 1948, who proved that $N(3) = N(4) = 6$. In 1966, van Lint and Seidel [802] showed that $N(5) = 10$, $N(6) = 16$, $N(7) \geq 28$, and explicitly posed the problem of determining $N(n)$ for all n, or at least prove general lower and upper bounds for it. The upper bound $N(n) \leq \frac{1}{2}n(n+1)$ was proved by Gerzon and published by Lemmens and Seidel [783] in 1973. In 2000, de Caen [236] proved that $N(n) \geq \frac{2}{9}(n+1)^2$ for every n in the form $n = 3 \cdot 2^{2t-1} - 1$ for integer t. This shows that $N(n)$ grows quadratically in n, up to a constant factor.

In 1973, Lemmens and Seidel [783] formulated the problem of estimating $N_\theta(n)$ for fixed θ, and proved that $N_\theta(n) = 2n - 2$ for $\theta = \arccos \frac{1}{3}$ and sufficiently large n. In 2018, Balla et al. [89] proved a stronger upper bound for $N_\theta(n)$ for all $\theta \neq \arccos \frac{1}{3}$. This implies that, for all large n, $N_\theta(n)$ is maximized at $\theta = \arccos \frac{1}{3}$.

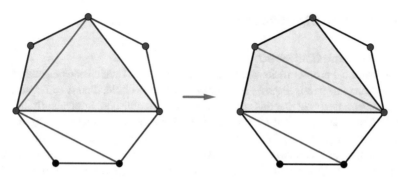

Fig. 5.5 The "flip" operation

Theorem 5.21 *For any angle* $\theta \in (0, \pi/2)$, $\theta \neq \arccos \frac{1}{3}$, *there is a constant* n_θ *such that* $N_\theta(n) \leq 1.93n$ *for all* $n \geq n_\theta$.

The authors also proved that for any set S of k fixed angles, there are at most $O(n^k)$ lines in $\mathbb{R}^n$ such that the angle between any two of them belongs to S. This confirms a conjecture of Bukh [226] made in 2016.

The Flip Distance Between Triangulations of a Convex Polygon
Let P be a convex polygon $A_1 A_2 \ldots A_n$ with n vertices and sides $A_1 A_2, \ldots, A_n A_1$. A line interval $A_i A_j$ which is not a side is called a *diagonal* of P with vertex set $\{A_i, A_j\}$. Two diagonals of P are *crossing* if they are non-disjoint but their vertex sets are disjoint. A set T of $n - 3$ pairwise non-crossing diagonals of P is called a *triangulation* of P. The name is justified by the fact that every triangulation T divides P into triangles.

Any diagonal e in any triangulation T is one of the diagonals of a unique convex quadrilateral whose four sides either belong to T or are sides on P. Let e' be the other diagonal of this quadrilateral. A *flip* is the operation that removes e from T and introduces e' into T, see Fig. 5.5. The result of a flip is another triangulation, T'. The *flip distance* between triangulations T_1 and T_2 of P is the minimal number of flips needed to transform T_1 into T_2.

In 1988, Sleator et al. [1135] proved that if $n > 12$, then the flip distance between any triangulations T_1 and T_2 of a convex n-gon P is at most $2n - 10$. The authors also proved that this bound is sharp provided that $n > N_0$, where N_0 is some large constant. The methods in [1135] provided no explicit bound on how large N_0 is, but the authors conjectured that the statement remains true with $N_0 = 12$. In 2014, Pournin [1006] proved this conjecture, by constructing, for every $n > 12$, a pair of explicit triangulations of a convex n-gon with flip distance $2n - 10$. Because the flip distances between any triangulations can be easily computed for $n \leq 12$, this result fully answers the question of how large the flip distance between triangulations of a convex n-gon can be.

Theorem 5.22 *For every* $n > 12$, *there exist triangulations* T_1 *and* T_2 *of an n-vertex convex polygon* P *such that the flipping distance between* T_1 *and* T_2 *is* $2n - 10$.

5.3   The Isoperimetric Inequality and Beyond

The Honeycomb Conjecture

An old classical problem asks what is the maximal area A which can be enclosed by a closed curve of fixed length L. If the curve encloses a circle, then $A = L^2/4\pi$, and it was long conjectured that this area is the largest possible one. Equivalently, a circle has the lowest length out of all plane curves which encloses a prescribed area. This conjecture was proved by Steiner [1159] in 1838 assuming that the solution exists. Since then, many rigorous unconditional proofs have been found. The same theorem is true in higher dimensions: the ball has the lowest perimeter out of all bodies with the given volume. This theorem is usually stated in terms of the inequality

$$P(S) \geq nV(S)^{\frac{n-1}{n}} V(B_1)^{\frac{1}{n}},$$

where $S \subset \mathbb{R}^n$ is a measurable set, $P(S)$ is its perimeter,[8] $V(\cdot)$ is the n-dimensional volume, and B_1 is the unit ball in $\mathbb{R}^n$. In the most general case (without any unnecessary additional conditions on the set S) this inequality was proved by De Giorgi [331] in 1958.

More than 2000 years ago (around 36 B.C.), Varro studied a related problem: what is the perimeter-optimal way to divide the plane into the cells of unit area? Varro proved that the regular hexagonal grid, depicted in Fig. 2.5, is better than the square grid or triangular grid. The statement that the regular hexagonal grid is better than *any* other way to divide the plane into the equal-area cells became known as the *honeycomb conjecture*.

To state this conjecture more formally in modern terms, we need some definitions. Let C be a partition of the plane into regions of equal unit area.[9] Let $B(0, r)$ be the ball with centre at the coordinate centre and radius r. Let

$$p(C) = \limsup_{r \to \infty} \frac{\text{Perimeter}(C \cap B(0, r))}{\text{Area}(C \cap B(0, r))}$$

be the average perimeter of C per unit area.

Let C^* be a partition of the plane into regular hexagons of unit area. C^* also has the names "regular hexagonal grid" and "honeycomb". The perimeter of any one hexagon in C^* is $2\sqrt[4]{12}$, hence $p(C^*) = \sqrt[4]{12}$. The honeycomb conjecture predicted that no partition C has lower $p(C)$. In 1943, Tóth [447] proved this conjecture under

---

[8] In the general case, $P(S)$ stands for "distributional perimeter" of S with a somewhat complicated definition. However, it coincides with the classical $(n-1)$-dimensional measure of ∂S when S has a smooth boundary.

[9] Formally, let G be a locally finite graph in $\mathbb{R}^2$, consisting of smooth curves, and such that $\mathbb{R}^2 \setminus G$ has infinitely many bounded connected components, all of unit area, and let C be the union of these bounded components.

the hypothesis that the cells are convex. In 2001, Hales [556] proved the conjecture unconditionally.

Theorem 5.23 *For any partition C of the plane into regions of equal unit area,* $p(C) \geq \sqrt[4]{12}.$

To prove Theorem 5.23, Hales first reduced the honeycomb conjecture to the problem of perimeter-optimal partitioning of a flat torus. Then the compactness of the torus was used in a crucial way to prove the so-called "hexagonal isoperimetric inequality" stating that a certain functional is minimized by a regular hexagon, and Theorem 5.23 was deduced from this fact.

The honeycomb conjecture is the two-dimensional version of the three-dimensional *Kelvin problem*. In 1887, Kelvin asked how to partition a 3-dimensional space into cells of equal volume while minimizing the area of the partitioning surface. Kelvin proposed a solution, which is a natural generalization of the hexagonal honeycomb in two dimensions, and conjectured that it is optimal. The Kelvin conjecture was widely believed for over 100 years. Remarkably and surprisingly, in 1993 Weaire and Phelan [1249] found a better way to partition space than Kelvin's way. This partitioning method became known as the *Weaire–Phelan structure*. Weaire and Phelan, however, did not provide a proof that their structure is optimal. Hence, the problem to find an optimal partition of 3-dimensional space remains open.

Enclosing and Separating Two Regions

The isoperimetric theorem states that a ball is the perimeter-minimizing way to enclose a given volume in $\mathbb{R}^d$. A related problem, studied by Plateau in the nineteenth century, is to find the way to enclose and separate *two* regions of prescribed volumes $V_1 > 0$ and $V_2 > 0$ in $\mathbb{R}^3$ with the least possible total surface area of the boundary. In 1896, Boys [196] assumed without proof that the optimal shape is the one which is usually formed by two soap bubbles.

A standard *double bubble* in $\mathbb{R}^3$ is a construction which consists of three spherical caps meeting along a common circle at 120-degree angles (if two of the caps have equal radii, the third one becomes a flat disc). It divides the space $\mathbb{R}^3$ into 3 regions: an infinite one and 2 finite ones with some volumes V_1 and V_2. The *double bubble conjecture* predicts that the volumes V_1 and V_2 cannot be enclosed and separated by a surface with lower area. In 1993, undergraduate students Foisy et al. [456] proved a two-dimensional version of this conjecture. In 2000, Hass and Schlafly [573] resolved the conjecture in $\mathbb{R}^3$ in the special case $V_1 = V_2$. In 2002, Hutchings et al. [633] proved the double bubble conjecture in general.

Theorem 5.24 *The standard double bubble provides the way to enclose and separate two regions of prescribed volumes $V_1 > 0$ and $V_2 > 0$ in $\mathbb{R}^3$ with the least possible total surface area of the boundary.*

The starting point of the proof of Theorem 5.24 is the 1976 result of Almgren [32, Theorem VI.2] implying that a perimeter-minimizing configuration exists. This allowed the authors to start with such an optimal configuration (let us call it C)

and use the optimality of C to prove its properties. The authors proved that the larger region in C is connected, while the smaller one has at most two connected components. Then they established more and more properties of C, culminating in the final conclusion that C must in fact be the standard double bubble.

Quantitative Isoperimetric Inequalities

The isoperimetric theorem states that a ball has the lowest perimeter out of all bodies with the given volume. In 1992, Hall proved a result stating that the perimeter of a set E may be close to optimal only if E is almost a ball.

More formally, the *isoperimetric deficit* $D(E)$ of a Borel set $E \subset \mathbb{R}^n$ with $0 < |E| < \infty$ is

$$D(E) = \frac{P(E) - P(B)}{P(B)},$$

where $P(\cdot)$ is the perimeter and B is a ball such that $|B| = |E|$. The *Fraenkel asymmetry* $\lambda(E)$ of E is

$$\lambda(E) = \min_{B \in \mathcal{B}} \frac{|E \Delta B|}{r(B)^n},$$

where $\mathcal{B}$ is the set of all balls B with $|B| = |E|$, $X \Delta Y = (X \setminus Y) \cup (Y \setminus X)$, and $r(B)$ is the radius of ball B. Then $D(E)$ measures how far a set E is from being optimal in the isoperimetric inequality, while $\lambda(E)$ measures how far it is from being a ball. Hall's theorem states that

$$\lambda(E) \leq C_n D(E)^{1/4}. \tag{5.7}$$

Theorems of this type are called *quantitative isoperimetric inequalities*. Hall further conjectured that the exponent $1/4$ in (5.7) can be improved to $1/2$, which would be optimal. In 2008, Fusco et al. [487] confirmed this conjecture.

Theorem 5.25 *For every $n \geq 2$ there exists a constant C_n such that the inequality*

$$\lambda(E) \leq C_n \sqrt[2]{D(E)}$$

holds for every Borel set E in $\mathbb{R}^n$ with $0 < |E| < \infty$.

The proof of Theorem 5.25 is based on reducing the problem, by using suitable geometric transformations, to the case of more and more symmetric sets. This idea is called "symmetrization", and is discussed in more detail in the next subsection.

Symmetrization Methods

To prove the isoperimetric theorem, Steiner [1159] invented in 1838 the powerful method of "symmetrization", which, since then, has become a major tool for proving many other geometric inequalities. For non-zero $h \in \mathbb{R}^n$, the set $H = \{x \in \mathbb{R}^n, \langle x, h \rangle = 0\}$ is called a *hyperplane* going through the origin. For every $x \in \mathbb{R}^n$

there exists a unique decomposition $x = y + th$, where $y \in H$, $t \in \mathbb{R}$, and we can consider the pair (y, t) as "coordinates" in $\mathbb{R}^n$. For fixed y and h, the set $\{x, x = y + th, -\infty < t < \infty\}$ is a line, which we denote by $y + \mathbb{R}h$. The *Steiner symmetrization* of a convex body K with respect to a hyperplane H is the body

$$S_H(K) = \left\{ (y, t) : K \cap (y + \mathbb{R}h) \neq \emptyset, \ |t| \leq \frac{1}{2} |K \cap (y + \mathbb{R}h)| \right\},$$

where $|K \cap (y + \mathbb{R}h)|$ denotes the length of line interval $K \cap (y + \mathbb{R}h)$.

In 1909, Carathéodory and Study [249] proved that for every convex body K there exists a sequence of Steiner symmetrizations of K that converges to a ball. This implies that if some property P of convex bodies (i) is true for a convex body K provided that it is true for $S_H(K)$, and (ii) is true for balls and bodies sufficiently close to balls, then P must be true for all convex bodies. Indeed, by the Carathéodory–Study theorem, there is a sequence $S_1, \ldots, S_m$ of Steiner symmetrizations such that $K' = S_m(\ldots S_1(K) \ldots)$ is close to a ball. Hence P is true for K' by (ii) and therefore it is true for K by m applications of (i).

This is a very general and powerful method for proving the properties of convex bodies. However, if P is a quantitative property with parameters, then (i) above may hold in the form "if P is true for for $S_H(K)$ then it is also true for K but possibly with worse parameters". In this case, every iteration of (i) results in some quantitative loss, and, to apply the method described above, it is very desirable to have a version of the Carathéodory–Study theorem with the number m of iterations as small as possible. In 1951, Hadwiger [551] proved an upper bound on m of the order $n^{n/2}$, where n is the dimension. In 1989, Bourgain et al. [191] made a dramatic improvement and proved that we can transform any convex body into a body somewhat close to a ball in just $m = cn \log n$ steps. In 2003, Klartag and Milman [721] reduced this bound to a linear one in the dimension.

Theorem 5.26 *For any $\epsilon > 0$, there exists a constant $c > 0$ such that for any $n \geq 2$ and any convex body $K \subset \mathbb{R}^n$, there exist $m \leq (2 + \epsilon)n$ Steiner symmetrizations $S_1, S_2, \ldots S_m$ such that the convex body $K' = S_m(S_{m-1}(\ldots S_2(S_1(K)) \ldots))$ contains a ball of radius r_1, but is contained in a ball of radius r_2, with $\frac{r_1}{r_2} = c$.*

To prove that $m \leq Cn$ symmetrizations suffices for some constant C, the authors managed to reduce the general problem to two special cases: K is the cube and K is the cross polytope (the convex hull of the points in $\mathbb{R}^n$ formed by permuting the coordinates $(\pm 1, 0, \ldots 0)$). The problem is easy for the cube. The case of the cross polytope is more difficult but the authors were able to solve it by an elementary argument. To prove that we can choose $C = 2 + \epsilon$ for any $\epsilon > 0$, the use of much deeper tools like Milman's "quotient of subspace theorem" was needed.

There are other methods of "symmetrization" that are widely used in mathematics. Minkowski symmetrization, introduced by Blaschke in 1916, is a prominent example. For any convex body S in $\mathbb{R}^n$, its *Minkowski symmetrization* $M_H(S)$ with respect to a hyperplane H going through the origin is the set of all midpoints of line segments AB with $A \in S$ and B being a reflection of some point in S with respect

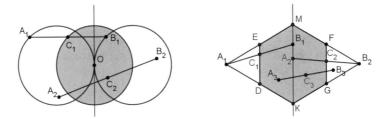

Fig. 5.6 Minkowski symmetrizations of a circle and a triangle

to H, see Fig. 5.6. In 1988, Bourgain et al. [192] proved that a sequence of $cn \log n$ Minkowski symmetrizations selected at random suffices to transform any convex body in $\mathbb{R}^n$ into an approximate ball. For random symmetrizations, this bound is the best possible. However, Klartag [718] proved in 2002 that with carefully selected not-completely-random symmetrizations, we can do this in only $5n$ steps.

Theorem 5.27 *For any $n \geq 2$ and any convex body $S \subset \mathbb{R}^n$, there exist $5n$ Minkowski symmetrizations $M_1, M_2, \ldots M_{5n}$ such that the convex body $S' = M_{5n}(M_{5n-1}(\ldots M_2(M_1(S)) \ldots))$ contains a ball of radius r_1, but is contained in a ball of radius r_2, with $\frac{r_1}{r_2} = \left(1 - c\frac{|\log\log n|}{\sqrt{\log n}}\right) / \left(1 + c\frac{|\log\log n|}{\sqrt{\log n}}\right)$, where c is a universal constant, independent of n and S.*

The basic idea in the proof of Theorem 5.27 is to replace the random symmetrizations studied in [192] by symmetrizations with respect to the vectors of a random orthogonal basis. This results in n symmetrizations. Then another random basis is chosen and the process is repeated, and so on.

The Log-Brunn–Minkowski Inequality

After the 1838 proof of Steiner [1159], many other proofs of the isoperimetric inequality have been developed. One of the simplest proofs uses the Brunn–Minkowski inequality, originally established by Brunn [217] in 1887 and Minkowski [899] in 1896. One of several equivalent formulations of the (plane version) of this inequality states that

$$S((1 - \lambda)K + \lambda L) \geq S(K)^{1-\lambda} S(L)^{\lambda}, \qquad \lambda \in [0, 1], \tag{5.8}$$

where $K, L \subset \mathbb{R}^2$ are convex bodies (compact convex sets with non-empty interiors), $S(\cdot)$ denotes the area, and $(1 - \lambda)K + \lambda L = \{x \in \mathbb{R}^2 : x = (1 - \lambda)a + \lambda b, a \in K, b \in L\}$. A large body of research is devoted to various strengthenings and generalizations of (5.8). Here we discuss one example.

The support function $h_K : \mathbb{R}^2 \to \mathbb{R}$ of a convex body $K \subset \mathbb{R}^2$ is given by

$$h_K(x) = \max_{y \in K} \langle x, y \rangle \quad \forall x \in \mathbb{R}^2, \tag{5.9}$$

where $\langle \cdot, \cdot \rangle$ is the inner product. For convex bodies K and L, and $\lambda \in [0, 1]$, the *log Minkowski combination*, $(1 - \lambda)K \oplus \lambda L$, is the set of all points $x \in \mathbb{R}^2$ such that $\langle x, u \rangle \leq h_K(u)^{1-\lambda} h_L(u)^\lambda$ for all $u \in \mathbb{R}^2$ such that $\langle u, u \rangle = 1$. In 2012, Böröczky, Lutwak, Yang and Zhang [172] proved the following result.

Theorem 5.28 *If K and L are origin-symmetric convex bodies in the plane, then the inequality*

$$S((1 - \lambda)K \oplus \lambda L) \geq S(K)^{1-\lambda} S(L)^\lambda \tag{5.10}$$

holds for all real $\lambda \in [0, 1]$.

Because it is known that $(1 - \lambda)K \oplus \lambda L \subseteq (1 - \lambda)K + \lambda L$, inequality (5.10) is stronger than (5.8). It is called the *log-Brunn–Minkowski inequality* and is conjectured to hold in all dimensions. Theorem 5.28 proves this conjecture in $\mathbb{R}^2$. In 2019, Yang and Zhang [1280] proved it in $\mathbb{R}^3$.

Isoperimetric Inequality in Heisenberg Groups

There are analogues of the isoperimetric inequalities in other areas of mathematics. Here we discuss one example, the vertical versus horizontal isoperimetric inequality in Heisenberg groups, which has applications in the embedding theory for metric spaces and in the analysis of complexity of algorithms for combinatorial problems.

For an element $g \in G$ of a group G, define $g^n = g \cdot g \cdots \cdot g$ (n times) and write g^{-n} for $(g^{-1})^n$. Also, let $[g, h] = ghg^{-1}h^{-1}$ denote the commutator of $g, h \in G$. For every $k \in \mathbb{N}$, the *k-th discrete Heisenberg group*, denoted $H_{\mathbb{Z}}^{2k+1}$, is the group with $2k + 1$ generators $a_1, b_1, \ldots, a_k, b_k, c$ and relations

$$[a_1, b_1] = \cdots = [a_k, b_k] = c \quad \text{and}$$

$$[a_i, a_j] = [b_i, b_j] = [a_i, b_j] = [a_i, c] = [b_i, c] = 1$$

for every distinct $i, j \in \{1, \ldots, k\}$. The *horizontal perimeter* of a finite subset $\Omega \subseteq H_{\mathbb{Z}}^{2k+1}$ is $|\partial_h \Omega|$, where

$$\partial_h \Omega = \{(x, y) \in \Omega \times (H_{\mathbb{Z}}^{2k+1} \setminus \Omega) : x^{-1}y \in S_k\},$$

where $S_k = \{a_1, b_1, a_1^{-1}, b_1^{-1}, \ldots, a_k, b_k, a_k^{-1}, b_k^{-1}\}$. The *vertical perimeter* of Ω is

$$|\partial_v \Omega| = \left(\sum_{t=1}^{\infty} \frac{|\partial_v^t \Omega|^2}{t^2} \right)^{\frac{1}{2}},$$

where

$$\partial_v^t \Omega = \{(x, y) \in \Omega \times (H_{\mathbb{Z}}^{2k+1} \setminus \Omega) : x^{-1}y \in \{c^t, c^{-t}\}\}.$$

In 2018, Naor and Young [923] proved an inequality which bounds $|\partial_v \Omega|$ in terms of $|\partial_h \Omega|$, and is called the *vertical vs horizontal isoperimetric inequality* in Heisenberg groups. See the original paper [923] for various applications of this result.

Theorem 5.29 *There exists a universal constant $C < \infty$ such that the inequality*

$$|\partial_v \Omega| \le \frac{C}{k}|\partial_h \Omega|$$

holds for every integer $k \ge 2$ and every finite subset $\Omega \subseteq H_{\mathbb{Z}}^{2k+1}$.

5.4 Topics in Plane Geometry

Convex Polygons Which Can Tile the Plane

We say that a bounded convex set K can *tile the plane* if the plane can be covered by identical copies of K (possibly translated and rotated) with no overlaps (except on the boundary) and no gaps. In the 1910s, Bieberbach (see [1042]) suggested to determine all the convex domains which can tile the whole plane. It is easy to see that if a bounded convex set K tiles the plane then K must actually be a convex m-gon, $m \ge 3$. It is also not difficult to verify that any triangle and any convex quadrilateral can tile the plane.

For a convex m-gon, denote by $\alpha_1, \ldots, \alpha_m$ the sizes of its angles (in radians) and by $l_1, l_2, \ldots, l_m$ the lengths of its sides, enumerated counterclockwise starting from a fixed vertex. In 1918, Reinhardt [1042] proved that no convex m-gon with $m > 6$ can tile the plane, and a convex hexagon can tile the plane if and only if it belongs to one of the following three types: (i) $\alpha_1 + \alpha_2 + \alpha_3 = 2\pi$, and $l_1 = l_4$; (ii) $\alpha_1 + \alpha_2 + \alpha_4 = 2\pi$, $l_1 = l_4$, and $l_3 = l_5$; or (iii) $\alpha_1 = \alpha_3 = \alpha_5 = \frac{2}{3}\pi$, $l_1 = l_2$, $l_3 = l_4$, and $l_5 = l_6$. After this, only the case of pentagons remained open.

Reinhardt [1042] discovered the following five types of pentagons that can tile the plane.

(1) $\alpha_1 + \alpha_2 + \alpha_3 = 2\pi$;
(2) $\alpha_1 + \alpha_2 + \alpha_4 = 2\pi$ and $l_1 = l_4$;
(3) $\alpha_1 = \alpha_3 = \alpha_4 = \frac{2}{3}\pi$, $l_1 = l_2$, and $l_4 = l_3 + l_5$;
(4) $\alpha_1 = \alpha_3 = \frac{1}{2}\pi$, $l_1 = l_2$, and $l_3 = l_4$;
(5) $\alpha_1 = \frac{1}{3}\pi$, $\alpha_3 = \frac{2}{3}\pi$, $l_1 = l_2$, and $l_3 = l_4$.

Some people believed that this list is complete, but Kershner [704] discovered three new types of such pentagons in 1968.

(6) $\alpha_1 + \alpha_2 + \alpha_4 = 2\pi$, $\alpha_1 = 2\alpha_3$, $l_1 = l_2 = l_5$, and $l_3 = l_4$;
(7) $2\alpha_2 + \alpha_3 = 2\alpha_4 + \alpha_1 = 2\pi$, and $l_1 = l_2 = l_3 = l_4$;
(8) $2\alpha_1 + \alpha_2 = 2\alpha_4 + \alpha_3 = 2\pi$, and $l_1 = l_2 = l_3 = l_4$.

In 1975, James [650] discovered the next type

(9) $\alpha_5 = \frac{\pi}{2}, \alpha_1 + \alpha_4 = \pi, 2\alpha_2 - \alpha_4 = 2\alpha_3 + \alpha_4 = \pi$, and $l_1 = l_2 + l_4 = l_5$.

In 1978, Schattschneider [1086] reported four new types of tiling pentagons discovered by mathematical amateur Marjorie Rice.

(10) $\alpha_2 + 2\alpha_5 = 2\pi, \alpha_3 + 2\alpha_4 = 2\pi$, and $l_1 = l_2 = l_3 = l_4$;
(11) $\alpha_1 = \frac{\pi}{2}, \alpha_3 + \alpha_5 = \pi, 2\alpha_2 + \alpha_3 = 2\pi$, and $2l_1 + l_3 = l_4 = l_5$;
(12) $\alpha_1 = \frac{\pi}{2}, \alpha_3 + \alpha_5 = \pi, 2\alpha_2 + \alpha_3 = 2\pi$, and $2l_1 = l_3 + l_5 = l_4$;
(13) $\alpha_1 = \alpha_3 = \frac{\pi}{2}, 2\alpha_2 + \alpha_4 = 2\alpha_5 + \alpha_4 = 2\pi, l_3 = l_4$, and $2l_3 = l_5$.

Then another tile

(14) $\alpha_1 = \frac{\pi}{2}, 2\alpha_2 + \alpha_3 = 2\pi, \alpha_3 + \alpha_5 = \pi$, and $2l_1 = 2l_3 = l_4 = l_5$

was reported by Stein [1156] in 1985, and another one

(15) $\alpha_1 = \frac{\pi}{3}, \alpha_2 = \frac{3\pi}{4}, \alpha_3 = \frac{7\pi}{12}, \alpha_4 = \frac{\pi}{2}, \alpha_5 = \frac{5\pi}{6}$, and $l_1 = 2l_2 = 2l_4 = 2l_5$

was found by Mann et al., announced in 2015, and published in a journal [836] in 2018. The pentagons listed above are depicted in Table 5.1. As an illustration, the tiling of the plane using the pentagon of type (15) is presented on Fig. 5.7.

In 2017, Rao [1026] posted online a preprint in which he gave a computer-assisted proof that this list is complete. This finished the solution of the celebrated *Bieberbach's problem*.

Theorem 5.30 *A convex pentagon can tile the whole plane by identical copies if and only if it belongs to one of the fifteen types listed above.*

The Lattice Problem of Steinhaus

In the 1950s, Steinhaus asked whether there exist two sets A and B in the plane such that every set congruent to A has exactly one point in common with B, except for the trivial cases $A = \mathbb{R}^2, |B| = 1$ and $B = \mathbb{R}^2, |A| = 1$. He also asked if this is possible with $A = \mathbb{Z}^2$. The first question was answered affirmatively by Sierpiński [1129] in 1958, while the second one remained open for decades. In 2002, Jackson and Mauldin [645] proved that the second question also has an affirmative answer.

A function $f : \mathbb{R}^2 \to \mathbb{R}^2$ is called an *isometry* if $d(f(x), f(y)) = d(x, y)$ for all $x, y \in \mathbb{R}^2$, where d is the usual Euclidean distance. For an isometry f, define $f(\mathbb{Z}^2) = \{y \in \mathbb{R}^2 \mid y = f(x), x \in \mathbb{Z}^2\}$.

Theorem 5.31 *There exists a set $S \subset \mathbb{R}^2$ such that $|S \cap f(\mathbb{Z}^2)| = 1$ for every isometry f.*

The proof of Theorem 5.31 heavily uses the axiom of choice, hence the existence of the set S has been proved without constructing any explicit example of such a set. In particular, the authors left open whether a set S satisfying the conditions of the theorem can be bounded, measurable, or connected.

Table 5.1 Pentagons that can tile the plane

Nr.

(1)–(3)

(4)–(6)

(7)–(9)

(10)–(12)

(13)–(15)

The Pyjama Problem

For any $\epsilon \in (0, 1/2)$, let S_ϵ be the set consisting of vertical strips of width 2ϵ around every integer x-coordinate. This set resembles the pattern on a pair of striped pyjamas, hence it has name "pyjama stripe". The problem asking whether it is possible to cover the whole plane by finitely many rotations of a pyjama stripe around the origin, see Fig. 5.8, is known as the pyjama problem. In 2015, Manners [837] answered this question in the affirmative.

More formally, for every $\epsilon > 0$, the "pyjama stripe" S_ϵ is the set of points $(x, y) \in \mathbb{R}^2$ such that $||x|| \le \epsilon$, where $||x||$ denotes the distance from the real number x to the nearest integer. For any $\phi \in [0, \pi)$, let T_ϕ be the rotation of $\mathbb{R}^2$

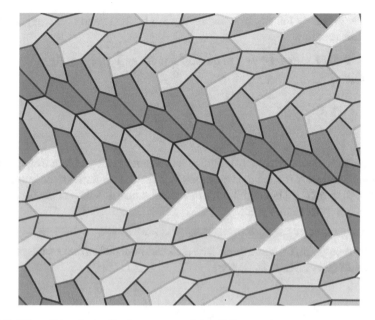

Fig. 5.7 Tiling of the plane using the pentagon of type (15)

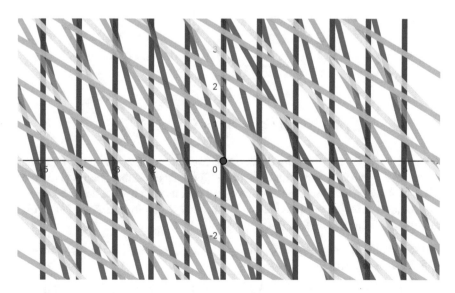

Fig. 5.8 Rotations of a pyjama stripe. The pyjama problem asks if finitely many such rotations can cover the plane

around the origin by the angle ϕ, and let $T_\phi(S_\epsilon)$ be the image of the pyjama stripe S_ϵ after this rotation.

Theorem 5.32 *For every $\epsilon > 0$ there exist a finite sequence $0 \leq \phi_1 < \cdots < \phi_k < \pi$ such that*

$$\bigcup_{i=1}^{k} T_{\phi_i}(S_\epsilon) = \mathbb{R}^2.$$

Theorem 5.32 was proved by reducing its statement to a problem from topological dynamics closely related to Furstenberg's $\times 2$, $\times 3$ Theorem proved in [482] (see the discussion before Theorem 3.5 for its formulation), and then using the methods from [482] to resolve this problem.

The "Capacity" of Compact Plane Sets

Let E be a polygon, or, more generally, a closed bounded subset of the plane. For any k points $A_1, \ldots A_k \in E$, define

$$p(E, A_1, A_2, \ldots, A_k) = \left(\prod_{i=1}^{k-1} \prod_{j=i+1}^{k} |A_i A_j| \right)^{\frac{2}{k(k-1)}}.$$

Then let

$$p_k(E) := \max_{A_1, A_2, \ldots, A_k \in E} p(E, A_1, A_2, \ldots, A_k),$$

and

$$p(E) = \lim_{k \to \infty} p_k(E).$$

The number $p(E)$ is called the *logarithmic capacity* (or the *transfinite diameter*) of the set E. In 1951, Pólya and Szegő [995] conjectured that the regular n-gon minimizes the logarithmic capacity among all n-gons with a fixed area and proved this conjecture for $n = 3$ and $n = 4$. In 2004, Solynin and Zalgaller [1145] confirmed this conjecture for all n.

Theorem 5.33 *For any n-gon E_n, $p(E_n) \geq p(E_n^*)$, where E_n^* is the regular n-gon with the same area as E_n.*

The proof of the cases $n = 3$ and $n = 4$ of Theorem 5.33 in [995] uses the method of Steiner symmetrization, but this method fails for $n \geq 5$ because such symmetrization may increase the number of sides of a polygon. Instead, the proof of Theorem 5.33 in [1145] uses a decomposition of the polygon E_n into triangles that are proportional in a certain sense. Although the logarithmic capacity is not an additive function, the authors showed that it satisfies a certain kind of "semiadditivity" for the decompositions they consider. This allows them to derive the desired lower bound for $p(E_n)$ using the logarithmic capacities of the triangles in the decomposition.

A related notion of "capacity" of compact sets on the plane is the analytic capacity defined via complex analysis. The *analytic capacity* of a compact subset E of the complex plane $\mathbb{C}$ is defined as

$$\gamma(E) = \sup |f'(\infty)|$$

where the supremum is taken over all analytic functions $f : C \setminus E \to C$ such that $|f| \leq 1$ on $C \setminus E$, and $f'(\infty) = \lim_{z \to \infty} z(f(z) - f(\infty))$.

In 1967, Vitushkin [1237] asked whether the analytic capacity is *semiadditive*, that is, whether $\gamma(E \cup F) \leq C(\gamma(E) + \gamma(F))$ for all compact sets E and F. Vitushkin showed that a positive answer to this question would have important applications to the theory of rational approximation. In 2003, Tolsa [1206] answered this question affirmatively.

Theorem 5.34 *There exists an absolute constant C such that $\gamma(E \cup F) \leq C(\gamma(E) + \gamma(F))$ for all compact sets E and F.*

More generally, Tolsa [1206] proved that semiadditivity holds for any countable collection of sets E_i with a compact union.

5.5 Topics in Space Geometry

A Body with Same-Volume Max Sections

For a convex body $K \subset \mathbb{R}^d$ with $0 \in \text{int}(K)$, and $u \in \mathbb{S}^{d-1}$, let

$$M_K(u) = \max_{t \in \mathbb{R}} \text{vol}_{d-1}\left(K \cap (u^\perp + tu)\right),$$

where $\mathbb{S}^{d-1}$ is the unit sphere in $\mathbb{R}^d$, $u^\perp$ is the hyperplane containing the origin and orthogonal to u, and $\text{vol}_{d-1}\left(K \cap (u^\perp + tu)\right)$ is the $d-1$-dimensional volume of the section of K by the affine hyperplane $u^\perp + tu$. It is known [492] that for origin-symmetric convex bodies $K_1, K_2 \subset \mathbb{R}^d$ the condition $M_{K_1}(u) = M_{K_2}(u) \,\forall u \in \mathbb{S}^{d-1}$ implies that $K_1 = K_2$. In particular, if $M_K(u) = c$ is a constant, then K must be a Euclidean ball in $\mathbb{R}^d$. In 1969, Klee [723] asked whether the same is true for general (not necessary origin-symmetric) convex bodies. In 2014, Nazarov, Ryabogin and Zvavitch [927] gave a negative answer to this question.

Theorem 5.35 *In all dimensions $d \geq 3$, there exists a convex body $K \subset \mathbb{R}^d$ that is not a Euclidean ball such that $M_K(u) = c_K$ for some constant c_K independent of u.*

Interestingly, the proof of Theorem 5.35 is very different for even and odd d. The proof for even d is easier and uses elementary algebra. The proof for odd d is much more difficult and uses some deep results, like, for example, the Borsuk–Ulam theorem [174], stating that every continuous function from $\mathbb{S}^m$ to $\mathbb{R}^m$ which maps antipodal[10] points to antipodal points, must map some point to 0.

A Geometric Problem with an Engineering Origin

The *d-dimensional skeleton* of a k-polytope (k-dimensional convex polytope) P is the set of all faces of the polytope of dimension at most d. In 2013, Tokuyama conjectured that it is always possible to find n points in the d-dimensional skeleton of an nd-dimensional convex polytope P so that their center of mass is a target point in P. In the author's words, the conjecture was originally motivated by the engineering problem of determining how counterweights can be attached to some 3-dimensional object to adjust its center of mass to reduce vibrations when the object moves. In 2015, Dobbins [356] proved this conjecture.

Theorem 5.36 *For any nd-polytope P and for any point $p \in P$, there are points $p_1, \ldots, p_n$ in the d-dimensional skeleton of P with barycenter p.*

On Inscribed Polyhedra

Let $S \subset \mathbb{R}^3$ be either the unit sphere (that is, the set of all points $x = (x_1, x_2, x_3) \in \mathbb{R}^3$ such that $x_1^2 + x_2^2 + x_3^2 = 1$), or hyperboloid (defined by the equation $x_1^2 + x_2^2 - x_3^2 = 1$), or cylinder (defined by $x_1^2 + x_2^2 = 1$ with x_3 free). We say that a convex polyhedron P is *inscribed* in S if $P \cap S$ is exactly the set of vertices of P.

A 1-*skeleton* of a convex polyhedron P is the set of vertices and edges of P, considered as a graph. In 1832, Steiner [1158] asked which graphs are 1-skeletons of

(a) an arbitrary convex polyhedron in $\mathbb{R}^3$;
(b) a convex polyhedron inscribed in the sphere;
(c) a convex polyhedron inscribed in the cylinder; and
(d) a convex polyhedron inscribed in the hyperboloid.

Question (a) was answered by a famous 1928 Theorem of Steinitz [1160], which states that a graph G is the 1-skeleton of a convex polyhedron in $\mathbb{R}^3$ if and only if G is planar and 3-connected.[11] In 1992, Hodgson et al. [611] gave a full characterization of possible 1-skeletons of convex polyhedra inscribed in the sphere, thus answering question (b). In 2020, Danciger, Maloni and Schlenker [314] solved the remaining parts (c) and (d) of Steiner's question by reducing them to the already solved part (b).

Theorem 5.37 *For a planar graph G, the following conditions are equivalent:*

(i) G is the 1-skeleton of some convex polyhedron inscribed in the cylinder;

---

[10] Two points on a sphere are called *antipodal* if they are in exactly opposite directions from the sphere's center.

[11] A graph G is called k-*connected* if it has more than k vertices and remains connected whenever fewer than k vertices are removed.

(ii) G is the 1-skeleton of some convex polyhedron inscribed in the hyperboloid; and

(iii) G is the 1-skeleton of some convex polyhedron inscribed in the sphere and G admits a Hamiltonian cycle.[12]

A *quadric surface* in $\mathbb{R}^3$ is the zero set of a quadratic irreducible polynomial in 3 variables. This is a generalization of conic sections (ellipses, parabolas, and hyperbolas) on the plane. It is known that, up to projective transformations, every quadric surface is either the sphere, or the cylinder, or the hyperboloid. Hence, Theorem 5.37 completes the classification of polyhedra in $\mathbb{R}^3$ that are inscribed in a quadric surface.

On the Covering Numbers

We say that convex body $K \subset \mathbb{R}^n$ is *symmetric with respect to the origin*, or just *symmetric* in short, if $x \in K$ implies that $-x \in K$. The set $K^\circ = \{y \in \mathbb{R}^n : \sup_{x \in K} \langle x, y \rangle \leq 1\}$, where $\langle \cdot, \cdot \rangle$ is the inner product, is called the *polar body* of K.

For two convex bodies K and T in $\mathbb{R}^n$, the *covering number* of K by T, denoted $N(K, T)$, is the minimal number of translates of T needed to cover K. In 1972, Pietsch [982] conjectured the existence of two constants $a, b \geq 1$ such that for any dimension n, and for any two symmetric convex bodies K and T in $\mathbb{R}^n$, we have

$$\log N(T^\circ, aK^\circ) \leq b \log N(K, T)$$

(this is one of many equivalent formulations of the conjecture). The conjecture originated in operator theory and became known as the "duality conjecture for entropy numbers". In 2004, Artstein et al. [54] proved this conjecture in the special but important case when either K or T is a Euclidean ball.

Theorem 5.38 *Let D be the Euclidean unit ball in $\mathbb{R}^n$. There exist two universal constants α and β such that for any dimension n and any symmetric convex body $K \subset \mathbb{R}^n$,*

$$N(D, \alpha^{-1}K^\circ)^{\frac{1}{\beta}} \leq N(K, D) \leq N(D, \alpha K^\circ)^\beta.$$

The starting point in the proof of Theorem 5.38 is an earlier work [55] of the same authors, in which they established the same result up to some factor depending on the body K. This result is then applied iteratively to different bodies, carefully selected in such a way that the resulting product of covering numbers shrinks, and the bound in Theorem 5.38 follows with absolute constants.

On Minkowski Valuations

In 1900, Hilbert [605] presented his famous list of problems. His 3rd problem asked whether, for any two polyhedra of equal volume, it is always possible to cut the first one into finitely many polyhedral pieces which can be reassembled to yield the

---

[12] A *Hamiltonian cycle* in a graph is a closed path visiting each vertex exactly once.

second one. In just a year, Dehn [336] proved that the answer is "No". The proof is given by constructing the "Dehn invariant", a function whose domain is the set of all polyhedra which is preserved by the operation of cutting and reassembling pieces, but takes different values for a cube and regular tetrahedron of the same volume. In 1965, Sydler [1171] proved that the Dehn invariant, together with the volume, tells the whole story: if two polyhedra P and Q have the same volume and the same Dehn invariant, it is always possible to get Q from P by cutting it into polyhedral pieces and reassembling.

The Dehn invariant is an example of a "valuation". Other examples of valuations are volume and surface area. An important class of valuations are Minkowski valuations. Let $\mathcal{K}^n$ and $\mathcal{P}^n$ denote the sets of convex bodies and convex polytopes in $\mathbb{R}^n$, respectively. An operator $Z : \mathcal{K}^n \to \mathcal{K}^n$ is called a *Minkowski valuation* if

$$Z(K_1) + Z(K_2) = Z(K_1 \cup K_2) + Z(K_1 \cap K_2),$$

whenever $K_1, K_2, K_1 \cup K_2 \in \mathcal{K}^n$, where "+" denotes *Minkowski addition* $A + B = \{a + b \mid a \in A, b \in B\}$. In 1957, Hadwiger [552] gave a complete characterization of continuous and rigid motion-invariant valuations. In 2005, Ludwig [821] characterized translation-invariant, $\mathrm{SL}(n)$-equivariant Minkowski valuations. An operator $Z : \mathcal{K}^n \to \mathcal{K}^n$ is called $\mathrm{SL}(n)$-*equivariant* if $Z(\phi K) = \phi Z(K)$ for all[13] $\phi \in \mathrm{SL}(n)$, and it is called *translation-invariant* if $Z(K + x) = Z(K)$ for all $x \in \mathbb{R}^n$. An example of a valuation with these properties is the *difference operator* $D : \mathcal{K}^n \to \mathcal{K}^n$ which transforms every $K \in \mathcal{K}$ into $D(K) = K + (-K)$. Ludwig [821] proved that this is essentially the only such example.

Theorem 5.39 *If* $Z : \mathcal{P}^n \to \mathcal{K}^n$, $n \geq 2$, *is a translation-invariant,* $\mathrm{SL}(n)$-*equivariant Minkowski valuation, then there exists a constant* $c \geq 0$ *such that* $Z(P) = cD(P)$ *for every* $P \in \mathcal{P}^n$.

An operator $Z : \mathcal{K}^n \to \mathcal{K}^n$ is called *homogeneous* of degree $r \in \mathbb{R}$ if $Z(sK) = s^r Z(K)$ for $s \geq 0$. In fact, Theorem 5.39 is a consequence of a more general result which gives a complete characterization of all homogeneous $\mathrm{SL}(n)$ equivariant Minkowski valuations, not necessary translation invariant.

5.6 Knot Theory

Transforming Knot Diagrams

A *knot* is a closed, non-self-intersecting curve in $\mathbb{R}^3$. Two knots K_1 and K_2 are *equivalent* if K_1 can be transformed smoothly, without intersecting itself, to coincide with K_2. A knot equivalent to a circle is called the *unknot*. Knots are usually represented on the plane using knot diagrams. A *knot diagram* is a projection of a

---

[13] Recall that $\mathrm{SL}(n)$ denotes the set of $n \times n$ matrices with determinant 1.

knot into a plane which (i) is injective everywhere, except at a finite number of crossing points, which are the projections of only two points of the knot, and (ii) records over/under information at each crossing. A knot diagram is *trivial* if it has no crossings.

A fundamental question in knot theory is: given two knot diagrams, how can we know if they represent the same knot? In 1927, Reidemeister [1038] proved that this is the case if and only if one diagram can be transformed into another one by a sequence of simple operations, which have acquired the name *Reidemeister moves*. Specifically, the Reidemeister moves of a knot diagram are (i) twist and untwist in either direction, or (ii) move one strand completely over another, or (iii) move a strand completely over or under a crossing, see Fig. 5.9. In particular, Reidemeister's theorem implies that a diagram represents the unknot if and only if it can be transformed into a trivial one by such moves. But how many moves are required for such a transformation for a diagram which initially had n crossings? The first explicit upper bound was obtained by Hass and Lagarias [572] in 2001.

Theorem 5.40 *There is a positive constant c such that for each $n \geq 1$, any knot diagram D of the unknot with n crossings can be transformed into a trivial knot diagram using at most 2^{cn} Reidemeister moves.*

In fact, Theorem 5.40 is true with the explicit constant $c = 10^{11}$. However, the bound 2^{cn} grows exponentially in n. An important open question was whether a result like Theorem 5.40 is possible with the bound growing polynomially in n. In 2015, Lackenby [753] answered this question affirmatively.

Theorem 5.41 *For every diagram D of the unknot with n crossings there is a sequence of at most $(236n)^{11}$ Reidemeister moves that transforms D into a trivial diagram. Moreover, every diagram in this sequence has at most $(7n)^2$ crossings.*

The proof of Theorem 5.41 uses many ideas from a 2006 paper of Dynnikov [377], who proved that every diagram D of the unknot with n crossings can be transformed to a trivial diagram such that all diagrams on the way have at most $2(n+1)^2$ crossings. This bound is even better than the bound $(7n)^2$ in Theorem 5.41. However, Dynnikov did not prove that the number of moves is polynomial in n, and this is the main new contribution of Theorem 5.41.

A "Nice" Representative of a Knot

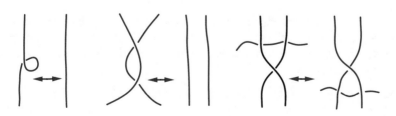

Fig. 5.9 The Reidemeister moves

If knots K_1 and K_2 are equivalent, we say that they "represent" the same knot. For example, the unknot can be represented by any closed plane curve without self-intersections, but is usually represented as a circle, which is intuitively the "nicest" way. More generally, the question of what is the "nicest" representative of an arbitrary knot is a well-studied question for various definitions of "niceness". One possible measure of "niceness" is the ropelength. The *thickness* of a knot representative K is

$$\tau(K) = \inf_{x,y,z \in K} r(x, y, z),$$

where the infimum is taken over all triples of pairvise distinct points x, y, z on K, and $r(x, y, z)$ is the radius of the circle passing through these points. The *ropelength* of K is the quotient of its length by its thickness. Given a family of equivalent knots, the ones with minimal ropelength in the family (such knots are also known as *ideal knots*) can serve as "nice-looking" representatives of the family. However, even the existence of such ropelength minimizers is a highly non-trivial question, because, in principle, there could be an infinite family of knot representatives whose ropelength approaches some limit from above but never reaches it. In 2002, Cantarella, Kusner and Sullivan [247] proved that this cannot be the case.

Theorem 5.42 *In any family of equivalent knot representatives, there exists a ropelength minimizer.*

With Theorem 5.42, we finally know that the ropelength criterion for selecting the "nicest" knot representative is feasible. However, determining the knots with minimal ropelength is a very non-trivial problem. Even the basic question "What is the minimal possible ropelength of a non-trivial knot?" remains open. Litherland [804] proved in 1999 that it is at least $5\pi \approx 15.71$. In the same paper [247] in which they proved Theorem 5.42, Cantarella, Kusner and Sullivan improved this lower bound to $2\pi(2 + \sqrt{2}) \approx 21.45$. In 2006, Denne et al. [346] improved it further to 31.32. Computer experiments suggests that the correct answer may be around 32.66.

An alternative way to measure the "niceness" of a knot representative is its *distortion*. The distortion of a knot representative γ is

$$\delta(\gamma) = \sup_{p,q \in \gamma} \frac{d_\gamma(p, g)}{d_{\mathbb{R}^3}(p, q)},$$

where d_γ denotes the arclength along γ and $d_{\mathbb{R}^3}$ denotes the Euclidean distance in $\mathbb{R}^3$. In 1983, Gromov [536] asked whether every knot K in $\mathbb{R}^3$ has a representative γ with $\delta(\gamma) < 100$. In 2011, Pardon [967] proved that this is not the case, with any constant in place of 100.

Theorem 5.43 *For any $C > 0$ there exists a knot K in $\mathbb{R}^3$ such that $\delta(\gamma) > C$ for any representative γ of K.*

Pardon [967] proved Theorem 5.43 by presenting explicit examples of families of knots with distortion going to infinity. For example, let $\gamma_{p,q}$ be the curve in $\mathbb{R}^3$ defined by the equations

$$x = r\cos(p\phi), \quad y = r\sin(p\phi), \quad z = -\sin(q\phi),$$

where $r = \cos(q\phi)+2$ and $0 \leq \phi < 2\pi$. The set $K_{p,q}$ of curves equivalent to $\gamma_{p,q}$ is called the (p, q)-*torus knot*. Pardon proved that the distortion of any representative of the (p, q)-torus knot is at least $\frac{1}{160}\min(p, q)$. Theorem 5.43 follows.

High-Dimensional Knot Theory
The classical knot theory is trivial in $\mathbb{R}^d$ for $d > 3$, because every pair of closed, non-self-intersecting curves in $\mathbb{R}^4$ can be transformed into each other smoothly and without intersecting itself. However, a meaningful theory was initiated by Fox [464] in 1962. The central notion of this theory is the notion of a slice knot. Let $B^4 = \{x \in \mathbb{R}^4 : |x| \leq 1\}$ be the unit ball in $\mathbb{R}^4$, and let $S^3 = \{x \in \mathbb{R}^4 : |x| = 1\}$ be its boundary. A knot $K \subset S^3$ is called (smoothly) *slice* if it bounds a smoothly embedded 2-dimensional disk in B^4.

We say that a knot K has n *crossings* if n is the minimal number of crossing a knot diagram of a knot equivalent to K may have. By 2005, all knots of under 13 crossings were classified according to whether they are slice knots or not, the only exception being the *Conway knot*. This is the knot with 11 crossings depicted in Fig. 5.10. The question of whether this knot is slice was unresolved for over 50 years. In 2020, Piccirillo [981] proved that the answer to this question is negative.

Theorem 5.44 *The Conway knot is not slice*

The standard tools used to prove that a given knot is not slice are various invariants. For example, imagine that we have a function f which assigns to each knot a real number, such that f from every slice knot is (for example) 0. Then proving that $f(K) \neq 0$ for a given knot K shows that it is not slice. However, if $f(K) = 0$, this gives no information whether K is slice or not: the invariant f does not work for K.

The problem with the Conway knot C is that it was conjectured to be not slice, but all the known invariants failed to certify this. Piccirillo [981] managed to construct a highly non-trivial knot C' with the property that the Conway knot C is slice if and only if C' is slice, and then tried to apply known invariants to C'. One of them,

Fig. 5.10 The Conway knot

called Rasmussen's s-invariant, happened to work, certified that C' is not slice, and Theorem 5.44 followed.

5.7 Minimal Surfaces

Simply Connected Complete Minimal Surfaces
Intuitively, a surface is a generalization of a plane, which is not necessarily flat, like a sphere or semi-sphere. Formally,[14] a (topological) surface S with boundary is a second-countable Hausdorff topological space in which every point has an open neighbourhood homeomorphic to some open subset of the closure of the upper half-plane in $\mathbb{C}$. If every point $x \in S$ has an open neighbourhood homeomorphic to an open subset of $\mathbb{C}$, then we say that the surface S has no boundary. For example, a semi-sphere is a surface with boundary, while a sphere has no boundary. A compact[15] surface without boundary is called *closed*.

---

[14] In this definition, $\mathbb{C}$ is the complex plane, "the closure of the upper half-plane" is the set $\{z \in \mathbb{C} : \mathrm{Im}(z) \geq 0\}$, "second-countable" is a topological space T in which there exists some countable collection $\mathcal{U} = \{U_i\}_{i=1}^{\infty}$ of open sets such that any open set in T can be written as a union of elements of some subfamily of $\mathcal{U}$; "Hausdorff" refers to a topological space where for any two distinct points there exist neighbourhoods of each which are disjoint from each other, and "homeomorphic" means the existence of a homeomorphism, a continuous function between topological spaces that has a continuous inverse function.

[15] Recall that topological space is called *compact* if each of its open covers has a finite subcover.

The definition of a surface is abstract and does not require that it is a subset of the Euclidean space $\mathbb{R}^3$. Intuitively, a surface S is called *embedded* in $\mathbb{R}^3$ if it can be placed into $\mathbb{R}^3$ without self-intersections. More formally, this means the existence of a homeomorphism from S onto a subset of $\mathbb{R}^3$, which carries the subspace topology inherited from $\mathbb{R}^3$. There are examples of surfaces, such as the real projective plane,[16] which cannot be embedded into $\mathbb{R}^3$.

The simplest example of a surface is the plane. Any open subset of the plane, such as the open unit disk, is also a surface. Intuitively, such examples are uninteresting. A correct formalization of the word "interesting" in this context in the notion of completeness. Intuitively, a surface is complete if, when we walk "straight" inside it, we never meet its "end", and can walk straight forever. By "walk straight" we mean follow a geodesic, a curve on a surface which locally minimizes length. More formally, a parametrized curve $\gamma : I \to M$ from an interval $I \subset \mathbb{R}$ to a surface M is a *geodesic* if there is a constant $v \geq 0$ such that for any $t \in I$ there is a neighborhood J of t in I such that for any $t_1, t_2 \in J$ we have $d(\gamma(t_1), \gamma(t_2)) = v |t_1 - t_2|$, where $d(x, y)$ is the length of the shortest path from x to y within M. A surface M is said to be *complete* when for every point $p \in M$, any parametrized geodesic $\gamma : [0, e) \to M$ of M, starting from $p = \gamma(0)$, may be extended to a parametrized geodesic $\gamma' : \mathbb{R} \to M$ defined on the entire real line $\mathbb{R}$. For example, the plane and the sphere are complete surfaces, while an infinite cone with removed vertex is not, because some geodesics can "hit" the vertex and then cannot be extended.

Another "uninteresting" example of a surface is a disjoint union of two parallel planes. A topological space (and in particular a surface) is called *connected* if it cannot be represented as the union of two or more disjoint non-empty open subsets. A surface S is called *simply connected* if for every two points $x \in S$ and $y \in S$ there is a path in S with endpoints x and y, and every two such paths can be continuously transformed into each other within S while keeping both endpoints x and y fixed. For example, an infinite cylinder is connected but not simply connected.

There are particularly nice and important examples of surfaces which are called minimal surfaces. This concept has many equivalent definitions, and here is one of them. A surface $M \subset \mathbb{R}^3$ is called a *minimal surface* if every point $p \in M$ has a neighbourhood S, with boundary B, such that S has minimal area out of all surfaces S' with the same boundary B. Minimal surfaces are among the most studied objects in geometry and topology. A trivial example of a minimal surface is the plane. One of the simplest non-trivial examples, described by Euler in 1774, is a helicoid. A *helicoid* (see Fig. 5.11a) is a surface which can be defined by equations $x = s \cos(\alpha t)$, $y = s \sin(\alpha t)$, $z = t$, where α is a constant, and s, t are real parameters, ranging from $-\infty$ to ∞.

An active research direction is to understand the possible shapes of complete minimal surfaces. For example, it is easy to see that planes and helicoids are simply connected. Since 1774, many other complete minimal surfaces embedded in $\mathbb{R}^3$

---

[16] The *real projective plane* can be defined as the unit square $[0, 1] \times [0, 1]$ with its sides identified by the following rules: $(0, y) \sim (1, 1 - y)$ for $0 \leq y \leq 1$ and $(x, 0) \sim (1 - x, 1)$ for $0 \leq x \leq 1$.

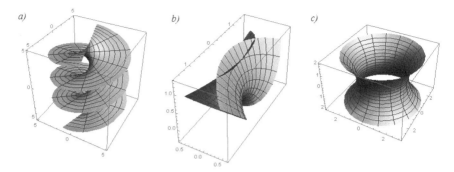

Fig. 5.11 Fragments of (**a**) a helicoid, (**b**) the Enneper surface, and (**c**) a catenoid

have been discovered, but none of them are simply connected. In 2005, Meeks and Rosenberg [870] proved[17] that no other such examples exist.

Theorem 5.45 *Any complete embedded simply connected minimal surface in $\mathbb{R}^3$ is either a plane or a helicoid.*

We remark that Theorem 5.45 is not true if we remove the "embedded" condition and allow the surface to have self-intersections. An example of a complete simply connected minimal surface with self-intersection is the *Enneper surface* depicted in Fig. 5.11b.

Surfaces of Genus 0

Intuitively, a surface is called orientable if it has two distinct "sides". An example of a surface which does not have this property is the *Möbius strip*, which can be defined as the square $[0, 1] \times [0, 1]$ with its top and bottom sides identified by the relation $(x, 0) \sim (1 - x, 1)$ for $0 \leq x \leq 1$. Formally, a surface S is called *orientable* if it does not contain a homeomorphic copy of the Möbius strip.

The *genus* of a connected orientable surface is the maximum number of cuttings that can be made along non-intersecting closed simple curves on the surface without making it disconnected. In particular, a surface has genus 0 if cutting it along any closed curve makes it disconnected. A *planar domain* is a connected surface that is non-compact and has genus 0.

Planes and helicoids are examples of minimal planar domains in $\mathbb{R}^3$. Another such example, discovered by Euler in 1744, is the *catenoid*. This is the surface obtained by rotating the catenary curve $y = \frac{a}{2}(e^{x/a} + e^{-x/a})$ in three-dimensional space around the x-axis, see Fig. 5.11c. Here, $a > 0$ is a real parameter. Another one-parameter family of minimal planar domains was discovered by Riemann in the 1860s. They are called the "Riemann minimal examples". We will not define

---

[17] In fact, Theorem 5.45 is proved in [870] with the extra hypothesis that the surface is "properly embedded", but this hypothesis can be removed by Theorem 5.47 (ii) discussed below.

these here but mention that any such surface can be rotated so that, after rotation, it intersects every horizontal plane in a circle or in a line.

An embedded surface $S \subset \mathbb{R}^3$ is called *properly embedded* in $\mathbb{R}^3$ if the pre-image of any compact subset of $\mathbb{R}^3$ is compact in the surface. In 2015, Meeks et al. [869] proved that, up to scaling and rigid motion, any properly embedded minimal planar domain M in $\mathbb{R}^3$ must be a plane, helicoid, catenoid, or one of the Riemann minimal examples. In particular, this theorem has the following easy-to-state corollary.

Theorem 5.46 *Every properly embedded minimal planar domain M in $\mathbb{R}^3$ can be rotated such that, after rotation, it intersects every horizontal plane in a circle or in a line.*

Surfaces with Finite Topology

In 1965, Calabi [238] conjectured that (a) a complete minimal surface in $\mathbb{R}^3$ must be unbounded, and, moreover, (b) it has an unbounded projection on every line, unless it is a plane. If we allow surfaces to have self-intersections, these conjectures turn out to be false: a counterexample to part (b) was given by Jorge and Xavier [665] in 1980, and to part (a) by Nadirashvili [918] in 1996. However, these counterexamples are not embedded, and Yau [1283] asked in 2000 whether Calabi's conjectures are true for embedded surfaces.

A surface M is said to have a *finite topology* if it is homeomorphic to a compact surface $\hat{M}$ (of finite genus k and empty boundary) minus a finite number of points $p_1, \ldots, p_j \in \hat{M}$. A closed set $E \subset M$ homeomophic to a neighbourhood of any removed point p_i is called an *end*[18] of M. So, a surface M is of finite topology if it has finite genus and a finite number of ends. Intuitively, the genus represents the number of holes in the surface while each end represents a topologically distinct way to move to infinity within the surface.

In 2008, Colding and Minicozzi [293] proved the following properties of complete embedded minimal surfaces with finite topology.

Theorem 5.47 *Let M be a complete embedded minimal surface with finite topology in $\mathbb{R}^3$. Then:*

(i) if M is contained in a halfspace of $\mathbb{R}^3$ then M is a plane;
(ii) M is properly embedded in $\mathbb{R}^3$.

Because every surface which cannot be covered by any halfspace is unbounded, and has an unbounded projection on every line, part (i) of Theorem 5.47 confirms both Calabi conjectures for embedded surfaces with finite topology. Part (ii) is important because it implies that many results which have been proved for properly embedded surfaces are true for all complete embedded surfaces. For example, Theorem 5.45 was originally proved in [870] for properly embedded surfaces only.

---

[18] Equivalently, an end of a topological space X is a function e which assigns to each compact set K in X some connected component with non-compact closure $e(K)$ of the complement $X - K$ in a compatible way, so that $K_1 \subseteq K_2 \Rightarrow e(K_1) \supseteq e(K_2)$.

Another corollary of Theorem 5.47 (ii) is that the plane, the helicoid and the catenoid are the only complete embedded minimal planar domains in $\mathbb{R}^3$ with finite topology (note that the Riemann minimal examples mentioned in the context of Theorem 5.46 do not count because they have an infinite number of ends).

Surfaces of Genus 1

An equivalent definition of minimal surfaces uses the notion of curvature. For any point x on a surface $M \subset \mathbb{R}^3$, we can build a vector u perpendicular to M, choose any plane P containing u (called a *normal plane*), and measure the curvature[19] $\kappa(x, P)$ at x of the curve which is the intersection of S and P. Let $k_1(x)$ and $k_2(x)$ be the minimum and maximum values of $\kappa(x, P)$ over all choices of normal plane P. The *mean curvature* of M at x is

$$H(x) = \frac{1}{2}(k_1(x) + k_2(x)). \tag{5.11}$$

A surface $M \subset \mathbb{R}^3$ is *minimal* if and only if its mean curvature is equal to zero at all points.

The product $K(x) = k_1(x) \cdot k_2(x)$ is called the *Gaussian curvature* of M at x. The integral of the Gaussian curvature over the whole surface is called the *total curvature* of M.

The helicoid is an example of a complete minimal surface in $\mathbb{R}^3$ with finite topology and infinite total curvature. Between 1774 and 2009, the helicoid was the only known surface with these properties. In 2009, Weber, Hoffman and Wolf [1250] proved the existence of at least one more such surface.

Theorem 5.48 *There exists a complete (and hence properly embedded) minimal surface in $\mathbb{R}^3$ with finite topology and infinite total curvature, which is not a helicoid.*

The authors proved Theorem 5.48 by constructing an explicit example of a surface with these properties. The surface they constructed is of genus one and has one end. It has got a name: the "embedded genus one helicoid".

Surfaces of Higher Genus

Let $M \subset \mathbb{R}^3$ be a complete embedded minimal surface in $\mathbb{R}^3$ with finite topology. If M is simply connected, then it must be the plane or the helicoid, see Theorem 5.45. More generally, if M is a planar domain (connected, non-compact and with genus 0) then it must be the plane, the helicoid or the catenoid, see Theorem 5.46. Are there other examples with higher genus? In 2009, Weber et al. [1250] constructed an example with one end and genus $g = 1$, see Theorem 5.48. In 2016, Hoffman, Traizet and White [614] proved that such surfaces exist for every genus g.

---

[19] Intuitively, the *curvature* of a plane curve γ at a point $x \in \gamma$ is $1/r$, where r is the radius of the circle which best approximates γ at x. More formally, if the curve is (locally) the graph of some twice differentiable function f, then its curvature at x_0 is $\kappa = \frac{f''(x_0)}{(1+f'(x_0)^2)^{3/2}}$.

Theorem 5.49 *For every positive integer g, there exists a connected, complete embedded minimal surface in $\mathbb{R}^3$ with one end and genus g.*

Theorem 5.49 was proved by first constructing surfaces with the listed properties embedded in space $\mathbb{S}^2 \times \mathbb{R}$, where $\mathbb{S}^2 = \{(x, y, z) \in \mathbb{R}^3 : x^2 + y^2 + z^2 = r^2\}$ is a sphere of radius r, and then going to the limit $r \to \infty$ to get a helicoid with the desired properties in $\mathbb{R}^3$. The most difficult part of the argument is to prove that the genus is preserved under this limit operation.

Theorem 5.49 states that a connected complete minimal surface in $\mathbb{R}^3$ may have an arbitrary genus. What about the number of ends? A classical 1990 conjecture of Hoffman and Meeks [613] states that if a complete connected minimal surface of genus g has a finite number k of ends, then $k \le g + 2$. However, before 2019, it was not known if there exists any upper bound on k depending only on g. In 2019, Meeks et al. [868] established the existence of such an upper bound.

Theorem 5.50 *For every positive integer g, there exists a constant C_g such that any complete minimal surface in $\mathbb{R}^3$ with genus g and finite topology has at most C_g ends.*

The Isoperimetric Inequality for Minimal Surfaces

An old classical theorem states that a circle has the lowest length out of all plane curves which encloses a prescribed area. Equivalently, if a plane curve of length L encloses area A, then

$$4\pi A \le L^2, \tag{5.12}$$

with equality if and only if the curve is a circle. This is called the *isoperimetric inequality on the plane*. In 1921 Carleman [250] proved that (5.12) remains correct for any minimal surface in $\mathbb{R}^3$ with area A and boundary length L, which is homeomorphic to a disk $D = \{(x, y) \in \mathbb{R}^2 : x^2 + y^2 \le 1\}$. Since then, the question of whether (5.12) remains true for any compact minimal surface with boundary, not necessary homeomorphic to a disk, became a famous open problem. In 1959, Reid [1037] proved (5.12) for all minimal surfaces in $\mathbb{R}^3$ with connected boundary. In 1984, Li et al. [790] proved the same for all minimal surfaces in $\mathbb{R}^3$ whose boundary has two connected components.

In 1974, Simon proved that the weaker inequality $2\pi A \le L^2$ is valid for all minimal surfaces. In 2003, Stone [1163] improved this to $2\sqrt{2}A \le L^2$. Finally, Brendle [205] in 2019 posted online a preprint in which he proved (5.12) for all minimal surfaces in $\mathbb{R}^3$ (and, more generally, in $\mathbb{R}^4$).

Theorem 5.51 *Let M be a compact minimal surface embedded in $\mathbb{R}^3$ (or, more generally, in $\mathbb{R}^4$) with area A and boundary length L. Then the isoperimetric inequality (5.12) holds.*

Minimal Surfaces in Other Spaces

Minimal surfaces can be studied in spaces other than $\mathbb{R}^3$. For example, one can study surfaces $M \subset \mathbb{S}^3$, where

$$\mathbb{S}^3 = \{(x_1, x_2, x_3, x_4) \in \mathbb{R}^4 : x_1^2 + x_2^2 + x_3^2 + x_4^2 = 1\} \tag{5.13}$$

in the unit 3-sphere in $\mathbb{R}^4$. For example, $\mathbb{R}^3$ does not contain closed minimal surfaces, while $\mathbb{S}^3$ does. The simplest examples of such surfaces are the *equator* $\{(x_1, x_2, x_3, x_4) \in \mathbb{S}^3 : x_4 = 0\}$ and the *Clifford torus*

$$\left\{(x_1, x_2, x_3, x_4) \in \mathbb{S}^3 : x_1^2 + x_2^2 = x_3^2 + x_4^2 = \frac{1}{2}\right\}. \tag{5.14}$$

The equator has genus 0, while the Clifford torus has genus 1. In 1970, Lawson [771] proved that, for any integer $g > 0$, there exists at least one closed embedded minimal surface in $\mathbb{S}^3$ with genus g. One may attempt to classify what such surfaces with given genus g may look like, at least for small g. In 1966, Almgren [31] proved that every closed minimal surface in $\mathbb{S}^3$ of genus 0 is congruent to the equator. In 1970, Lawson [772] conjectured that the Clifford torus is the only closed embedded minimal surface in $\mathbb{S}^3$ of genus 1. In 2013, Brendle [203] confirmed this conjecture.

Theorem 5.52 *Every closed embedded minimal surface in $\mathbb{S}^3$ of genus 1 is congruent to the Clifford torus.*

5.8 General Surfaces

The Willmore Conjecture

While the theory of minimal surfaces had a golden age at the beginning of the twenty-first century, with a large number of impressive results, there are of course major developments in the theory of surfaces which are not necessary minimal. One example is the resolution of the *Willmore conjecture*.

For a closed orientable surface $S \subset \mathbb{R}^3$, its *Willmore energy* is

$$W(S) = \int_S H^2(x)\mathrm{d}x,$$

where $H(x)$ is the mean curvature of S at x defined in (5.11). The Willmore energy is a way to measure the "total curvature" of a surface. It has nice mathematical properties and appears naturally in some physical contexts. It is known that $W(S) \geq 4\pi$ for every closed surface S, with equality if S is an embedded round sphere. In 1965, Willmore [1265] conjectured that the stronger lower bound $W(S) \geq 2\pi^2$ holds for every closed surface of genus 1. This bound is the best possible, because equality holds for the torus T^* obtained by rotating in $\mathbb{R}^3$ a circle of radius 1 with center at distance $\sqrt{2}$ from the axis of revolution. Equivalently, T^* can be defined as

$$\{(x, y, z) \in \mathbb{R}^3 : x = (\sqrt{2} + \cos u) \cos v, \ y = (\sqrt{2} + \cos u) \sin v, \ z = \sin u, \ u, v \in \mathbb{R}\}.$$

In 1982, Li and Yau [791] proved that $W(S) \geq 8\pi > 2\pi^2$ for every genus 1 closed surface $S \subset \mathbb{R}^3$ with self-intersections, after which Willmore's conjecture remained open only for embedded surfaces. In 2014, Marques and Neves [847] confirmed this remaining case of the conjecture.

Theorem 5.53 *If $S \subset \mathbb{R}^3$ is a closed embedded orientable surface of genus 1, then $W(S) \geq 2\pi^2$.*

Marques and Neves [847] deduced Theorem 5.53 from the more general result that the Clifford torus (5.14) minimizes both the functional $V(S) = \int_S (1 + H^2(x))dx$ and the surface area $A(S)$ over all closed surfaces $S \subset \mathbb{S}^3$ of genus $g \geq 1$ embedded into the 3-sphere $\mathbb{S}^3$. This last result was proved using the newly developed "min-max theory" of minimal surfaces, see [847] for further details.

Self-Similar Shrinkers

A surface $S \subset \mathbb{R}^3$ is called a *self-similar shrinker* if it satisfies the equation

$$H(x) = \frac{1}{2}\langle x, u(x)\rangle$$

for every $x \in S$, where $H(x)$ is the mean curvature (5.11), $u(x)$ is the unit vector perpendicular to S at x, and $\langle x, y\rangle$ is the inner product. The classication of self-similar shrinkers is an important problem with implications for the analysis of singularities. The simplest example of a compact self-similar shrinker in $\mathbb{R}^3$ is the round sphere of radius 2 centred at the origin. A well-known conjecture states that this is in fact the only example of a compact, embedded self-similar shrinker in $\mathbb{R}^3$ of genus 0. In 1992, Angenent [45] constructed an example of a compact embedded self-similar shrinker in $\mathbb{R}^3$ of genus 1. In 2015, Drugan [362] constructed a self-similar shrinker of genus 0 but with self-intersections. This implies that none of the assumptions of the conjecture can be removed. In 2016, Brendle [204] proved the conjecture.

Theorem 5.54 *The only compact, embedded self-similar shrinker in $\mathbb{R}^3$ of genus 0 is the round sphere of radius 2 centred at the origin.*

As a first step in the proof of Theorem 5.54, Brendle proved that if M is a compact embedded self-similar shrinker in $\mathbb{R}^3$ of genus 0 then the intersection of M with any plane P which passes through the origin must be a piecewise smooth Jordan curve. Then this fact is used to prove that M is star-shaped.[20] From this, the author deduced

---

[20] A set $S \subset \mathbb{R}^n$ is called a *star-shaped* if there exists an $x_0 \in S$ such that for all $x \in S$ the line segment from x_0 to x lies in S.

that the mean curvature (5.11) does not change sign on M. However, Huisken [630] proved in 1990 that the round sphere is the only compact self-similar shrinker with this last property.

5.9 Topology in Higher Dimensions

The Poincaré Conjecture

Many of the definitions given for surfaces in Sect. 5.7 can be straightforwardly extended to higher dimensions. A k-dimensional topological manifold M without boundary (or *k-manifold* for short) is a second-countable Hausdorff topological space in which every point has an open neighbourhood homeomorphic to some open subset of $\mathbb{R}^k$. A compact k-manifold without boundary is called *closed*. A k-manifold M is called *simply connected* if for every two points $x \in M$ and $y \in M$ there is a path in M with endpoints x and y, and every two such paths can be continuously transformed into each other within M while keeping both endpoints x and y fixed. Topology studies k-manifolds up to homeomorphisms.

A surface is a k-manifold for $k = 2$. A fundamental result in the topology of surfaces is the classification theorem of closed surfaces. The *connected sum* of two surfaces S_1 and S_2 is the surface obtained by removing a disk from each of them and gluing them along the boundary components that result. The *classification theorem of closed surfaces* states that any connected closed surface is homeomorphic to either the 2-sphere or the connected sum of g tori for $g \geq 1$, or the connected sum of k real projective planes for $k \geq 1$. The simplest special case of this theorem states that every simply connected closed surface is homeomorphic to the 2-sphere. In 1904, Poincaré [989], in an attempt to build a similar theory for 3-manifolds, conjectured that every simply connected closed 3-manifold is homeomorphic to the 3-sphere (5.13). This statement became a fundamental conjecture in topology and in the whole of mathematics. In 2000, the Clay Mathematics Institute included the Poincaré conjecture in the list of seven Millennium Prize Problems [252], for which they offered a million-dollar prize for the first correct solution.

In 1960, Smale [1136] proved that the high-dimensional analogue of the conjecture is true for k-manifolds in dimensions $k \geq 5$. In 1982, Freedman [471] proved the same for $k = 4$, but the original case $k = 3$ remained open. In 1982, Thurston [1201] formulated a more general conjecture, called the *geometrization conjecture*, which would imply a classification method for 3-manifolds in a similar way as the classification theorem for closed surfaces described above. In 2002 and 2003, Perelman posted online three papers [973–975] in which he proved the Thurston's geometrization conjecture, and in particular the Poincaré conjecture. This result is considered by many as *the* greatest theorem of the twenty-first century. In particular, as of 2020, the Poincaré conjecture remains the only solved Millennium Prize Problem from the list [252].

Theorem 5.55 *Every simply connected closed 3-manifold is homeomorphic to the 3-sphere.*

Triangulations of Surfaces and Manifolds

A (two-dimensional, orientable) *triangulation T* is a way to glue together a collection of oriented triangles, called the *faces*, along their edges in a connected way that matches the orientations. If the number of triangles is finite, then T is homeomorphic to an orientable topological surface S. Let us define the genus of T as the genus of S. The triangulation T is called *rooted* if it has a distinguished oriented edge, called the *root edge*.

Triangulations are important in many areas of mathematics and physics, for example, they play a crucial role in the theory of random surfaces. Many surfaces studied in physics, like the surface of the ocean, behave so unpredictably that the best way to model them is to assume that they are "random" and use the tools of probability theory. However, how do we formally define the term "random surface"? Here, triangulations can help. For every n, there is a finite number of triangulations with n faces, and we can select one uniformly at random. Then the (appropriately defined) limit as $n \to \infty$ is a good model of a random surface.

To estimate the probability that a random triangulation has any property P, we need to estimate the number of n-face triangulations with property P and divide by the total number of n-face triangulations. Hence, estimating the numbers n-face triangulations with various properties has a central importance in this theory. In particular, what is the number $\tau(n, g)$ of rooted triangulations with n faces and genus g? In 1986, Bender and Canfield [115] established the asymptotic growth of $\tau(n, g)$ when $n \to \infty$ and g is fixed. However, the important case when n and g both go to infinity remained open. In 2019, Budzinski and Louf posted online a preprint [221] in which they estimated $\tau(n, g)$ up to sub-exponential factors in the case when the ratio g/n converges to a constant.

To formulate the result, we need some notation. Let $\lambda_c = \frac{1}{12\sqrt{3}}$. For any $\lambda \in (0, \lambda_c]$, let $h \in (0, \frac{1}{4}]$ be such that $\lambda = \frac{h}{(1+8h)^{3/2}}$, and let

$$d(\lambda) = \frac{h \log \frac{1+\sqrt{1-4h}}{1-\sqrt{1-4h}}}{(1+8h)\sqrt{1-4h}}.$$

For any $\theta \in (0, \frac{1}{2})$, let $\lambda(\theta)$ be the unique solution to the equation $d(\lambda) = \frac{1-2\theta}{6}$, and let

$$f(\theta) = 2\theta \log \frac{12\theta}{e} + \theta \int_2^{1/\theta} \log \frac{1}{\lambda(1/t)} dt, \qquad 0 < \theta < \frac{1}{2}.$$

Also, let $f(0) = \log 12\sqrt{3}$ and $f(1/2) = \log \frac{6}{e}$.

Theorem 5.56 *Let g_n be a sequence such that $0 \leq g_n \leq \frac{n+1}{2}$ for every n and $\lim_{n\to\infty} \frac{g_n}{n} = \theta \in [0, \frac{1}{2}]$. Then*

$$\tau(n, g_n) = n^{2g_n} \exp(f(\theta)n + o(n)) \quad \text{as} \quad n \to \infty.$$

The key step in the proof of Theorem 5.56 is the derivation of the *bounded ratio lemma*, which states that $\frac{\tau(n,g)}{\tau(n-1,g)}$ stays bounded as long as $\frac{g}{n}$ is bounded away from $\frac{1}{2}$. This lemma is then used to adapt the well-known argument of Angel and Schramm [44] who proved a similar result in the genus 0 case.

Theorem 5.56 counts two-dimensional orientable triangulations. More generally, triangulations can be non-orientable and can be studied in an arbitrary dimension. We say that points in $u_0, u_1, \ldots, u_d \in \mathbb{R}^n$ (where $n \geq d$) are *affine independent* if the vectors $u_1 - u_0, \ldots, u_d - u_0$ are linearly independent. A *d-simplex T* in $\mathbb{R}^n$ is the convex hull of a set S of $d + 1$ affine independent points in $\mathbb{R}^n$. Clearly, T is a d-dimensional polytope with $d + 1$ vertices. The convex hull of any non-empty subset $S' \subseteq S$ is called a *face* of T. A *simplicial complex K* is a set of simplexes such that (i) every face of a simplex from K is also in K, and (ii) the non-empty intersection of any two simplexes in K is a face of both of them. The *dimension* of a simplicial complex K in $\mathbb{R}^n$ is the maximal dimension of a simplex in it.

A simplicial complex K is called *locally finite* if each vertex of K belongs only to finitely many simplexes in K. A *simplicial triangulation* is a homeomorphism to a locally finite simplicial complex. Triangulations can be used to define random d-dimensional manifolds, and have many other applications. For example, a (finite) triangulation is a convenient way to represent a manifold M for a computer program, which can then compute various parameters of M, like genus. Without triangulation, it is unclear how to "enter" a manifold as a (finite) input a computer can understand. Hence, the question of existence of triangulations is of central importance in topology.

In 1925, Radó [1016] proved that every closed two-dimensional surface has a simplicial triangulation. In 1926, Kneser [726] asked whether the same is true for every closed k-dimensional topological manifold. The conjectured affirmative answer to this question is known as the *triangulation conjecture* in dimension d. In 1952, Moise [901] proved the conjecture for $d = 3$. However, Casson proved in the 1980s that the conjecture is false in dimension $d = 4$, see [21]. Before 2016, the conjecture remained open in higher dimensions. In 2016, Manolescu [838] proved that it is false for every dimension $d \geq 5$.

Theorem 5.57 *For every $d \geq 5$, there exists a closed d-dimensional topological manifold that does not admit a simplicial triangulation.*

If a manifold M has a simplicial triangulation K with finitely many simplexes, how many simplexes of every dimension can it contain? This basic question turns out to be surprisingly difficult even if M is a sphere. For example, the 2-dimensional sphere has a triangulation (tetrahedron) with 4 vertices, 6 edges, and 4 two-dimensional faces, and one may ask for which triples of positive integers (v, e, r) it has a triangulation with v vertices, e edges, and r faces? More generally, the *f-vector* of a d-dimensional simplicial complex K is the vector $f(K) = (f_0(K), f_1(K), \ldots, f_d(K))$, where $f_m(K)$ is the number of m-dimensional simplexes in K. Then, what is the set $\mathcal{F}$ of possible f-vectors of simplicial complexes homeomorphic to a sphere? The famous "g-conjecture for spheres" predicted that $\mathcal{F} = \mathcal{F}_P$, where $\mathcal{F}_P$ is the set of possible f-vectors of simplicial polytopes (convex polytopes whose facets are all simplexes). Because the set $\mathcal{F}_P$ was fully characterized by the works of Billera and Lee [148] and Stanley [1153] in 1980, the g-conjecture gives the full description of the set $\mathcal{F}$. Because every simplicial polytope is a triangulation of a sphere, the inclusion $\mathcal{F}_P \subseteq \mathcal{F}$ is obvious. However, the inclusion $\mathcal{F} \subseteq \mathcal{F}_P$, stating that for every simplicial triangulation of a sphere there exists a simplicial polytope with the same f-vector, turned out to be a very deep and difficult question, and it remained open until 2018.

Another deep conjecture about the possible number of simplexes in a d-dimensional simplicial complex K in $\mathbb{R}^n$ was the *Grünbaum–Kalai–Sarkaria conjecture*. If $d = 1$ and $n = 2d = 2$, then K is just a planar graph. If a connected simple planar graph has v vertices, e edges, and divides the plane into r regions (or, equivalently, if v, e, and r are the number of vertices, edges, and two-dimensional faces of a convex polyhedron in $\mathbb{R}^3$) then

$$v - e + r = 2.$$

This formula is known as *Euler's formula*, it was stated by Descartes [348] in 1621 and Euler [434] in 1758 and rigorously proved by Legendre [779] in 1794. Euler's formula easily implies that $e \leq 3v$ (and in fact $e \leq 3v - 6$ if $v \geq 3$). In 1970, Grünbaum [540] asked if a similar inequality can be proved for any d-dimensional simplicial complex K in $\mathbb{R}^{2d}$. In the 1990s, Kalai [673] and Sarkaria [1080] made a precise conjecture that if K contains $f_m(K)$ m-dimensional simplexes then

$$f_d(K) \leq (d + 2) f_{d-1}(K).$$

The multiplier $d + 2$ in this inequality is known to be the best possible one.

In particular, the case $d = 2$ of the conjecture states that if K is a 2-dimensional simplicial complex (set of triangles glued along their edges) in $\mathbb{R}^4$, then the number of triangles in K is at most 4 times the number of edges. Even this $d = 2$ case remained open up to 2018.

In 2018, Adiprasito [8] posted online a preprint in which he proved the g-conjecture for spheres, and also proved the Grünbaum–Kalai–Sarkaria conjecture for all d.

Theorem 5.58

(a) *The set of possible f-vectors of a simplicial complex homeomorphic to a sphere is the same as the set of possible f-vectors of simplicial polytopes.*

(b) *For any d-dimensional simplicial complex K in $\mathbb{R}^{2d}$, one has $f_d(K) \leq (d + 2) f_{d-1}(K)$.*

Both parts (a) and (b) of Theorem 5.58 are corollaries of a deep theorem proved in [8] called the "hard Lefschetz theorem for toric varieties" whose exact formulation is too technical to be stated here. We only mention that it is part of Hodge theory, which has also been used to prove, for example, Theorems 2.39 and 2.40 discussed earlier in this book.

Theorem 5.58 discusses possible numbers of simplexes in the triangulation of a sphere. A related question is to understand what is the minimal number of vertices a triangulation of a given manifold M can have. This question is particularly important and well-studied in the case when M is the real projective space[21] $\mathbb{RP}^d$. It is easy to construct a triangulation of $\mathbb{RP}^d$ with the number of vertices exponential in d, but can we do better? In 2020, Adiprasito, Avvakumov and Karasev posted online a preprint [9] in which they answered this long-standing open question by constructing a triangulation of $\mathbb{RP}^d$ with sub-exponentially many vertices.

Theorem 5.59 *For all positive integers d, there exists a triangulation of $\mathbb{RP}^{d-1}$ with at most*

$$\exp\left(\left(\frac{1}{2} + o(1)\right)\sqrt{d}\log d\right)$$

vertices.

The proof of Theorem 5.59 is short and elegant. At its heart is the construction of a convex centrally symmetric d-dimensional polytope P such that (i) all the vertices of P lie on the unit sphere; (ii) for any vertex $A \in P$, if $F, F' \subset P$ are faces with $A \in F$ and $-A \in F'$, then $F \cap F' = \emptyset$, and (iii) the number of vertices of P is less than $\exp\left(\left(\frac{1}{2} + o(1)\right)\sqrt{d}\log d\right)$. This polytope is then used to construct the triangulation as in Theorem 5.59.

---

[21] See the discussion before Theorem 4.32 for the definition of projective space.

Chapter 6
Probability and Statistics

6.1 Random Graphs

Properties of Erdős–Rényi Graphs

The *Erdős–Rényi graph* $G(n, p)$ with edge density $p = p(n)$ is a random graph with n vertices such that every possible edge is included independently with probability p. This is the classical model of random graphs introduced by Erdős and Rényi [415] in 1960. We have already discussed the chromatic number of such graphs, see Theorem 2.36 and the discussion before it.

In 1966, Erdős and Rényi [416] proved that if n is even and tends to infinity, then $G(n, p)$ has a perfect matching with probability tending to 1 if $\lim_{n\to\infty} \frac{p(n)}{(\log n)/n} = +\infty$, but the same probability tends to 0 if $\lim_{n\to\infty} \frac{p(n)}{(\log n)/n} = 0$. More generally, if the vertices of an n-vertex graph G can be partitioned into n/k disjoint copies of a k-vertex graph H, we say that G has an *H-factor*. A natural and well-studied question is for what minimal edge density p the random graph $G(n, p)$ has an H-factor with high probability, provided that n is divisible by k and tends to infinity. The Erdős–Rényi theorem solves this question for $H = K_2$, where K_n denotes the complete n-vertex graph. In 1991, Łuczak and Rucinńki [819] solved it if H is an arbitrary tree. However, the problem remained open even if $H = K_3$ is a triangle.

In 1993, Alon and Yuster [38] formulated a conjecture which, if true, would solve this problem for an arbitrary graph H, up to an ϵ term in the exponent. In 2008, Johansson et al. [660] proved this conjecture. For a graph H on at least two vertices, let

$$d^*(H) = \max_{H'\subseteq H} \frac{e(H')}{v(H') - 1},$$

where $v(H')$ and $e(H')$ is the number of vertices and edges in graph H', respectively.

B. Grechuk, *Landscape of 21st Century Mathematics*,
https://doi.org/10.1007/978-3-030-80627-9_6

Theorem 6.1 *For any graph H on at least two vertices, and any $\epsilon > 0$, the probability that the Erdős–Rényi random graph with $n = k \cdot v(H)$ vertices and edge density $p = n^{-\frac{1}{d^*(H)} + \epsilon}$ can be partitioned on k subgraphs isomorphic to H tends to 1 as $k \to \infty$.*

It is known that the same probability tends to 0 if $\lim_{n \to \infty} \left(p(n)/n^{-\frac{1}{d^*(H)}} \right) = 0$, hence Theorem 6.1 is the best possible up to an ϵ term.

Another interesting research direction about Erdős–Rényi graphs is the study of their Turán properties. We say that a graph is F-*free* if it does not contain a copy of a graph F as a subgraph. For a graph G, we write $ex(G, F)$ for the maximal number of edges that an F-free subgraph of G may have. The number

$$\pi(F) = \lim_{n \to \infty} \frac{ex(K_n, F)}{e(K_n)}$$

is known as the *Turán density* of the graph F.

The function $ex(G, F)$ was first studied by Mantel [839] in 1907, and then popularized by Erdős [395] and Turán [1217]. In particular, in 1941 Turán [1217] determined $ex(K_n, K_k)$ for all integers n and k. For a general graph G, it is known that

$$ex(G, F) \geq \pi(F)e(G).$$

In 1997, Kohayakawa et al. [729] conjectured that if G is a random graph with sufficiently high edge density $p(n)$, then this inequality is essentially tight. Specifically, the opposite inequality

$$ex(G, F) \leq (\pi(F) + \epsilon)e(G)$$

holds with high probability, provided that $p(n)$ grows at least as $Cn^{-1/m(F)}$, where $m(F)$ denotes the maximum value of $\frac{e(H)-1}{v(H)-2}$ over all subgraphs H of F with $v(H) \geq 3$. In 2016, Schacht [1084] proved this conjecture.

Theorem 6.2 *For every graph F with $e(F) \geq 1$ and $v(F) \geq 3$ and every $\epsilon > 0$, there exists a $C = C(\epsilon) > 0$ such that for every sequence of probabilities $q_n \geq Cn^{-1/m(F)}$, we have*

$$\lim_{n \to \infty} \mathbb{P}\left[ex(G(n, q_n), F) \leq (\pi(F) + \epsilon) e(G(n, q_n)) \right] = 1.$$

In fact, Theorem 6.2 is just one of many corollaries of a general methodology developed in [1084] which allow the transfer of many classical extremal results from the deterministic to the probabilistic setting. For example, this methodology gives an alternative proof of Theorem 2.8.

Combinatorial Problems on Random Graphs

Let K_n be a complete n-vertex graph such that each edge e has some non-negative weight w_e. The famous *travelling salesman problem* (TSP) asks to compute a cycle that visits each vertex precisely once (called a *Hamiltonian cycle*) with the minimum possible sum of the edge weights. The model in which the edge costs w_e are random, chosen independently from a fixed distribution μ on the non-negative real numbers, is known as the *mean field model of distance*. Let L_n be the (random) minimum sum of the edge weights of a Hamiltonian cycle in this model. A central problem in this area is to determine the limiting behaviour of L_n as n grows. This problem has been extensively studied when μ is either the uniform distribution on the interval $[0, 1]$ or the exponential distribution with mean 1.

In 1986, Mézard and Parisi [887, 888] used a non-rigorous analysis with a physical origin (called *replica analysis*) which allowed them to derive the conjectured analytic formula for the limit L^* of L_n for any distribution μ such that

$$\lim_{t \to 0^+} \frac{\mathbb{P}(w_e < t)}{t} = 1. \tag{6.1}$$

Specifically, for any $x > 0$, let $y = y(x)$ be the unique real number such that

$$\left(1 + \frac{x}{2}\right) e^{-x} + \left(1 + \frac{y}{2}\right) e^{-y} = 1.$$

Let $L^* = \frac{1}{2} \int_0^\infty y(x) \mathrm{d}x = 2.0415....$ Mézard and Parisi [887, 888] predicted that L_n converges in probability to L^*. In 2010, Wästlund [1247] rigorously confirmed this prediction.

Theorem 6.3 *For any μ satisfying* (6.1), *and for any $\epsilon > 0$,*

$$\lim_{n \to \infty} \mathbb{P}\left[\left|L_n - L^*\right| > \epsilon\right] = 0.$$

Another famous combinatorial problem which has been intensively studied for graphs with random weights is the *minimum matching problem*. A *perfect matching* in a graph is a set of edges of which each vertex is incident to exactly one. In a graph G with weight $w_e \geq 0$ assigned to every edge e, the minimum matching problem asks for the perfect matching of minimum total weight $M(G)$. Let $K_n(\mu)$ be a complete n-vertex graph with n even, in which weights $w_e \geq 0$ are selected independently at random from a distribution μ. In 1985, Mézard and Parisi [886] used replica analysis to predict that if μ satisfies (6.1), then $M(K_n(\mu))$ converges in probability to $\pi^2/12$. In 2001, Aldous [26] rigorously confirmed this prediction.

If the distribution μ has density $\rho(t)$, $t \geq 0$, then (6.1) is equivalent to $\lim_{t \to 0^+} \rho(t) = 1$. Such distributions are said to have pseudo-dimension 1. More generally, for any real $d > 0$, we say that the distribution μ with density $\rho(t)$, $t \geq 0$ has *pseudo-dimension d* if[1]

$$\lim_{t \to 0^+} \frac{\rho(t)}{d \cdot t^{d-1}} = 1.$$

In 2012, Wastlund [1248] established the asymptotic behaviour of $M(K_n(\mu))$ for distributions of pseudo-dimension $d \geq 1$.

Theorem 6.4 *For any $d \geq 1$ there exists a number $\beta(d)$ such that for any μ of pseudo-dimension d and any $\epsilon > 0$,*

$$\lim_{n \to \infty} \mathbb{P}\left[\left| \frac{M(K_n(\mu))}{n^{1-1/d}} - \beta(d) \right| > \epsilon \right] = 0.$$

The proof of Theorem 6.4 also implies a method for computing $\beta(d)$ numerically for any given $d \geq 1$.

The main idea of the proof is to introduce a new two-person game called *Exploration*, which is played on graphs with random edge lengths. The author proved that on some graphs this game has a game-theoretical value, which is well-defined with probability 1, and this fact can be used to rigorously justify the replica prediction in the minimum matching problem. Variants of this game can be used to study other problems on random graphs, such as the TSP.

Replica analysis was initiated by Mézard and Parisi [886] and has been used to non-rigorously predict the limiting behaviour of a large variety of other parameters of random graphs, including, for example, the independence ratio in random regular graphs. The *independence ratio* $\mathrm{IR}(G)$ of an n-vertex graph G is the size of its maximum independent set divided by n. For positive integers n, d with nd even, let $G_{n,d}$ denote the uniformly random d-regular graph on n vertices, sampled as follows: start with n isolated vertices each equipped with d labelled half-edges, and form the graph by taking a uniformly random matching on the nd half-edges. For fixed d, let $\mathrm{IR}_n = \mathrm{IR}(G_{n,d})$.

In 2005, Hartmann and Weigt [569] and Rivoire [1048] used replica analysis methods to predict that IR_n converges in probability to the limit $\alpha^*(d)$ given by the following formula. For any integer $d > 0$, consider the functions

$$\lambda_d(q) = q \frac{1 - (1-q)^{d-1}}{(1-q)^d}, \quad \alpha_d(q) = q \frac{1 - q + dq/(2\lambda_d(q))}{1 - q^2(1 - 1/\lambda_d(q))},$$

---

[1] Such distributions are used to approximate the euclidean models, where the vertices of a graph are distributed randomly in the unit cube in $\mathbb{R}^d$ and the edge lengths are given by euclidean distance.

and $f_d(q)$, defined by

$$-\log\left[1 - q\left(1 - \frac{1}{\lambda_d(q)}\right)\right] - \left(\frac{d}{2} - 1\right)\log\left[1 - q^2\left(1 - \frac{1}{\lambda_d(q)}\right)\right] - \alpha_d(d)\log\lambda_d(q).$$

Let $q^*(d)$ be the largest $q \leq 2(\log d)/d$ such that $f_d(q) = 0$, and let $\alpha^*(d) = \alpha_d(q^*(d))$.

In 2016, Ding et al. [352] rigorously confirmed this prediction for all sufficiently large d.

Theorem 6.5 *There exists a constant d_0 such that for all $d \geq d_0$ the independence ratio IR_n of the random d-regular graph $G_{n,d}$ converges in probability to $\alpha^*(d)$.*

Uniform Spanning Forests

Recall that a *forest* is a graph with no cycles and a *tree* is a connected forest. A *spanning forest* on a graph G is a (non-induced) subgraph H of G which (i) is a forest, and (ii) contains at least 1 edge incident to each vertex of G. If, moreover, H is a tree, then it is called a *spanning tree*. If G is finite, it is easy to select a spanning forest (or spanning tree) uniformly at random. However, for some applications we need a model for a *uniform spanning forest* (USF) on an infinite graph, such as the graph $\mathbb{Z}^d$ with vertices $x = (x_1, \ldots, x_d)$ with integer coordinates, where x and y are connected by an edge if $\sum_{i=1}^{d} |x_i - y_i| = 1$.

In 1991, Pemantle [971] introduced the following model for a USF on $\mathbb{Z}^d$. Let $\mathbb{Z}_n^d$ be the (finite, induced) subgraph of $\mathbb{Z}^d$ consisting of vertices $x = (x_1, \ldots, x_d)$ with $|x_i| \leq n$ for all i. Let H_n^d be a spanning tree on $\mathbb{Z}_n^d$ selected uniformly at random. For any set C of edges in $\mathbb{Z}^d$, let $P_n(C)$ be the probability that all these edges are included in H_n^d, and let $P(C) = \lim_{n \to \infty} P_n(C)$. A (random) subgraph H^d of $\mathbb{Z}^d$ such that any set C of edges of $\mathbb{Z}^d$ is included in H^d with probability $P(C)$ is called a USF on $\mathbb{Z}^d$.

For $x, y \in \mathbb{Z}^d$, let $N(x, y)$ be the minimum number of edges outside a USF in a path joining x and y in $\mathbb{Z}^d$. Pemantle proved that, for $d \leq 4$ (but not for $d \geq 5$), a USF is connected (and thus is a spanning tree) almost surely. In other words, $N(x, y) = 0$ a.s. for all $x, y \in \mathbb{Z}^d$ if and only if $d \leq 4$. In 2004, Benjamini et al. [117] proved a more general theorem which computes $\max_{x,y \in \mathbb{Z}^d} N(x, y)$ in all dimensions. In particular, it shows that $N(x, y) \leq 1$ a.s. for all $x, y \in \mathbb{Z}^d$ if and only if $d \leq 8$.

Theorem 6.6 *Almost surely,*

$$\max_{x,y \in \mathbb{Z}^d} N(x, y) = \left\lfloor \frac{d-1}{4} \right\rfloor,$$

where $\lfloor x \rfloor$ denotes the largest integer not exceeding the real number x.

6.2 Random Matrices

The Probability of Singularity

Recall that an $n \times n$ matrix A is called *invertible* (or *non-singular*), if there exists an $n \times n$ matrix B (usually denoted A^{-1} and called the *inverse matrix* of A) such that $AB = BA = I_n$, where I_n is the $n \times n$ identity matrix. Otherwise A is called *singular* (or *degenerate*). A square matrix A is singular if and only if its determinant $\det(A)$ is equal to zero.

One of the important questions in the theory of random matrices is to estimate the probability that a random $n \times n$ matrix A is singular. One of the easiest models of random matrices is the *Bernoulli matrix* M_n, defined as the $n \times n$ matrix whose entries are independent identically distributed (i.i.d.) random variables, equal to $+1$ or -1 with probabilities $1/2$. Let

$$P_n = \mathbb{P}(\det(M_n) = 0)$$

be the probability that M_n is singular.

Because the equality of two columns or two rows of M_n (up to sign) is an obvious sufficient condition for singularity, it is clear that

$$P_n \geq (1 + o(1))n^2 2^{1-n} = \left(\frac{1}{2} + o(1)\right)^n.$$

It has been long conjectured that in fact

$$P_n = \left(\frac{1}{2} + o(1)\right)^n,$$

that is, the equality of rows or columns is the *main* reason for the singularity. This conjecture is stated explicitly, for example, by Odlyzko [943] in 1988, but in fact has been known for considerably longer. In 1967, Komlós [732] proved that $\lim_{n \to \infty} P_n = 0$. In 1995, Kahn et al. [672] gave the first exponential upper bound $P_n = O(0.999^n)$. In 2006, Tao and Vu [1184] improved this to $P_n = O(0.958^n)$. In 2007, the same authors [1185] achieved the next significant improvement.

Theorem 6.7 $P_n \leq \left(\frac{3}{4} + o(1)\right)^n$ *as* $n \to \infty$.

In 2010, Bourgain et al. [193] improved the upper bound on P_n further to $P_n \leq \left(\frac{1}{\sqrt{2}} + o(1)\right)^n$. Finally, in 2020 Tikhomirov [1204] proved the upper bound $P_n \leq \left(\frac{1}{2} + o(1)\right)^n$, which matches the lower bound up to the $o(1)$ term and proves the conjecture.

Theorem 6.8 $P_n = \left(\frac{1}{2} + o(1)\right)^n$ *as* $n \to \infty$.

In fact, Tikhomirov [1204] considered a more general model, where, for $p \in (0, 1)$, $B_n(p)$ is the random $n \times n$ matrix with independent entries b_{ij} such that $\mathbb{P}(b_{ij} = 1) = p$ and $\mathbb{P}(b_{ij} = 0) = 1 - p$. He proved that, for any fixed $p \in (0, 1/2]$, $B_n(p)$ is singular with probability $(1 - p + o(1))^n$. In 2020, Jain et al. [649] posted online a preprint in which they derived estimates of similar and better accuracy for the singularity probability of square matrices whose entries are independent copies of an arbitrary non-constant real-valued random variable ξ with finite support, see [649] for details.

Norm of the Inverse
For many applications, one is interested not only in whether a random matrix A is invertible, but also "how well" it is invertible. The quantitative characterization of invertibility of A is the norm of its inverse matrix A^{-1}. The *norm* of an $n \times n$ matrix A is

$$\|A\| = \sup_{x \in \mathbb{R}^n} \frac{|Ax|}{|x|}.$$

where $|x|$ is the Euclidean norm of $x \in \mathbb{R}^n$. It is common to say that A is "well invertible" if $\|A^{-1}\|$ is not too large. We will assume the convention that $\|A^{-1}\| = \infty$ if A is singular.

The norms $\|A\|$ and $\|A^{-1}\|$ are connected with singular values of the matrix A. For any $n \times n$ matrix A, possibly with complex entries, let A^* denote the conjugate transpose of A. Then the eigenvalues of A^*A are real and non-negative. The square roots of these eigenvalues are called *singular values* of A. Let $s_{\min}(A)$ and $s_{\max}(A)$ be the smallest and the largest singular values, respectively. Then $s_{\max}(A) = \|A\|$ and $s_{\min}(A) = 1/\|A^{-1}\|$. In particular, a matrix A is singular if and only if $s_{\min}(A) = 0$ and is considered to be "well invertible" if $s_{\min}(A)$ is not too small.

Theorem 6.8 estimates the probability that $\|M_n^{-1}\| < \infty$ for the $n \times n$ matrix M_n with i.i.d. Bernoulli entries. In 1963, von Neumann [934] predicted that $\|A^{-1}\|$ should be proportional to $\sqrt{n}$ with high probability for a large class of random $n \times n$ matrices A. Before 2008, this was open even for $A = M_n$. In this special case, Spielman and Teng [1147] made in 2002 a more precise conjecture that

$$\mathbb{P}\left(s_{\min}(M_n) \leq \frac{\epsilon}{\sqrt{n}}\right) = \mathbb{P}\left(\|M_n^{-1}\| \geq \frac{\sqrt{n}}{\epsilon}\right) \leq C\epsilon + c^n \tag{6.2}$$

for any $\epsilon \geq 0$ and some universal constants $C > 0$ and $c \in (0, 1)$.

In[2] 2009, Tao and Vu [1189] proved that the norm $\|M_n^{-1}\|$ is polynomially bounded with probability at least $1 - n^{-C}$ for any $C > 0$.

---

[2] The first draft of this paper was posted online in 2005.

Theorem 6.9 *For any constant $C > 0$, there are constants B and N such that for any $n \geq N$ the inequality $\|M_n^{-1}\| \leq n^B$ holds with probability at least $1 - n^{-C}$.*

Theorem 6.9 is a corollary of the inverse Littlewood–Offord theory, developed by the authors. Let $S = \pm v_1 \pm \cdots \pm v_n$, where the v_i are (non-random) integers and the signs are selected independently from the same distribution. The *Littlewood–Offord problem* asks to maximize the probability $\mathbb{P}[S = 0]$ subject to some constraints on the v_i. Tao and Vu [1189] developed a methodology for solving an inverse problem: given that $\mathbb{P}[S = 0]$ is large, what can be said about $\{v_i\}$? They proved that in this case $\{v_i\}$ is (almost) contained in a generalized arithmetic progression, and demonstrated some applications of this fact, including Theorem 6.9.

In 2008, Rudelson [1067] proved a version of Theorem 6.9 with $C \leq 1/2$, but with the explicit dependence of B from C and for a more general class of random matrices. We next formulate this result. A random variable X is called *subgaussian* if there exists a constant b such that

$$\mathbb{P}(|X| > t) \leq 2e^{-t^2/b^2} \quad \text{for all} \quad t > 0.$$

The minimal $b > 0$ such that this holds is called the *subgaussian moment* of X. For an integer $n > 0$, let $A_{n,X}$ be an $n \times n$ matrix whose entries a_{ij} are i.i.d. with the same distribution as X.

Theorem 6.10 *Let X be a subgaussian random variable with mean 0 and variance 1. Then there are positive constants C_1, C_2, and C_3 such that for any integer $n > 0$, and any $\epsilon > C_1/\sqrt{n}$, the inequality $\|A_{n,X}^{-1}\| \leq C_2 \cdot n^{3/2}/\epsilon$ holds with probability greater than $1 - \epsilon/2 - 4e^{-C_3 n}$.*

Theorem 6.10 implies Theorem 6.9 for $C \leq 1/2$, but in general Theorems 6.9 and 6.10 do not imply each other. Later in 2008, Rudelson and Vershynin [1069] proved a much more general result which implies both Theorems 6.9 and 6.10, and much more.

Theorem 6.11 *For any $b > 0$ there exist constants $C = C(b) > 0$ and $c = c(b) \in (0, 1)$ such that for any subgaussian random variable X with mean 0, variance 1, and subgaussian moment at most b, any integer $n > 0$, and any $\epsilon \geq 0$, the probability that $\|A_{n,X}^{-1}\| \geq \sqrt{n}/\epsilon$ is at most $C\epsilon + c^n$.*

The estimate in Theorem 6.11 is optimal up to the values of constants C and c. Theorem 6.11 implies Theorem 6.10. It also implies that Theorem 6.9 is true, and, moreover, we can take $B = C + \frac{1}{2}$ in it. In the special case $A_{n,X} = M_n$, Theorem 6.11 implies the Spielman–Teng conjecture (6.2). In addition, Rudelson and Vershynin [1069] confirmed the von Neumann prediction mentioned above under the assumption that the fourth moment of X is bounded.

Similar to Theorem 6.9, Theorem 6.11 was derived from the analysis of the Littlewood–Offord problem. Rudelson and Vershynin [1069] studied a more general version of this problem, in which, for i.i.d. random variables X_i and real numbers v_i, the task is to estimate the probability p that $\sum_{i=1}^n v_i X_i$ is close to some given number v. The authors showed that this happens when the v_i are essentially contained in an arithmetic progression of length $1/p$, and deduced Theorem 6.11 from there.

Theorems 6.9, 6.10 and 6.11 study the distribution of the norm of the inverse (or, equivalently, of the smallest singular value), for the matrices with independent entries. The same problem is important in other models of random matrices, for example, for matrices with random unitary perturbation. A matrix U is called *unitary* if $U^*U = UU^* = I$, where I is the identity matrix. The *unitary group* $U(n)$ is the group of all $n \times n$ unitary matrices with matrix multiplication as the group operation. There is a unique probability measure μ on $U(n)$ with the following property: if U is a random unitary matrix with distribution μ, then so are AU and UA, for any fixed matrix $A \in U(n)$. This probability measure μ is called the *Haar measure* on $U(n)$. A random matrix U uniformly distributed with respect to this measure can be constructed explicitly as follows. Let $\mathbb{C}^n$ be the space of vectors $z = (z_1, \ldots, z_n)$ with n complex coordinates and inner product $\langle z, w \rangle = \sum_{i=1}^n z_i \bar{w}_i$, where $\bar{w}_i$ is the complex conjugate of w_i. For a finite subset $W \subset \mathbb{C}^n$, let $W^\perp = \{z \in \mathbb{C}^n \mid \langle z, w \rangle = 0, \forall w \in W\}$. Let $S = \{z \in \mathbb{C}^n, \langle z, z \rangle = 1\}$ be the unit sphere in $\mathbb{C}^n$. Let u^1 be the vector selected uniformly at random from S, and, then, inductively, for any $i = 2, \ldots, n$, let u^i be the vector selected uniformly at random from $S \cap \{u^1, \ldots, u^{i-1}\}^\perp$. Then the matrix U with columns $u^1, \ldots, u^n$ is uniformly distributed with respect to the Haar measure.

In 2014, Rudelson and Vershynin [1070] proved that a random unitary perturbation of a fixed matrix D is well invertible with high probability.

Theorem 6.12 *Let D be an arbitrary fixed $n \times n$ matrix, $n \geq 2$. Let U be a random matrix uniformly distributed over the unitary group $U(n)$ with respect to the Haar measure. Then*

$$\mathbb{P}\left(s_{\min}(D + U) \leq t\right) \leq t^c n^C$$

for all $t > 0$ and some absolute positive constants C, c.

The Distribution of the Largest Eigenvalue

In the previous section we mostly discussed estimates for the norm $\|A^{-1}\|$ of the inverse of a random square matrix A. In many applications, it is also important to estimate the norm $\|A\|$ of A itself. Equivalently, we can estimate $\|A\|^2$, which is equal to the largest eigenvalue of $A'A$, where A' denotes the transpose of a matrix A with real entries.

A special but important case is when the entries of A are independent and follow the standard normal distribution $N(0, 1)$. In a series of papers in 1993–1996, Tracy and Widom [1210–1212] proved that the maximal eigenvalue of $A'A$, appropriately scaled, converges to an explicit distribution is now known as the *Tracy–Widom law*. In 2001, Johnstone [664] generalized this result to the case when A is a rectangular matrix.

To formulate the result, we need some notation. Let

$$G(x) = \frac{1}{\pi} \int_0^\infty \cos\left(\frac{t^3}{3} + xt\right) dt,$$

and let $q(x)$ be the (unique) function such that

$$q''(x) = xq(x) + 2q^3(x) \quad \text{and} \quad \lim_{x \to +\infty} \frac{q(x)}{G(x)} = 1.$$

The probability distribution with cumulative distribution function

$$F(s) = \exp\left(-\frac{1}{2}\int_s^\infty (q(x) + (x-s)q^2(x))dx\right), \quad s \in \mathbb{R}$$

is called the *Tracy–Widom law of order* 1.

Let A be an $n \times p$ matrix whose entries are independent and follow $N(0, 1)$. Let l_{np} be the largest eigenvalue of matrix $A'A$. Let

$$\mu_{np} = \left(\sqrt{n-1} + \sqrt{p}\right)^2 \quad \text{and} \quad \sigma_{np} = \left(\sqrt{n-1} + \sqrt{p}\right)\left(\frac{1}{\sqrt{n-1}} + \frac{1}{\sqrt{p}}\right)^{1/3}.$$

Theorem 6.13 *If* $n = ak$, $p = bk$, *where* $a \geq b$ *are constants and* $k \to \infty$, *then* $\frac{l_{np} - \mu_{np}}{\sigma_{np}}$ *converges in distribution to the Tracy–Widom law of order* 1.

The Distribution of All Eigenvalues: The Hermitian Case

In the previous sections we discussed the distributions of the minimal and maximal singular values of a random matrix A, which are the (square roots of) the minimal and maximal eigenvalues of the matrix $X = A^*A$, where A^* is the conjugate transpose of A. The matrix X has the property that $X^* = X$, and such matrices are called *Hermitian*. More generally, one may ask for the distribution of *all* eigenvalues of a random Hermitian matrix X. A standard model for random Hermitian matrices is given by Wigner Hermitian matrices. A *Wigner Hermitian* $n \times n$ *matrix* M_n is a Hermitian matrix whose upper diagonal entries are i.i.d. complex random variables with mean 0 and variance 1. A central result of this theory is the *Wigner semi-circular law*. Let ρ be the function with support on $[-2, 2]$ given by

$$\rho(x) = \frac{\sqrt{4 - x^2}}{2\pi}, \quad -2 \leq x \leq 2.$$

Note that the graph of ρ is a semi-circle. The semi-circular law states that if M_n is a Wigner Hermitian matrix, then, for any real number x,

$$\lim_{n \to \infty} \frac{1}{n} |\{1 \leq i \leq n, \lambda_i(M_n/\sqrt{n}) \leq x\}| = \int_{-2}^x \rho(y)dy,$$

where the convergence is in probability, $|\cdot|$ denotes the number of elements in a finite set, and $\lambda_i(X)$ denotes the i-th eigenvalue of a matrix X. This fundamental theorem was proved by Wigner [1261] in 1958 for matrices whose entries have Gaussian distribution, and by Pastur [968] in 1972 in general.

The Wigner semi-circular law describes the global distribution of the eigenvalues, but cannot be used to obtain the local eigenvalue statistics, such as, for example, the distribution of gaps between consecutive eigenvalues. In 2011, Tao and Vu [1187] developed a methodology to establish the existence of limiting distribution for such gaps. To formulate their result rigorously, we need some notation. Let M_n be the $n \times n$ Hermitian matrix with upper triangular complex entries $\psi_{ij} = \xi_{ij} + \tau_{ij}\sqrt{-1}$, $1 \le i < j \le n$, and diagonal real entries ξ_{ii}, $1 \le i \le n$, where

(i) for $1 \le i < j \le n$, ξ_{ij} and τ_{ij} are i.i.d. copies of a real random variable ξ with mean zero and variance $1/2$;
(ii) for $1 \le i \le n$, ξ_{ii} are i.i.d. copies of a real random variable ξ' with mean zero and variance 1; and
(iii) ξ and ξ' have exponential decay, i.e., there exist constants C and C_0 such that $\mathbb{P}(|\xi| \ge t^C) \le e^{-t}$ and $\mathbb{P}(|\xi'| \ge t^C) \le e^{-t}$ for all $t > C_0$.

The distributions ξ and ξ' are called the *atom distributions* of M_n.

Let $\lambda(M_n\sqrt{n}) = (\lambda_1, \ldots, \lambda_n)$ be the vector of eigenvalues λ_i of the matrix $M_n\sqrt{n}$, arranged in non-increasing order: $\lambda_1 \le \lambda_2 \le \cdots \le \lambda_n$. Let

$$S_n(s; \lambda(M_n\sqrt{n})) = \frac{1}{n}|\{1 \le i \le n - 1 : \lambda_{i+1} - \lambda_i \le s\}|.$$

Theorem 6.14 *Let M_n be a random Hermitian matrix satisfying the conditions (i)–(iii) above, whose atom distributions ξ, ξ' have support on at least three points, and $s > 0$. Then the limit*

$$G(s) = \lim_{n \to \infty} \mathbb{E}[S_n(s; \lambda(M_n\sqrt{n}))]$$

exists and depends only on s but not on ξ, ξ'.

Theorem 6.14 states that expected value $\mathbb{E}[S_n(s; \lambda(M_n\sqrt{n}))]$ converges to the same limit $G(s)$, no matter how ξ and ξ' are distributed. Hence, we can compute $G(s)$ for ξ and ξ', say, normally distributed, and the resulting $G(s)$ works for all distributions. In fact, Theorem 6.14 is just one of many corollaries of the much more general "Four moment theorem" proved in [1187], which, informally, states that many properties of the eigenvalues of a random Hermitian matrix are determined by the first four moments of the atom distributions.

The Distribution of All Eigenvalues: The Non-Hermitian Case
The semi-circular law and Theorem 6.14 study the distribution eigenvalues of random Hermitian matrices. Here, we discuss the non-Hermitian case, in which the eigenvalues are not necessary real. Based on numerical experiments made in

the 1950s, it was conjectured that if A is an $n \times n$ matrix whose entries are i.i.d. complex random variables with mean 0 and variance 1, then the eigenvalues of A, appropriately scaled, become uniformly distributed in the unit disk of the complex plane when $n \to \infty$. This statement became known as the *circular law conjecture*. In 1967, Mehta [872] proved this conjecture for random matrices with Gaussian distribution of entries. Then many authors proved it for more and more general distributions. Finally, Tao and Vu [1186] proved the general case of the conjecture in 2010.

To formulate this result rigorously, we need some notation. Given an $n \times n$ matrix A, let

$$\mu_A(x, y) := \frac{1}{n}|\{1 \leq i \leq n, \ Re(\lambda_i) \leq x, \ Im(\lambda_i) \leq y\}|$$

be the empirical spectral distribution (ESD) of its eigenvalues $\lambda_i \in \mathbb{C}, i = 1, \ldots, n$. We interpret μ_A as a discrete probability measure on $\mathbb{C}$. Let A_n be a sequence of random matrices. For a probability measure μ_∞ on $\mathbb{C}$, and continuous compactly supported function $f : \mathbb{C} \to \mathbb{R}$, let

$$\Delta(\mu_\infty, f, n) = \int_{\mathbb{C}} f(z) d\mu_{1/\sqrt{n}A_n}(z) - \int_{\mathbb{C}} f(z) d\mu_\infty.$$

We say that the ESD of $\frac{1}{\sqrt{n}}A_n$ *converges in probability* to μ_∞ if

$$\lim_{n \to \infty} \mathbb{P}\left(|\Delta(\mu_\infty, f, n)| \geq \epsilon\right) = 0$$

for every f and for every $\epsilon > 0$. Also, we say that the ESD of $\frac{1}{\sqrt{n}}A_n$ converges to μ_∞ *almost surely* if, with probability 1, $\Delta(\mu_\infty, f, n)$ converges to zero for all f.

Theorem 6.15 *Let A_n be the $n \times n$ random matrix whose entries are i.i.d. complex random variables with mean zero and variance one. Then, the ESD of $\frac{1}{\sqrt{n}}A_n$ converges (both in probability and in the almost sure sense) to the uniform distribution on the unit disk.*

Theorem 6.15 was proved by establishing the universality principle, which states that the limiting distribution of the ESD of $\frac{1}{\sqrt{n}}A_n$ does not depend on the distribution of matrix entries, as soon as they have the same variance. This implies that the limiting law for a general distribution is the same as for the Gaussian distribution, which is the unit disk by Mehta's theorem [872].

The circular law, established in Theorem 6.15, describes the global distribution of the eigenvalues, and cannot be used to control their local properties, for example, to determine how many of the eigenvalues are real. In 1994, Edelman et al. [381] answered this question for real random matrices whose entries are i.i.d. and have normal distribution. In 2015, Tao and Vu [1188] extended this result to matrices with a more general distribution of the entries. Specifically, for a positive integer n, let M_n be an $n \times n$ matrix whose entries ξ_{ij} are jointly independent real random variables which

(i) have exponential decay, i.e., $\mathbb{P}(|\xi_{ij}| \geq t) \leq C \exp(-t^c)$ for some constants $C, c > 0$ (independent of n) and all i, j, and
(ii) satisfy $\mathbb{E}[\xi_{ij}^k] = E[Z^k]$ for $k = 1, 2, 3, 4$ and all i, j, where Z is the random variable having standard normal distribution.

Note that the entries ξ_{ij} are not required to be identically distributed.

Theorem 6.16 M_n has $\sqrt{\frac{2n}{\pi}} + o(\sqrt{n})$ real eigenvalues asymptotically almost surely.

In fact, Theorem 6.16 is just one of many corollaries of a much more general theorem, which, very roughly speaking, states that many properties of the eigenvalues of a random matrix with independent entries depend only on the first four moments of the entries, see [1188] for details.

Other Models of Random Matrices
In addition to the simple models of random matrices discussed above, there are other models, often arising from physical applications. One of them is the random band matrix model, which is used as a tractable mean-field model for large quantum systems.

Fix a sequence $w_1, w_2, \ldots, w_n, \ldots$ of positive numbers. For every positive integer n, let A_n be a random $n \times n$ matrix whose entries a_{ij} are such that

(i) $a_{ij} = a_{ji}$ for all i and j;
(ii) $a_{ij} = 0$ if either $i = j$ or $\min(|i - j|, n - |i - j|) > w_n$; and
(iii) all a_{ij} such that $0 < \min((i - j), n - (i - j)) \leq w_n$ are selected independently at random, each equal to 1 or -1 with equal probabilities.

Random matrices A_n selected in accordance to these rules are called *random band matrices*.

An important research direction in the study of random band matrices is to understand the limiting behaviour of their norms, after appropriate scaling. In 2008, Khorunzhiy [710] proved that if $\lim_{n \to \infty} \frac{w_n}{\log^{3/2} n} = +\infty$, then $\|A_n/2\sqrt{2w_n}\|$ converges in distribution to 1. He also conjectured that the same conclusion remains true under the weaker assumption $\lim_{n \to \infty} \frac{w_n}{\log n} = +\infty$. This assumption is the weakest one could hope for, because Bogachev et al. [161] proved in 1991 that the conclusion of the conjecture fails if $\lim_{n \to \infty} \frac{w_n}{\log n} = 0$. In 2010, Sodin [1141] confirmed the Khorunzhiy conjecture.

Theorem 6.17 If $\lim_{n \to \infty} \frac{w_n}{\log n} = +\infty$ then $\|A_n/2\sqrt{2w_n}\|$ converges in distribution to 1.

Most of the theorems discussed above study some specific model of random matrices, e.g. matrices whose entries are independent, and perform an asymptotic analysis as $n \to \infty$. In practice, one may need to work with random matrices whose entries may be dependent, and may need to analyse a finite number of such matrices of a fixed size. Can some meaningful conclusion be derived in this generality?

For random variables, we have asymptotic results like the central limit theorem, but also many non-asymptotic results. One important example of the latter is the following classical inequality proved by Hoeffding [612] in 1963. It states that if $X_1, \ldots, X_n$ are independent random variables such that $a_i \leq X_i \leq b_i$ almost surely, and $\overline{X} = \frac{1}{n} \sum_{i=1}^{n} X_i$, then

$$\mathbb{P}\left(\overline{X} - \mathbb{E}[\overline{X}] \geq t\right) \leq \exp\left(-\frac{2n^2 t^2}{\sum_{i=1}^{n} (b_i - a_i)^2}\right).$$

This inequality allows us to bound the probability that the sum of independent random variables exceeds some threshold. For many applications, it is important to have a similar result for the sum of Hermitian matrices, where we bound the size of the maximal eigenvalue $\lambda_{\max}$ of the sum. In 2012, Tropp [1213] established a matrix version of the Hoeffding inequality.

Consider a finite sequence $A_1, \ldots, A_n$ of fixed Hermitian $d \times d$ matrices. Define $\sigma^2 = \|\sum_{k=1}^{n} A_k^2\|$. Let $X_1, \ldots, X_n$ be a sequence of independent, random, Hermitian $d \times d$ matrices satisfying $\mathbb{E}[X_k] = 0$ and $X_k \preceq A_k$ almost surely, where we write $A \preceq B$ if the matrix $B - A$ is positive semidefinite.

Theorem 6.18 *For $X_1, \ldots, X_n$ as above, and for all $t \geq 0$,*

$$\mathbb{P}\left(\lambda_{\max}\left(\sum_{k=1}^{n} X_k\right) \geq t\right) \leq d \cdot e^{-t^2/8\sigma^2}.$$

In fact, Tropp also extended many other classical inequalities to the matrix setting, see [1213] for details.

6.3   From Independence to Convex Sets

The Central Limit Theorem and Beyond
A central result in probability theory is the *central limit theorem* (CLT), stating that if $X_1, X_2, \ldots$ is a sequence of independent and identically distributed random variables with finite mean $\mu = \mathbb{E}[X_i]$ and finite variance $\sigma^2 = \mathrm{Var}[X_i]$, then $\sqrt{n}\left(\frac{1}{n}\sum_{i=1}^{n} X_i - \mu\right)$ converges in distribution to the normal distribution $N(0, \sigma^2)$. This theorem was first stated, for the special case when X_i follows the Bernoulli distribution, by de Moivre in 1733. The general case was proved by Lyapunov [823] in 1901.

If $\mu = 0$, the CLT implies that the sequence

$$Y_n = \frac{1}{\sqrt{n}} \sum_{i=1}^{n} X_i$$

converges to the normal distribution, regardless of the shape of the distribution of the individual X_i. In 1978, Lieb [794] conjectured that this convergence is "monotone", in the sense that Y_{n+1} is "closer" to the normal distribution than Y_n. The word "closer" in this conjecture is formalized in terms of Shannon entropy [1110]. For a random variable X with probability density function f and support $S \subset \mathbb{R}$, its *Shannon entropy* is

$$\text{Ent}(X) = -\int_S f(x) \log f(x) dx.$$

In the 1940s, Shannon [1113] outlined the proof that $\text{Ent}(Y_2) \geq \text{Ent}(Y_1)$. In 1959, Stam [1152] gave a rigorous proof of this fact. Inductively, this implies that $\text{Ent}(Y_1) \leq \text{Ent}(Y_2) \leq \text{Ent}(Y_4) \leq \cdots \leq \text{Ent}(Y_{2^k}) \leq \ldots$. This motivated Lieb [794] to conjecture that in fact $\text{Ent}(Y_n) \leq \text{Ent}(Y_{n+1})$ for all $n \geq 1$. Because the normal distribution has the largest entropy among all distributions with given mean and variance, this conjecture can be interpreted as "Y_{n+1} is closer to the normal distribution than Y_n". However, before 2004, the Lieb conjecture remained open even for $n = 2$. In 2004, Artstein et al. [53] proved the conjecture for all n.

Theorem 6.19 *For any sequence $X_1, X_2, \ldots$ of independent and identically distributed square-integrable random variables, the entropy of the normalized sum* $\text{Ent}\left(\frac{X_1 + \cdots + X_n}{\sqrt{n}}\right)$ *is a non-decreasing function of n.*

The main difficulty in the proof of Theorem 6.19 is that it is not clear how to "build" the sum of $n + 1$ independent copies of a random variable from the sum of n such copies. The authors overcome this difficulty by first noting that the inequality $\text{Ent}(Y_n) \leq \text{Ent}(Y_{n+1})$ is equivalent to

$$J(Y_n) \geq J(Y_{n+1}), \tag{6.3}$$

where J is the *Fisher information*, defined, for a random variable X with smooth positive density f, as $J(X) = \int_{\mathbb{R}} \frac{(f'(x))^2}{f(x)} dx$. Next the authors derived a new formula for the Fisher information, which is particularly suitable for proving inequalities like (6.3).

The central limit theorem has various extensions and generalizations, which allow the X_i to not be identically distributed and even allow some (limited) form of dependence between X_i. In 2007, Klartag [720] proved a version of the CLT for the case when the random vector $X = (X_1, \ldots, X_n)$ is uniformly distributed in a convex body[3] in $\mathbb{R}^n$. This is a remarkable result because in this case the components X_i of X may be far from being independent.

---

[3] Recall that a *convex body* in $\mathbb{R}^n$ is a compact, convex set $K \subset \mathbb{R}^n$ with a non-empty interior.

Theorem 6.20 *There exists a sequence ϵ_n converging to 0 for which the following holds. For any convex body $K \subset \mathbb{R}^n$, there exist a unit vector θ in $\mathbb{R}^n$, $t_0 \in \mathbb{R}$, and $\sigma > 0$ such that*[4]

$$\sup_{A \subset \mathbb{R}} \left| \mathbb{P}\{\langle X, \theta \rangle \in A\} - \frac{1}{\sqrt{2\pi}\sigma} \int_A e^{-\frac{(t-t_0)^2}{2\sigma^2}} \, dt \right| \leq \epsilon_n,$$

where X is a random vector uniformly distributed in K, and the supremum runs over all measurable sets $A \subset \mathbb{R}$.

The starting point in the proof of Theorem 6.20 is the 1978 observation of Sudakov [1167] that if X is a random vector in $\mathbb{R}^n$ with mean zero and identity covariance matrix, then the random variable $\langle X, \theta \rangle$ is approximately Gaussian in almost all directions θ if and only if the norm $|X|$ is close to $\sqrt{n}$ with high probability. Hence, the problem reduces to obtaining the concentration of mass estimates in the spirit of Theorem 6.22 discussed below. Klartag [720] proved a version of Theorem 6.22 which works for all (rather than some) $c > 0$, at the cost of a much weaker estimate for the probability. However, this weaker estimate suffices to deduce Theorem 6.20.

The Gaussian Correlation Inequality

In Theorem 6.20, the condition that X is uniformly distributed in a convex set is used as a replacement for the condition that the components of X are independent. Another famous example when convexity is used as a replacement of independence is the famous *Gaussian correlation inequality*. By definition, events A and B are *independent* if $\mu(A \cap B) = \mu(A) \cdot \mu(B)$, where μ is the probability measure. The *Gaussian correlation conjecture* predicted that inequality

$$\mu(A \cap B) \geq \mu(A) \cdot \mu(B) \tag{6.4}$$

holds for all convex and centrally symmetric sets $A, B \subset \mathbb{R}^n$ if μ is the n-dimensional Gaussian probability measure on $\mathbb{R}^n$, centred at the origin.[5] A special case of the conjecture was formulated by Dunnett and Sobel [373] in 1955, the general case was stated by Gupta et al. [541] in 1972. Despite the simple-looking formulation and efforts of many researchers, the conjecture remained open for decades. In 2014, Royen [1066] proved the conjecture, and it is now called the Gaussian correlation inequality.

---

[4] Here, $\langle \cdot, \cdot \rangle$ denotes the inner product in $\mathbb{R}^n$.

[5] A symmetric $n \times n$ matrix Σ is called *positive definite* if $x^T \Sigma x > 0$ for all non-zero $x \in \mathbb{R}^n$. Given a positive definite $n \times n$ matrix Σ, let $f : \mathbb{R}^n \to \mathbb{R}$ be the function given by $f(x) = (2\pi)^{-n/2} \det(\Sigma)^{-1/2} \exp(-\frac{1}{2} x^T \Sigma^{-1} x)$, $x \in \mathbb{R}^n$. For a subset $A \subset \mathbb{R}^n$, let $\mu(A) = \int_A f(x) dx$.

Theorem 6.21 *Let μ be the n-dimensional Gaussian probability measure on $\mathbb{R}^n$, centred at the origin. Then inequality (6.4) holds for all convex and centrally symmetric sets $A, B \subset \mathbb{R}^n$.*

The proof of Theorem 6.21 is short and works, more generally, for multivariate gamma distributions. The inequality (6.4) in this generality reduces to the proof of positivity of a certain function $F(\tau)$ for all $\tau \in (0, 1]$. The parameter τ "measures" the dependency between A and B, with $\tau = 0$ corresponding to the case when A and B are independent, in which case (6.4) holds as an equality and $F(0) = 0$. Hence, it suffices to prove that $F'(\tau) > 0$ on $(0, 1)$. The author then used the Laplace transform to represent $F'(\tau)$ as a sum of non-negative terms, not all 0, and Theorem 6.21 follows.

The Concentration of Mass and Beyond

A convex body $K \subset \mathbb{R}^n$ is called *isotropic* if it has volume 1, its centre of mass is at the origin, and there exists a positive constant L_K such that

$$\int_K \langle x, \theta \rangle \mathrm{d}x = L_K^2$$

for every θ in the unit sphere $\mathbb{S}^{n-1}$. Any convex body can be made isotropic by an affine transformation of $\mathbb{R}^n$. The fundamental *concentration of mass principle* states that, in high dimensions, "most" of the mass of an isotropic convex body K is concentrated "near" a sphere with radius $\sqrt{n}$. In the language of the probability theory, this means that if x is selected uniformly at random inside K, then $||x||_2 = \sqrt{\langle x, x \rangle}$ is "close" to $\sqrt{n}$ with high probability. In 2006, Paouris [965] proved that the probability that $||x||_2$ is much higher that $\sqrt{n}$ is exponentially small.

Theorem 6.22 *There exists an absolute constant $c > 0$ such that if x is selected uniformly at random inside an isotropic convex body K in $\mathbb{R}^n$, then*

$$\mathbb{P}(||x||_2 \geq c\sqrt{n} L_K t) \leq \exp(-\sqrt{n}t)$$

for every $t \geq 1$.

A starting point in the proof of Theorem 6.22 is an earlier result of Paouris [964] stating that the statement of the theorem is equivalent to the inequality

$$\left(\int_K ||x||_2^q \mathrm{d}x \right)^{1/q} \leq c\sqrt{n} L_K, \quad \text{for every } 2 \leq q \leq \sqrt{n}.$$

This inequality was proved by analysing the properties of the L_q-centroid bodies of K, which are the bodies $Z_q(K)$ defined by the support function[6]

$$h_{Z_q(K)}(y) = \left(\int_K |\langle x, y \rangle|^q \, dx \right)^{1/q}.$$

The statement of Theorem 6.22 was conjectured by Bobkov and Nazarov [159, 160] in 2003, who proved it for a special case. The theorem has many applications. In particular, the famous *hyperplane conjecture* (also known as the *slicing conjecture*) predicts that any convex body in $\mathbb{R}^n$ of unit volume has a hyperplane section whose $(n-1)$-dimensional volume is at least a universal constant. This conjecture is equivalent to the inequality $L_K \leq c$ for some constant c independent of n. As partial progress, Bourgain [181] proved in 1991 that $L_K \leq c \sqrt[4]{n} \log n$ for some absolute constant $c > 0$. In combination with the result of Klartag [719], Theorem 6.22 implies a better estimate $L_K \leq c \sqrt[4]{n}$.

This bound remained unimproved for almost 15 years. In 2020, Chen [271] posted online a preprint in which he proved that $L_K \leq n^{f(n)}$ for some function $f : [0, \infty) \to [0, \infty)$ such that $\lim_{n \to \infty} f(n) = 0$. This is a dramatic improvement! In fact, Chen proved a stronger result which we now formulate.

A uniform distribution in a convex body is a special case of the well-studied family of *log-concave* distributions. A probability measure μ on $\mathbb{R}^n$ is called *absolutely continuous* if there exists a function $g : \mathbb{R}^n \to [0, \infty)$ (called a *density*) such that $\mu(A) = \int_A g(x) dx$ for every measurable $A \subset \mathbb{R}^n$. If the set $D = \{x \in \mathbb{R}^n \,|\, g(x) > 0\}$ is convex, and $-\log(g(x))$ is a convex function on D, then μ is called *log-concave*. A density g (and the corresponding probability measure μ) is called *isotropic* if

$$\int_{\mathbb{R}^n} x \, g(x) \, dx = 0 \quad \text{and} \quad \int_{\mathbb{R}^n} \langle x, \theta \rangle \, g(x) \, dx = 1 \quad \text{for every} \quad \theta \in \mathbb{S}^{n-1}.$$

The *boundary measure* of $A \subset \mathbb{R}^n$ is

$$\mu(\partial A) = \liminf_{\epsilon \to 0^+} \frac{\mu(\{x \in \mathbb{R}^n : d(x, A) \leq \epsilon\}) - \mu(A)}{\epsilon},$$

where $d(x, A)$ is the minimum Euclidean distance between x and A. The *isoperimetric constant* of μ, also known as the *Cheeger constant*, is defined by

$$\psi_\mu = \inf_{A \subseteq \mathbb{R}^n} \frac{\mu(\partial A)}{\min\{\mu(A), \mu(\mathbb{R}^n \setminus A)\}}.$$

---

[6] See (5.9) for the definition of the support function.

In 1995, Kannan et al. conjectured the existence of some universal constant $c > 0$, independent from the dimension, such that $\psi_\mu \geq c$ for any isotropic log-concave μ in any dimension n. This statement became known as the *KLS conjecture* and has many corollaries. In particular, it implies the hyperplane conjecture discussed above. In 2013, Eldan [383] introduced a new proof technique called *stochastic localization* to prove that $\psi_\mu \geq cn^{-1/3}(\log n)^{-1/2}$ for some constant $c > 0$. In 2017, Lee and Vempala [777] proved that

$$\psi_\mu \geq cn^{-1/4}. \tag{6.5}$$

Chen's main result in [271] improves this bound dramatically.

Theorem 6.23 *There is a function* $f : [0, \infty) \to [0, \infty)$ *such that* $\lim_{n \to \infty} f(n) = 0$ *and*

$$\psi_\mu \geq n^{-f(n)}$$

for every positive integer n *and every isotropic log-concave probability measure* μ *on* $\mathbb{R}^n$. *Moreover, we can take* $f(n) = c\sqrt{\frac{\log \log n}{\log n}}$ *for some absolute constant* $c > 0$.

Lee and Vempala [777] proved their previous bound (6.5) by using Eldan's stochastic localization method in combination with a new technique called *bootstrapping*, which is applied recursively and uses a bound on ψ_μ to achieve an even better bound. Using Eldan's bound as an input, they obtained the bound (6.5). They also outlined an idea how to use (6.5) to obtain a better bound, and so on, but they could not justify one crucial step in this recursive argument. Chen managed to justify the missing step, make the argument work, and the iterative application of the Lee–Vempala bootstrapping procedure leads to the bound in Theorem 6.23.

We next discuss the connection of the KLS conjecture with concentration properties of Lipschitz functions. Theorem 6.22 states that the function $f(x) = ||x||_2$ is "well-concentrated". Similar results has been derived for a wider class of functions. By the triangle inequality, $||x||_2 - ||y||_2 \leq ||x - y||_2$. More generally, a function f is called 1-*Lipschitz* if $|f(x) - f(y)| \leq ||x - y||$ for all $x, y \in \mathbb{R}^n$. A large body of literature is devoted to concentration properties of arbitrary 1-Lipschitz functions with respect to an arbitrary log-concave probability measure μ on $\mathbb{R}^n$.

Given a function $f : \mathbb{R}^n \to \mathbb{R}$, we write $E_\mu(f) = \int_{\mathbb{R}^n} f d\mu$, and $||f||_{L^1(\mu)} = E_\mu(|f|)$. We say that μ satisfies *first-moment concentration* if there exists a $D > 0$ such that for every 1-Lipschitz f we have

$$||f - E_\mu(f)||_{L^1(\mu)} \leq \frac{1}{D}.$$

In 2009, Milman [895] proved that this weak-looking condition implies a much stronger concentration property, similar to the one in Theorem 6.22.

Theorem 6.24 *For every absolutely continuous log-concave probability measure μ on $\mathbb{R}^n$ satisfying first-moment concentration, and every 1-Lipschitz $f : \mathbb{R}^n \to \mathbb{R}$, the inequality*

$$\mu(|f - E_\mu(f)| \geq t) \leq c \exp(-CDt)$$

holds for all $t > 0$, where c and C are universal positive constants.

Let $x \in \mathbb{R}^n$ be selected at random with respect to a measure μ, and let f be some useful function we want to compute. In general, because x is random, $f(x)$ is also random and unpredictable. However, if the conclusion of Theorem 6.24 (known as *exponential concentration*) holds, then we know that $f(x) \approx E_\mu(f)$ with very high probability.

In 1983, Gromov and Milman [535] proved that

$$\mu(|f - E_\mu(f)| \geq t) \leq c \exp(-Ct\psi_\mu) \tag{6.6}$$

for every log-concave probability measure μ on $\mathbb{R}^n$ and every 1-Lipschitz $f : \mathbb{R}^n \to \mathbb{R}$. The improved lower bound on ψ_μ in Theorem 6.23 automatically implies a better concentration bound in (6.6). Moreover, if the KLS conjecture is true, then *every* 1-Lipschitz f has the exponential concentration property as in Theorem 6.24. See the survey of Lee and Vempala [778] for more implications of this conjecture.

Separability of Random Points
One important consequence of mass concentration is linear separability. Let X be a random point cloud consisting of n points x_i, $i = 1, \ldots, n$, sampled independently and identically from a Gaussian distribution in $\mathbb{R}^d$ with non-singular covariance matrix. The set X is called *linearly separable* (or 1-convex) if every point in it can be separated from other points by a hyperplane.

If you select n random points on the plane, then, if n is not too small, you will likely find a point which is in the convex hull of the other points, and therefore cannot be separated from them by a line. In 2009, Donoho and Tanner [358] proved that (at least for the Gaussian distribution), the situation in high dimension is dramatically different.

Theorem 6.25 *For any $\epsilon > 0$, with overwhelming probability for large $d < n$, each subset of X of size at most $\frac{d}{2e \log(n/d)}(1 - \epsilon)$ can be separated from other points of X by a hyperplane.*

Theorem 6.25 implies that, with very high probability, a set of n random points in $\mathbb{R}^d$ is linearly separable even if n grows exponentially fast in terms of d. This result has many applications. For example, Gorban et al. [505, 521] proved the same result for a large family of non-Gaussian distributions and demonstrated how to use such theorems (known as *stochastic separation theorems*) to develop efficient error-correction mechanisms in artificial intelligence systems.

6.4   Discrete Time Stochastic Processes, Random Walks

Simple Random Walks on the Plane

Let $X_1, X_2, \ldots, X_n, \ldots$ be a sequence of independent vector-valued random variables, each taking values $(1, 0)$, $(0, 1)$, $(-1, 0)$ or $(0, -1)$ with equal probabilities. The sequence $S_n = \sum_{k=1}^{n} X_k$, $n = 1, 2, \ldots$, is called a *simple random walk* in $\mathbb{Z}^2$. This is one of the simplest examples of a stochastic process, but some basic questions about it turned out to be surprisingly hard to answer. For example, Erdős and Taylor [424] asked in 1960 how many times does a simple planar random walk revisit the most frequently visited point in the first n steps. More formally, for each pair of integers (x, y) and positive integer n, let $T(n, x, y)$ be the number of positive integers $k \leq n$ such that $S_k = (x, y)$. Then let $T^*(n) = \max_{x,y} T(n, x, y)$ be the number of times that the random walk visits the most visited point. Erdős and Taylor [424] proved that the limit

$$\lim_{n \to \infty} \frac{T^*(n)}{(\log n)^2},$$

if it exists, is between $\frac{1}{4\pi}$ and $\frac{1}{\pi}$, and conjectured that the limit actually exists and is equal to $\frac{1}{\pi}$ almost surely. This conjecture remained open for over 40 years, until was proved by Dembo et al. [342] in 2001.

Theorem 6.26 $\frac{T^*(n)}{(\log n)^2}$ *converges to* $\frac{1}{\pi}$ *almost surely as* $n \to \infty$.

Theorem 6.26 was proved by first confirming a conjecture of Perkins and Taylor [976] for an analogous problem for the 2-dimensional Wiener process,[7] and then using the fact that the simple random walk approximates the Wiener process in the limit.

Another basic question one may ask about simple random walks is the probability that two such walks S and S' intersect. For a simple random walk S on $\mathbb{Z}^2$ and integer $k \geq 0$, denote by $S[0, k]$ the (random) set of vertices S visited after k steps. If the probability

$$\mathbb{P}[S[0, k] \cap S'[0, k] = \emptyset]$$

decays proportionally to $k^{-\gamma}$ for some constant $\gamma > 0$, then γ is called the *intersection exponent*. In 1988, Duplantier and Kwon [374] used ideas from theoretical physics to predict that, for two simple random walks on the plane, $\gamma = 5/8$. In 2001, Lawler et al. [770] provided the first rigorous mathematical proof of this prediction.

---

[7] See Section 6.5 below for the definition of the Wiener process.

Theorem 6.27 *There exists a constant c > 0 such that the inequality*

$$c^{-1}k^{-5/8} \le \mathbb{P}[S[0,k] \cap S'[0,k] = \emptyset] \le ck^{-5/8}$$

holds for all $k \ge 1$, where S and S' are two independent simple random walks starting from neighbouring vertices in $\mathbb{Z}^2$.

The starting point in the proof of Theorem 6.27 is the 2000 result of Lawler and Puckette [769] stating that the probability $\mathbb{P}[S[0,k] \cap S'[0,k] = \emptyset]$ in Theorem 6.27 is, up to a multiplicative constant, equal to the probability that the trajectories of two 2-dimensional Wiener processes intersect. The authors then proved that the intersection exponent for n such processes is $\gamma_n = \frac{1}{24}(4n^2 - 1)$. With $n = 2$, this gives $5/8$.

Another interesting question is to estimate how long in takes before all the points in a certain region will be visited. In 1921, Pólya [991] proved that a simple random walk on the plane eventually visits all points with integer coordinates with probability one.[8] But how long do we need to wait until it visits, for example, all points in the disk centred at the origin with radius n? If we denote this time by T_n, Kesten and Révész proved the existence of constants $0 < a < b < \infty$ such that for every $t > 0$

$$e^{-b/t} \le \liminf_{n\to\infty} \mathbb{P}(\log T_n \le t(\log n)^2) \le \limsup_{n\to\infty} \mathbb{P}(\log T_n \le t(\log n)^2) \le e^{-a/t}.$$

In 1993, Lawler [767] proved that this holds with $a = 2$, $b = 4$, and stated a conjecture, which he attributed to Kesten, that the limit actually exists and is equal to $e^{-4/t}$. In 2004, Dembo et al. [343] confirmed this conjecture.

Theorem 6.28 *If T_n denotes the time it takes for a simple random walk in $\mathbb{Z}^2$ to completely cover the disc of radius n, then*

$$\lim_{n\to\infty} \mathbb{P}(\log T_n \le t(\log n)^2) = e^{-4/t}.$$

Similar to the proofs of Theorems 6.26 and 6.27, the proof of Theorem 6.28 first reduces the problem to the corresponding question about 2-dimensional Wiener processes. The key tool in the proof of the latter result is to consider such processes on many scales simultaneously. The authors called this idea a "multi-scale refinement of the classical second moment method".

Simple Random Walks on Graphs

Simple random walks can be studied, more generally, on arbitrary graphs. Let $G = (V, E)$ be a finite connected graph. A simple random walk on G is the process which starts at some vertex v at time $t = 0$, and then at times $t = 1, 2, \dots$ moves to a vertex selected uniformly at random among the neighbours of v in G. Let τ_{cov}

---

[8] Interestingly, this property is no longer true for random walks in $\mathbb{R}^d$ for $d \ge 3$.

be the first time at which every vertex of G has been visited, and let $\mathbb{E}_v[\tau_{\text{cov}}]$ be the expectation of τ_{cov} (which depends on the initial vertex v). The number

$$t_{\text{cov}}(G) = \max_{v \in V} \mathbb{E}_v[\tau_{\text{cov}}]$$

is called the *cover time* of G.

In 1996, Winkler and Zuckerman [1269] made a conjecture which relates the cover time to another important graph parameter, the blanket time. For any $v \in V$, let $N_v(t)$ be the (random) number of times the random walk has visited v up to time t. For $\delta \in (0, 1)$, let $\tau_{\text{bl}}(\delta)$ be the first time $t \geq 1$ at which

$$N_v(t) \geq \delta t \frac{\deg(v)}{2|E|}$$

holds for all $v \in V$. It is known that as $t \to \infty$, the proportion of time a random walk spends at any vertex v converges to $\frac{\deg(v)}{2|E|}$. Hence, $\tau_{\text{bl}}(\delta)$ is the first time at which all vertices have been visited at least a δ fraction of the expected number of visits. The number

$$t_{\text{bl}}(G, \delta) = \max_{v \in V} \mathbb{E}_v[\tau_{\text{bl}}(\delta)]$$

is called the *δ-blanket time* of G.

It is clear that $t_{\text{cov}}(G) \leq t_{\text{bl}}(G, \delta)$ by definition. Winkler and Zuckerman [1269] conjectured that the blanket and cover times are within an $O(1)$ factor. In 2012, Ding et al. [351] confirmed this conjecture.

Theorem 6.29 *For every $\delta \in (0, 1)$, there exists a $C = C(\delta)$ such that for every connected graph G, one has $t_{\text{bl}}(G, \delta) \leq C t_{\text{cov}}(G)$.*

The proof of Theorem 6.29 is based on a strong connection, established by the authors, between cover times, blanket times, and the *majorizing measures theorem for Gaussian processes* proved by Talagrand [1177] in 1987, see [351] for details. In the same paper [351], the authors also showed how to efficiently compute the cover time of an arbitrary graph to within an $O(1)$ factor. This answers a question asked by Aldous and Fill in 1994.

Theorems 6.28 and 6.29 study the question of how long to wait until all vertices of a graph are visited by a random walk. A related question is for how long a fixed vertex can be avoided. Here, we will discuss this question in the special but important case when the underlying graph G is a bounded-degree planar graph.

More formally, assuming that the initial vertex v_0 of the simple random walk on G is selected uniformly at random from the vertices of G, denote by $\phi(T, G)$ the probability that the walk does not return to v_0 for T steps. Let $\mathcal{G}(D)$ be the set of all finite planar graphs with degrees of all vertices bounded by D, and let

$$\phi_D(T) = \sup_{G \in \mathcal{G}(D)} \phi(T, G).$$

In 2001, Benjamini and Schramm [120] proved that $\phi_D(T) \to 0$ as $T \to \infty$, but left the rate of decay of $\phi_D(T)$ as an open question. It is known that $\phi_D(T) \geq \frac{c}{\log T}$ for some constant $c > 0$. In 2013, Gurel-Gurevich and Nachmias [545] proved an upper bound on $\phi_D(T)$ in the same form, thus answering the question of Benjamini and Schramm up to a constant factor.

Theorem 6.30 *For any $D \geq 1$, there exists a constant $C = C(D) < \infty$ such that* $\phi_D(T) \leq \frac{C}{\log T}$ *for any $T \geq 2$.*

The proof of Theorem 6.30 is based on a deep theory of distributional limits of sequences of finite graphs, initiated by Benjamini and Schramm [120] in 2001. Informally, the limit of a sequence G_n is a random infinite graph U with a distinguished vertex u, called a root, such that neighbourhood of a random vertex v in G_n converges in an appropriate sense to the neighbourhood of u in U. The authors proved that the graph U is almost surely recurrent, that is, a random walk on it visits its starting position infinitely often with probability one. This result has many corollaries, including Theorem 6.30.

Random Walks on Groups

Random walks can also be studied on groups. Let G be an infinite group with finite generating set S. Let $S^{-1} = \{g \in G \mid g^{-1} \in S\}$, and let $|g|$ be the smallest integer $k \geq 0$ such that $g \in G$ has representation $g = \prod_{i=1}^{k} s_i$ with each $s_i \in S \cup S^{-1}$. A *symmetric finitely supported probability measure* on G is a function $\mu : G \to [0, 1]$ such that the set $\{g \in G \mid \mu(g) > 0\}$ is finite, $\sum_{g \in G} \mu(g) = 1$, and $\mu(g^{-1}) = \mu(g)$ for all $g \in G$. Let $X_1, X_2, \ldots$ be a sequence of i.i.d. random variables with distribution μ, that is, such that $\mathbb{P}[X_i = g] = \mu(g)$ for all $g \in G$ and all i. A *random walk* on G with step distribution μ is the sequence $W_1, W_2, \ldots$ given by $W_n = \prod_{i=1}^{n} X_i$.

A fundamental quantity associated with the random walk W_n is its speed function

$$L_\mu(n) = \mathbb{E}[|W_n|] = \sum_{g \in G} |g| \mathbb{P}[W_n = g],$$

and a central question is how fast $L_\mu(n)$ can grow with n. It is straightforward that the growth of $L_\mu(n)$ can be at most linear. On the other hand, Lee and Peres [776] proved in 2013 that $L_\mu(n) \geq c_\mu \sqrt{n}$ for some constant c_μ. In 2015, Brieussel [212] proved that $L_\mu(n)$ can grow as n^γ for any $\gamma \in [1/2, 1]$. More formally, the *upper exponent* of a function f is $\gamma = \limsup_{n \to \infty} \frac{\log f(n)}{\log n}$. If $\lim_{n \to \infty} \frac{\log f(n)}{\log n}$ exists, γ is called the *exponent* of f. Brieussel proved that for every $\gamma \in [1/2, 1]$ there exists a random walk with upper exponent of the speed function (or "upper speed exponent" for short) γ.

Another important function related to the random walk W_n is its *entropy*

$$H_\mu(n) = - \sum_{g \in G} \mathbb{P}[W_n = g] \log \mathbb{P}[W_n = g],$$

where we adopt the convention that $0 \log 0 = 0$. It is known that the speed exponent γ and entropy exponent $\theta = \lim\limits_{n \to \infty} \frac{\log H_\mu(n)}{\log n}$ of any random walk, if they exist, must satisfy

$$\theta \leq \gamma \leq \frac{1}{2}(\theta + 1). \tag{6.7}$$

Conversely, Amir [40] proved that for any real numbers $\gamma, \theta \in [3/4, 1]$ satisfying (6.7), there exists a random walk with speed exponent γ and entropy exponent θ. He also conjectured that the same is true for a large range $\gamma, \theta \in [1/2, 1]$. In 2015, Brieussel and Zheng posted online a preprint [212], in which they proved this conjecture.

Theorem 6.31 *For any* $\gamma \in [1/2, 1]$ *and* $\theta \in [1/2, 1]$ *satisfying (6.7), there exist a finitely generated group G and a symmetric probability measure μ of finite support on G, such that the random walk on G with step distribution μ has speed exponent γ and entropy exponent θ.*

In fact, Brieussel and Zheng proved much more precise results describing how $L_\mu(n)$ and $H_\mu(n)$ can grow. We write $f(n) \sim g(n)$ if $\frac{1}{C} f(n) \leq g(n) \leq Cf(n)$, $\forall n \geq 1$ for some constant C. Brieussel and Zheng proved that for any non-decreasing function $f : [1; \infty) \to [1; \infty)$ such that $f(1) = 1$ and $x/f(x)$ is non-decreasing, there exists a random walk such that $L_\mu(n) \sim \sqrt{n} f(\sqrt{n})$. The implied constant C is universal. The same statement is true for $H_\mu(n)$, and similar results have been obtained for many other parameters of a random walk W_n, for example, for the return probability $\mathbb{P}[W_{2n} = e]$.

On k-Dependent q-Colourings
An important research direction in probability theory is the study of stochastic processes with little or no dependence between random variables at distant locations. A k-dependent process indexed by integers is one of the simplest examples of this phenomenon. We say that a stochastic process $(X_i)_{i \in \mathbb{Z}}$ is k-*dependent* if the random sequences $(\ldots, X_{i-2}, X_{i-1})$ and $(X_{i+k}, X_{i+k+1}, \ldots)$ are independent of each other, for each $i \in \mathbb{Z}$.

Among the simplest examples of stochastic processes are those which take only a finite number of values. A stochastic process $(X_i)_{i \in \mathbb{Z}}$ is called a q-*colouring* (of $\mathbb{Z}$) if each X_i takes values in $\{1, \ldots, q\}$, and almost surely we have $X_i \neq X_{i+1}$ for all $i \in \mathbb{Z}$. In 2008, Schramm proved that no stationary[9] 1-dependent 3-colouring of the integers exists, and asked whether there exists a k-dependent q-colouring for some pair of positive integers (k, q). In 2016, Holroyd and Liggett [618] determines all such pairs.

---

[9] A process X is *stationary* if $(X_i)_{i \in \mathbb{Z}}$ and $(X_{i+1})_{i \in \mathbb{Z}}$ are equal in law.

Theorem 6.32 *There exist a stationary* 1-*dependent* 4-*colouring of* $\mathbb{Z}$*, and a stationary* 2-*dependent* 3-*colouring of* $\mathbb{Z}$.

A process X is called an *r-block-factor* (of an independent and identically distributed process) if

$$X_i = f(U_{i+1}, U_{i+2}, \ldots, U_{i+r}), \quad \forall i$$

for some independent and identically distributed $(U_i)_{i \in \mathbb{Z}}$ and some measurable function f. An r-block-factor is obviously stationary and $r - 1$-dependent. Ibragimov and Linnik [636] proved that, conversely, every stationary and $r - 1$-dependent Gaussian processes is an r-block-factor, but claimed without proof that this is not true in the non-Gaussian case. In 1989, Aaronson et al. [1] indeed proved the existence of a family of 1-dependent processes that are not 2-block-factors, but their construction is intricate and counterintuitive. Because it is known that no r-block-factor q-coloring exists for any r and q, Theorem 6.32 gives the first natural examples of finitely dependent stationary processes that are not block factors.

To prove Theorem 6.32, it suffices to construct an example of a stochastic process with the given properties. To describe a stochastic process, it suffices to define the probabilities that $X_1, \ldots, X_n$ take any given values. The authors first derived a (complicated) expression for these probabilities, but the challenge was to prove that this expression always takes non-negative values. By searching for recursions satisfied by this expression, the authors then derived a much simpler equivalent expression, from which the non-negativity follows easily.

6.5 Continuous Time Stochastic Processes

The Wiener Process

One of the most famous and classical examples of a continuous time stochastic process is the d-dimensional Wiener process, which is a mathematical formalization of Brownian motion, the random motion of particles in a liquid or gas. More formally, a *one-dimensional Wiener process* is a real-valued stochastic process $\{W(t)\}_{t \geq 0}$ satisfying the following properties

(i) *independent increments*: for any $0 \leq t_1 < t_2 < \cdots < t_n < \infty$, the random variables $W_{t_2} - W_{t_1}, W_{t_3} - W_{t_2}, \ldots, W_{t_n} - W_{t_{n-1}}$ are mutually independent;

(ii) *stationary increments*: if $0 \leq s \leq t$, random variables $W_t - W_s$ and $W_{t-s} - W_0$ have the same probability law;

(iii) *continuity*: the function $t \to W(t)$ is continuous in t with probability 1.

It is well-known that properties (i)–(iii) imply that[10]

(iv) the increment $W(s + t) - W(s)$ has the normal distribution with mean rt and variance $\sigma^2 t$, where r and σ are constant real numbers.

A one-dimensional Wiener process is called *standard* if $r = 0$, $\sigma = 1$, and $W_0 = 0$ almost surely. A *standard d-dimensional Wiener process* is a vector-valued stochastic process $W(t) = (W_1(t), \ldots, W_d(t))$ whose components $W_i(t)$ are independent, standard one-dimensional Wiener processes.

For $a > 0$ and $t \geq 0$, let

$$W^a(t) = \{x \in \mathbb{R}^d \mid \exists s \in [0, t] : \rho(x, W(s)) \leq a\},$$

where ρ is the distance in $\mathbb{R}^d$. $W^a(t)$ is known as a *Wiener sausage*. It was first described by Spitzer [1150] in 1964, and has applications to modelling physical processes like heat conduction. Let $|W^a(t)|$ denote its (d-dimensional) volume. Spitzer proved that

$$\mathbb{E}|W^a(t)| \sim \begin{cases} 2\pi t / \log t, & \text{for } d = 2, \\ 2\pi at, & \text{for } d = 3, \\ K_{a,d} \cdot t, & \text{for } d \geq 3, \end{cases}$$

where $K_{a,d} > 0$ is a constant depending on a and d, $\mathbb{E}$ denotes mathematical expectation, and by $f(t) \sim g(t)$ we mean $\lim_{t \to \infty} \frac{f(t)}{g(t)} = 1$. An interesting research direction is to determine with what probability $|W^a(t)|$ is "close" to its expected value.

Let a be fixed and define $V_t = |W^a(t)|$. Starting from the 1975 work of Donsker and Varadhan [360], many researchers have derived very good upper bounds on the probabilities of the form $\mathbb{P}(V_t \leq f(t)\mathbb{E}[V_t])$ for some $f(t)$ such that $\lim_{t \to \infty} f(t) = 0$. These are known as *probabilities of large deviations*. In 2001, van den Berg et al. [131] investigated the much more delicate question of finding upper bounds for probabilities of the form $\mathbb{P}(V_t \leq b\,\mathbb{E}[V_t])$ for constant b, which are called *moderate deviation probabilities*. They proved that such probabilities decay as t^{-C} for $d = 2$, and as $\exp\left(-Ct^{(d-2)/d}\right)$ for $d \geq 3$.

Theorem 6.33 *For every $a > 0$, $b \in (0, K_{a,d})$,*

$$\lim_{t \to \infty} \frac{1}{t^{(d-2)/d}} \log \mathbb{P}(|W^a(t)| \leq bt) = -C_{a,d,b}, \quad d \geq 3,$$

---

[10] Historically, a *Wiener process* is defined as a process satisfying (i)–(iv), a process satisfying (i) and (ii) is called a *Levy process*, and the fact that (i)–(iii) imply (iv) is stated as "The Wiener process is the only continuous Levy process".

where $C_{a,d,b} > 0$ is some constant depending on a, d, and b. Similarly, for every $a > 0$, $b \in (0, 2\pi)$,

$$\lim_{t \to \infty} \frac{1}{\log t} \log \mathbb{P}(|W^a(t)| \le bt / \log t) = -C_b, \quad d = 2.$$

A related research direction is to understand the distribution of the intersection volume of two Wiener sausages. Let

$$S^a(t) = W_1^a(t) \cap W_2^a(t)$$

be the intersection of two Wiener sausages corresponding to two independent Wiener processes starting at the origin, see Fig. 6.1, and let $|S^a(t)|$ be its (d-dimensional) volume. Let a be fixed and define $V_t = |S^a(t)|$. In 2004, van den Berg et al. [132] proved that, for large t and constant b, the probability $\mathbb{P}(V_{bt} \ge t)$ decays as $\exp\left(-Ct^{(d-2)/d}\right)$ for $d \ge 3$. In other words, the probability that the Wiener sausages' intersection volume is large (larger than t after time bt) decays exponentially fast in t, with constant in the exponent $\frac{d-2}{d}$. This gives us a good idea of what the distribution of V_t looks like.

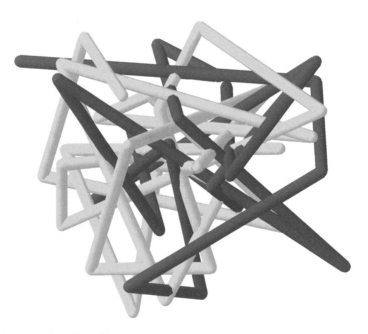

Fig. 6.1 Intersection of two Wiener sausages

Theorem 6.34 *For every a > 0 and b > 0,*

$$\lim_{t \to \infty} \frac{1}{t^{(d-2)/d}} \log \mathbb{P}(|S^a(bt)| \geq t) = -C_{a,d,b}, \quad d \geq 3$$

where $C_{a,d,b} > 0$ is some constant depending on a, d, and b. Similarly,

$$\lim_{t \to \infty} \frac{1}{\log t} \log \mathbb{P}(|S^a(bt)| \geq t/\log t) = -C_b, \quad d = 2.$$

The idea of the proof of Theorem 6.34 is to wrap the Wiener sausages around a torus of an appropriate size, show that the changes to the intersection volume caused by this procedure are negligible in the limit, and then use the methodology developed in an earlier work [131], including Theorem 6.33, to compute the intersection volume on the torus.

The Poisson Process
Another fundamental continuous time stochastic process is the *Poisson process*, which is used to count random events arriving at some constant rate μ. More formally, a *counting process* is a stochastic process $\{N(t)\}_{t \geq 0}$ whose values are non-negative, integer, and non-decreasing. A counting process is called a *Poisson process with rate $\mu > 0$* if

(i) $N(0) = 0$;
(ii) the process $\{N(t)\}_{t \geq 0}$ has independent increments;
(iii) the increments $N(s + t) - N(s)$ are Poisson distributed with mean μt. In other words, for any integer $k \geq 0$,

$$\mathbb{P}(N(s + t) - N(s) = k) = \frac{e^{-\mu t}(\mu t)^k}{k!}.$$

In 1991, Cheng and Lebowitz [272] observed a surprising connection between Poisson process and the sequence of squares of distances from a fixed point on the coordinate plane to all points with integer coordinates. More formally, for fixed real numbers $\alpha, \beta \in [0, 1]$, let $\Lambda_{\alpha,\beta}$ be the infinite sequence $0 \leq \lambda_1 \leq \lambda_2 \leq \ldots$ formed by the numbers $(m - \alpha)^2 + (n - \beta)^2$ for integer m, n, arranged in non-decreasing order. Cheng and Lebowitz observed numerically that this non-random sequence shares many properties with the random sequence of event times in the Poisson process with rate π. For example, they observed that these sequences seem to have the same limiting behaviour as the pair correlation function. For the interval $[a, b] \subset \mathbb{R}$, the *pair correlation function* is

$$R_2[a, b](\lambda) := \frac{1}{\pi \lambda}|\{j \neq k \mid \lambda_j \leq \lambda, \; \lambda_k \leq \lambda, a \leq \lambda_j - \lambda_k \leq b\}|,$$

where $|S|$ denotes the number of elements in set S. If λ_i are the event times in the Poisson process with rate π, then $R_2[a, b](\lambda)$ shows how often the time interval between the events is greater than a but less then b, and it is known that $\lim_{\lambda \to \infty} R_2[a, b](\lambda) = \pi(b - a)$ almost surely. Numerical experiments of Cheng and Lebowitz showed that $R_2[a, b](\lambda)$ for sequence $\Lambda_{\alpha, \beta}$ seems to have the same limiting behaviour.

In 2003, Marklof [845] rigorously confirmed this numerical observation for almost all pairs (α, β). An irrational number α is called *Diophantine* if there exist constants $k, C > 0$ such that $\left| \alpha - \frac{p}{q} \right| > \frac{C}{q^k}$ for every pair of integers p and $q \neq 0$. It is known that (Lebesgue) almost all real numbers are Diophantine. Real numbers $\alpha_1, \alpha_2, \dots \alpha_k$ are *linearly independent over* $\mathbb{Q}$ if there are no rational numbers $r_1, r_2, \dots r_k$, not all 0, such that $\sum_{i=1}^{k} r_i \alpha_i = 0$.

Theorem 6.35 *If $\alpha, \beta, 1$ are linearly independent over $\mathbb{Q}$, α is Diophantine, and $a < b$, then*

$$\lim_{\lambda \to \infty} R_2[a, b](\lambda) = \pi(b - a).$$

A Process for Modelling Diffusion

We next describe a stochastic process which is often used to model diffusion. We start with its one-dimensional version. At time $t = 0$, let us put, for each $x \in \mathbb{Z}$, a particle in site x, independently and with the same probability σ. Then, for each particle z, independently generate a random variable t_z from the exponential distribution with parameter 1, wait for time t_z, and then jump to an adjacent site right or left, with probabilities p and $q = 1 - p$, respectively, provided that the target site is unoccupied (and otherwise stay). This action is then repeated and performed in parallel for all particles. We call this the *asymmetric simple exclusion process* (ASEP).

For $x \in \mathbb{R}$, denote by $[x]$ the largest integer in $[0, x]$ if $x \geq 0$, and the smallest integer in $[x, 0]$ if $x \leq 0$. For $v \in \mathbb{R}$, let

$$J^{(v)}(t) = J_+^{(v)}(t) - J_-^{(v)}(t),$$

where $J_+^{(v)}(t)$ is the number of particles that began in $(-\infty, 0]$ at time 0 but lie in $[[vt] + 1, \infty)$ at time t, and $J_-^{(v)}(t)$ is the number of particles that began in $[1, \infty)$ at time 0 but lie in $(-\infty, [vt]]$ at time t. The random variable $J^{(v)}(t)$ is interpreted as the total net particle current seen by an observer moving at speed v during time interval $[0, t]$. The mean of $J^{(v)}(t)$ is easy to estimate, but the problem of estimating its variance is more delicate. In 1994, Ferrari and Fontes [452] proved that

$$\lim_{t \to \infty} \frac{\text{Var}(J^v(t))}{t} = \sigma(1 - \sigma)|(p - q)(1 - 2\sigma) - v|,$$

which implies that the variance grows linearly in time, except when $v = (p - q)(1 - 2\sigma)$, which is called the *characteristic speed*. $J^{(v)}(t)$ for this v is called the *current across the characteristic*, and the problem of estimating the growth rate of

$\text{Var}(J^v(t))$ in this case remained open. In 2010, Balázs and Seppäläinen [85] solved this problem and proved that the variance grows as $t^{2/3}$.

Theorem 6.36 *Let $v = (p - q)(1 - 2\sigma)$. Then there exists positive constants t_0 and C such that*

$$C^{-1}t^{2/3} \leq \text{Var}(J^{(v)}(t)) \leq Ct^{2/3},$$

for all $t \geq t_0$.

To study diffusion in higher dimensions we need to define a d-dimensional version of ASEP. Here we discuss the case $d = 2$, the definition in higher dimensions is similar. At time $t = 0$, let us put, for each $x \in \mathbb{Z}^2$, a particle in site x, independently and with probability σ. Then, for each particle p, independently generate a random variable t_p from an exponential distribution, wait for time t_p, and then jump to an adjacent site up, down, right, or left with probabilities p_1, p_2, p_3, or p_4, respectively,[11] provided that the target site is unoccupied (and otherwise stay). This action is then repeated and performed in parallel for all particles. We call this the *asymmetric simple exclusion process (ASEP)* on the plane.

For concreteness, let us assume that $\sigma = 1/2$, $p_1 = p_2 = 1/4$, $p_3 = 1/2$, $p_4 = 0$. For any $x \in \mathbb{Z}^2$, let $\eta_x(t)$ be the number of particles (which can be 0 or 1) in position x at time t. Next define

$$D(t) = \frac{4}{t} \sum_{x \in \mathbb{Z}^2} x_1^2 \cdot \text{cov}(\eta_x(t), \eta_{(0,0)}(0)),$$

where x_1 is the first coordinate of x, and $\text{cov}(X, Y) = \mathbb{E}[XY] - \mathbb{E}[X]\mathbb{E}[Y]$ is the covariance between random variables X and Y. The function $D(t)$ is known as the *diffusion coefficient*, and understanding how fast $D(t)$ grows with t is crucial for estimating the speed of a diffusion process. In 2004, Yau [1281] proved that $D(t)$ grows approximately as $(\log t)^{2/3}$ for large t. This growth rate is a corollary of the following theorem.

Theorem 6.37 *There exists a constant $\gamma > 0$ such that, for sufficiently small $\lambda > 0$,*

$$\lambda^{-2}|\log \lambda|^{2/3}e^{-\gamma|\log\log\log\lambda|^2} \leq \int_0^\infty e^{-\lambda t}t D(t)\mathrm{d}t \leq \lambda^{-2}|\log \lambda|^{2/3}e^{\gamma|\log\log\log\lambda|^2}.$$

The proof of Theorem 6.37 is based on an earlier work of Landim et al. [761] which connects the problem of estimating the growth of $D(t)$ to the analysis of the so-called "Green function" of the dynamics. In [761], the Green function was estimated to degree three, which implies a lower bound $D(t) > (\log t)^{1/2}$. Yau

---

[11] If $p_1 = p_2 = p_3 = p_4 = 1/4$, and the mean of the exponential distribution is μ, this is called a *continuous time simple random walk on $\mathbb{Z}^2$, with jump rates μ.*

[1281] estimated the Green function to a much higher degree, and combined this with the results in [761] to deduce Theorem 6.37.

A Process for Modelling the Spread of Infection

There are many other important stochastic processes used for modelling fundamental processes in physics, chemistry, medicine, etc. Here we discuss a process for modelling the spread of infection. At time 0, let us put n_x "A-particles" at each $x \in \mathbb{Z}^d$, where n_x are i.i.d. random variables following a Poisson distribution with mean μ. We also put a finite number of "B-particles" in arbitrary (non-random) positions on $\mathbb{Z}^d$. In this model, the B-particles model organisms affected by an infection, while the A-particles model unaffected organisms. As time progresses, each A-particle and each B-particle performs a continuous time simple random walk on $\mathbb{Z}^d$, with the same jump rate D. The walks are independent, except that when a B-particle and an A-particle coincide, the latter turns into the former. Let $B'(t)$ be the set of all locations $x \in \mathbb{Z}^d$ visited by a B-particle during $[0, t]$, and let $B(t) = B'(t) + [-1/2, 1/2]^d$. Figure 6.2a depicts region $B(t)$ is the simple case when $B'(t)$ is the set of 4 points $(0, 0)$, $(0, 1)$, $(1, 1)$, and $(1, 0)$. In general, region $B(t)$ is the region affected by infection by time t. In 2008, Kesten and Sidoravicius [705] proved that in the limit as $t \to \infty$ the shape of this region converges to a non-random shape growing at a linear rate.

Theorem 6.38 *There exists a nonrandom, compact, convex set $B_0 \subset \mathbb{R}^d$ such that for all $\epsilon > 0$, the inclusion*

$$(1 - \epsilon) B_0 \subset \frac{1}{t} B(t) \subset (1 + \epsilon) B_0$$

holds almost surely as $t \to \infty$.

The conclusion of Theorem 6.38 is illustrated in Fig. 6.2b.

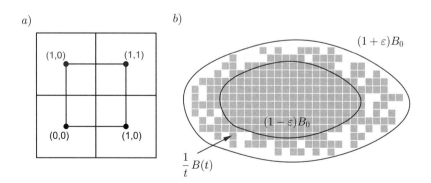

Fig. 6.2 The shape of the region affected by infection

The Internal DLA Process

The *internal diffusion limited aggregation process* (*internal DLA process* for short) was proposed by Meakin and Deutch [867] in 1986 as a model of industrial chemical processes. For each integer time $n \geq 1$, construct a random set $A(n) \subset \mathbb{Z}^2$ such that (i) $A(1)$ consists of one point $(0, 0)$, and (ii) for each $n \geq 1$, let $A(n + 1)$ be $A(n)$ plus the first point at which a simple random walk from the origin hits $\mathbb{Z}^2 \setminus A(n)$. For any real $t \geq 1$, let $A(t) = A(\lfloor t \rfloor)$, where $\lfloor t \rfloor$ is the largest integer not exceeding t.

Meakin and Deutch found numerically that, for large n, $A(n)$ becomes close to a disk with at most logarithmic fluctuations. In 1992, Lawler et al. [768] proved that the asymptotic shape of $A(n)$ is indeed a disk. In 2012, Jerison et al. [654] confirmed that the fluctuations are indeed at most logarithmic. For $r > 0$, let $B(r) = \{z = (z_1, z_2) \in \mathbb{Z}^2 \mid z_1^2 + z_2^2 < r^2\}$.

Theorem 6.39 *For every $\gamma > 0$, there exists a constant $a = a(\gamma)$ such that*

$$\mathbb{P}\left[B(r - a \log r) \subset A(\pi r^2) \subset B(r + a \log r) \right] \geq 1 - r^{-\gamma}$$

for all sufficiently large r.

In 2013, the same authors [655] proved an analogue of Theorem 6.39 in $\mathbb{R}^d$ in $d \geq 3$, with $\log r$ replaced by $\sqrt{\log r}$. This means that the higher dimensional DLA process has even smaller fluctuations.

6.6 Bernoulli Convolutions

Let $\lambda, p \in (0, 1)$ be real numbers, and let $\xi_0, \xi_1, \xi_2, \ldots$ be a sequence of i.i.d. random variables with $\mathbb{P}(\xi_n = 1) = p$ and $\mathbb{P}(\xi_n = -1) = 1 - p$ for all n. By a *generalized Bernoulli convolution* we mean a random variable

$$X_{\lambda, p} = \sum_{n=0}^{\infty} \xi_n \lambda^n.$$

If $p = 1/2$, $X_{\lambda, 1/2}$ is called a *Bernoulli convolution*. Bernoulli convolutions were introduced by Jessen and Wintner [658] in 1935, and have been intensively studied since then. The main research question is for which values of λ the Bernoulli convolution is absolutely continuous.[12] In 1939, Erdős [396] observed that $X_{\lambda, 1/2}$ is not absolutely continuous for $\lambda < 1/2$, and also for $\lambda \in E$ for some non-empty

---

[12] Recall that $X_{\lambda, p}$ is called *absolutely continuous* if there exists a function $f : \mathbb{R} \to \mathbb{R}$ such that $\mathbb{P}(a \leq X_{\lambda, p} \leq b) = \int_a^b f(x) dx$ whenever $a \leq b$.

set[13] $E \subset [1/2, 1)$. However, Solomyak [1142] proved in 1995 that the set E has Lebesgue measure 0. Moreover, Shmerkin [1118] proved in 2014 that $\dim(E) = 0$, where dim is the Hausdorff dimension defined in (1.23).

Despite this, before 2019 very few explicit examples of λ-s for which $X_{\lambda, p}$ is absolutely continuous were known, and no such examples were known for $p \neq 1/2$. In 2019, Varjú [1224] proved a theorem which provides a lot of such examples. For example, it implies that $X_{\lambda, p}$ is absolutely continuous for $\lambda = 1 - 10^{-50}$ and $\frac{1}{4} \leq p \leq \frac{3}{4}$.

To formulate the theorem, we need a few more definitions. The *Mahler measure* of any polynomial $P(x) = a \prod_{i=1}^{d} (x - z_i)$ is $M(P) = a \prod_{j:|z_j|>1} |z_j|$. A real number r is called *algebraic* if $P(r) = 0$ for some polynomial P with rational coefficients. The (unique) monic polynomial P_r of smallest degree with this property is called the *minimal polynomial* of r. The Mahler measure of r is $M_r = M(P_r)$.

Theorem 6.40 *For every $\epsilon > 0$ and $p \in (0, 1)$, there is an (explicitly computable) constant $c = c(\epsilon, p) > 0$ such that for any algebraic number λ satisfying $1 > \lambda > 1 - c \min(\log M_\lambda, (\log M_\lambda)^{-1-\epsilon})$, the Bernoulli convolution $X_{\lambda, p}$ is absolutely continuous.*

Let ν_λ denote the probability distribution of the Bernoulli convolution $X_{\lambda, 1/2}$. In addition to studying when ν_λ is absolutely continuous, researchers are interested in when ν_λ obeys the weaker property of being full-dimensional. In 2009, Feng and Hu [450] proved that the limit

$$\lim_{r \to 0+} \frac{\log \nu_\lambda([x - r, x + r])}{\log r}$$

exists and is constant for ν_λ-almost every x. This constant is called the *dimension* of ν_λ and is denoted $\dim(\nu_\lambda)$. We say that ν_λ is *full-dimensional* if $\dim(\nu_\lambda) = 1$. Solomyak [1142] proved in 1995 that, for almost all $\lambda \in (1/2, 1)$, ν_λ is absolutely continuous and therefore full-dimensional. This implies that the set E_1 of exceptional λ's for which $\dim(\nu_\lambda) < 1$ has Lebesgue measure 0. In 2014, Hochman [608] proved the stronger result that E_1 must have packing dimension 0.

---

[13] In fact, Erdős provided explicit examples of numbers in the set E: the inverses of the Pisot numbers. A *Pisot number* is an algebraic integer greater than 1 all of whose conjugate elements have absolute value less than 1, where the algebraic numbers are called *conjugate* if they have the same minimal polynomial. For example, the positive root $\theta = 1.3247...$ of the equation $x^3 - x - 1 = 0$ is a Pisot number, hence $\theta^{-1} \in E$.

The *box dimension* of a set S of real numbers is $\mathrm{bdim}(S) = \limsup\limits_{r \to 0+} \frac{\log N_S(r)}{\log(1/r)} = 0$, where $N_S(r)$ is the minimal number of intervals of length r needed to cover S. The *packing dimension* of S is

$$\mathrm{pdim}(S) = \inf\left\{ \sup_n \mathrm{bdim}(S_n) : S \subseteq \bigcup_{n=1}^{\infty} S_n \right\}.$$

Theorem 6.41 *The set* $E_1 = \{\lambda \in (1/2, 1) : \dim(\nu_\lambda) < 1\}$ *has packing dimension* 0.

In 2019, Varjú [1225] made a step further and proved that the set E_1 is a subset of algebraic numbers and is therefore countable.

Theorem 6.42 *For all transcendental* $\lambda \in (1/2, 1)$, $\dim(\nu_\lambda) = 1$.

The proof of Theorem 6.42 is based on an earlier work of Hochman, who, in the same paper [608] where he proved Theorem 6.41, also proved Theorem 6.42 conditional on one conjecture. The conjecture predicts the existence of a universal constant $C > 0$ such that the inequality $|\xi_1 - \xi_2| > C^{-n}$ holds for any two roots ξ_1, ξ_2 of any polynomials P_1, P_2 of degree at most n with coefficients $-1, 0$ and 1. The intuitive justification is that there are only 3^{n+1} such polynomials, each having at most n roots, hence, for large C, the average distance between any such roots is much larger than C^{-n}. However, the conjecture turns out to be difficult and remains open. Instead, Varjú [1225] proved a weaker property in the same spirit, and showed that even this weaker property, when plugged into the result of Hochman [608], implies Theorem 6.42.

Theorem 6.42 implies that the set E_1 is a subset of algebraic numbers. The inclusion is proper: Theorem 6.40 implies that there are some algebraic $\lambda \in (1/2, 1)$ for which ν_λ is absolutely continuous, and in particular $\dim(\nu_\lambda) = 1$. Hence, the exact characterization of the set E_1 remains an open question.

6.7 Percolation Theory

Percolations on $\mathbb{Z}^d$

Let $G = (V, E)$ be an infinite graph and let $p \in [0, 1]$. The *Bernoulli(p) bond percolation* on $G = (V, E)$ is a subgraph of G to which each edge of G is included independently with probability p. The *Bernoulli(p) site percolation* on $G = (V, E)$ is a subgraph of G to which each vertex of G is included independently with probability p, and each edge (u, v) is included if and only if both vertices u and v are.

Bond and site percolations are important models in mathematics for studying processes such as diffusion. The study of percolations is motivated by physical applications and mathematical beauty. An important special case is when G is the graph with vertex set $\mathbb{Z}^d$ such that x, $y \in \mathbb{Z}^d$ are connected by an edge if and only if $||x - y|| = 1$. Let G_p be a bond percolation in $\mathbb{Z}^d$. Given $r > 0$, let $P_d(p, r)$ be the probability that there exists a path in G_p from the origin 0 to some vertex $x \in \mathbb{Z}^d$ with $||x|| > r$. It is known that for every $d \geq 2$ there exists a $p_d \in [0, 1]$ such that $L_d(p) = \lim_{r \to \infty} P_d(p, r) = 0$ for $p < p_d$ but $L_d(p) = 1$ for $p > p_d$. This implies that, with probability 1, G_p has no infinite connected components if $p < p_d$ but has at least one such component if $p > p_d$.

What about the case $p = p_d$? The number p_d is called the *critical probability*, the percolation with $p = p_d$ is called the *critical* percolation, and its study is particularly challenging. It is believed that $L_d(p_d) = 0$ for all $d \geq 2$, and this is considered to be one of the most challenging open problems in probability theory. In dimension $d = 2$, Kesten [706] proved in 1980 that $p_2 = 1/2$ and $L_2(p_2) = L_2(1/2) = 0$. In 1990, Hara and Slade [563] proved that $L_d(p_d) = 0$ for $d \geq 19$.

Intuition from physics suggest that not only $L_d(p_d) = 0$, but in fact $P_d(p_d, r) \approx r^{-1/\rho}$ for some $\rho = \rho(d) > 0$, which is called the *one-arm exponent* for critical bond percolation in $\mathbb{Z}^d$. In 2011, Kozma and Nachmias [745], refining the result of Hara and Slade, proved that for all $d \geq 19$ the one-arm exponent indeed exists and is equal to 1/2. We write $f_d(r) \approx g_d(r)$ if $C^{-1} f_d(r) \leq g_d(r) \leq C f_d(r)$ for some constant $C > 0$ which might depend on d but not on r.

Theorem 6.43 $P_d(p_d, r) \approx r^{-2}$ *for every* $d \geq 19$.

We next say a few words about the proof. Theorem 6.43 estimates, for the critical bond percolations on $\mathbb{Z}^d$ with $d \geq 19$, the probability that there exists a path from the origin 0 to some vertex $x \in \mathbb{Z}^d$ with $||x|| > r$. The authors argued that this may happen if either (i) the connected component $C(0)$ containing the origin is large, or (ii) $C(0)$ is small but contains a large number of points in $[r/3, 2r/3]$, or (iii) $C(0)$ is small and contains a small number of points in $[r/3, 2r/3]$. To estimate the probability of (i), they used the theorem of Hara and Slade [563] stating that $\mathbb{P}(|C(0)| > n) \approx 1/\sqrt{n}$, and in particular $\mathbb{P}(|C(0)| > cr^4) \approx r^{-2}$. To bound the probabilities in (ii) and (iii) they use the 2008 result of Hara [564] stating that the probability that 0 is connected to any fixed point x is $\approx |x|^{2-d}$. From, this, the authors managed to deduce the "recurrence" relation of the form

$$P_d(p_d, r) \leq \frac{c}{r^2} + (r/3)^2 P_d(p_d, r/3)^2 + \frac{1}{20} P_d(p_d, r/3),$$

from which the upper bound in Theorem 6.43 follows by an inductive argument.

Percolations on General Quasi-Transitive Graphs

Given an infinite graph $G = (V, E)$ and parameter $p \in [0, 1]$, a fundamental question in percolation theory is "what does the Bernoulli(p) (bond or site) percolation on G look like?". Most importantly, one is interested in the following aspects of this question.

(i) Infinite component structure: for which p does the percolation on G have an infinite component? If the infinite component exists, is it unique?

(ii) Finite component size: what is the size of a typical finite component?

To study question (i), the following definition is fundamental. For a given graph $G = (V, E)$, let $p_c(G)$ be the infimum of all $p \in [0, 1]$ such that the Bernoulli(p) (bond or site) percolation on G has an infinite connected component almost surely, and let $p_u(G)$ be the infimum of all p for which this infinite connected component is unique almost surely. It is clear that $p_c(G) \leq p_u(G)$ by definition, and part (i) essentially reduces to the question: for which graphs G is this inequality strict? For many families of graphs, the following "monotonicity in p" property has been proved. If $p_c(G) = p_u(G)$, then there are no infinite components for all $p < p_c(G)$, a unique one for all $p > p_c(G)$, and the only open question is what happens at the critical case $p = p_c(G)$, see Theorem 6.43 and the discussion before it. In contrast, if $0 < p_c(G) < p_u(G) < 1$, then there are three non-empty phases for percolation on G. In the first phase, $p \in (0, p_c]$, there are no infinite connected components; in the second phase, $p \in (p_c, p_u)$, there are infinitely many of them; and in the third phase, $p \in [p_u, 1)$, there is a unique infinite connected component.

In 1987, Aizenman et al. [18] proved that the bond percolation on $\mathbb{Z}^d$ has at most one infinite connected component for every $d \geq 1$, and in particular $p_c(\mathbb{Z}^d) = p_u(\mathbb{Z}^d)$. In 1992, Gandolfi et al. [491] generalized this result to all amenable quasi-transitive graphs, which we now define. An *automorphism* of a graph G is a bijection $\sigma : V \to V$ such that pair of vertices (u, v) is an edge if and only if $(\sigma(u), \sigma(v))$ is an edge. For vertices $u, v \in V$, let us write $u \sim v$ if there is an automorphism σ such that $\sigma(u) = v$. An (infinite) graph G is called *quasi-transitive* if its vertex set V can be partitioned into finitely many classes $V_1, \ldots, V_k$ such that $u \sim v$ if and only if vertices u and v belong to the same class V_i. If this is true with $k = 1$ (that is, $u \sim v$ for all $u, v \in V$), then the graph G is called *transitive*. A graph $G = (V, E)$ is called *amenable* if, for every $\epsilon > 0$, there is a finite set $A \subset V$, such that $|\partial A| \leq \epsilon |A|$, where ∂A is the set of vertices outside A connected by edge to at least one vertex in A.

In 1996, Benjamini and Schramm [118] initiated a systematic study of bond percolations on general quasi-transitive graphs. In particular, they conjectured that if G is a connected, locally finite[14], quasi-transitive graph, then $p_c(G) < p_u(G)$ if and only if G is nonamenable. As discussed above, if G is amenable then it has at most one infinite connected component, and $p_c(G) = p_u(G)$, hence only the "if" direction of this conjecture is open.

In 2001, Benjamini and Schramm [119] studied the case when the graph G is hyperbolic. An infinite transitive graph $G = (V, E)$ has one *end* if, after deletion of any finite set of vertices (and the corresponding edges), the remaining graph has exactly one infinite connected component. Transitive, nonamenable, planar graphs with one end are known as "planar hyperbolic graphs".

---

[14] A graph G is called *locally finite* if every vertex $v \in V$ has finite degree.

Theorem 6.44 *For any planar hyperbolic graph G we have $0 < p_c(G) < p_u(G) < 1$, for Bernoulli bond or site percolations on G.*

Theorem 6.44 can be extended to show that $p_c(G) < p_u(G)$ holds for all nonamenable planar quasi-transitive graphs. This verifies the "if" direction of the 1996 Benjamini–Schramm conjecture under the additional assumption of planarity. This assumption, however, is quite restrictive. In 2020, Hutchcroft [632] verified this conjecture under a different hypothesis of non-unimodularity. Let $S_u v$ be the set of vertices $w \in V$ such that there is an automorphism σ such that $\sigma(u) = u$ and $\sigma(v) = w$. A graph G is called *unimodular* if $|S_u v| = |S_v u|$ whenever $u \sim v$, and nonunimodular otherwise.

Theorem 6.45 *For Bernoulli bond percolation on any connected, locally finite, quasi-transitive, nonunimodular graph G, we have $p_c(G) < p_u(G)$.*

Interestingly, the case of unimodular graphs is usually considered to be *easier* to analyse than the nonunimodular case, because a powerful tool called the mass-transport principle works better in the unimodular case for obtaining proofs by contradiction. In particular, this is the way Theorem 6.44 was proved. Hutchcroft [632] proved Theorem 6.45 by a completely different, combinatorial argument, thus resolving the seemingly more difficult case of the Benjamini–Schramm conjecture first. This gives a hope that Theorem 6.45 will become a crucial step in the eventual full resolution of the conjecture.

We next discuss question (ii), about the sizes of finite components. Let $|K_v|$ be the size (number of edges) in the connected component containing vertex v in the bond percolation. Then we need to estimate, for $n > 0$, the probability $\mathbb{P}[n \le |K_v| < \infty]$ that the component is finite but has at least n edges. It is known that in every connected, locally finite graph G, and for every $p \in (0, 1)$, the quantity

$$\xi(p) := -\limsup_{n \to \infty} \frac{1}{n} \log \mathbb{P}[n \le |K_v| < \infty]$$

does not depend on the choice of the vertex v. It is clear that $\xi(p) \ge 0$, and an important question to investigate is whether it is 0 or strictly positive. Indeed, if $\xi(p) > 0$, then there exists a constant $c = c(p) > 0$ such that

$$\mathbb{P}[n \le |K_v| < \infty] \le e^{-cn}$$

for every $n \ge 1$. In other words, for every vertex v, the probability that v belongs to a large finite component is exponentially small, hence the typical components are small.

Let G be a connected, locally finite, transitive graph. Then it follows from works of Menshikov [881], Aizenman and Barsky [17] and Aizenman and Newman [19] in the 1980s that $\xi(p) > 0$ for every $0 \le p < p_c(G)$. It is also not difficult to prove (see Corollary 3.5 in [599]) that $\xi(p) = 0$ for every $p_c(G) \le p < 1$ if G is amenable. In 2019, Hermon and Hutchcroft posted online a preprint [599] in which they investigated the remaining non-critical case when G is nonamenable and $p_c(G) < p$.

Theorem 6.46 *Let $G = (V, E)$ be a connected, locally finite, nonamenable, quasi-transitive graph. Then $\xi(p) > 0$ for every $p_c(G) < p \leq 1$.*

Theorem 6.46 was proved by showing the finiteness of the *truncated exponential moment* $m(p) := \mathbb{E}_p[e^{t|K_v|} I_{|K_v| < \infty}]$, where I_A denotes the indicator function of the event A. The well-known *Russo's formula* was used to represent the derivative $\frac{d}{dp} m(p)$ as a difference of two non-negative terms, which have been analysed separately, and then $\frac{d}{dp} m(p)$ was bounded using martingale methods. From this, the authors deduced the finiteness of $m(p)$, and Theorem 6.46 follows.

In combination with the previous results, Theorem 6.46 finishes the analysis of positivity of $\xi(p)$ for all $p \neq p_c(G)$. It remains to investigate only the critical case $p = p_c(G)$, which is the most difficult to analyse.

Dynamical Percolations

The *triangular grid* is the (infinite) graph G whose vertex set $V \subset \mathbb{R}^2$ is given by

$$V = \{(k + m/2, \sqrt{3}m/2) \mid k, m \in \mathbb{Z}\},$$

and two vertices are connected by an edge if and only if the distance between them is 1. For the site percolation on the triangular grid G, it is known that the critical probability $p_c(G) = 1/2$, and that, with probability 1, the critical percolation has no infinite connected components.

This motivates the study of the following dynamic model of the critical site percolation on G. Assume that each vertex of G, independently of the other vertices, (i) is coloured black or white at time $t = 0$, with probability $1/2$ for each colour, (ii) generate a random variable T according to an exponential distribution with parameter $\lambda = 1/2$, wait for time T, and switch the colour, and (iii) repeat part (ii) again and again. This is called the *dynamical critical site percolation* on the triangular grid G. If, at some moment $t = t_0$, the graph G has an infinite connected subgraph consisting of only white vertices, we say that *percolation occurs*. Let $\mathcal{T}$ be the set of all times $t \in [0, 1]$ at which percolation occurs.

The fact that the (static) critical site percolation on G has no infinite connected components with probability 1 implies that at any fixed moment $t_0 > 0$, the probability that percolation occurs at time $t = t_0$ is equal to 0. However, because there are uncountably many "candidate" time moments t_0, this does not imply that $\mathcal{T}$ is the empty set. In 1997, Häggström et al. [553] proved that this set is empty almost surely if $G = \mathbb{Z}^d$ with $d \geq 19$. However, Schramm and Steif [1097] proved in 2010 that the situation is completely opposite for the planar triangular grid.

Theorem 6.47 *Almost surely, the set of times $\mathcal{T}$ such that dynamical critical site percolation on the triangular grid has an infinite connected component is nonempty.*

The proof of Theorem 6.47 is based on new quantitative "noise sensitivity" results. Intuitively, these results mean that the configurations of black and white points at time moments t_0 and $t_0 + \epsilon$ are almost independent even if $\epsilon > 0$ is small. This gives the system an "opportunity" to "try" a very large number of different configurations over short time intervals, and eventually "find" some configurations at which percolation occurs.

6.8   Compressed Sensing and Related Problems

Compressed Sensing
A central problem in statistics is to recover some useful information from the data one observes (observations). This problem becomes easier if the number of observations is large. However, in many applications, the number of observations is small, even far smaller than the number of variables or parameters we want to compute. In this section we list some important research developments which allows us to deal with such situations.

The central ingredient of this story is the methodology of *compressed sensing*, which allows us to recover a sparse signal from a surprisingly low number of measurements. A cornerstone of modern compressed sensing is the theorem proved by Donoho [359] in 2006, which we formulate below.

Let ψ_i, $i = 1, 2, \ldots, m$, be an orthonormal basis in $\mathbb{R}^m$, and let

$$X = \left\{ x \in \mathbb{R}^m \mid \left(\sum_i |(x, \psi_i)|^p \right)^{1/p} \leq R \right\}$$

(where $0 < p \leq 1$ and $R > 0$ are constants) be the set of vectors whose coefficients (x, ψ_i) in this basis are not too large. Let $I_n : X \rightarrow \mathbb{R}^n$ be of the form $I_n(x) = ((x, \xi_1), \ldots, (x, \xi_n))$ for some vectors $\xi_i \in \mathbb{R}^m$, $A_n : \mathbb{R}^n \rightarrow \mathbb{R}^m$ be an arbitrary operator, and let

$$E(n, m, p, R) = \inf_{A_n, I_n} \sup_{x \in X} ||x - A_n(I_n(x))||_2.$$

Let us call vectors $x \in X$ l_p-*sparse*. Such vectors arise in numerous applications, for example, x may represent an image (e.g. a medical image) with m pixels. The vectors $\xi_i \in \mathbb{R}^m$ represent measurements, $I_n(x)$ is the result of n measurements, $A_n : \mathbb{R}^n \rightarrow \mathbb{R}^m$ is the procedure of reconstructing x from the measurements, $\sup_{x \in X} ||x - A_n(I_n(x))||_2$ is the worst-case error of this procedure, and $E(n, m, p, R)$ is the minimal worst-case error we can achieve using a non-adaptive procedure, that is, the same for all x. Donoho [359] proved that we can achieve $O(N^{\frac{1}{2} - \frac{1}{p}})$ error, where $N = \frac{n}{\log(m_n)}$, using just $n = N \log m_n$ non-adaptive measurements. This methodology has dramatic implications in, for example, fast reconstruction of medical images. By the end of 2020, the paper [359] had over

28,000 citations in Google Scholar, which makes it one of the most cited (if not *the* most cited) mathematical papers of the twenty-first century.

Theorem 6.48 *For any $0 < p \leq 1$, $A > 0$ and $\gamma > 1$ there is a constant $C_p = C_p(A, \gamma)$ such that*

$$E(n, m_n, p, R) \leq C_p R \left(\frac{n}{\log(m_n)} \right)^{\frac{1}{2} - \frac{1}{p}},$$

whenever $m_n \sim An^\gamma$, $n \to \infty$.

Theorem 6.48 proves the existence of a recovery procedure with very good properties. Later in 2006, Candès et al. [245] provided a concrete efficient algorithm (called l_1-*regularization*) which does the job, even for inaccurate measurements.

We next formulate this result. A vector $x_0 \in \mathbb{R}^m$ is called *S-sparse* if it is supported on a set T_0 with $|T_0| \leq S$. For $x_0 \in \mathbb{R}^m$ let $y = Ax_0 + e$, where A is an $n \times m$ matrix ($n < m$) and $e \in \mathbb{R}^n$. Vectors $x_0 \in \mathbb{R}^m$ represent, for example, images or signals in various applications. The vector $y = Ax_0 + e$ is the result of n measurements (given by rows of A) with error e. Given y, we then try to recover x_0. This can be done by solving the following optimization problem, which is called the l_1-*regularization problem*:

$$\min_{x \in \mathbb{R}^m} ||x||_1 \quad \text{subject to} \quad ||Ax - y||_2 \leq \epsilon, \tag{6.8}$$

where $\epsilon \geq 0$ is a parameter, and $||x||_1 = \sum_{i=1}^{m} |x_i|$ is the l_1 norm. Let x^* be a solution to (6.8). Candès et al. [245] developed sufficient conditions which guarantee that x^* is close to x_0. In particular, if the measurements are exact ($e = 0$) then (6.8) with $\epsilon = 0$ gives $x^* = x_0$.

To state these sufficient conditions, we need some more notation. For $T \subset \{1, \ldots, m\}$, let A_T be an $n \times |T|$ submatrix of A formed from the columns of A corresponding to the indices in T. For $S > 0$, let δ_S be the smallest number such that the inequalities

$$(1 - \delta_S)||c||_2^2 \leq ||A_T c||_2^2 \leq (1 + \delta_S)||c||_2^2 \tag{6.9}$$

hold for all T with $|T| \leq S$ and all $c \in \mathbb{R}^{|T|}$.

Theorem 6.49 *Let A be an $n \times m$ matrix ($n < m$), and let S be such that $\delta_{3S} + 3\delta_{4S} < 2$. Then for any S-sparse $x_0 \in \mathbb{R}^m$ and any $e \in \mathbb{R}^n$ such that $||e||_2 \leq \epsilon$, we have $||x^* - x_0||_2 \leq C_S \cdot \epsilon$, where the constant C_S depends only on δ_{4S}. For example, $C_S \approx 8.82$ for $\delta_{4S} = \frac{1}{5}$ and $C_S \approx 10.47$ for $\delta_{4S} = \frac{1}{4}$.*

Crucially, the authors showed that the condition $\delta_{3S} + 3\delta_{4S} < 2$ in Theorem 6.49 holds for almost all matrices A with unit-normed columns if $n = O(S \log m)$, hence, with high probability, we can recover x_0 after just $O(|T_0| \log m)$ random measurements.

The error $C_S \cdot \epsilon = O(||e||_2)$ in Theorem 6.49 is essentially optimal, provided that the measurement error e is arbitrary (random or non-random). In statistics, one usually assumes that the error e is random. In 2007, Candès and Tao [242] proved that in this case one can recover x_0 with much better accuracy with high probability by using the Dantzig selector, a quantity which can be efficiently computed using linear programming.

For $x_0 \in \mathbb{R}^m$ let $y = Ax_0 + e$, where A is $n \times m$ matrix $(n < m)$ and $e \in \mathbb{R}^n$ is a random error (noise) whose components e_i are i.i.d. and follow the normal distribution $N(0, \sigma^2)$. The *Dantzig selector* x^* is a solution to the problem

$$\min_{x \in \mathbb{R}^m} ||x||_1 \quad \text{subject to} \quad ||A^T(y - Ax)||_\infty \leq \lambda_m \cdot \sigma,$$

where $\lambda_m > 0$ is a constant, A^T is the transpose of the matrix A, and $||z||_\infty = \max_{1 \leq i \leq m} |z_i|$.

Let A_T and δ_S be as defined in (6.9). Also, for $S, S' > 0$, such that $S + S' \leq m$, let $\theta_{S,S'}$ be the smallest number such that the inequality

$$|(A_T c, A_{T'} c')| \leq \theta_{S,S'} ||c||_2 ||c'||_2$$

holds for all disjoint sets $T, T' \subset \{1, \ldots, m\}$ with $|T| \leq S$ and $|T'| \leq S'$ and all $c \in \mathbb{R}^{|T|}, c' \in \mathbb{R}^{|T'|}$.

Theorem 6.50 *Let A be an $n \times m$ matrix $(n < m)$, S be such that $\delta_{2S} + \theta_{S,2S} < 1$, $x_0 \in \mathbb{R}^m$ be S-sparse, and $x^* \in \mathbb{R}^m$ be the Dantzig selector with $\lambda_m = \sqrt{2(1 + a) \log m}$ for some $a > 0$. Then the inequality*

$$||x^* - x_0||_2 \leq C\lambda_m \sqrt{S}\sigma,$$

where $C = 4/(1 - \delta_{2S} - \theta_{S,2S})$, holds with probability exceeding $1 - (\sqrt{\pi \log m} \cdot m^a)^{-1}$.

The error in Theorem 6.50 is optimal up to the constant and $\sqrt{\log m}$ factors.

The proofs of Theorems 6.48–6.50 are not very difficult. Once results like this are formulated, they are not very difficult to prove. In applied mathematics and statistics, the main challenge is often to *formulate* the methods which work well. After this, the formal proof of the mathematical properties of such methods are sometimes straightforward, and even if not, the methods can be successfully used in practice even without such proofs.

Recovery from Magnitude Measurements

In Theorems 6.48–6.50, measurements of a vector x are modelled as (x, z_i) for some vectors z_i. In some applications, we measure a complex vector $x \in \mathbb{C}^n$, and only (the squares of) the absolute values $|(x, z_i)|$ of complex numbers (x, z_i) can be measured. Such measurements are called the *magnitude measurements* of x. If the vectors z_i are fixed and given, the problem of recovering x from such measurements can be computationally infeasible. In contrast, Candès et al. [246] proved in 2013 that we can select z_i at random, and recover x efficiently with high probability after a number of measurements not much higher than the dimension.

Specifically, let $x \in \mathbb{C}^n$, and let $z_1, \ldots z_m$ be vectors independently and uniformly sampled from the unit sphere in $\mathbb{C}^n$. We call the numbers $b_i = |(x, z_i)|^2$, $i = 1, 2, \ldots, m$ the *magnitude measurements* of x.

Theorem 6.51 *There exist positive constants c_0 and γ such that an arbitrary $x \in \mathbb{C}^n$ can be exactly recovered from $m \geq c_0 n \log n$ magnitude measurements in polynomial time with probability at least $1 - 3e^{-\gamma m/n}$.*

Theorem 6.51 was proved by providing an explicit and efficient procedure, a so-called "trace-norm minimization problem" that reduces to a semidefinite program, which provides the recovery from magnitude measurements as described in the theorem.

Low-Rank Matrix Recovery

A related problem, which is also central in many applications, is the recovery of missing entries in a low-rank matrix. For example, if the rows of the matrix represent people who use some platform to watch films online, the columns are films, and the matrix entries represent the scores users gave to films, then most of the entries will be missing because most users haven't seen most films. We may assume that the matrix is (close to being) low-rank, because the scores may essentially depend on a limited number of features. In this example, it would be very useful to "recover the missing entries", because this would give predictions of which scores users would give to which films, and these predictions are crucial to develop individual *recommendations* of which films are worth watching.

In 2009, Candès and Recht [244] proved that if we have only a small portion of random entries of a random matrix of rank r, we can recover all other entries with high probability, by solving a simple convex optimization problem, which can be done fast using existing software.

What exactly is meant by a "random matrix of rank r" needs some clarification. For positive integers n_1, n_2, and $0 < r \leq \min\{n_1, n_2\}$, let $u_1, \ldots u_r$ be a family of orthonormal vectors in $\mathbb{R}^{n_1}$ sampled uniformly at random among all such families. Let $v_1, \ldots v_r$ be a similar random family of orthonormal vectors in $\mathbb{R}^{n_2}$. Then

$$M = \sum_{i=1}^{r} \sigma_i u_i v_i^T,$$

where σ_i are some (non-random) positive numbers, is an $n_1 \times n_2$ matrix of rank r, which we call a random matrix from the *random orthogonal model*. Also, recall that a *singular value* of a matrix X is an eigenvalue of XX^T, and denote by $||X||_*$ the sum of all singular values of X.

Theorem 6.52 *Let M be an $n_1 \times n_2$ matrix of rank r sampled from the random orthogonal model, and put $n = max(n_1, n_2)$. Suppose we observe m entries of M with locations Ω sampled uniformly at random. Then there are constants C and c such that if $m \geq Crn^{5/4} \log n$, the minimizer of the problem*

$$\min \|X\|_* \quad subject\ to \quad X_{ij} = M_{ij}, \ (i, j) \in \Omega$$

is unique and equal to M with probability at least $1 - cn^{-3} \log n$.

Principal Component Analysis

In many applications, real life data have the form of a matrix $M = L_0 + S_0$, where L_0 is a low-rank matrix and S_0 is some perturbation. A fundamental problem, known as "principal component analysis" is to recover L_0 from M. Most known methods work if S_0 is small, but this assumption often fails in applications. In 2011, Candès et al. [243] proved that L_0 can be efficiently recovered if S_0 does not have too many non-zero entries, which, however, can be arbitrary large. The theorem is formulated here for square matrices L_0, S_0 but can be extended to rectangular matrices as well.

More formally, let $\mu > 0$ be a fixed real number. Let $e_i \in \mathbb{R}^n$ be a vector with i-th component 1 and others 0. Let L_0 be a real $n \times n$ matrix of rank r whose singular value decomposition[15] $L_0 = UDV^T$ satisfies (i) $\max_i \|U^T e_i\| \leq \frac{\mu r}{n}$, (ii) $\max_i \|V^T e_i\| \leq \frac{\mu r}{n}$, and (iii) $\max_{i,j} |w_{ij}| \leq \frac{\sqrt{\mu r}}{n}$, where w_{ij} are entries of the matrix UV^T. Conditions (i)–(iii) are needed to ensure that L_0 is not sparse. The number of non-zero entries of D is equal to r. The sum of all entries of D is denoted $\|L_0\|_*$ and is called the *nuclear norm* of L_0.

Next, let Σ be a fixed $n \times n$ matrix with ± 1 entries Σ_{ij}. Choose an integer $0 < m < n^2$. Let S_0 be an $n \times n$ matrix with entries $[S_0]_{ij}$ whose support set Ω is uniformly distributed among all sets of cardinality m, and such that the sign of $[S_0]_{ij}$ is Σ_{ij} for all $(i, j) \in \Omega$. Let $M = L_0 + S_0$.

Let $\hat{L}$, $\hat{S}$ be the solution to the optimization problem

$$\min_{L,S} \left(\|L\|_* + \frac{1}{\sqrt{n}} \|S\|_1 \right) \quad subject\ to \quad L + S = M.$$

Theorem 6.53 *Suppose that L_0 and S_0 are as above. Then, there are positive constants c, c_r, c_s such that with probability at least $1 - cn^{-10}$ (over the choice of support of S_0), we have $\hat{L} = L_0$ and $\hat{S} = S_0$ provided that $\operatorname{rank}(L_0) \leq c_r n \mu^{-1} (\log n)^{-2}$ and $m \leq c_s n^2$.*

---

[15] A matrix U is called *orthogonal* if $U^T U = UU^T = I$, where I is the identity matrix. The *singular value decomposition* of a square matrix L is the representation $L = UDV^T$ where D is a diagonal matrix and U, V are orthonormal matrices.

6.9   Other Theorems Related to Probability

The "majority is stablest" Theorem

Imagine that there is an election with 2 candidates and n voters. The votes for the candidates may be encoded as -1 and 1, hence the result of the election is a vector in $\{-1, 1\}^n$. The "voting rule" is a way to determine the winner of the election for each possible result, which is a function $f : \{-1, 1\}^n \to \{-1, 1\}$. Such functions are called *boolean functions*. For example, the most popular rule "who gets more votes wins" corresponds to the *majority function*

$$\mathrm{Maj}_n(x) = \mathrm{sign}\left(\sum_{i=1}^n x_i\right), \quad x = (x_1, x_2, \ldots, x_n) \in \{-1, 1\}^n,$$

where, for simplicity, we may assume that n is odd to guarantee that $\sum_{i=1}^n x_i \neq 0$. Can we find a better voting rule? To answer this, we first need to define criteria for what is counted as "better". First of all, the rule should be fair, which can be formalized as $E[f] = 0$, where

$$E[f] = \frac{1}{2^n} \sum_{x \in \{-1,1\}^n} f(x)$$

is the average value of f. Another obvious criterion is that any single voter should have a small influence on the election result. One more desirable criterion is stability, which informally means that miscounting a small number of votes has only small chance of changing the result. In 2010, Mossel et al. [914] proved that the majority rule is the best one with respect to these criteria.

We next formulate this result rigorously. For $i \in [n] := \{1, 2, \ldots, n\}$, let $N_i(f)$ be the number of x, such that $f(x) \neq f(x_1, \ldots, x_{i-1}, -x_i, x_{i+1}, \ldots, x_n)$. The ratio $\mathrm{Inf}_i(f) = N_i(f)/2^n$ is called the *influence* of i. Equivalently,

$$\mathrm{Inf}_i(f) = \sum_{S \subset [n], i \in S} \hat{f}(S)^2,$$

where

$$\hat{f}(S) = \frac{1}{2^n} \sum_{x \in \{-1,1\}^n} \left[f(x) \prod_{i \in S} x_i \right].$$

For $0 \leq \rho \leq 1$, the *noise stability* of f is

$$S_\rho(f) = \sum_{S \subset [n]} \rho^{|S|} \hat{f}(S)^2.$$

For example, for the majority function, we have

$$\lim_{n \to \infty} S_\rho(\text{Maj}_n) = \frac{2}{\pi} \arcsin(\rho).$$

In 2007, Khot et al. [712] conjectured that the majority function has (in the limit $n \to \infty$ and up to ϵ) the highest noise stability among all functions with $E[f] = 0$ and low influence. This statement was known as the "Majority Is Stablest" conjecture, and has applications to the analysis of voting systems and to algorithmic complexity for some combinatorial problems. In 2010, Mossel et al. [914] proved this conjecture.

Theorem 6.54 *For any* $0 \le \rho \le 1$ *and* $\epsilon > 0$ *there exists a* $\delta > 0$ *such that if* $f : \{-1, 1\}^n \to \{-1, 1\}$ *satisfies* $E[f] = 0$ *and* $\text{Inf}_i(f) \le \delta$ *for all* i, *then*

$$S_\rho(f) \le \frac{2}{\pi} \arcsin(\rho) + \epsilon.$$

Theorem 6.54 is a corollary of a powerful "invariance principle for multilinear polynomials with low influences" developed by the authors, which also has other applications, see [914] for details.

The OSSS Inequality

Let $[n] = \{1, 2, \ldots, n\}$, and let $X = \{0, 1\}^n$ be the set of vectors $w = (w_1, \ldots, w_n)$ with each w_i being either 0 or 1. Assume that we select each $w \in X$ with some probability $\mathbb{P}(w) \ge 0$, such that $\sum_{w \in X} \mathbb{P}(w) = 1$. For $A \subseteq X$, define $\mathbb{P}[A] = \sum_{w \in A} \mathbb{P}(w)$. For any function $f : X \to \mathbb{R}$, define the *expectation* $E_\mathbb{P}(f) = \sum_{w \in X} \mathbb{P}(w) f(w)$ and *variance* $\text{Var}_\mathbb{P}(f) = E_\mathbb{P}(f^2) - E_\mathbb{P}(f)^2$. For $f, g : X \to \mathbb{R}$, let $\text{Cov}_\mathbb{P}(f, g) = E_\mathbb{P}(fg) - E_\mathbb{P}(f) E_\mathbb{P}(g)$.

A *decision tree* T is an algorithm that takes $w \in X$ as an input, and then tries to determine $f(w)$ by querying the values of w_i, $i \in [n]$, one bit after the other. The bit it queries at step k may depend on the values of the bits obtained at steps $1, \ldots, k - 1$. The algorithm stops when it can determine $f(\omega)$ from the bits it has. Let $\delta_i(f, T)$ be the probability that T, starting from a random input ω, queries bit w_i before it stops.

In 2005, O'Donnell et al. [945] proved that if each bit w_i of $w \in X$ is selected independently, then the inequality

$$\text{Var}_\mathbb{P}(f) \le \sum_{i=1}^{n} \delta_i(f, T) \text{Cov}_\mathbb{P}(f, w_i) \tag{6.10}$$

holds for any boolean function f and any decision tree T calculating f. This result became known as the *OSSS inequality*. In 2019, Duminil-Copin et al. [371] proved a version of this inequality in the case when bits of w may depend on each other. Instead, the authors imposed the condition that the function f is increasing and the probability $\mathbb{P}$ is monotonic. A function $f : X \to \mathbb{R}$ is called *increasing* if

$f(x) \geq f(y)$ whenever $x \geq y$ coordinate-wise. A probability $\mathbb{P}$ is called *monotonic* if for any $i \in [n]$, any $F \subset [n]$, and any $x, y \in \{0, 1\}^F$ satisfying $x \leq y$, $\mathbb{P}[w_j = x_j, \forall j \in F] > 0$ and $\mathbb{P}[w_j = y_j, \forall j \in F] > 0$, we have

$$\mathbb{P}[w_i = 1 | w_j = x_j, \forall j \in F] \leq \mathbb{P}[w_i = 1 | w_j = y_j, \forall j \in F].$$

Theorem 6.55 *Inequality* (6.10) *holds for every increasing function* $f : X \to [0, 1]$, *any monotonic probability* $\mathbb{P}$, *and any decision tree* T.

The authors demonstrated applications of Theorem 6.55 to mathematical physics, specifically to the analysis of so-called lattice spin models and their random-cluster representations.

Chapter 7
Algorithms and Complexity

7.1 Exact Algorithms

Primality Testing

A fundamental algorithmic task in number theory is the *primality testing problem*: given an integer $k > 1$, determine whether it is prime or composite. A naive method would be to try all possible divisors from 2 to $\sqrt{k}$, but the number of operations required for this method grows exponentially in the number of digits of k. Is there a faster method?

The primality testing problem is an example of *decision problem*, which is a yes-or-no question on an infinite set of possible (finite) inputs. We say that a decision problem belongs to class P if it has a polynomial-time algorithm, that is, an algorithm which solves the problem, always gives the correct answer, and, for any input S of length n, performs at most Cn^m operations, where C and m are some constants. For the primality testing problem, this means determining whether an integer k is prime or composite in $O((\log k)^m)$ operations.

In 1976, Miller [891] developed a polynomial-time algorithm for primality testing assuming the *Extended Riemann Hypothesis*. In 1980, Rabin [1013] modified Miller's algorithms and obtained an unconditional but randomized algorithm: it works in polynomial time but may output an incorrect answer with some negligibly small probability. The Miller–Rabin test is extensively used in practice, but leaves the question whether the problem is in class P open.

In 1983, Adleman et al. [11] presented an algorithm for primality testing which does not involve any randomness (such algorithms are called *deterministic*) and runs in time $(\log k)^{O(\log\log\log k)}$. Finally, Agrawal et al. [12] presented a deterministic algorithm which runs in polynomial time, thus proving that the problem belongs to the class P.

Theorem 7.1 *There exists a deterministic polynomial-time algorithm for the primality testing problem.*

© The Author(s), under exclusive license to Springer Nature Switzerland AG 2021
B. Grechuk, *Landscape of 21st Century Mathematics*,
https://doi.org/10.1007/978-3-030-80627-9_7

The proof is based on the easy-to-check fact that if k is prime then, for all positive integers a and r

$$\exists f, g \in \mathbb{Z}(x): \quad (x+a)^k - (x^k + a) = (x^r - 1)g(x) + kf(x), \tag{7.1}$$

where $\mathbb{Z}(x)$ is the set of polynomials with integer coefficients. Conversely, the authors showed how to deterministically select $r = r(k)$ and a set $\mathcal{A} = \mathcal{A}(k)$ of integers such that if (7.1) holds for this r and all $a \in \mathcal{A}$, then k must be either a prime or a prime power. Moreover, the selected $r(k)$ is bounded by a polynomial in $\log k$, which implies that condition (7.1) can be checked in polynomial time. The size of the set $\mathcal{A}$ is also polynomial in $\log k$. Finally, it is known that testing whether an integer is a perfect power can be done in polynomial time, and Theorem 7.1 follows.

The Graph Isomorphism Problem

One of the fundamental algorithmic problems in graph theory is the *graph isomorphism problem*, which asks to determine whether two given finite graphs are isomorphic. The isomorphism problems for many other objects, such as groups, can be reduced to it. It has been intensively studied for decades, and efficient algorithms for many families of graphs are known. However, despite substantial efforts, the $2^{O(\sqrt{n \log n})}$ time algorithm discovered by Babai and Luks [72] in 1983 remained the fastest one working for general n-vertex graphs. In 2015, Babai posted online a preprint [69] in which he developed a much faster algorithm.

Theorem 7.2 *The graph isomorphism problem for n-vertex graphs can be solved in time $O(2^{(\log n)^c})$ for some fixed $c > 0$.*

Algorithms with running time $O(2^{(\log n)^c})$ are called *quasipolynomial time algorithms*.

The starting point of the Babai algorithm is similar to many previous graph isomorphism algorithms. Take $k = (\log n)^c$ vertices on both graphs and colour them with k different colours. The idea is to look for an isomorphism which matches all colours. We next colour the remaining vertices based on how many neighbours of each colours they have, and proceed recursively. If, as a result, there are few vertices in each colour class, we can just try all possible colour-preserving candidate isomorphisms. If we have found the isomorphism, we are done, otherwise try a different choice of the initial k points, and so on. Note that there are at most $n^k = n^{(\log n)^c}$ initial guesses to check. The problem is that, for some highly symmetric graphs, we may end up with many vertices coloured the same way. However, Babai proved that a modification of this method works for all graphs except for the specific family called *Johnson graphs*. Because the isomorphism for Johnson graphs can be efficiently tested using different methods, this proves Theorem 7.2.

Theorem 7.2 motivates a large body of further research. We will mention one example. We say that a graph H is a *topological subgraph* of a graph G if H can be obtained from G by deleting vertices and edges as well as dissolving degree two vertices (that is, deleting a vertex of degree two and connecting its two neighbours). In 2015, Grohe and Marx [533] proved that the graph isomorphism problem can be solved in polynomial time for graphs which exclude some fixed h-vertex graph as a topological subgraph. This is a very general class of graphs, which includes, for example, graphs of bounded degree, and graphs excluding any fixed graph as a minor[1]. The running time of the Grohe–Marx algorithm is $O(n^{f(h)})$ for some function f. In 2020, Neuen posted online a preprint [930] in which he presented a common generalization of this result and Theorem 7.2: an isomorphism test that runs in time $O(n^{(\log h)^c})$ on all n-vertex graphs excluding some h-vertex graph as a topological subgraph.

P, NP, and the Dichotomy Conjecture

NP is the set of decision problems for which the problem instances where the answer is "yes" have proofs verifiable in polynomial time[2]. For example, the graph isomorphism problem is clearly in NP, because the proof can be an isomorphism itself. A central conjecture in theoretical computer science, and one of the central conjectures in the whole of mathematics, is that P$\neq$NP, that is, not all problems in NP can be solved in polynomial time. This is one of the problems for which a million dollars is offered for a solution [252].

We say that problem A is *(Turing) reducible* to a problem B if there is an algorithm that solves problem A using a polynomial number of calls to a subroutine for problem B, and in polynomial time outside of those subroutine calls. We say that problem B is NP-*hard* if every problem A in the class NP is reducible to B. It follows from the definition that if P$\neq$NP then there can be no polynomial time algorithm for solving B. Because P$\neq$NP is a widely believed conjecture, the proof that a problem is NP-hard is regarded as a convincing evidence of its intractability. If an NP-hard problems belongs to the class NP, it is called NP-*complete*.

The first example of an NP-complete problem was presented by Cook [301] in 1971. After this, to prove that a new problem B is NP-hard, it suffices to take any known NP-hard problem A and reduce it to B. Many thousands of problems have been proved to be NP-hard in this way. By now, for a majority of natural computational problems, we either have a polynomial algorithm or a proof of NP-hardness. An ideal scenario would be to have such a classification for all problems. However, Ladner [755] proved in 1975 that if P$\neq$NP, then there exist problems in NP which are neither in P nor NP-complete. Such problems are called NP-*intermediate*. The graph isomorphism problem is one of the most famous examples of a natural problem which is believed to be NP-intermediate.

---

[1] See the discussion before Theorem 2.51 for the definition of minor.

[2] Here is a more formal definition. Let $\Sigma = \{0, 1\}$, Σ^n be the set of words $\sigma_1 \ldots \sigma_n$ with each σ_i being 0 or 1, and let $\Sigma^* = \bigcup_{n=0}^{\infty} \Sigma^n$. Let $|x|$ denote the length of any word $x \in \Sigma^*$. Let P_2 be the set of functions $F(z, x) : \Sigma^* \times \Sigma^* \to \Sigma$, for which there is a polynomial P and a program on a computer which computes F in at most $P(|z| + |x|)$ operations. Let NP be the set of functions $f : \Sigma^* \to \Sigma$ for which there exists a polynomial P, and a function $F \in P_2$, such that $f(x) = 1$ if and only if $\exists z \in \Sigma^{P(|x|)}$ such that $F(z, x) = 1$.

However, there is a large class of problems in which the full classification "either in P or NP-complete" turned out to be possible. To introduce this class, we need a few more definitions. Let Σ be a finite set. Let $X = \{x_1, \ldots x_n\}$ be a set of n variables taking values in Σ. An *assignment* $a = (a_1, \ldots, a_n) \in \Sigma^n$ means that we assign to each variable x_i the value a_i. A *relation* R is a subset of Σ^k, where k is a positive integer. A *constraint* C (corresponding to R) is a k-element subset $x_{i_1}, \ldots x_{i_k}$ of X. We say that an assignment $a \in \Sigma^n$ satisfies the constraint C if $(a_{i_1}, \ldots a_{i_k}) \in R$. Let Γ be a finite set of relations. A *constraint satisfaction problem* CSP(Γ) is: given Σ, X, a finite set $R_1, \ldots, R_m$ of relations such that each $R_i \in \Gamma$, and a finite set $C_1, C_2, \ldots, C_t$ of constraints (each constraint C_j corresponds to some relation R_i), is there an assignment $a \in \Sigma^n$ that satisfies all the constraints? A large variety of computational problems can be expressed as a CSP(Γ) for some Γ. In 1993, Feder and Vardi [440] conjectured that every such problem is either in P or NP-complete. This conjecture became known as the *Dichotomy conjecture*. In 2017, Bulatov [228] posted online a preprint with the proof of this conjecture. An independent proof has been given by Zhuk [1292].

Theorem 7.3 *For every fixed Γ, the problem* CSP(Γ) *is either solvable in polynomial time or is NP-complete.*

Counting Problems

So far we have considered only decision problems, which are problems with a "Yes" or "No" answer. For example, the graph isomorphism problem (see Theorem 7.2) asks whether two given graphs G and H are isomorphic. More generally, we may ask to count *how many* isomorphisms between G and H there are.[3] As another example, there is a decision problem to determine whether there exists a perfect matching in a given graph G, and the corresponding counting problem asks how many such matchings are there.

Note that the graph isomorphism problem is in the class NP because any isomorphism f is a "certificate" that G and H are isomorphic, and the correctness of this certificate can be verified in polynomial time. The counting version asks to count such certificates. Similarly, any perfect matching in a graph G is a certificate that the graph has a perfect matching. More generally, for any problem in class NP, we may formulate a counting problem asking to count the corresponding certificates. The class of all such counting problems is denoted #P. The class of counting problems solvable in polynomial time is called FP, and, similarly to P$\neq$NP, there is a widely believed conjecture that #P$\nsubseteq$FP. We say that a counting problem is #P-*hard* if any problem in #P reduces to it in polynomial time, and, similar to NP-hardness for decision problems, #P-hardness is convincing evidence that a counting problem is intractable.

---

[3] More formally, the problem asks how many bijections f there are between vertex sets of G and H such that vertices u, v in G are connected by an edge if and only if $f(u)$ and $f(v)$ are connected by an edge in H.

If a counting version of a problem is in FP, the corresponding desicion problem is clearly in P, because we can just count certificates and check whether the result is 0. However, the converse direction does not hold. For example, there is a well-known polynomial-time algorithm for checking whether a graph G has a perfect matching. On the other hand, Valiant [1223] proved in 1979 that the problem of counting perfect matchings in G is #P-hard.

Theorem 7.3 states that every constraint satisfaction problem CSP(Γ) is either solvable in polynomial time or is NP-complete. Every such problem has a natural counting version, denoted #CSP(Γ), in which the goal is to *count* the number of assignments satisfying all the constraints. A large variety of counting problems from all areas of mathematics, including the problem of counting perfect matchings mentioned above, are special cases of #CSP(Γ) for some Γ. By analogy with Theorem 7.3, one may ask whether every #CSP(Γ) is either in FP or #P-complete. In 2013, Bulatov [227] confirmed that this is indeed the case.

Theorem 7.4 *For every fixed Γ, the problem #CSP(Γ) is either solvable in polynomial time or is #P-complete.*

While Theorem 7.4 is a great achievement, its proof provides no explicit algorithm which, for a given set of relations Γ, would decide whether the problem #CSP(Γ) is tractable. Also in 2013, Dyer and Richerby [376] provided such an algorithm.

Theorem 7.5 *There is an algorithm which, given set of relations Γ as an input, decides whether the counting problem #CSP(Γ) is in FP or is #P-complete.*

Since 2013, Theorems 7.4 and 7.5 have been extended and generalized in various ways. In particular, Cai and Chen [237] established in 2017 a version of Theorem 7.4 for CSPs with (possibly complex) weights. See [237] for the exact formulation of this result and further details.

The Complexity of Multiplication
Integer multiplication is one of the basic operations in mathematics. The usual school method requires $O(n^2)$ operations to compute the product of two n-digit integers. In 1963, Karatsuba [681] presented a method which can do such multiplication in $O(n^{\log_2 3}) \approx O(n^{1.585})$ elementary operations. This was improved to $O(n^{1+\epsilon})$ for any $\epsilon > 0$ by Toom [1207] in 1963, and to $O(n \log n \log \log n)$ by Schönhage and Strassen [1096] in 1971.

Schönhage and Strassen further conjectured the existence of an algorithm which can do the same task in $O(n \log n)$ operations. However, the running time of their algorithm remained unimproved for over 35 years. In 2009, Fürer [481] presented a faster method, which almost achieves the conjectural $O(n \log n)$ running time.

The *iterated logarithm* $\log^*$ is the unique function which has the properties (i) $\log^* n = 0$ if $n \leq 1$ and (ii) $\log^* n = 1 + \log^*(\log n)$ if $n > 1$.

Theorem 7.6 *Two n-digit integers can be multiplied in time $n \log n \, 2^{O(\log^*(n))}$.*

Theorem 7.6 opened the road for further progress. The bound in the theorem can be written as $O(n \log n\, K^{\log^*(n)})$ for some constant K. In a series of papers, algorithms with better and better values of K have been developed, with $K = 4$ achieved by Harvey and van der Hoeven [570] in 2019. Later in 2019, the same authors posted online a preprint [571] in which they described the long-conjectured $O(n \log n)$ algorithm.

Theorem 7.7 *Two n-digit integers can be multiplied in time $O(n \log n)$.*

To prove Theorem 7.7, the authors first presented a relatively easy multiplication algorithm which works in $O(n \log n)$ time subject to a plausible but unproved hypothesis about the least prime in an arithmetic progression.[4] The authors then demonstrated how to use similar ideas to develop a second algorithm, which is much more technical, but runs in time $O(n \log n)$ unconditionally. The algorithm uses a new "Gaussian resampling" technique to reduce the multiplication problem to the evaluation of several multidimensional discrete Fourier transforms over the complex numbers, for which there are known fast algorithms. This allows the authors to prove a recurrence inequality in the form

$$M(n) < \frac{Kn}{n'} M(n') + O(n \log n), \qquad n' = n^{\frac{1}{d}+o(1)},$$

where $M(n)$ is the complexity of n-bit integer multiplication, $d > 0$ is an integer parameter of the algorithm we are allowed to choose, and $K > 0$ is an absolute constant independent from d. If we choose any $d > K$, this inequality implies that $M(n) = O(n \log n)$.

Because integer multiplication has been used in a huge number of other algorithms, Theorem 7.7 immediately implies faster algorithms for many problems, including, for example, finding the greatest common divisor of two n-bit integers, conversion of an n-digit integer from one base to another (e.g. from decimal to binary), faster approximation of transcendental constants like π and functions like e^x, and so on.

A related but more difficult problem is to multiply $n \times n$ matrices. Matrix multiplication is a basic operation which lies at the core of many algorithms, so doing it faster makes a big impact. A trivial algorithm for multiplying $n \times n$ matrices takes $O(n^3)$ operations. In 1968, Strassen [1165] described a method for multiplying two 2×2 matrices using only 7 entry multiplications. The recursive application of this method implies that two $n \times n$ can be multiplied in time $O(n^{\log_2 7}) \approx O(n^{2.81})$. Then many authors developed faster and faster methods, but, after the 1990 algorithm of Coppersmith and Winograd [303], requiring $O(n^{2.375477\cdots})$ operations, progress

---

[4] For relatively prime positive integers a and r, let $P(a, r)$ be the smallest prime p such that $p \equiv a \pmod{r}$, and let $P(r) = \max_a P(a, r)$. In 1944, Linnik [801] proved that $P(r) = O(r^L)$ for some constant L. In 2011, Xylouris [1279] proved that we can take $L = 5.18$. It is widely believed that the same is true for any $L > 1$, and this would imply the $O(n \log n)$ bound for the first algorithm in [571].

stopped for over 20 years. Finally, the Coppersmith–Winograd algorithm was improved by Davie and Stothers [326] in 2013.

Theorem 7.8 *Two $n \times n$ matrices can be multiplied in time $O(n^{2.37368\cdots})$.*

Theorem 7.8 opened the road for a further chain of successive improvements. We mention a 2020 preprint of Alman and Williams [30] achieving $O(n^{2.3728596\cdots})$. However, all the recent improvements of the exponent are relatively small. A big open question is whether the bound can be improved to $O(n^{2+\epsilon})$ for any $\epsilon > 0$.

The Frobenius Coin Problem and Beyond

Given positive coprime integers $a_1, a_2, \ldots, a_n$, denote by $S(a_1, a_2, \ldots, a_n)$ the (always finite) set of positive integers which cannot be represented as $\sum_{i=1}^{n} m_i a_i$ for some non-negative integers $m_1, m_2, \ldots m_n$. The famous *Frobenius coin problem* asks to compute the largest integer in $S(a_1, a_2, \ldots, a_n)$. The problem is known to be NP-hard if n is part of the input. However, Kannan [679] presented in 1992 a polynomial algorithm to solve it for every fixed n. In 2003, Barvinok and Woods [103] proved that, for every fixed n, the number of elements in $S(a_1, a_2, \ldots, a_n)$ can also be computed efficiently.

Theorem 7.9 *There exists an algorithm, which, given $a_1, a_2, \ldots, a_n$, computes the number of integers in $S(a_1, a_2, \ldots, a_n)$ in polynomial time for any fixed n.*

The proof of Theorem 7.9 is based on the powerful method of generating functions. To every set S of positive integers we can associate the generating function

$$f(S, x) = \sum_{m \in S} x^m.$$

If $S = S(a_1, a_2, \ldots, a_n)$ is as in Theorem 7.9, Barvinok and Woods developed an efficient method to compute a short and convenient formula for $f(S, x)$. Then the generating function of the complement $\bar{S}$ of S is

$$f(\bar{S}, x) = \frac{1}{1 - x} - f(S, x),$$

and the number of elements in $\bar{S}$ is just $\lim_{x \to 1} f(\bar{S}, x)$.

The Algorithmic Lovász Local Lemma

If a finite number of events in a probability space are mutually independent, and each event has probability less than 1, then there is a positive probability that none of these events happen. In 1975, Erdős and Lovász [413] proved a version of this result which allows for a limited dependence among the events. This method is known as the *Lovász local lemma*, and allows us to non-constructively prove the existence of certain combinatorial objects meeting a prescribed collection of criteria, even if these objects occur with exponentially small probability. An important problem in the theory of algorithms was to develop an efficient procedure to actually *find* objects whose existence can be proved in this way. In 1991, Beck [109] demonstrated how to do this under some restrictive additional conditions. Then many researchers proved versions of Beck's result with less and less restrictive conditions. This line of research culminated in the 2010 theorem of Moser and Tardos [913], who developed the algorithmic version of the Lovász local lemma in general.

To formulate the result, we need some notation. Let $\mathcal{P}$ be a finite collection of mutually independent random variables in a fixed probability space Ω. Let S be a finite family of events in Ω determined by $\mathcal{P}$. For $A \in S$, let $v(A)$ be the (unique) minimal subset of $\mathcal{P}$ that determines A. For $A \in S$, let $\Gamma_S(A)$ be the set of $B \in S$ such that $A \neq B$ but $v(A) \cap v(B) \neq \emptyset$. We say that an evaluation of the variables in $\mathcal{P}$ *violates* $A \in S$ if it makes A happen.

Theorem 7.10 *If there exists an assignment of reals $x : S \to (0, 1)$ such that*

$$\forall A \in S : \mathbb{P}[A] \leq x(A) \prod_{B \in \Gamma_S(A)} (1 - x(B)),$$

then there exists an assignment of values to the variables in $\mathcal{P}$ not violating any of the events in S. Moreover, this assignment can be found by a randomized algorithm with expected number of resampling steps at most $\sum_{A \in S} \frac{x(A)}{1 - x(A)}$.

Theorem 7.10 without the "moreover" part is (a slightly modified version of) the original Lovász local lemma. The "moreover" part provides the long-desired constructive version of this lemma. In 2013, Chandrasekaran et al. [261] obtained a version of Theorem 7.10 with a deterministic algorithm in the "moreover" part at the cost of assuming slightly more restrictive bounds on $\mathbb{P}[A]$.

Instance Optimal Algorithms

Traditionally, the efficiency of algorithms are evaluated based on the worst-case running time. However, given algorithm A for some problem, it is usually very difficult to prove that its worst-case running time is optimal. And even if it is, there still may exist a different algorithm B, with worse worst-case running time, but performing much faster than A in most instances.

In 2003, Fagin et al. [437] introduced the notion of *instance optimality*, the strongest possible notion of algorithm optimality one can imagine, and presented an example of an instance optimal algorithm for a practically important problem of aggregation in multimedia databases.

Assume that each object in a database has m grades, one for each of m attributes. For each attribute, there is a sorted list, which lists each object and its grade under that attribute, sorted by grade (highest grade first). Each object is assigned an overall grade, which is a fixed monotone function of the attribute grades, such as min or average. The problem is to determine the top k objects with the highest overall grades.

Theorem 7.11 *There is an algorithm to solve the problem above which is instance optimal, that is, optimal (within a constant factor) for any database.*

Cake-Cutting Protocols

For some algorithmic problems, even the existence of any finite algorithm, not necessary efficient, is already a major advance. Here we discuss one example.

Assume that you want to share a cake with a friend, such that no-one thinks the other one has a better piece. The definition of "better" can be complicated, because the cake may be non-uniform (heterogeneous), and smaller pieces may be preferred because they have more cream, chocolate, etc. However, here is a simple method you can use. One person can divide the cake into two pieces which she/he evaluates as the same, the second person chooses the preferred piece out of these two, and the remaining piece goes to the first person. In this way, no participant thinks that the other one gets a strictly better piece. There is a similar method to divide the cake among 3 friends, discovered independently by Selfridge and Conway around 1960. In 1995, Brams and Taylor [197] proposed a method which works for any number n of friends, but it can require an arbitrarily large number of cuts, even for $n = 4$. In 2016, Aziz and Mackenzie [67] developed a method for $n = 4$ friends which takes a bounded number of steps. In 2020, the same authors [68] presented a method which works in a bounded number of steps for every fixed n.

We now formulate the result rigorously. Let us call the interval $[0, 1]$ a *cake*, and any finite union of subintervals of $[0, 1]$ a *piece of cake*. Let X be the set of all pieces of cake. Let $N = \{1, \ldots, n\}$ be the set of *agents*, each agent i having its own valuation function $V_i : X \to [0, \infty)$ such that

(i) $V_i(X \cup X') = V_i(X) + V_i(X')$ for all disjoint $X, X' \in X$, and
(ii) for every $X \in X$ and $0 \le \lambda \le 1$, there exists an $X' \subset X$ in X with $V_i(X') = \lambda V_i(X)$.

A *cake allocation* is a partition $[0, 1] = \bigcup_{i=1}^{n} X_i$ into disjoint pieces $X_i \in X$. We say that agent i is *envious* if $V_i(X_i) < V_i(X_j)$ for some j. A *discrete protocol* is an algorithm which returns an allocation after a sequence of queries which either

(1) for given $x \in [0, 1]$ and $r \ge 0$, ask agent i to return a point $y \in [0, 1]$ such that $V_i([x, y]) = r$ (CUT query), or
(2) for given $x, y \in [0, 1]$, ask agent i to return $V_i([x, y])$ (EVALUATE query).

A protocol is called *envy-free* if no agent is envious if he/she follows the protocol truthfully, even if other agents lie or cooperate. A protocol is bounded if the number of queries required to return a solution is bounded by a constant $C = C(n)$ depending only on n and not on V_i.

Cake cutting is a metaphor for the allocation of a heterogeneous divisible good among multiple agents with possibly different preferences over different parts of the good. The existence of a discrete and bounded envy-free cake cutting protocol was the central open problem in this area. As mentioned above, this problem has been solved by Aziz and Mackenzie for $n = 4$ in [67] and then for general n in [68].

Theorem 7.12 *There exists a discrete and bounded envy-free cake cutting protocol for any number of agents.*

The number of queries $C(n)$ guaranteed by the proof of Theorem 7.12 grows quite fast with n. If we define $T_1(n) = n$, $T_2(n) = n^n$, and then inductively $T_{k+1}(n) = n^{T_k(n)}$, then the proof implies the upper bound $C(n) \leq T_6(n)$. The protocol is therefore impractical even for $n = 4$. However, once the existence of a bounded protocol is established, there is a hope that the upper bound for $C(n)$ will be improved in future works. The current best lower bound is $C(n) \geq an^2$ for some constant a, see Procaccia [1008], so even a protocol with a polynomial number of queries may exist.

7.2 Approximation Algorithms and Complexity

The Sparsest Cut Problem

For problems which are hard to solve exactly, one may aim for an algorithm which can efficiently find an approximate solution. One example when this is possible is the *sparsest cut problem*, a fundamental combinatorial problem both for theory and applications. This problem asks to divide the vertices of a graph G into two classes in such a way that the ratio r of the number of edges between the classes divided by the number of vertices in the smaller class is minimized. The optimal ratio $\alpha(G)$ is also known as the *edge expansion* of the graph. It is NP-hard to compute $\alpha(G)$ for a given graph G exactly. However, Leighton and Rao [781] developed in 1999 a polynomial time $O(\log n)$-approximation algorithm for this problem, that is, an algorithm which produces a cut with r at most $O(\alpha(G) \log n)$. In 2009, Arora et al. [51] improved the approximation ratio from $O(\log n)$ to $O(\sqrt{\log n})$.

Theorem 7.13 *There exist a polynomial-time algorithm that, given any graph G with n vertices, produces a cut with ratio $r = O(\alpha(G)\sqrt{\log n})$.*

The proof of Theorem 7.13 uses the theory of embeddings into metric spaces (see Theorem 3.80), and the phenomenon of measure concentration (see, for example, Theorem 6.22). The idea is to consider a more general version of the problem where G is a weighted graph with non-negative "length" associated to each edge, and then embed the vertices of the weighted graph into some metric space such that the distance between any pair of points is close to the length of the corresponding edge. Then the graph vertices are partitioned according to the "clusters" they form in this space. This can be done efficiently using a semidefinite programming technique. Theorem 3.80 gives an embedding with better distortion, which translates into the sparsest cut algorithms with better approximation ratio.

The Travelling Salesman Problem

Let $G = (V, E)$ be a graph in which a "length" $w_{ij} \geq 0$ is assigned to every edge $e \in E$ connecting vertices i and j. What is the shortest length of a closed tour in G visiting every vertex? We may assume that G is connected, otherwise such a tour does not exist. We may further assume that G is complete and that edge lengths satisfy the triangle inequality

$$w_{ij} + w_{jk} \geq w_{ik}, \qquad \forall i, j, k \in V, \qquad (7.2)$$

because otherwise we can go from i to k through j. Once (7.2) is satisfied, we may then assume that the tour visits every vertex exactly once, and ask for the shortest hamiltonian cycle in G. This problem is called the (metric) *travelling salesman problem* (TSP). In the 1970s, Christofides [278] and Serdyukov [1102] developed a polynomial time algorithm which finds a hamiltonian cycle of length at most $\frac{3}{2}l(G)$, where $l(G)$ is the length of the shortest one. In 2015, Karpinski et al. [686] proved that it is NP-hard to approximate $l(G)$ to any factor better than $\frac{123}{122}$, hence the best possible approximation ratio r^* for a polynomial time algorithm lies in the interval $\left[\frac{123}{122}, \frac{3}{2}\right]$. Because the Christofides–Serdyukov algorithm had been unimproved for over 40 years, some people started to conjecture that $r^* = \frac{3}{2}$. In 2020, Karlin et al. [682] posted online a preprint which disproves this conjecture.

Theorem 7.14 *There exists a constant $\epsilon > 0$ and a randomized algorithm that outputs a TSP tour with expected length at most $\frac{3}{2} - \epsilon$ times the length of the optimal tour. In fact, one can take $\epsilon = 10^{-36}$.*

The algorithm in Theorem 7.14 is a modification of the Christofides-Serdyukov algorithm. The latter first finds a minimum spanning tree[5] in G and then adds the minimum cost matching on the odd degree vertices of the tree. The new algorithm does the same but chooses a *random* spanning tree from a suitable distribution. Theorem 7.14, as formulated, has no practical importance, because the improvement in the approximation ratio is tiny. However, the algorithm itself may be practical, because for most graphs it finds a TSP tour with much better approximation ratio than $\frac{3}{2} - 10^{-36}$. In fact, there are no known examples of graphs G on which this ratio is worse than $\frac{4}{3}$. Also, Theorem 7.14 is very important theoretically, because it shows that the approximation ratio $\frac{3}{2}$ is not optimal and (hopefully) opens the road for further progress.

The Set-Covering Problem

In the *set-covering problem*, the input is a collection of finite sets $P_1, P_2, \ldots, P_m$ and positive numbers $c_1, c_2, \ldots, c_m$. Let $I = \bigcup_{j=1}^{m} P_j$ be the union of these sets. A subset $J \subset \{1, \ldots, m\}$ is called a *cover* if $\bigcup_{j \in J} P_j = I$. The *cost of cover J* is $c(J) = \sum_{j \in J} c_j$. The SET COVER problem asks to find a cover J of minimum cost. This is a well-known NP-hard problem, even if all $c_j = 1$.

In 1979, Chvátal [282] presented a greedy algorithm which finds in polynomial time a cover J with cost $c(J) \leq (\log n + 1)c^*$, where $n = |I|$ and c^* is the cost of the optimal cover. Since then, it was a well-known open question to either improve the approximation ratio of this algorithm or prove that it is NP-hard to do so. In 1998,

---

[5] That is, a spanning tree of minimal total length of all edges. See the discussion before Theorem 6.6 for the definition of spanning tree.

Feige [444] proved that any polynomial time algorithm returning a cover J with $c(J) \leq ((1 - \epsilon) \log n)c^*$ for $\epsilon > 0$ would imply that any problem in NP with input size n can be solved in time $n^{O(\log\log n)}$. Finally, in 2014 Dinur and Steurer [355] proved the same result with $n^{O(1/\epsilon)}$ instead of $n^{O(\log\log n)}$. This is a polynomial time algorithm for any fixed $\epsilon > 0$, hence it implies the following theorem.

Theorem 7.15 *For every $\epsilon > 0$, it is NP-hard to approximate SET COVER to within a $(1 - \epsilon) \log n$ factor, where n is the size of the instance.*

Approximating the Permanent

Some problems which are NP-hard to solve exactly can be efficiently solved approximately with *any* given accuracy! A famous example is the problem of computing the permanent. Let A be an $n \times n$ matrix with integer entries $a(i, j)$, $i = 1, \ldots, n$, $j = 1, \ldots, n$. The *permanent* of A is

$$\text{per}(A) = \sum_{\sigma} \prod_{i=1}^{n} a(i, \sigma(i)),$$

where the sum is over all permutations σ of $\{1, 2, \ldots, n\}$. The evaluation of the permanent of a matrix is an old and well-studied problem going back to at least the 1812 memoirs of Binet and Cauchy. In 1979, Valiant [1223] proved that computation of the permanent exactly is #P-hard, even if all $a(i, j)$ are either 0 or 1.[6] However, Jerrum et al. [656] presented in 2004 an algorithm which, for any matrix A with non-negative entries, computes (with high probability) an arbitrarily close approximation to its permanent in time that depends polynomially on the input size and the desired error. Given Valiant's result, this is the best we could hope for.

In general, a *polynomial-time approximation scheme* (PTAS) is an algorithm which, given a parameter $\epsilon > 0$ and an instance of an optimization problem of length n, produces in time $P_\epsilon(n)$ a solution which differs from the optimal one by a factor of $1 + \epsilon$. Here, P_ϵ is a polynomial which depends on ϵ. If the running time is bounded by a polynomial in both n and ϵ^{-1}, we say that the problem has a *fully polynomial-time approximation scheme* (FPTAS). If the corresponding algorithm is randomized, we call it a *fully-polynomial randomized approximation scheme* (FPRAS). More concretely, a FPRAS for the permanent computation is a randomized algorithm which, given an $n \times n$ matrix A and $\epsilon \in (0, 1]$ as an input, runs in time polynomial in n and ϵ^{-1}, and outputs a (random) number Z such that

$$\mathbb{P}[\exp(-\epsilon)Z \leq \text{per}(A) \leq \exp(\epsilon)Z] \geq \frac{3}{4}.$$

The probability $\frac{3}{4}$ here can be made arbitrary close to 1 by running the algorithm several times and averaging the outputs.

---

[6] See the discussion before Theorem 7.4 for the definition of #P-hardness

Theorem 7.16 *There exists a fully polynomial randomized approximation scheme (FPRAS) for the permanent computation of an arbitrary $n \times n$ matrix A with non-negative entries.*

The proof of Theorem 7.16 first treats the important special case when all entries $a(i, j)$ are either 0 or 1, and then extends to the general case of arbitrary non-negative entries. The starting point is the 1986 observation of Jerrum et al. [657] that the (approximate) permanent computation for 0-1 matrices is equivalent to sampling perfect matchings from a bipartite graph G (almost) uniformly at random. The latter problem is solved using the "Markov chain Monte Carlo" method, first proposed by Broder [214] in 1986, which considers a random walk on the space of all perfect and "near-perfect" matchings (that is, matchings with 2 unmatched vertices) on G, such that, no matter where we start the walk, its position after polynomially many steps is nearly uniformly random. The method works unless the number of perfect matchings is negligible compared to the number of near-perfect ones. Jerrum et al. [656] developed a modified version of this algorithm, in which each near-perfect matching gets a weight in such a way that the number of weighted near-perfect matchings becomes not much higher than the number of perfect matchings.

Inapproximability of Max Clique and Chromatic Number
Theorem 7.16 gives an example of an important problem which is hard to solve exactly but easy to solve approximately. For some other problems, however, even finding an approximate solution with any non-trivial accuracy is hard. This is the case, for example, for the max clique and chromatic number problems.

Given a graph G, let $c(G)$ be the size of the largest clique in G, and $\chi(G)$ be the chromatic number of G (the smallest number of colours needed to colour the vertices of G so that no two adjacent vertices share the same colour). The MAX CLIQUE problem is: given a graph G, compute $c(G)$. The CHROMATIC NUMBER problem is: given a graph G, compute $\chi(G)$. These problems are known to be NP-hard to solve exactly, so it makes sense to look for approximations. We say that an algorithm approximates MAX CLIQUE (respectively, CHROMATIC NUMBER) to within $f(n)$ if, given a graph G with n vertices, it returns a number k such that $\frac{c(G)}{f(n)} \le k \le f(n)c(G)$ (respectively, $\frac{\chi(G)}{f(n)} \le k \le f(n)\chi(G)$).

In 2007, Zuckerman [1294] proved that both these problems are NP-hard to approximate to within $n^{1-\epsilon}$ for any $\epsilon > 0$. Note that the trivial "algorithm" outputting n for any n-vertex graph is an approximation to within n, so the theorem states that we cannot improve this trivial "algorithm" by ϵ in the exponent.

Theorem 7.17 *(i) for any $\epsilon > 0$, it is NP-hard to approximate MAX CLIQUE to within $n^{1-\epsilon}$, and (ii) for any $\epsilon > 0$, it is NP-hard to approximate CHROMATIC NUMBER to within $n^{1-\epsilon}$.*

Essentially all proofs that a problem A is NP-hard proceed by choosing an NP-hard problem B and presenting a polynomial time algorithm for B assuming the existence of such algorithm for A. In 1999, Håstad [577] used the hypothetical $n^{1-\epsilon}$-approximation algorithm for MAX CLIQUE to develop a *randomized* algorithm for an NP-hard problem. Using Håstad's result, Feige and Kilian [445] developed a randomized algorithm for an NP-hard problem assuming the existence of an $n^{1-\epsilon}$-approximation algorithm for CHROMATIC NUMBER. The proof of Theorem 7.17 proceeds by derandomizing these algorithms. To do this, the authors use techniques based on random walks on expander graphs, as well as sum-product theorems, see Theorem 2.23 and related discussion. This is a nice example of the unity of mathematics.

The MAXEkSAT Problem

Another example for which the approximation accuracy of a trivial algorithm cannot be beaten is the MAXEkSAT problem. Let $x_1, x_2, \ldots, x_n$ be boolean variables. A *k-clause* is any expression of the form $y_1 \vee y_2 \vee \cdots \vee y_k$, where each y_i is either x_j or $\neg x_j$ for some j. The problem which asks, given m k-clauses, to decide if it is possible to satisfy them all is a famous problem which is called the k-SAT problem, and is known to be NP-hard for any $k \geq 3$. In fact, this is the problem which was *first* proved to be NP-hard by Cook [301] in 1971. The problem of deciding what is the maximal number of clauses we can satisfy is known as MAXEkSAT. It is easy to satisfy a $1 - 2^{-k}$ fraction of clauses by random assignment. In 2001, Håstad [578] proved that it is NP-hard to approximate MAXEkSAT with any constant ratio better than $1 - 2^{-k}$. In other words, no polynomial time algorithm can beat in the approximation ratio the trivial random assignment algorithm.

Theorem 7.18 *For any $\epsilon > 0$ and any fixed $k \geq 3$, it is NP-hard to, given m k-clauses, distinguish the case when all clauses are satisfiable from the case when at most $(1 - 2^{-k} + \epsilon)m$ of them are satisfiable.*

The starting point in the proof of Theorem 7.18 is the celebrated *PCP theorem* of Arora et al. [50] which states that for every "Yes" instance of every decision problem in NP there is a proof (that this is indeed a "Yes" instance) that can be checked by a randomized algorithm which uses a logarithmic number of random bits, provided that the algorithm is allowed to communicate with the prover (ask questions back) a constant number of rounds. See the discussion before Theorem 7.47 below for more details about this prover-verifier model. The PCP theorem turned out to be an extremely useful tool in proving that a problem is NP-hard to approximate. For example, it immediately implies the existence of some constant $C < 1$ such that it is NP-hard to approximate MAXE3SAT with any constant ratio better than C. By refining this technique, Håstad [578] proved that one can take $C = 1 - 2^{-3} = \frac{7}{8}$ in this result, which proves Theorem 7.18 for $k = 3$. Then generalization to any $k \geq 3$ follows easily.

The Edge-Deletion Problem

The problem we discuss next is the *edge-deletion problem*. How many edges do we need to remove from a given graph G to make it, for example, triangle-free, or 3-colourable? More generally, a graph property is *monotone* if it is closed under removal of vertices and edges. The edge-deletion problem is: given a monotone property P and a graph G, compute the smallest number $E_P(G)$ of edge deletions that are needed in order to turn G into a graph satisfying P.

In 2009, Alon et al. [37] proved that we can quickly solve any such problem approximately with additive error ϵn^2 for any fixed $\epsilon > 0$ but not with error $n^{2-\delta}$ for any $\delta > 0$. Before 2009, an ϵn^2 approximation algorithm was not known even for the triangle-free property, and the NP-hardness result was not known even for computing $E_P(G)$ *exactly*.

Theorem 7.19 *For any fixed $\epsilon > 0$ and any monotone property P, there is a deterministic algorithm which, given a graph G with n vertices and m edges, approximates $E_P(G)$ in linear time $O(n + m)$ to within an additive error of ϵn^2. On the other hand, if all bipartite graphs satisfy P, then for any fixed $\delta > 0$ it is NP-hard to approximate $E_P(G)$ to within an additive error of $n^{2-\delta}$.*

The ϵn^2-approximation algorithm is based on the idea that every graph G can be approximated by a small weighted graph W such that the solution of a certain (easier-to-solve) problem on W is approximately $E_P(G)$. As a step towards the proof that this idea actually works, the authors developed an efficient algorithmic version of the famous Szemerédi regularity lemma [1174], which we have briefly discussed earlier in this book in the context of Theorem 2.56. The proof of the NP-hardness result in Theorem 7.19 is even more involved, and uses deep tools from external graph theory as well as analysis of eigenvalues of the graph adjacency matrices.

The Unique Games Conjecture

We next discuss the accuracy of efficient approximation for the famous minimum vertex cover problem. A *vertex cover* of a graph G is a set S of vertices of G such that for each edge (u, v) of G either $u \in S$ or $v \in S$ (or both). Let $\tau(G)$ be the size of the smallest vertex cover of G. The *minimum vertex cover problem* is: given a graph G, compute $\tau(G)$. Its approximation with factor $c > 1$ is: given a graph G, compute a number k such that $\tau(G) \leq k \leq c\tau(G)$.

Because the minimum vertex cover problem is well-known to be NP-hard to solve exactly, it is natural to look for approximate algorithms. An old simple algorithm[7] gives an approximation with factor $c = 2$, and this is the best known. In the opposite direction, Håstad [578] proved in 2001 that the problem is NP-hard to approximate to within any factor $c < \frac{7}{6} = 1.1666....$ In 2005, Dinur and Safra [354] proved the same result with the better factor $10\sqrt{5} - 21 = 1.3606...$ in place of $\frac{7}{6}$.

Theorem 7.20 *The minimum vertex cover problem is NP-hard to approximate to within any constant factor smaller than $10\sqrt{5} - 21 = 1.3606...$*

In 2002, Khot [711] made a fundamental conjecture, now known as the *Unique Games Conjecture*, which, if true, would imply better hardness of approximation results for a wide range of combinatorial problems, including the minimum vertex cover problem. For fixed parameters $0 < c \leq s < 1$ and an integer m, consider

---

[7] Find any maximal matching M in G by a greedy algorithm and return a vertex cover C that consists of all endpoints of the edges in M.

the following problem. The input is a system of k linear equations in n variables $x_1, x_2, \ldots, x_n$ over $\mathbb{Z}_m$, each equation having the form $a \cdot x_i + b \cdot x_j = c, a, b, c \in \mathbb{Z}_m$, such that either

(i) there is an assignment satisfying at least sk of the equations, or
(ii) every assignment satisfies fewer than ck of the equations.

The problem $UG(c, s, m)$ is to decide which of the cases (i) or (ii) is true. The *Unique Games Conjecture* (UGC) predicts that for every $0 < c \leq s < 1$ there exists an m such that $UG(c, s, m)$ is NP-hard.

In 2008, Khot and Regev [714] proved that the approximation ratio $c = 2$ of the trivial algorithm for the minimal vertex cover is the best possible, unless the UGC is false.

Theorem 7.21 *Assume the Unique Games Conjecture. Then the minimum vertex cover problem is NP-hard to approximate to within any constant factor smaller than* 2.

The proof of Theorem 7.21 combines the PCP theorem [50] with the combinatorial methods developed in [354] to prove Theorem 7.20, to derive that the conclusion of Theorem 7.21 follows from a stronger form of the UGC. The authors then proved that this stronger form is equivalent to the original conjecture.

By Theorem 7.21, UGC implies that the known approximation algorithm for the minimal vertex cover has optimal accuracy. Similar results have been proved for a large class of other important optimization problems, see Theorem 7.23 below. Because of this, resolving the UGC in one way or another is considered as one of the most important open problems in the theory of approximation algorithms. In 2018, Khot et al. [713] made dramatic progress in the positive direction by proving the conjecture for $s < 1/2$.

Theorem 7.22 *For every $0 < c \leq s < 1/2$ there exists an m such that* $UG(c, s, m)$ *is NP-hard.*

The proof of Theorem 7.22 is long and is distributed over many papers. In the penultimate paper [353] in this series, Dinur et al. completed the proof of Theorem 7.22, conditional on a combinatorial hypothesis they introduced. The hypothesis states that "pseudorandom sets in the Grassmann graph have near-perfect expansion". The *Grassmann graph* $J_q(n, k)$ is the graph whose vertices are the k-dimensional subspaces of an n-dimensional vector space over a finite field with q elements, and two vertices are connected by an edge if and only if the intersection of the corresponding subspaces is $k - 1$-dimensional. A set $S \subset J_q(n, k)$ has *near-perfect expansion* if its edge expansion (2.22) is very close to 1, see [353] for the exact definition. The final paper [713] in the series proves this hypothesis and thus completes the unconditional proof of Theorem 7.22.

Theorem 7.22 implies new unconditional NP-hardness results for many optimization problems whose hardness was earlier known only conditionally on the UGC. In particular, it implies that the minimum vertex cover problem is NP-hard to approximate to within any constant factor smaller than $\sqrt{2} = 1.4142...$, which is an improvement in comparison with Theorem 7.20.

Optimal Approximation Algorithms for Every CSP
Theorem 7.21 states that if the Unique Games Conjecture (UGC) is true then the known approximation algorithm for the minimum vertex cover problem achieves the best possible approximation ratio, unless P $=$ NP. In 2008, Raghavendra [1018] proved the same result for a large family of problems, called the constraint satisfaction problems (CSPs). See the discussion before Theorem 7.3 for the definition of this class of problems. There we have discussed the exact versions of CSPs, where the problem is to determine whether all constraints can be satisfied. Each CSP also has a optimization version in which the goal is to maximize the number of satisfying constraints. Raghavendra [1018] proved the following impressive result.

Theorem 7.23 *Assume the Unique Games and* P $\neq$ NP *conjectures. Then there exists an (explicitly presented) polynomial-time algorithm which for every CSP and every* $\epsilon > 0$ *obtains an approximation ratio which is within* ϵ *of the optimal ratio a polynomial-time algorithm can achieve.*

Theorem 7.3 provides a dichotomy which CSPs are solvable in polynomial time and which are NP-hard to solve exactly. Theorem 7.4 does the same for (exact) counting versions of CSPs. Finally, Theorem 7.23 provides the best possible approximation algorithm for such problems we might have, conditional on the UGC. This explains why the UGC is the central conjecture in the theory of approximation algorithms.

Problems on Graphs Arising from Social Network Analysis
The analysis of social networks leads to a variety of problems on graphs which look a bit unnatural from the point of view of abstract graph theory but are central for understanding social networks. One example is determining the betweenness centrality of graph vertices. Let $\sigma_{s,t}$ be the total number of shortest paths from vertex s to vertex t in an undirected graph G. Let $\sigma_{s,t}(v)$ the number of such shortest paths which contain the vertex v. The number

$$C_B(v) = \sum_{s \neq v \neq t} \frac{\sigma_{s,t}(v)}{\sigma_{s,t}}$$

is called the *betweenness centrality* of the vertex v. This concept was introduced by Freeman [472] in 1977, but became especially important with the development of social networks, where the betweenness centrality is a measure of "power": high $C_B(v)$ allows the person v to, for example, decide which information to share and which to not, and information not shared by v will be spread significantly more slowly.

However, before 2001, the fastest algorithm to compute $C_B(v)$ for all vertices of an n-vertex graph worked in $O(n^3)$ steps. In 2001, Brandes [199] provided a significant improvement for graphs with not too many edges, which is often the case in applications.

Theorem 7.24 *There is an algorithm which computes the betweenness centrality of all vertices of a graph with n vertices and m edges in time $O(nm)$.*

Another problem arising from social network analysis is the influence maximization problem. Let G be a directed graph, in which each directed edge $w \rightarrow v$ has a weight $b_{vw} \in [0, 1]$. Each vertex v has a threshold θ_v, and the thresholds θ_v are chosen independently and uniformly at random from the interval $[0, 1]$. At time 0, we select some set A of vertices and mark them as "active". At times $t = 1, 2, \ldots,$ vertex v becomes active if

$$\sum_{w \rightarrow v,\, w \text{ active}} b_{vw} \geq \theta_v,$$

and if a vertex becomes active it stays active forever. This process ends if no activations occur from round t to $t + 1$. The *influence* $\sigma(A)$ of a set A of vertices is the expected number of active vertices at the end of the process. The *influence maximization problem* asks, given a graph G, weights b_{vw}, and parameter k, to find a k-vertex set A with maximal $\sigma(A)$.

In social network analysis, the "activation process" is when people start to use some product or service because their friends do, and the influence maximization problem asks which set A of individuals should be initially convinced to use it, to maximize the "cascade effect". The problem is NP-hard to solve exactly, but Kempe et al. [701] proved in 2015 that it can be efficiently solved approximately, with the approximation factor arbitrary close to $1 - 1/e = 0.63...$, where e $= 2.718...$ is the base of the natural logarithm.

Theorem 7.25 *For any $\epsilon > 0$, there is a polynomial-time algorithm approximating the maximum influence to within a factor of $1 - 1/e - \epsilon$.*

The proof of Theorem 7.25 does not introduce any new algorithms. Instead, the authors proved that an already known natural greedy algorithm achieves the stated approximation guarantee. The proof is based on the analysis of submodular functions, which we define in the next subsection.

Maximization of a Submodular Function

Given a collection $\mathcal{F}$ of subsets of $[n] = \{1, \ldots, n\}$, the *max k-cover problem* is the problem of selecting k subsets from $\mathcal{F}$ such that their union has maximum cardinality. This problem is well-known to be NP-hard to solve exactly, and one is looking for approximation algorithms, where the approximation ratio is the ratio between the cardinality of the union of the k subsets selected by the algorithm and the optimal solution. There is an easy polynomial time algorithms which achieves approximation ratio $1 - 1/e = 0.632....$ In 1998, Feige [444] proved that this is the best possible: approximating the max k-cover problem with any constant ratio better than this is NP-hard.

The max k-cover problem is a special case of the following more general problem. Let 2^X denotes the set of all subsets of a set X. A function $f : 2^X \to \mathbb{R}_+$ is called *monotone* if $f(A) \le f(B)$ whenever $A \subseteq B$. We also assume that $f(\emptyset) = 0$. A function f is called *submodular* if

$$f(A \cup B) + f(A \cap B) \le f(A) + f(B) \quad \text{for all} \quad A, B \subseteq X.$$

Consider the optimization problem

$$\max_{|S| \le k} f(S). \tag{7.3}$$

The max k-cover problem is the special case of (7.3) with $X = \mathcal{F}$, and $f(S)$ being the cardinality of union of sets in S. A far-reaching generalization of (7.3) is the optimization problem

$$\max_{S \in I} f(S), \tag{7.4}$$

where I is a family of independent sets in a (finite) matroid[8] (X, I). In addition to the max k-cover, problem (7.4) includes a number of other interesting and useful combinatorial optimization problems as special cases.

There is a simple greedy algorithm to solve problem (7.4) that achieves approximation ratio $1/2$. In 2011, Calinescu et al. [239] improved it to $1 - 1/e$. By Feige's theorem, this approximation ratio is the best possible, even in the special case of the max k-cover problem, unless $P = NP$.

Theorem 7.26 *There is a randomized algorithm giving a $(1 - 1/e)$-approximation (in expectation) to the problem (7.4) for any monotone submodular function f (given by a value oracle) and an arbitrary matroid (given by a membership oracle).*

The Uncapacitated Facility Location Problem
The next problem we discuss is the (metric) *uncapacitated facility location problem* (UFL). The problem is: given a set F of potential facility locations, each location $i \in F$ with a facility cost f_i, a set C of clients and a metric d over $F \cup C$, the goal is to find a subset $A \subseteq F$ of locations of open facilities, so that the total cost

$$S = \sum_{i \in A} f_i + \sum_{j \in C} d(j, i_j)$$

---

[8] Recall that a *finite matroid* is a pair (X, I), where X is a finite set and $I \subset 2^X$ is a family of its subsets, such that (i) $\emptyset \in I$; (ii) if $A \in I$ and $A' \subset A$ then $A' \in I$; and (iii) if $A \in I$, $B \in I$, and $|A| > |B|$, then there exists an $x \in A \setminus B$ such that $B \cup \{x\} \in I$. See Theorems 2.40 and 2.70 for some properties of matroids.

is minimized, where i_j is the closest open facility to client j. This practically important problem is NP-hard to solve exactly. Moreover, Sviridenko [1170] proved in 2002 that it is NP-hard to approximate it within any constant factor better than 1.463....

In 1997, Shmoys et al. [1120] gave the first constant-factor approximation algorithm for this problem with ratio 3.16. After a number of subsequent improvements, Byrka and Aardal [235] achieved an approximation ratio of 1.5. Building on this work, Li [792] in 2013 improved the approximation ratio further to 1.488.

Theorem 7.27 *There exists a randomized polynomial-time algorithm for the metric UFL problem which gives a solution whose expected cost is at most* 1.488 *times the cost of the optimal solution.*

The previous algorithm of Byrka and Aardal [235] is a parametrized algorithms with parameter γ which we are allowed to choose. The approximation ratio 1.5 was achieved by running, depending on the instance, either this algorithm with $\gamma \approx 1.6774$, or an earlier algorithm developed by Jain et al. [648] in 2002. Theorem 7.27 has been proved by noticing that running the Byrka–Aardal algorithm with γ selected at *random* from a suitable distribution improves the approximation ratio to 1.488.

7.3   Linear Programming and Polytopes

The Simplex Method

One of the fundamental problems in optimization is the linear programming problem: given an $m \times n$ matrix A and vectors $b \in \mathbb{R}^m$ and $c \in \mathbb{R}^n$ as an input, find $x \in \mathbb{R}^n$ which maximizes the linear function $f(x) = c^T x$ subject to constraints $Ax \leq b$ and $x \geq 0$. In fact, all the coefficients are usually assumed to be rational numbers.

The feasible region defined by the constraints is a (possibly unbounded) convex polyhedron P. Because the objective function f is linear, it either diverges to infinity or attains its maximum value on at least one of the vertices (extreme points) of P. Moreover, for every vertex A on which f is not maximal, there is an edge e incident to A such that f increases along e. If e is a finite edge AB, we can move from A to B and increase f. By repeating this procedure, we can find the maximum of f. This method was discovered by Dantzig in 1947 (see [316]) and is called the *simplex method*. There are variations of this method with different ways to choose an edge to move at each step, if f increases along several edges. In 1972, Klee and Minty [725] gave an example of a linear program on which the version of simplex method originally suggested by Dantzig takes exponential time. The existence of a version of the simplex method which always runs in polynomial time remains an open question.

If a polyhedron P is bounded, it is called a *polytope*. The *combinatorial distance* between vertices u and v of P is the minimum number of steps needed to go from u to v, where a step consists in traversing an edge. The *(combinatorial) diameter* of P is the maximum combinatorial distance between a pair of its vertices. In 1957, Hirsch conjectured that a d-dimensional polytope with n facets cannot have diameter greater than $n - d$. If this conjecture were true, the simplex method could, at least in principle, move from any initial vertex to the vertex at which f attains its maximum, in at most $n - d$ steps. The conjecture was well-believed and proved in some special cases. However, Santos [1078] proved in 2012 that in general this conjecture is false.

Theorem 7.28 *There exists a 43-dimensional polytope with 86 facets and diameter at least 44.*

The proof of Theorem 7.28 is based on the properties of spindles. A d-*spindle* is a d-dimensional polytope P which has two special vertices u and v such that every facet of P contains exactly one of these vertices. The *length* of a spindle is the combinatorial distance between u an v. Santos proved that if there exists a d-spindle P with n facets and length l, then there also exists an $n - d$-spindle P' with $2n - 2d$ facets and with length at least $l + n - 2d$. In particular, if $l > d$, then P' violates the Hirsch conjecture. Santos then gave an example of a 5-spindle of length 6, which finishes the proof of Theorem 7.28.

Even with the Hirsch conjecture being false, it is still possible that the diameter of any polytope is bounded by a polynomial in n and d. As partial progress, Kalai and Kleitman [674] proved in 1992 that the diameter is at most $n^{\log_2 d + 1}$, while Barnette [101] established in 1969 an upper bound for the diameter which grows linearly in n for every *fixed* d. However, even if a short walk between any pair of vertices exists, it does not imply that there exist a version of the simplex method which can actually *find* such walks.

On the other hand, even the versions of the simplex method which are known to be exponentially slow in the worst case demonstrate a surprisingly good performance on all instances arising in practice. In 1979, Khachiyan [707] made a theoretical breakthrough by proving that a different method, called the *ellipsoid algorithm*, always solves the linear program in time polynomial in the number of bits needed to represent it. However, the ellipsoid algorithm is almost never used in practice because the simplex method demonstrates much better performance in all practical instances.

Starting with the work of Borgwardt [170] in 1977, many researchers demonstrated that the simplex method runs in expected polynomial time on various distributions of random inputs. However, this still does not explain its good performance in practice, because practical instances are usually very different from random ones. In 2004, Spielman and Teng [1148] invented a new method for analysing the algorithms: *smoothed analysis*. Intuitively, an algorithm has polynomial smoothed complexity if it runs in polynomial time on a small random perturbation of *any* instance. More formally, let A be an algorithm which solves some problem whose input x is a vector in $\mathbb{R}^m$ for some m, with norm $||x||$. Let

$C_A(x)$ be the number of steps A works at input x. The worst-case complexity of A is $f(n) = \max_{x \in X_n} C_A(x)$, where X_n is the set of all possible inputs of bit-length n. The *smoothed complexity* of A is

$$f(n, \sigma) = \max_{x \in X_n} E_g[C_A(x + (\sigma||x||)g)],$$

where $(\sigma||x||)g$ is a vector of Gaussian random variables of mean 0 and standard deviation $\sigma||x||$, and E_g is the corresponding expectation. We say that an algorithm A has *polynomial smoothed complexity* if $f(n, \sigma)$ is bounded from above by a polynomial in n and $1/\sigma$. Because the practical instances are usually noisy, a polynomial smoothed complexity provides an adequate theoretical explanation for the good practical performance of an algorithm. The main result of Spielman and Teng [1148] establishes this for the simplex method.

Theorem 7.29 *The simplex algorithm for the linear programming problem has polynomial smoothed complexity.*

Extension Complexity
There are also different methods of solving the linear programming problem. For example, the *interior point method*, introduced by Karmarkar [683] in 1984, runs in polynomial-time in the worst case, and at the same time is competitive with the simplex method in practical instances. These methods make linear programming a popular tool in optimization: if any problem of interest can be expressed as a linear program with polynomially many variables and constraints, then it can be solved in polynomial time. Sometimes, even if the naive implementation of a combinatorial problem as a linear program requires exponentially many constraints, one can use a few extra variables to reduce the number of constraints down to a polynomial. Formally, the *extension complexity* of any polytope P is the smallest number of inequalities necessary to describe a higher-dimensional polytope Q that can be linearly projected onto P. If the solution set of a combinatorial problem can be described as a polytope P with polynomial extension complexity, then the problem can be reduces to a "small" linear program, and then solved efficiently.

In 2012, Fiorini et al. [454] proved that the polytopes describing the solutions sets of some well-known NP-hard problems grow faster than polynomially in n. This is a rare example when we have a formal proof that some powerful method (in this case linear programming) cannot efficiently solve a problem. It allows us to reject a sequence of flawed P = NP proofs, which claimed to solve some NP-hard problems in exactly this way.

In 2017, Rothvoss [1063] proved that even the polytope for the perfect matching problem, which is a well-known problem in P, has exponential extension complexity. This resolves a long-standing open question in the field. Let $G = (V, E)$ be a graph with vertex set V and edge set E. A set $M \subseteq E$ of edges of G is called a *matching* if no two edges in M share a common vertex. A matching is called *perfect* if $|M| = |V|/2$. Let us enumerate the edges of G from 1 to $|E|$. For any subset $S \subset E$, let $\chi_S \in \mathbb{R}^{|E|}$ be the characteristic vector of S, that is, the i-th coordinate of χ_S is 1 if the edge number i belongs to S, and 0 otherwise. The *perfect matching polytope* $P(G)$ in a graph G is the convex hull of all characteristic vectors of perfect matchings in G.

Theorem 7.30 *For all even n, the extension complexity of the perfect matching polytope in the complete n-vertex graph is at least 2^{cn} for some constant c.*

Systems of Linear Equations

An important special case of linear programming is the task of solving a system of linear equations, which can be written in matrix form as

$$Ax = b, \tag{7.5}$$

where A is an $m \times n$ matrix, $x \in \mathbb{R}^n$ represents n variables, and $b \in \mathbb{R}^m$ is a vector of constants. If $m = n$ and the matrix A is invertible, then the system (7.5) has a unique solution which is given by $x = A^{-1}b$. Note, however, that this formula has only theoretical importance, because solving the system (7.5) is an easier problem than finding the inverse matrix A^{-1}.

If the matrix A is not invertible, the exact solution to (7.5) may not exist, and one may look for an approximation. Moore [905] in 1920 and Penrose [972] in 1955 developed a way to write down an approximate solution to (7.5) using a special kind of "pseudo-inverse matrix" which always exists and is unique. For any $m \times n$ matrix A with real or complex entries, its *Moore–Penrose inverse* $A^\dagger$ is the (unique) $n \times m$ matrix $A^\dagger$ such that (i) $AA^\dagger A = A$, (ii) $A^\dagger AA^\dagger = A$, and (iii) matrices $AA^\dagger$ and $A^\dagger A$ are Hermitian (symmetric in the real case). We say that $\tilde{x} \in \mathbb{R}^n$ is an approximate solution to (7.5) with accuracy ϵ if $\|\tilde{x} - A^\dagger b\|_A \leq \epsilon \|A^\dagger b\|_A$, where $\|x\|_A = \sqrt{x^T A x}$.

A special case of (7.5) which often arises in practice is when the matrix A with entries a_{ij} is real, symmetric, and *weakly diagonally dominant*, that is, $a_{ii} \geq \sum_{j \neq i} |a_{ij}|$ for all i. The task of solving such systems arises as a subroutine in many applications, e.g. in various problem on graphs or when solving elliptic partial differential equations by numerical methods. The systems arising in some applications are very large, so methods running in quadratic time (or even in time $k^{1+\epsilon}$ for some $\epsilon > 0$ where k is the length of input) may be impractical. In 2014, Spielman and Teng [1149] made a breakthrough by showing that symmetric diagonally dominant linear systems can by solved with any given accuracy in time linear in the input size, if we ignore the logarithmic factors. In later works, this theorem has been used to obtain substantial asymptotic improvements in the running time for various important problems in algorithmic graph theory.

Theorem 7.31 *There exists a randomized algorithm that on inputting a symmetric, weakly diagonally dominant $n \times n$ matrix A with m non-zero entries, vector $b \in \mathbb{R}^n$, and $\epsilon > 0$, produces an $\tilde{x}$ such that $\|\tilde{x} - A^\dagger b\|_A \leq \epsilon \|A^\dagger b\|_A$ in expected time $O(m \log^c n \log(1/\epsilon))$ for some constant c.*

In the original proof of Theorem 7.31 the authors did not attempt to optimize the constant c, preferring to present the simplest possible proof instead. However, in a subsequent work [742], Koutis et al. proved that the constant c is Theorem 7.31 can be chosen arbitrarily close to 1.

7.4   Numerical Algorithms

Numerical Solutions to Polynomial Equations

A fundamental problem in mathematics is numerical computation of roots of polynomials in one variable. While an exact formula for roots expressed in radicals does not exist for general polynomials of degree $d \geq 5$, there are methods which can find the roots numerically with any given accuracy. One of the most efficient and popular methods for doing this task is Newton's method.

Let P_d be the set of polynomials of degree d in one complex variable such that all their roots have absolute values less than 1. The last condition can be easily ensured at the preprocessing step and is not a restriction. For $P \in P_d$ and $z_0 \in \mathbb{C}$, consider the sequence

$$z_0, \; z_1 = f_P(z_0), \; z_2 = f_P(z_1), \ldots, \quad \text{where} \quad f_P(z) = z - \frac{P(z)}{P'(z)}.$$

If this sequence converges to a root z^* of P, we say that z_0 is in the *basin* of z^*. This is called *Newton's method* for finding roots, and it is usually very efficient. However, its convergence depends on the choice of the initial point z_0.

In 2001, Hubbard et al. [626] proved the existence of a small set S_d, which depends only on d, such that if we start Newton's method from all points of S_d, we are guaranteed to find all the roots of any polynomial $P \in P_d$.

Theorem 7.32 *For every $d \geq 2$, there is a set S_d consisting of at most $1.11d \cdot \log^2 d$ points in $\mathbb{C}$ with the property that for every polynomial $P \in P_d$ and every root z^* of P, there is a point $s \in S_d$ in the basin of z^*.*

The proof of Theorem 7.32 is constructive, and the set S_d with the stated properties is explicitly described. In fact, it consists of approximately $0.26 \log d$ circles, and in each circle there are about $4.16d \log d$ points selected to cover the circle approximately uniformly.

Stable Roots Computation and the Condition Number

Because in many numerical studies the coefficients of polynomials are known only approximately, the central issue is to understand how a slight perturbation of the coefficients affects the roots. In 1993, Shub and Smale [1126] introduced the condition number $\mu(P)$ of a polynomial $P : \mathbb{C} \to \mathbb{C}$ such that for polynomials with small $\mu(P)$ the root computation is stable with respect to perturbations of the coefficients. For the polynomial $P(z) = \sum_{i=0}^{N} a_i z^i$ with complex coefficients a_i, its

Weil norm is

$$||P|| = \sqrt{\sum_{i=0}^{N} \binom{N}{i}^{-1} |a_i|^2},$$

where $\binom{N}{i} = \frac{N!}{i!(N-k)!}$ are binomial coefficients, and its *(normalized) condition number* is

$$\mu(P) = \max_{z \in \mathbb{C}: P(z)=0} \left(\sqrt{N} \frac{||P||(1+|z|^2)^{N/2-1}}{|P'(z)|} \right).$$

To find the roots of a polynomial P, Shub and Smale suggested to choose another polynomial P_0 of the same degree with known roots and analyse the roots of $P_t = (1-t)P_0 + tP$. In 2011, Bürgisser and Cucker [232] proved that this method can efficiently find the roots of almost all polynomials P provided that the condition number of P_0 is small. To apply this method, we need to have an explicit family of polynomials of every degree N with as small a condition number as possible. Shub and Smale [1125] proved that a random polynomial P of degree N has $\mu(P) \leq N$ with probability at least $1/2$, but finding an *explicit* family of polynomials with such condition number was an important open problem. In 2019, Beltrán et al. posted online a paper [113] in which they constructed an explicit sequence of polynomials with an even better upper bound on μ.

Theorem 7.33 *There exists a constant C and an explicit formula, which, given any integer $N > 0$, produces a polynomial P_N of degree N such that*

$$\mu(P_N) \leq C\sqrt{N}.$$

In the proof of Theorem 7.33, the polynomial P_N is defined by specifying the set of its roots. The authors first constructed a set of N points on the unit sphere $\mathbb{S}^2 \subset \mathbb{R}^3$ with suitable geometric properties, and then the images of these points through the stereographic projection onto the complex plane are the zeros of P_N. The authors also proved the existence of a constant $c > 0$ such that $\mu(P) \geq c\sqrt{N}$ for every polynomial P of degree N. Hence, the sequence of polynomials constructed in Theorem 7.33 has the optimal condition numbers, up to a constant factor.

Minimization of a Polynomial in Several Variables
A more complicated but also very important problem with numerous applications is to find the global minimum of a real polynomial in several variables. The problem is NP-hard even for polynomials of degree 4, so it is natural to look for some approximation methods. In 2001, Lasserre [766] proved a theorem connecting this problem to the so-called *linear matrix inequality (LMI) problem*, which can be solved efficiently. The LMI problem is the problem of minimizing $\sum_{i=1}^{m} c_i y_i$ subject to the constraint that the matrix $A_0 + \sum_{i=1}^{m} A_i y_i$ is positive semidefinite, where y_i are real variables, c_i are given real numbers, and A_i are given square matrices.

Theorem 7.34 *Let $P : \mathbb{R}^n \to \mathbb{R}$ be a polynomial of even degree with (unknown) global minimum $p^* = P(x^*)$. If the non-negative polynomial $P(x) - p^*$ is a sum of squares of polynomials, then x^* can be found from the LMI problem as above, with c_i and A_i being efficiently computable from the coefficients of P.*

Because the LMI problem can be solved efficiently, Theorem 7.34 provides an efficient method to find x^* in the special case when $P(x) - p^*$ is a sum of squares. Using Theorem 7.34, Lasserre [766] also showed how to find an approximate minimum for general P by solving a finite sequence of LMI problems. For polynomial minimization problems arising in practice, this sequence is often not too long, and the resulting method is efficient.

Computing the Nash Equilibrium

Another fundamental problem is computing the m-player Nash equilibrium. This is the central problem in game theory with applications to, for example, economics. Informally, a *Nash equilibrium* is a situation in a non-cooperative game in which no player can gain anything by changing his/her strategy. A famous theorem of Nash [924] from 1950 implies that Nash equilibrium always exists. Since then, it was a long-standing open question whether or not it can be computed efficiently. On the one hand, no-one had presented a polynomial-time algorithm, even for $m = 2$ players. On the other hand, it is difficult to use standard NP-completeness theory for decision problems to prove the hardness results for Nash equilibrium, because, by Nash's theorem, the decision version of this problem is trivial, always with a "Yes" answer. To study such problems, Papadimitriou [966] introduced in 1994 a different complexity class, called PPAD. We do not provide a formal definition here, but mention that (i) the problem of computing Nash-equilibrium belongs to this class, and (ii) the conjecture that PPAD contains problems with no polynomial-time algorithms is well-believed. A problem A in class PPAD is called PPAD-*complete* if every other problem in PPAD has a polynomial-time reduction to A. Hence, PPAD-complete problems cannot have a polynomial time algorithm, unless the well-believed conjecture stated in (ii) fails.

In 2009, Daskalakis et al. [319] made a breakthrough by showing that the problem of computing the m-player Nash equilibrium is PPAD-complete for every $m \geq 3$. This provides strong evidence for the intractability of this problem, except in the important case of $m = 2$ players. In the same year, Chen et al. [270] extended the analysis of [319] to the remaining $m = 2$ case, thus making a final step in settling the complexity of computing the Nash equilibrium.

Formally, let $\mathbb{P}^n$ be the set of vectors $x = (x_1, \ldots, x_n) \in \mathbb{R}^n$ such that all $x_i \geq 0$ and $\sum_{i=1}^n x_i = 1$. The problem of computing a two-player Nash equilibrium is: given two $m \times n$ matrices A and B with rational entries, compute vectors $x^* \in \mathbb{P}^m$ and $y^* \in \mathbb{P}^n$ such that $(x^*)^T A y^* \geq x^T A y^*$ and $(x^*)^T B y^* \geq (x^*)^T B y$ for all $x \in \mathbb{P}^m$ and $y \in \mathbb{P}^n$.

Theorem 7.35 *The problem of computing a two-player Nash equilibrium is PPAD-complete.*

The proof of the three-player version of Theorem 7.35 in [319] starts with the PPAD-complete problem 3-DIMENSIONAL BROUWER, a 3-dimensional discrete version of the problem of computing Brouwer's fixed point[9]. This problem is then reduced to the analysis of *degree-3 graphical games*, an important class of games proposed by Kearns et al. [697] in 2001. Finally, the graphical game problem has been reduced to computing four-player, and then three-player Nash equilibrium. However, there seems to be no way to reduce graphical game to two-player Nash. Chen et al. resolved this difficulty by reducing 3-DIMENSIONAL BROUWER directly to two-player Nash, without using graphical games as an intermediate step.

The authors also proved that the problem of computing a two-player Nash equilibrium does not have a fully polynomial approximation scheme unless PPAD is contained in P. In other words, ϵ-approximate Nash equilibrium cannot be computed in time polynomial in N and $1/\epsilon$, where N is the input size. On the positive side, Lipton et al. [803] presented in 2003 an $N^{O(\log N/\epsilon^2)}$-time algorithm for this approximate computation.

The Saddle-Point Problems
In the special case $B = -A$, a two-player Nash equilibrium can be computed efficiently. In this case, the Nash equilibrium problem can be formulated as a saddle-point problem. Such problems also arise in many other applications, for example in image denoising and segmentation. In 2011, Chambolle and Pock [259] developed a simple algorithm that can solve a large class of saddle-point problems. Practical experiments suggest that the convergence is in fact fast, which makes the algorithm widely applicable.

By a *saddle-point problem* we mean a problem of the form

$$\min_{x \in \mathbb{R}^n} \max_{y \in \mathbb{R}^m} H(x, y), \tag{7.6}$$

where $H : \mathbb{R}^n \times \mathbb{R}^m \to \mathbb{R}$ is some function. A pair (x^*, y^*) is called a *saddle-point* of problem (7.6) if $H(x, y^*) \geq H(x^*, y^*)$ for all $x \in \mathbb{R}^n$ and $H(x^*, y) \leq H(x^*, y^*)$ for all $y \in \mathbb{R}^m$. Chambolle and Pock developed a method to solve this problem for a large family of functions H which we define next.

Let $\langle x, y \rangle$ denote the inner product in $\mathbb{R}^n$, and let $||x|| = \sqrt{\langle x, x \rangle}$. Let $G : \mathbb{R}^n \to [0, +\infty]$ and $F : \mathbb{R}^m \to [0, +\infty]$ be proper convex lower-semicontinuous (l.s.c.) functions, and let

$$F^*(y) = \sup_{x \in \mathbb{R}^m} (\langle x, y \rangle - F(x))$$

be the convex conjugate of F. Let $K : \mathbb{R}^n \to \mathbb{R}^m$ be a continuous linear operator with norm $||K|| = \max_{x \in \mathbb{R}^n} \frac{||Kx||}{||x||}$, and let $K^* : \mathbb{R}^m \to \mathbb{R}^n$ be such that $\langle Kx, y \rangle =$

---

[9] *Brouwer's fixed-point theorem* states that for any compact convex set $K \subset \mathbb{R}^n$ and any continuous function $f : K \to K$ there is a point $x_0 \in K$ such that $f(x_0) = x_0$.

$\langle x, K^*y \rangle$ for all $x \in \mathbb{R}^n$, $y \in \mathbb{R}^m$. Finally, assume that $H(x, y)$ in (7.6) can be represented in the form $H(x, y) = \langle Kx, y \rangle + G(x) - F^*(y)$.

We next describe a method for computing a saddle-point of (7.6) developed in [259]. For any $V : \mathbb{R}^n \to [0, +\infty]$, $y \in \mathbb{R}^n$, and $\tau > 0$, define $d_{V,\tau}(x) = \arg \min_{x \in \mathbb{R}^n} \left(\frac{\|x-y\|^2}{2\tau} + V(x) \right)$. For $\tau > 0$, $\sigma > 0$, $\theta \in [0, 1]$, initial points $x^0 \in \mathbb{R}^n$, $y^0 \in \mathbb{R}^m$, and $\bar{x}^0 = x^0$, define the sequences $\{x^n\}$, $\{y^n\}$, $\{\bar{x}^n\}$ iteratively as follows:

$$y^{n+1} = d_{F^*,\sigma}(y^n + \sigma K \bar{x}^n),$$

$$x^{n+1} = d_{G,\tau}(x^n - \tau K^* y^{n+1}),$$

$$\bar{x}^{n+1} = x^{n+1} + \theta(x^{n+1} - x^n),$$

for every $n \geq 0$.

Theorem 7.36 *If $\theta = 1$ and $\tau\sigma\|K\|^2 < 1$, then there exist a saddle-point (x^*, y^*) of (7.6) such that $\lim_{n\to\infty} x^n = x^*$ and $\lim_{n\to\infty} y^n = y^*$.*

In 2016, the same authors showed that the algorithm presented in Theorem 7.36 is capable of solving an even more general class of problems, see their paper [260] for details.

Minimizing the Total Variation
Another important optimization problem arising in various applications including image denoising, zooming, and the computation of the mean curvature motion of interfaces, is the problem of minimizing the total variation. In 2004, Chambolle [258] presented an algorithm for constructing a sequence of approximations which is guaranteed to converge to the optimal solution of this problem.

Formally, for $N \times N$ matrix u with entries $u_{i,j}$ define its *total variation* as

$$J(u) = \sum_{i=1}^{N} \sum_{j=1}^{N} \sqrt{(u_{i+1,j} - u_{i,j})^2 + (u_{i,j+1} - u_{i,j})^2},$$

where $u_{N+1,j} = u_{N,j}$ and $u_{i,N+1} = u_{i,N}$ by convention. For a given $N \times N$ matrix g and $\lambda > 0$, consider the optimization problem

$$\min_u \left(\frac{\|u - g\|}{2\lambda} + J(u) \right) \tag{7.7}$$

which looks for a matrix u close to g but with small total variation $J(u)$. For example, if g represents a noisy image, an optimal solution u^* of (7.7) represents an image which preserves the essential features of g but removes noise.

Theorem 7.37 *There is an algorithm for constructing a sequence u_n of matrices which is guaranteed to converge to an optimal solution u^* of (7.7).*

Theorem 7.37 does not provide a formal estimate for how fast u_n converges to u^*, but numerical experiments in [258] show that this algorithm works quite fast in practical applications.

7.5 Average-Case Algorithms and Complexity

The Satisfiability Threshold Conjecture

If a problem is NP-hard in the worst case, one may ask if we can at least efficiently solve a random instance of it with high probability. Here, we discuss this question for the k-SAT problem. Recall that a k-*clause* is a disjunction of k Boolean variables, some of which may be negated. For a set V_n of n Boolean variables, let $C_k(V_n)$ denote the set of all $2^k n^k$ possible k-clauses on V_n. A random k-CNF formula $F_k(n, m)$ is formed by selecting uniformly, independently and with replacement m k-clauses from $C_k(V_n)$ and taking their conjunction. For each $k \geq 2$, let r_k be the supremum of r such that $F_k(n, rn)$ is satisfiable with probability tending to 1 as $n \to \infty$. Conversely, let r_k^* be the infimum of r such that $F_k(n, rn)$ is unsatisfiable with probability tending to 1 as $n \to \infty$. Obviously, $r_k \leq r_k^*$. The *Satisfiability Threshold Conjecture* predicts that in fact $r_k = r_k^*$ for all $k \geq 3$. If this conjecture is true, then the trivial algorithm outputting "Yes" if $\frac{m}{n} < r_k$ and "No" if $\frac{m}{n} > r_k$ would solve the large random k-SAT instances reasonably well.

In 1997, Monasson and Zecchina [902] used non-rigorous methods originating from statistical mechanics to predict that the asymptotic growth of r_k is $2^k \log 2$. Also in 1997, Dubois and Boufkhad [366] proved the upper bound $r_k^* \leq 2^k \log 2$. In 2004, Achlioptas and Peres [4] proved the asymptotically matching lower bound on r_k. This confirms the prediction of Monasson and Zecchina, and also implies that $r_k = r_k^*(1 - o(1))$, which is an asymptotic form of the Satisfiability Threshold Conjecture.

Theorem 7.38 *There exists a sequence δ_k convergent to 0 such that for all $k \geq 3$,*

$$r_k \geq 2^k \log 2 - (k + 1)\frac{\log 2}{2} - 1 - \delta_k.$$

Most of the previous proofs of the lower bounds for r_k were algorithmic: that is, researchers analysed some specific algorithm for finding satisfying assignments and proved that if r is smaller than a certain value $f(k)$, then this algorithm succeeds on random input $F_k(n, rn)$ with probability tending to 1 as $n \to \infty$. This proves that $r_k \geq f(k)$. However, no such method has been able to produce a lower bound better than $c2^k/k$ for some constant c. In contrast, the proof of Theorem 7.38 is non-algorithmic, and instead analyses the "geometry" of the set of satisfying assignments. This leads to the optimal asymptotic lower bound for r_k, and also produces specific lower bounds for small values of k. For example, Achlioptas and Peres proved that $r_4 \geq 7.91$, improving the previous algorithmic lower bound $r_4 \geq 5.54$.

Systems of Polynomial Equations

Another fundamental NP-hard problem is solving systems of polynomial equations. In 1998, Smale [1137] presented a list of problems he considered as the most important ones for the twenty-first century. Smale's 17th problem asks for the algorithm which could compute a solution of a random system of polynomial equations with polynomial average running time. In 2009, Beltrán and Pardo [114] developed a probabilistic algorithm which achieves exactly this.

Let $n > 0$ be a positive integer. Let H be the space of all systems $f = [f_1, \ldots, f_n] : \mathbb{C}^n \to \mathbb{C}^n$ on n polynomial equations in n unknowns. For $f \in H$, define

$$V_{\mathbb{C}}(f) = \{x \in \mathbb{C}^n : f_i(x) = 0,\ 1 \leq i \leq n\} \subset \mathbb{C}^n,$$

the set of solutions of f. Also, denote by D the maximal degree of a polynomial in f, and N the number of bits needed to write down the system f (that is, write down all coefficients of all polynomials in f). Note that $N \geq 2^{\min(n, D)}$.

Theorem 7.39 *There exists an (explicit) probabilistic algorithm that computes an approximate solution $x_0 \in V_{\mathbb{C}}(f)$ for any given $f \in H$ with any given accuracy (with probability 1), such that the average number of arithmetic operations of this algorithm is polynomial in N.*

In 2011, Bürgisser and Cucker [232] extended Theorem 7.39 in several directions. First, they developed a different version of the Beltrán–Pardo probabilistic algorithm and proved that this new version not only has polynomial average running time, but in fact has polynomial smoothed complexity, as defined in Theorem 7.29. This implies that *any* (non-random) system of polynomial equations can be solved fast if we allow a small random perturbation of the coefficients.

Theorem 7.40 *There is a (randomized) algorithm for computing an approximate solution to a system of polynomial equations which has polynomial smoothed complexity.*

Second, Bürgisser and Cucker [232] developed a *deterministic* algorithm for the same problem with running time $N^{O(\log \log N)}$. This was achieved by derandomization of the Beltrán–Pardo algorithm. In 2017, Lairez [758] found an alternative derandomization, which runs in time polynomial in N. This provides a deterministic solution to Smale's 17th problem.

Because the input size N is large in many applications, even a polynomial running time is often impractical. In 2020, Lairez [759] developed a probabilistic algorithm for the same problem whose average running time is $O(n^6 D^4 N)$. Because $N \geq 2^{\min(n, D)}$, this running time grows as $N^{1+o(1)}$ when $\min(n, D) \to \infty$. Such algorithms are called *quasilinear algorithms*.

Theorem 7.41 *There exists an (explicit) probabilistic algorithm that computes an approximate solution $x_0 \in V_{\mathbb{C}}(f)$ for any given $f \in H$ with any given accuracy (with probability 1), such that the average number of arithmetic operations of this algorithm is $O(n^6 D^4 N)$.*

The proof of Theorem 7.41 is based on the idea of rigid continuation paths. We start by computing a solution to each equation separately. Then we apply rigid motions of the equations to make the solutions to all equations match. In this way we obtain a new system of equations for which we know a solution. Finally, we continuously return the equations to the original position and keep track of the movement of the solution.

Cryptography

In the sections above we discussed some problems which are hard in the worst-case but "easy" for an average (random) input. Proving the existence of problems which have no polynomial algorithm even on average is a more difficult task than "just" proving that P $\neq$ NP. Even more difficult is to prove the existence of *one-way functions*, functions which are easy to compute on every input but are hard to invert given the image of a random input. For example, if we assume that integer factorization is hard, then the function which takes to two prime numbers p and q (say, in binary) as input and returns their product as output, is an obvious candidate for being a one-way function. Such functions are important because of their applications in cryptography, where they are used to create encryption schemes such that any message is easy to encrypt but difficult to decrypt. Much of current cryptography is based on the difficulty of integer factorization, but this problem is not NP-hard, and in fact is solvable in polynomial time by a quantum computer, which we will discuss later in this book.

Another important research direction in cryptography is the existence of a *fully homomorphic encryption (FHE) scheme*. An FHE scheme is a procedure that allows one to evaluate circuits over encrypted data without being able to decrypt. The notion of FHE scheme was introduced by Rivest et al. [1047] in 1978. An FHE would allow us to perform arbitrary computations on encrypted data without decrypting them, and would therefore lead to the possibility of more secure cloud computing. For example, a program could help with preparation of tax return forms using encrypted financial information. However, the existence of an FHE was a long-standing open problem, and many experts believed that it does not exist. In 2009, Gentry [495] proved that an FHE exists and can be actually constructed. This caused a revolution in cryptography.

Theorem 7.42 *Fully homomorphic encryption schemes exist.*

7.6   The Complexity Zoo

While the P vs NP question is the main open problem in theoretical computer science, there are many other interesting questions to study in the theory of computing, which naturally leads to the introduction of a huge variety of different complexity classes. For example, with respect to the running time, problems which have a (deterministic) algorithm running in time at most $2^{P(n)}$, where P is a polynomial and n is the length of input, forms a class called EXP. It is clear from the definition that P $\subset$ EXP and the inclusion is known to be strict. Also, it is not difficult to check that NP $\subseteq$ EXP. However, there are problems in EXP (think about the question "is a given chess position on the $n \times n$ board the winning one for white?") for which there is no obvious way to verify the "Yes" answer in polynomial time. Problems for which the "Yes" answer can at least be verified in exponential time form a (huge) complexity class NEXP.

The classes above are defined for deterministic algorithms. Theorem 7.10 provides a *randomized* algorithm to solve an important family of problems. Decision problems for which there is a randomized polynomial-time algorithm which correctly solves them with high probability form a complexity class called BPP. A central open question in the theory of randomized algorithms is whether P is equal to BPP.

We have also discussed the class #P of counting problems in the connection with Theorem 7.4. There are many more complexity classes, and the collection of them all is called the *complexity zoo*. A small portion of these classes will be discussed in the next subsections.

Parallel Algorithms

Modern computers have a large number of processors, and it is very useful to develop algorithms which can "parallelize" the task and allow all the processors to work in parallel. Formally, we say that a decision problem belongs to class NC if there is a deterministic parallel algorithm that solves it in polylogarithmic time using polynomially many processors. It follows from the definition that NC $\subseteq$ P. The class NC includes many important problems, for example, integer addition, multiplication, and division. For some other problems in P the question of existence of a good deterministic parallel algorithm is highly non-trivial. For example, this question has long been open for the problem of finding a perfect matching[10] in a planar graph.

There were many partial results indicating that the planar graph perfect matching problem should be in NC. First, Karp et al. [685] developed in 1986 a *randomized* parallel algorithm for this problem which works in polylogarithmic time on polynomially many processors. The class of problems which have such an algorithm is called "Random NC", or RNC. Second, Vazirani [1229] proved in 1989 that

---

[10] Recall that a *perfect matching* in a graph is a set of edges such that each vertex is incident to exactly one edge in the set.

the seemingly harder problem of *counting* perfect matchings in planar graphs is in NC. Third, Miller and Naor [892] developed an NC algorithm for finding a perfect matching in *bipartite* planar graphs. Despite this, NC algorithm for this problem on general planar graphs remained elusive until 2020, when such an algorithm was finally developed by Anari and Vazirani [42].

Theorem 7.43 *There is an NC algorithm which, given a planar graph, returns a perfect matching in it, if it has one.*

The proof of Theorem 7.43 is based on the method, first described by Mahajan and Varadarajan [827] in 2000, which shows how to use an NC algorithm for counting perfect matchings to find one. Unfortunately, Mahajan and Varadarajan were only able to do this for bipartite planar graphs. Anari and Vazirani proved Theorem 7.43 by extending this method to general planar graphs.

Memory-Efficient Algorithms
So far we have evaluated algorithms based solely on their running times. However, there are other important criteria. For example, efficient memory use is one of the fundamental requirements for algorithms, almost as important and well-studied as running time. If the decision problem has a deterministic algorithm which solves it using memory $O(\log n)$, where n is the size of the (read-only) input, we say that this problem belongs to the class *log-space*. In 2008, Reingold [1040] proved that the *undirected connectivity problem*, an important problem naturally arising in many applications, belongs to this class. This problem has the following formulation: given as input an undirected graph G with n vertices and two vertices s and t in G, find out whether or not these two vertices are connected by a path in G.

Theorem 7.44 *The undirected connectivity problem belongs to the class log-space.*

The proof of Theorem 7.44 is heavily based on the theory of expander graphs in general and Theorem 2.50 in particular. There was a known randomized memory-efficient algorithm for this problem, and expander graphs can be used to de-randomize such algorithms because they share many properties of random graphs, but, on the other hand, can be constructed deterministically as in Theorem 2.50.

In addition to log-space, there are many other complexity classes which are based on memory use. The most natural of them is the class PSPACE of problems which can be solved by an algorithm which uses memory polynomial in the input size.

The Polynomial Hierarchy
The NP-complete SAT problem asks to decide whether there exists a vector of boolean variables $x = (x_1, \ldots, x_n)$ such that $F(x)$ is true where F is a SAT formula. Let us now exchange the "Yes" and "No" answers and ask whether it is true that a given SAT formula is not satisfiable. In this formulation, the problem is not believed to belong to NP, because it is unclear how to give a proof of such un-satisfiability verifiable in polynomial time. Problems in class NP with "Yes" and "No" answers exchanged form a class called coNP. While an NP problem can be written as $\exists x F(x)$, a coNP problem asks to verify that $\forall x F'(x)$, where $F'(x)$ is the statement that x does *not* satisfy SAT formula F.

We can then consider more general problems in the form $\exists x \, \forall y \, F(x, y)$, and so on, with any number of existential and universal quantifiers. These problems form a sequence of nested complexity classes, which form a polynomial hierarchy, see below for the formal definition. It is widely believed that not only P$\neq$NP, but also NP$\neq$coNP, and, more generally, all the complexity classes in the polynomial hierarchy are different. However, proving this is even more difficult than proving P$\neq$NP, and therefore is well out of reach of the current methods.

Such separation problems are easier for the complexity classes with oracles, which we now define. Let $S = \{0, 1\}$, S^n be the set of words $\sigma_1 \ldots \sigma_n$ with each σ_i being 0 or 1, and let $S^* = \bigcup_{n=0}^{\infty} S^n$. Let $|x|$ denote the length of any word $x \in S^*$. Let $A : S^* \rightarrow S$ be an arbitrary function, which we call an *oracle*. Imagine a computer which can do everything a regular computer can do, but, in addition, can instantly compute $A(x)$ for any $x \in S^*$. Let P_k^A be the set of functions $F(x_1, \ldots, x_k)$ in k variables $x_i \in S^*$, for which there is a polynomial P and a program on such a computer which computes F in at most $P\left(\sum_{i=1}^{k} |x_i|\right)$ operations. For any integer $k \geq 0$, let Σ_k^A be the set of functions $f : S^* \rightarrow S$ for which there exists a polynomial P and function $F \in P_{k+1}^A$ such that $f(x) = 1$ if and only if

$$\exists x_1 \in S^{P(|x|)} \, \forall x_2 \in S^{P(|x|)} \ldots \exists x_k \in S^{P(|x|)} \text{ such that } F(x_1, x_2, \ldots, x_k, x) = 1,$$

where the last quantifier can be $\exists$ or $\forall$ depending on whether k is odd or even. It is clear from the definition that $\Sigma_0^A \subseteq \Sigma_1^A \subseteq \Sigma_2^A \subseteq \ldots$, and we denote by $PH^A = \bigcup_{k=0}^{\infty} \Sigma_k^A$ their union. If the oracle A is (for example) a constant function, it does not help with the calculation and can be ignored. In this case we will denote the corresponding complexity classes as just Σ_k, $k = 0, 1, 2 \ldots$, and $PH = \bigcup_{k=0}^{\infty} \Sigma_k$. Note that $\Sigma_0 = P$, $\Sigma_1 = NP$, and so on.

It is widely believed that $\Sigma_k \subsetneq \Sigma_{k+1}$ for all k, but this is wide open. On the other hand, Baker et al. [84] proved the existence of an oracle A such that $P^A \subsetneq NP^A$, and cited a question of Meyer asking whether there exists an oracle A such that $\Sigma_k^A \subsetneq \Sigma_{k+1}^A$ for all k. In 1986, Håstad [576] gave a positive answer to this question.

In 1981, Bennett and Gill [121] proved that $P^A \subsetneq NP^A$ is not just with respect to *some* oracle A but in fact with respect to *almost all* oracles, in the sense that it is true with probability 1 if the oracle A is selected at random. Many authors including Håstad [576] conjectured that, for a random oracle A, we also have $\Sigma_k^A \subsetneq \Sigma_{k+1}^A$ for all k. In 2017, Håstad et al. [579] confirmed this conjecture.

Theorem 7.45 *With probability 1, a random oracle A satisfies $\Sigma_k^A \subsetneq \Sigma_{k+1}^A$ for all integers $k \geq 0$.*

The Power of Quantum Computers

One of the major technological open questions of today is whether it is possible to build a full-scale quantum computer. We will not give a formal definition of this device here, but mention that when/if it will be built, it will be able to efficiently solve some problems for which no efficient algorithm on a regular/classical computer is currently known. A notable example is the integer factorization problem, for which a polynomial-time quantum algorithm has been developed by Shor [1123] in 1994. Examples like this lead to the widely believed conjecture that BQP $\not\subseteq$ P, where BQP is the class of decision problems solvable by a quantum computer in polynomial time. Moreover, it is believed that BQP $\not\subseteq$ PH, where PH is the union of all complexity classes in the polynomial hierarchy, see Theorem 7.45 above for the definition.

Proving separation results like BQP $\not\subseteq$ PH is well out of reach of the current techniques. Such problems often become easier with an oracle, for example, it is not known whether P $\neq$ PH, but it is known that $P^A \neq PH^A$ for some oracle A, and in fact for *most* oracles (again see Theorem 7.45). In 1993, Bernstein and Vazirani conjectured (at least verbally) the existence of an oracle A such that $BQP^A \not\subseteq PH^A$. In 2019, Raz and Tal [1028] confirmed this conjecture.

Theorem 7.46 *There exists an oracle A such that $BQP^A \not\subseteq PH^A$.*

We now say a few words about the proof. Let D and D' be two probability distributions over a finite set X. We say that an algorithm A *distinguishes between D and D' with advantage ϵ* if

$$\left| \mathbb{P}_{x \sim D}[A \text{ accepts } x] - \mathbb{P}_{x' \sim D'}[A \text{ accepts } x'] \right| = \epsilon.$$

Raz and Tal constructed a distribution D over strings $\{\pm 1\}^{2N}$ such that (i) there exists a quantum algorithm that makes one query to the input, and runs in time $O(\log N)$, that distinguishes between D and the uniform distribution with advantage at least $C/\log N$ for some constant C, but, on the other hand (ii) no Boolean circuit of quasi-polynomial (in N) size and constant depth[11] distinguishes between D and the uniform distribution with advantage better than polynomial in $\log N$. Theorem 7.46 follows from this fact via well-known reductions.

The Power of Quantum Provers

There are problems which are difficult to solve, but the solution is at least easy to verify once presented to you. Think, for example, about integer factorization. It is difficult to factorize a 10,000-digit number, but, if you could send it to an all-powerful but untrusted machine called a "prover", and get back the factorization, it would be easy to verify that the answer is correct. In a similar way, you may verify the "Yes" answer for any problem from class NP. If you are allowed to have many rounds of interactions with the prover, can toss secret coins, and are ready to accept a small probability of error, you can in fact verify a much wider class of problems. A classic example is: to verify the prover's claim that two graphs G and H are *not* isomorphic, you can randomly perturb their vertices and ask the prover to guess which graph is a perturbation of G. You can then repeat this procedure n times. If the graphs are indeed not isomorphic, the prover will be able to answer

---

[11] See, for example, [1258] for the missing definitions of Boolean circuit, its size and depth.

correctly each time. If, however, they are isomorphic, the prover will have at least one mistake with probability $1 - 2^{-n}$. In fact, Shamir [1108] proved in 1992 that the solution of every decision problem from the class PSPACE can be verified in this way. If you have two provers which cannot communicate with each other, you can cross-validate their answers, and, as proved by Babai et al. [70] in 1991, can verify in polynomial time the "Yes" answer to any problem from the class NEXP, which is believed to be much lager than PSPACE.

In 2020, Ji et al. posted online a preprint [659] in which they proved that if we assume that provers are quantum and allow them to share entangled qubits, then we can verify the "Yes" answer to all decidable problems and even to some undecidable ones!

We now need to give some definitions. Let $\Sigma = \{0, 1\}$, Σ^n be the set of words $\sigma_1 \ldots \sigma_n$ with each σ_i being 0 or 1, and let $\Sigma^* = \bigcup_{n=0}^{\infty} \Sigma^n$. Let $|x|$ denote the length of any word $x \in \Sigma^*$. A *language* is any set of words $L \subset \Sigma^*$. A language L is called *recursively enumerable* if there is a computer program such that its output is a full list $x_1, x_2, \ldots$ of the words in L (in any order). Note that if L is infinite, such a program will run forever. The set of all recursively enumerable languages is denoted RE.

Imagine two quantum computers, which we call provers, that are all-powerful, that is, can do any (even an infinite!) number of operations in, say, one second. The provers can share any number of entangled qubits, but any other communication between them is prohibited. Also, we have a standard (not quantum and not all-powerful) computer, called a verifier, which can do everything a regular computer can do, but, in addition, can ask any questions to the provers. Let MIP* be the set of all languages L for which there exists a polynomial P and a computer program on such a computer, which, for any input $x \in \Sigma^*$, and with probability greater than (say) $\frac{3}{4}$:

  (i) performs at most $P(|x|)$ operations, then stops and output Yes or No,
 (ii) outputs Yes if $x \in L$ and the provers give correct answers, and
(iii) outputs No if $x \notin L$ even if the provers are trying to cheat.

It is not difficult to see that MIP* is a subset of RE. Ji et al. [659] proved that in fact these classes coincide.

Theorem 7.47 MIP* = RE

A language L is called *decidable* if there is a program which, for any input $x \in \Sigma^*$, runs for a finite time and then correctly outputs whether $x \in L$ or not. The set RE contains all languages which are decidable and some languages which are not. Theorem 7.47, however, states that any language L in RE, even undecidable, belongs to MIP*. This means that, by interaction with two all-powerful quantum provers which share entangled qubits, we can check whether $x \in L$ after performing at most $P(|x|)$ operations. The theorem has important consequences in pure mathematics. In particular, it implies that *Connes' embedding conjecture*, a central conjecture in the theory of von Neumann algebras, is false. See [659] for details.

Theorem 7.47 was proved by constructing a verification protocol as described above for a specific problem H called the *halting problem*, the problem of determining whether a given computer program on a given input will ever halt or run forever. It is known that this problem is RE-complete, in the sense that any other problem in RE reduces to it, hence $H \in$ MIP* implies RE $\subseteq$ MIP*, and Theorem 7.47 follows.

7.7 Decidable and Undecidable Problems

The Word Problem

We have, until now, mostly discussed which problems can be solved *efficiently*. However, there are problems for which there is no algorithm at all, even if we are ready to wait any finite amount of time. Such problems are called *undecidable*. In 1937, Turing [1218] famously proved that the *halting problem*, the problem of determining whether a given computer program on a given input will ever halt or run forever, is undecidable.

Other famous examples of undecidable problems are the *word problems* in monoids and groups. Let S be a set and let $\cdot$ be a binary operation $S \times S \rightarrow S$. We say that pair $(S, \cdot)$ form a *monoid* if

(i) $(a \cdot b) \cdot c = a \cdot (b \cdot c)$ holds for all $a, b, c \in S$, and
(ii) there exists an element e in S, called the *identity element*, such that $e \cdot a = a \cdot e = a$ holds for all $a \in S$.

A monoid is called an *inverse monoid* if for every $x \in S$ there exists a $y \in S$ (called the *inverse* of x) such that $x = x \cdot y \cdot x$ and $y = y \cdot x \cdot y$. Note that every group is an inverse monoid with the extra property $x \cdot y = y \cdot x = e$ whenever y is an inverse of x.

Recall that a group G is called *finitely presented group* if it has a presentation $\langle S | R \rangle$ with a finite set of generators S and a finite set of relations R. If, moreover, the set R consists of just one relation, the group G is called a *one-relator group*. A *word* in a group G is any product of generators and their inverse elements. The *word problem* in G is the problem to decide whether a given word represents an identity element. The same definitions extends to inverse monoids and (with obvious modifications) to monoids.

In the middle of the twentieth century, Markov and Post independently proved that there are finitely presented monoids for which the word problem is undecidable. In 1958, Novikov [942] proved that the same is true for groups. This raises the question: for which special classes of monoids and groups is the word problem decidable? In particular, Magnus [826] proved in 1932 that the one-relator groups have decidable word problem. Can this result be extended to a (harder) word problem for inverse monoids? In 2020, Gray [519] gave a negative answer to this long-standing open question.

Theorem 7.48 *There exists a one-relator inverse monoid with undecidable word problem.*

We next concentrate on the case of groups. Their word problem is undecidable in general but solvable for some special classes of groups, like one-relator groups [826], finitely presented simple groups [1130], etc. For groups for which the word problem is solvable, we may ask *how efficiently* it is solvable. For example, for which groups is the word problem in P? For which groups it is in NP? Can it be NP-complete? In 2002, Sapir et al. [1079] answered the last question affirmatively.

Theorem 7.49 *There exists a finitely presented group for which the word problem is NP-complete.*

An even more ambition problem is to characterize *all* finitely generated groups with the word problem in, for example, the class NP. This question is closely related to the growth of the isoperimetric function. For any word a representing the identity element e in a finitely generated group G, let $T(a)$ be the minimal number of applications of the relators required to transform a into e. A function f is called an *isoperimetric function* of G if $T(a) \leq f(|a|)$ for any a representing e, where $|a|$ is the length of word a. We say that an isoperimetric function f is *polynomial* if $f(n) = cn^k$ for some constants c and k.

If a group G is finitely presented and has a polynomial isoperimetric function, then its word problem is obviously in NP. The converse to this statement is not true: there are groups H with word problem in NP but such that every isoperimetric function of H grows very fast. However, Birget et al. [151] proved in 2002 that every such group H can be embedded into a different group G which is finitely presented and has a polynomial isoperimetric function. This provides the desired characterization.

Theorem 7.50 *The word problem of a finitely generated group H is in NP if and only if H is a subgroup of a finitely presented group G with a polynomial isoperimetric function.*

Other Undecidable Problems in Groups

Novikov's theorem [942] that the word problem in groups is undecidable immediately implies the undecidability of some other fundamental problems for groups, including, for example, the *conjugacy problem*, which asks whether two given words represent conjugate elements[12] in a group G, and the *group isomorphism problem*, which asks to determine whether two given finite group presentations present isomorphic groups.

The converses to these statements are not true in a strong sense. Miller [893] constructed in 1971 the first examples of finitely generated groups with decidable word problem but undecidable conjugacy problem. Moreover, Macintyre constructed groups G with decidable word problem such that any group H that has a subgroup

---

[12] Recall that two elements a and b of a group G are *conjugate* if there is an element $g \in G$ such that $b = g^{-1}ag$.

isomorphic to G must have undecidable conjugacy problem. In other words, such groups G cannot be embedded in any group with decidable conjugacy problem.

Macintyre's examples have torsions (non-identity elements of finite order), and the construction used them in an essential way. Motivated by this, Collins asked in the 1970s whether there exists any torsion-free group G with the same property. This question was open for decades, until it was resolved by Darbinyan in a preprint [317] posted online in 2017.

Theorem 7.51 *There exists a finitely generated (in fact, finitely presented) torsion-free group G with decidable word problem such that G cannot be embedded into any finitely generated group with decidable conjugacy problem.*

The starting point of the proof of Theorem 7.51 is the existence of recursively enumerable recursively inseparable sets. Recall that a set $S \subset \mathbb{N}$ of natural numbers is called *recursive* is there is an algorithm which, given $k \in \mathbb{N}$, runs in finite time and correctly decides whether $k \in S$. A set S is *recursively enumerable* if there is a computer program that lists all elements of S. Two disjoint sets $S_1, S_2 \subset \mathbb{N}$ are called *recursively inseparable* if there is no recursive set $T \subset \mathbb{N}$ such that $S_1 \subseteq T$ and $S_2 \subset \mathbb{N} \setminus T$. In 1958, Smullyan [1140] proved the existence of recursively enumerable recursively inseparable sets. Darbinyan used these sets to construct a group G satisfying the conditions of Theorem 7.51.

In addition to the word, conjugacy, and isomorphism problems, there are many other basic decision problems about finitely generated and finitely presented groups, and the decidability status of some of them remains open or has been resolved only recently. A famous example is the problem of deciding whether or not a group has a proper subgroup of finite index.[13] In 2015, Bridson and Wilton [211] proved that this problem is also undecidable.

Theorem 7.52 *There is no algorithm that can determine whether or not a finitely presented group has a proper subgroup of finite index.*

Computability of Real Numbers
A related notion to decidability is computability of, for example, real numbers. A real number x is called *computable* if we can approximate it to within any given accuracy, that is, if there exists an algorithm which, for a given integer n, computes integers a_n and b_n such that $\left| x - \frac{a_n}{b_n} \right| \leq \frac{1}{n}$. Because there are uncountably many real numbers and only countably many algorithms, almost all real numbers are not computable. However, important real numbers naturally arising in mathematics and its applications, such as $\pi = 3.14159...$ and $e = 2.71828...$, are obviously computable, and it is difficult to imagine a real number which naturally arises mathematics or its applications but is not computable. In 2010, however, Hochman and Meyerovitch [609] provided examples of real numbers which are not

---

[13] Recall that a (left) coset of a subgroup H of group G with respect to $g \in G$ is the set $\{gh : h \in H\}$. We say that subgroup H is of *finite index* if there are only finitely many different cosets of H in G.

computable but quite meaningful, because they arise from the study of so-called shifts of finite type, which are important objects of study in symbolic dynamics.

Let $F \subset \mathbb{Z}^d$ be a finite set of $|F|$ points in $\mathbb{Z}^d$, and let Σ be a finite set of $|\Sigma|$ colours. Let L be any subset of $|F|^{|\Sigma|}$ colourings of F. A *translate* of F by $x \in \mathbb{Z}^d$ is the set $\{x + y \mid y \in F\}$. A colouring of $\mathbb{Z}^d$ in colours from Σ is called *admissible* for L if all translates of F are coloured according to a pattern belonging to L. The *shift of finite type* (SFT) defined by L is the set X of all admissible colourings of $\mathbb{Z}^d$.

For every positive integer n, let F_n be the set of points $(y_1, \ldots, y_d) \in \mathbb{Z}^d$ such that $1 \leq y_i \leq n$ for all $i = 1, 2, \ldots, d$. For the SFT X, let $f_X(n)$ be the number of different colourings of F_n which can appear in some colouring $x \in X$. The (topological) entropy of X is then defined as

$$h(X) = \lim_{n \to \infty} \frac{1}{n^d} \log_2 f_X(n).$$

Theorem 7.53 *There exists an SFT X such that the real number $y = h(X)$ is not computable.*

Theorem 7.53 was proved as a corollary of a more general result stating that a real number $h \geq 0$ is the entropy of some SFT if and only if it is *right recursively enumerable*, which means that there exists a computable sequence of rational numbers converging to h from above. This result immediately implies Theorem 7.53, because it is well-known that the set of right recursively enumerable real numbers is strictly larger than the set of computable real numbers.

Chapter 8
Logic and Set Theory

The vast majority of recent theorems related to logic and set theory are too technical to be included in this book. However, there are a few exceptions.

8.1 Squaring the Circle with and Without the Axiom of Choice

The *axiom of choice* is one of the fundamental axioms that lies in the foundation of modern mathematics. It states that given any set S of mutually disjoint nonempty sets, there exists a set X that contains exactly one element in common with each set $Y \in S$. The axiom look completely "obvious", but is known to have some paradoxical-looking consequences. For example, Banach and Tarski [93] used it to show that one can decompose a unit ball in $\mathbb{R}^3$ into a finite number of disjoint pieces, move and rotate the pieces, and construct two unit balls!

Tarski proved that this paradox does not work in $\mathbb{R}^2$: if we start with a measurable set A, partition it into (possibly unmeasurable) pieces $A_1 \cup \cdots \cup A_m$, and then rearrange the pieces (by translation and rotation) to get a measurable set B, then the sets A and B must have the same area. In 1925, Tarski asked whether any sets A and B of the same area can be converted into each other in this way. Specifically, he asked if a disk can be transformed in this way to a square of the same area. This problem became known as the *Tarski's circle squaring problem*.

In 1990, Laczkovich [754] answered this question affirmatively: a disk can indeed be partitioned into a finite number of pieces which can then be rearranged to form a square! The pieces in the Laczkovich proof are very complicated, and in particular non-measurable. Laczkovich left the problem of whether or not the disk can be squared with measurable pieces as an interesting open question. In 2017, Grabowski et al. [512] answered this question affirmatively. Moreover, they showed that this can be done by only translating the pieces, even without rotation.

B. Grechuk, *Landscape of 21st Century Mathematics*,
https://doi.org/10.1007/978-3-030-80627-9_8

More formally, we call two sets $A, B \subset \mathbb{R}^2$ *equidecomposable by translations* if there are partitions $A = A_1 \cup \cdots \cup A_m$ and $B = B_1 \cup \cdots \cup B_m$ such that $B_i = A_i + v_i$, $i = 1, \ldots, m$, for some vectors $v_1, \ldots, v_m \in \mathbb{R}^2$. If, moreover, all A_i (and thus B_i) are Lebesgue measurable, we say that A and B are equidecomposable by translations *with Lebesgue measurable parts*.

Theorem 8.1 *A disk and a square of the same area on the plane are equidecomposable by translations with Lebesgue measurable parts.*

Of course, there is nothing special about the disk and square, and the statement of Theorem 8.1 remains correct in all dimensions, and for any pair of bounded sets with non-empty interiors of the same Lebesgue measure subject to the technical condition that the boundaries of the sets have dimensions lower that the sets.

Some people may find the positive resolution of Tarski's circle squaring problem to be quite impressive. Others may argue that "if the axiom of choice can be used to transform a unit ball into two such balls, then it is not at all a surprise that it can be used to transform a disk into a square". Indeed, the proof of Theorem 8.1 uses the axiom of choice in a substantial way, and the pieces used in the decompositions are, while measurable, still quite complicated. In particular, they are not Borel sets. A subset $A \subset \mathbb{R}^2$ is a *Borel set* if it can be formed from open sets through the operations of countable union, countable intersection, and set difference.

In 2017, Marks and Unger [846] proved a version of Theorem 8.1 in which all pieces A_i and B_i used in the partitions are Borel sets. Moreover, the proof of their theorem does not use the axiom of choice. If such a proof is possible, we say that the problem has a *constructive solution*. The existence of a constructive solution to Tarski's circle squaring problem answers a question of Wagon [1244, Appendix C, Question 2.a] asked in 1985.

Theorem 8.2 *A disk and a square of the same area on the plane are equidecomposable by translations with Borel parts.*

In 1963, Dubins et al. [365] introduced the notion of *scissors congruence* in $\mathbb{R}^2$, which is equidecomposition using pieces whose boundaries consist of a single Jordan curve. They proved that a disk is not scissors congruent to a square, and, moreover, is not scissors congruent to any other convex set other than a disk. Hence, the Borel parts used in Theorem 8.2 are essentially the "nicest" pieces one could hope for.

8.2   Defining $\mathbb{Z}$ in $\mathbb{Q}$

The 10th problem in the Hilbert's famous list [605] asks for an algorithm which, given a polynomial equation in several variables with integer coefficients, decides whether it has an integer solution. In 1970, Matiyasevich [853] proved that there is no such algorithm.

Some researchers think that Hilbert was actually interested in the question of whether a polynomial equation with rational coefficients has a rational solution. This question is known as *Hilbert's 10th problem over* $\mathbb{Q}$ and remains an open question of central importance. A possible reason why Hilbert formulated the question over the integers is because he expected a positive solution, and a positive solution over $\mathbb{Z}$ would imply a positive solution over $\mathbb{Q}$. Indeed, any equation $P(x_1, \ldots, x_n) = 0$, where x_i are rational variables, can be written as

$$P\left(\frac{y_1}{z_1}, \ldots, \frac{y_n}{z_n}\right) = 0$$

where y_i and z_i are integer variables, which can then be reduced to a polynomial equation over the integers by multiplying by the common denominator.[1]

The argument above is possible because the condition $x \in \mathbb{Q}$ can be expressed as $\exists y, z \in \mathbb{Z} : x = \frac{y}{z}$. If we could express the condition that $t \in \mathbb{Z}$ as

$$t \in \mathbb{Q} \quad \text{and} \quad \exists x_1, \ldots, \exists x_n \in \mathbb{Q} : P(t, x_1, \ldots, x_n) = 0 \tag{8.1}$$

for some polynomial P with integer coefficients, then the argument as above would reduce solving polynomial equations over $\mathbb{Z}$ to solving them over $\mathbb{Q}$. By Matiyasevich's theorem, we could then conclude that Hilbert's 10th problem over $\mathbb{Q}$ has a negative answer.

In 1949, Robinson [1052] developed a formula like (8.1) but with a mixture of existential quantifiers $\exists$ with the universal quantifiers $\forall$. Specifically, she proved that there exists a polynomial $P(t, x_1, x_2, y_1, \ldots, y_7, z_1, \ldots, z_6)$ with integer coefficients such that a rational number t is an integer if and only if

$$\forall x_1, x_2 \in \mathbb{Q} \, \exists y_1, \ldots, y_7 \in \mathbb{Q} \, \forall z_1, \ldots, z_6 \in \mathbb{Q} \, P(t, x_1, x_2, y_1, \ldots, y_7, z_1, \ldots, z_6) = 0.$$

In 2009, Poonen [1001] found a simper formula with just two universal and seven existential quantifiers. In 2016, Koenigsmann [728] developed a similar formula which uses universal quantifiers only.

Theorem 8.3 *There exists a positive integer n, and a polynomial $P(t, x_1, \ldots, x_n)$ in $n + 1$ variables with integer coefficients, such that, for any $t \in \mathbb{Q}$, we have $t \in \mathbb{Z}$ if and only if*

$$\forall x_1, \ldots, \forall x_n \in \mathbb{Q} : P(t, x_1, \ldots, x_n) \neq 0.$$

The proof of Theorem 8.3 uses elementary tools like the Hasse principle for quadratic forms (which we discussed earlier in this book in connection with Theorem 1.26) and the law of quadratic reciprocity to transform some existential formulas into universal formulas. Theorem 8.3 is interesting in its own, but is

---

[1] We also need to take care about the additional condition $z_i \neq 0$, but this is not a difficult issue.

also considered as a major step towards proving (8.1) and resolving Hilbert's 10th problem over $\mathbb{Q}$.

In 2020, Prunescu [1009] proved a version of (8.1) for *exponential* diophantine equations, that is, equations of the form

$$P(t_1, \ldots, t_n) = 0,$$

where P is a polynomial with integer coefficients and t_j are expressions of the form α^β, where α and β can be variables or integers. Specifically, Prunescu proved that a rational number t is a positive integer if and only if

$$\exists x, y \in \mathbb{Q}: \ x > 0 \ \text{and} \ y - x > 0 \ \text{and} \ x^y = y^x \ \text{and} \ ty = (t+1)x. \quad (8.2)$$

The proof is short and follows from the old result that the only solutions to the equation $x^y = y^x$ in rational numbers x, y such that $0 < x < y$ are $x_n = (1 + \frac{1}{n})^n$, $y_n = (1 + \frac{1}{n})^{n+1}$ for a positive integer n. In combination with Matiyasevich's theorem, (8.2) implies that *there is no algorithm to decide whether a system of exponential diophantine equations has a solution in rational numbers.*

8.3   Comparing Cardinalities of Sets

We say that two (possibly infinite) sets A and B have the same *cardinality*, and write $|A| = |B|$, if there is a one-to-one correspondence (bijection) between their elements. We write $|A| \leq |B|$ if there exists a bijection from A to a subset of B. If $|A| \leq |B|$ but not $|A| = |B|$, we write $|A| < |B|$ and say that the set A has cardinality strictly less than the cardinality of the set B. In 1874, Cantor, using a simple diagonal argument, proved that the cardinality of the set of integers (denoted $\aleph_0$) is strictly less than the cardinality of the set of real numbers (denoted $2^{\aleph_0}$). He conjectured that there is no set whose cardinality is strictly between $\aleph_0$ and $2^{\aleph_0}$, and this statement became known as the *continuum hypothesis*. When Hilbert formulated his famous list [605] of problems for the twentieth century, he included the continuum hypothesis as problem number one.

In 1908, Zermelo [1290] proposed a system of axioms for set theory. This system, enriched by the axiom of choice and by one more crucial axiom suggested by Fraenkel [465] in 1921, became the most common foundation of the whole of mathematics. It is now called *Zermelo–Fraenkel set theory with the axiom of choice* (ZFC for short). In 1931, Gödel [501] proved his famous *incompleteness theorem*, stating that for every formal axiomatic system capable of modelling basic arithmetic, including ZFC,[2] there are statements (including statements about natural

---

[2] Here and further we assume without mention that ZFC is consistent, that is, cannot prove false statements.

numbers) which the system can neither prove nor disprove. Such statements are called *independent* of the axiomatic system.

In 1963, Cohen [285, 286] proved that the continuum hypothesis is independent from ZFC. Hence, there is no chance to either prove or disprove it, unless we assume without proof some new axioms. Since 1963, many other statements in set theory have been proved to be independent. In fact, almost every question in the form "is the cardinality of two given sets equal?" turned out to be either relatively easy or independent of ZFC.

However, there is one notable exception which we now describe. Let $A \subseteq^* B$ mean that the set $\{x : x \in A, x \notin B\}$ is finite. Let $\mathbb{N}^\mathbb{N}$ be the family of all infinite sequences of natural numbers, and let $D \subset \mathbb{N}^\mathbb{N}$. We say that D has a *pseudo-intersection* if there is an infinite $A \subseteq \mathbb{N}$ such that $A \subseteq^* B$ for all $B \in D$. We say that D has the *strong finite intersection property* (s.f.i.p. for short) if every non-empty finite subfamily of D has infinite intersection. D is called *well-ordered* by $\subseteq^*$ if $A \subseteq^* B$ or $B \subseteq^* A$ for all $A, B \in D$. D is called a *tower* if it is well-ordered by $\subseteq^*$ and has no pseudo-intersection. Let $\mathbf{p}$ be the smallest cardinality of $D \subset \mathbb{N}^\mathbb{N}$ which has the s.f.i.p. but has no pseudo-intersection. Let $\mathbf{t}$ be the smallest cardinality of $D \subset \mathbb{N}^\mathbb{N}$ which is a tower.

Clearly, both $\mathbf{p}$ and $\mathbf{t}$ are at least $\aleph_0$ and no more than $2^{\aleph_0}$. It is easy to see that $\mathbf{p} \leq \mathbf{t}$, since a tower has the s.f.i.p. In 1936, Hausdorff [580] proved that $\aleph_1 \leq \mathbf{p}$, where $\aleph_1$ is the smallest cardinality of an uncountable set. In 1948, Rothberger [1062] proved that $\mathbf{p} = \aleph_1$ implies $\mathbf{p} = \mathbf{t}$. This originates from the problem of whether we can prove $\mathbf{p} = \mathbf{t}$ unconditionally. For many decades, researchers could not prove that this equality is true, could not prove that it is false, but also could not prove that it is independent of ZFC.

In 2016, Malliaris and Shelah [834] resolved this long-standing open problem by showing that the equality $\mathbf{p} = \mathbf{t}$ is actually true, and this can be proved in ZFC.

Theorem 8.4 *The cardinal numbers $\mathbf{p}$ and $\mathbf{t}$ are equal.*

8.4 Characterization of Random Real Numbers

When we say that some property is true for a "typical" (or random) real number, we mean that it holds for all real numbers except possibly for a set of measure 0. For example, random real numbers are irrational. But can we formally define a set $U \subset \mathbb{R}$ consisting of precisely "random" numbers? By considering a binary expansion, the concept of a random real number is equivalent to the concept of a random sequence of 0s and 1s. One of the most natural formalizations of this concept is the notion of "computational randomness" suggested by Schnorr [1093] in 1971. The idea is that a player receives symbols in order, and, based on the string σ she currently observes and the amount $M(\sigma)$ of money she currently has, can make a bet $q \in [0, M(\sigma)]$ on what the next bit is. She then gets q if she is right and loses q otherwise. These rules imply the "fairness" condition

$$M(\sigma 0) + M(\sigma 1) = 2M(\sigma). \qquad (8.3)$$

The string is then called *random* if there is no computable betting strategy which can win an arbitrarily high sum of money. More formally, let Σ be the set of all finite strings σ consisting of bits (each bit is 0 or 1). A function $M : \Sigma \to \mathbb{Q}^+$, where $\mathbb{Q}^+$ is the set of non-negative rational numbers, is called *computable* if there is a computer program which returns $M(\sigma)$ on input σ. A *martingale* is a function $M : \Sigma \to \mathbb{Q}^+$ such that (8.3) holds for all $\sigma \in \Sigma$. For every infinite string Z of bits, let Z_n denote the first n bits. We say that M *succeeds* on Z if the sequence $\{M(Z_n)\}_{n=1}^{\infty}$ is unbounded. We say that an infinite string Z is *computably random* if no computable martingale succeeds on Z. A real number $z \in [0, 1)$ is called *computably random* if its binary expansion is computably random.

In 2016, Brattka et al. [200] characterized the set U of computably random real numbers in $[0, 1)$ via differentiability of computable functions. Formally, a *Cauchy name* for a real number x is a sequence $\{q_n\}_{n=1}^{\infty}$ of rational numbers such that (i) $|q_n - q_k| \le 2^{-n}$ for each $k \ge n$ and (ii) $\lim_{n\to\infty} q_n = x$. A real number $x \in \mathbb{R}$ is called *computable* if it has a computable Cauchy name. A sequence $\{x_n\}_{n=1}^{\infty}$ of real numbers is *computable* if there exists a computable double sequence $\{q_{n,k}\}_{n,k\in\mathbb{N}}$ of rational numbers such that, for each n, the sequence $\{q_{n,k}\}_{k=1}^{\infty}$ is a Cauchy name for x_n. A function $f : [0, 1) \to \mathbb{R}$ is called *computable* if

(a) for each computable sequence of real numbers $\{x_n\}_{n=1}^{\infty} \subset [0, 1)$, the sequence $\{f(x_n)\}_{n=1}^{\infty}$ is computable, and
(b) there is a computable function $h : \mathbb{N} \to \mathbb{N}$ such that, for each $n \in \mathbb{N}$, $|x - y| < 2^{-h(n)}$ implies $|f(x) - f(y)| < 2^{-n}$.

Now we are ready to state the main result of [200].

Theorem 8.5 *Let $z \in [0, 1)$. Then z is computably random if and only if each computable non-decreasing function $f : [0, 1) \to \mathbb{R}$ is differentiable at z.*

References

1. Aaronson, J., Gilat, D., Keane, M., de Valk, V.: An algebraic construction of a class of one-dependent processes. Ann. Probab. **17**(1), 128–143 (1989)
2. Achlioptas, D., Friedgut, E.: A sharp threshold for k-colorability. Random Struct. Algorithm. **14**(1), 63–70 (1999)
3. Achlioptas, D., Naor, A.: The two possible values of the chromatic number of a random graph. Ann. Math. **162**(3), 1335–1351 (2005)
4. Achlioptas, D., Peres, Y.: The threshold for random k-SAT is $2^k \log 2 - o(k)$. J. Am. Math. Soc. **17**(4), 947–973 (2004)
5. Adamczewski, B., Bugeaud, Y.: On the complexity of algebraic numbers, I. Expansions in integer bases. Ann. Math. **165**(2), 547–565 (2007)
6. Addario-Berry, L., Dalal, K., McDiarmid, C., Reed, B.A., Thomason, A.: Vertex-colouring edge-weightings. Combinatorica **27**(1), 1–12 (2007)
7. Addario-Berry, L., Dalal, K., Reed, B.A.: Degree constrained subgraphs. Discrete Appl. Math. **156**(7), 1168–1174 (2008)
8. Adiprasito, K.: Combinatorial Lefschetz theorems beyond positivity. Preprint (2018). arXiv:1812.10454
9. Adiprasito, K., Avvakumov, S., Karasev, R.: A subexponential size $\mathbb{RP}^n$. Preprint (2020). arXiv:2009.02703
10. Adiprasito, K., Huh, J., Katz, E.: Hodge theory for combinatorial geometries. Ann. Math. **188**(2), 381–452 (2018)
11. Adleman, L.M., Pomerance, C., Rumely, R.S.: On distinguishing prime numbers from composite numbers. Ann. Math. **117**, 173–206 (1983)
12. Agrawal, M., Kayal, N., Saxena, N.: PRIMES is in P. Ann. Math. **160**(2), 781–793 (2004)
13. Aharoni, R.: König's duality theorem for infinite bipartite graphs. J. Lond. Math. Soc. **2**(1), 1–12 (1984)
14. Aharoni, R.: Menger's theorem for countable graphs. J. Combin. Theory Ser. B **43**(3), 303–313 (1987)
15. Aharoni, R., Berger, E.: Menger's theorem for infinite graphs. Invent. Math. **176**(1), 1–62 (2009)
16. Ahlgren, S., Boylan, M.: Arithmetic properties of the partition function. Invent. Math. **153**(3), 487–502 (2003)
17. Aizenman, M., Barsky, D.J.: Sharpness of the phase transition in percolation models. Comm. Math. Phys. **108**(3), 489–526 (1987)
18. Aizenman, M., Kesten, H., Newman, C.M.: Uniqueness of the infinite cluster and continuity of connectivity functions for short and long range percolation. Comm. Math. Phys. **111**(4), 505–531 (1987)

© The Author(s), under exclusive license to Springer Nature Switzerland AG 2021
B. Grechuk, *Landscape of 21st Century Mathematics*,
https://doi.org/10.1007/978-3-030-80627-9

19. Aizenman, M., Newman, C.M.: Tree graph inequalities and critical behavior in percolation models. J. Stat. Phys. **36**(1), 107–143 (1984)
20. Akbari, S., Maimani, H.R., Yassemi, S.: When a zero-divisor graph is planar or a complete r-partite graph. J. Algebra **270**(1), 169–180 (2003)
21. Akbulut, S., McCarthy, J.D.: Casson's Invariant for Oriented Homology Three-Spheres: An Exposition, Mathematical Notes, vol. 36. Princeton University Press (2014)
22. Alberti, G., Csörnyei, M., Preiss, D.: Differentiability of Lipschitz functions, structure of null sets, and other problems. In: Proceedings of the International Congress of Mathematicians 2010 (ICM 2010) (In 4 Volumes) Vol. I: Plenary Lectures and Ceremonies Vols. II–IV: Invited Lectures, pp. 1379–1394. World Scientific (2010)
23. Aldaz, J.M.: Remarks on the Hardy–Littlewood maximal function. Proc. Roy. Soc. Edinburgh Sect. A **128**(1), 1–9 (1998)
24. Aldaz, J.M.: The weak type (1, 1) bounds for the maximal function associated to cubes grow to infinity with the dimension. Ann. Math. **173**(2), 1013–1023 (2011)
25. Aldaz, J.M., Varona, J.L.: Singular measures and convolution operators. Acta Mathematica Sinica English Series **23**(3), 487–490 (2007)
26. Aldous, D.J.: The $\zeta(2)$ limit in the random assignment problem. Random Struct. Algorithm. **18**(4), 381–418 (2001)
27. Alexander, J., Hirschowitz, A.: Polynomial interpolation in several variables. J. Algebraic Geom. **4**(2), 201–222 (1995)
28. Allen, P., Böttcher, J., Griffiths, S., Kohayakawa, Y., Morris, R.: The chromatic thresholds of graphs. Adv. Math. **235**, 261–295 (2013)
29. Allouche, J.P.: Sur la conjecture de "Syracuse–Kakutani–Collatz". Séminaire de théorie des nombres de Bordeaux **8**, 1–16 (1978)
30. Alman, J., Williams, V.V.: A refined laser method and faster matrix multiplication. Preprint (2020). arXiv:2010.05846
31. Almgren, F.J.: Some interior regularity theorems for minimal surfaces and an extension of Bernstein's theorem. Ann. Math. **84**, 277–292 (1966)
32. Almgren, F.J.: Existence and regularity almost everywhere of solutions to elliptic variational problems with constraints, Memoirs of the American Mathematical Society, vol. 165. American Mathematical Soc. (1976)
33. Alon, N.: Eigenvalues and expanders. Combinatorica **6**(2), 83–96 (1986)
34. Alon, N.: Restricted colorings of graphs. Surv. Comb. **187**, 1–33 (1993)
35. Alon, N., Krivelevich, M.: The concentration of the chromatic number of random graphs. Combinatorica **17**(3), 303–313 (1997)
36. Alon, N., Makarychev, K., Makarychev, Y., Naor, A.: Quadratic forms on graphs. Invent. Math. **163**(3), 499–522 (2006)
37. Alon, N., Shapira, A., Sudakov, B.: Additive approximation for edge-deletion problems. Ann. Math. **170**(1), 371–411 (2009)
38. Alon, N., Yuster, R.: Threshold functions for h-factors. Combin. Probab. Comput. **2**(2), 137–144 (1993)
39. Ambrosio, L., Cabré, X.: Entire solutions of semilinear elliptic equations in $\mathbb{R}^3$ and a conjecture of De Giorgi. J. Am. Math. Soc. **13**(4), 725–739 (2000)
40. Amir, G.: On the joint behaviour of speed and entropy of random walks on groups. Groups Geom. Dyn. **11**(2), 455–467 (2017)
41. Anari, N., Liu, K., Gharan, S.O., Vinzant, C.: Log-concave polynomials III: Mason's ultra-log-concavity conjecture for independent sets of matroids. Preprint (2018). arXiv:1811.01600
42. Anari, N., Vazirani, V.V.: Planar graph perfect matching is in NC. J. ACM **67**(4), 1–34 (2020)
43. Anderson, D.F., Frazier, A., Lauve, A., Livingston, P.S.: The zero-divisor graph of a commutative ring. II. In: Ideal Theoretic Methods in Commutative Algebra (Columbia, MO, 1999), Lecture Notes in Pure and Appl. Math., vol. 220, pp. 61–72. Dekker (2001)

44. Angel, O., Schramm, O.: Uniform infinite planar triangulations. Comm. Math. Phys. **241**(2), 191–213 (2003)
45. Angenent, S.B.: Shrinking doughnuts. In: Nonlinear Diffusion Equations and Their Equilibrium States, vol. 3, pp. 21–38. Springer (1992)
46. Apostol, T.M.: Modular functions and Dirichlet series in number theory. Graduate Texts in Mathematics, vol. 41. Springer Science & Business Media (2012)
47. Arestov, V., Babenko, A.: On Delsarte scheme of estimating the contact numbers. Trudy Matematicheskogo Instituta imeni VA Steklova **219**, 44–73 (1997)
48. Aron, R., Gurariy, V., Seoane, J.: Lineability and spaceability of sets of functions on $\mathbb{R}$. Proc. Am. Math. Soc. **133**(3), 795–803 (2005)
49. Arora, S., Lee, J., Naor, A.: Euclidean distortion and the sparsest cut. J. Am. Math. Soc. **21**(1), 1–21 (2008)
50. Arora, S., Lund, C., Motwani, R., Sudan, M., Szegedy, M.: Proof verification and the hardness of approximation problems. J. ACM **45**(3), 501–555 (1998)
51. Arora, S., Rao, S., Vazirani, U.: Expander flows, geometric embeddings and graph partitioning. J. ACM **56**(2), 1–37 (2009)
52. Artin, E.: Über die zerlegung definiter funktionen in quadrate. Abhandlungen aus dem mathematischen Seminar der Universität Hamburg **5**(1), 100–115 (1927)
53. Artstein, S., Ball, K., Barthe, F., Naor, A.: Solution of Shannon's problem on the monotonicity of entropy. J. Am. Math. Soc. **17**(4), 975–982 (2004)
54. Artstein, S., Milman, V., Szarek, S.J.: Duality of metric entropy. Ann. Math. **159**(3), 1313–1328 (2004)
55. Artstein, S., Milman, V.D., Szarek, S.J.: More on the duality conjecture for entropy numbers. Comptes Rendus Mathematique **336**(6), 479–482 (2003)
56. Artstein-Avidan, S., Milman, V.: The concept of duality in convex analysis, and the characterization of the Legendre transform. Ann. Math. **169**(2), 661–674 (2009)
57. Atar, R., Burdzy, K.: On Neumann eigenfunctions in lip domains. J. Am. Math. Soc. **17**(2), 243–265 (2004)
58. Atkin, A.O.L., O'Brien, J.N.: Some properties of $p(n)$ and $c(n)$ modulo powers of 13. Trans. Am. Math. Soc. **126**(3), 442–459 (1967)
59. Atkin, A.O.L., Swinnerton-Dyer, P.: Some properties of partitions. Proc. Lond. Math. Soc. **3**(1), 84–106 (1954)
60. Avila, A., Forni, G.: Weak mixing for interval exchange transformations and translation flows. Ann. Math. **165**(2), 637–664 (2007)
61. Avila, A., Jitomirskaya, S.: The ten martini problem. Ann. Math. **170**(1), 303–342 (2009)
62. Avila, A., Jitomirskaya, S., Marx, C.A.: Spectral theory of extended Harper's model and a question by Erdős and Szekeres. Invent. Math. **210**(1), 283–339 (2017)
63. Avila, A., Leguil, M.: Weak mixing properties of interval exchange transformations and translation flows. Preprint (2016). arXiv:1605.03048
64. Avila, A., Lyubich, M.: Hausdorff dimension and conformal measures of Feigenbaum Julia sets. J. Am. Math. Soc. **21**(2), 305–363 (2008)
65. Avila, A., Moreira, C.G.: Statistical properties of unimodal maps: the quadratic family. Ann. Math. **161**(3), 831–881 (2005)
66. Azbel, M.Y.: Energy spectrum of a conduction electron in a magnetic field. Sov. Phys. JETP **19**(3), 634–645 (1964)
67. Aziz, H., Mackenzie, S.: A discrete and bounded envy-free cake cutting protocol for four agents. In: Proceedings of the Forty-Eighth Annual ACM Symposium on the Theory of Computing, pp. 454–464 (2016)
68. Aziz, H., Mackenzie, S.: A bounded and envy-free cake cutting algorithm. Commun. ACM **63**(4), 119–126 (2020)
69. Babai, L.: Graph isomorphism in quasipolynomial time. Preprint (2015). arXiv:1512.03547
70. Babai, L., Fortnow, L., Lund, C.: Non-deterministic exponential time has two-prover interactive protocols. Comput. Complexity **1**(1), 3–40 (1991)

71. Babai, L., Hetyei, G., Kantor, W.M., Lubotzky, A., Seress, A.: On the diameter of finite groups. In: Proceedings 1990 31st Annual Symposium on Foundations of Computer Science, pp. 857–865. IEEE (1990)

72. Babai, L., Kantor, W.M., Luks, E.M.: Computational complexity and the classification of finite simple groups. In: 24th Annual Symposium on Foundations of Computer Science (SFCS 1983), pp. 162–171. IEEE (1983)

73. Babai, L., Seress, Á.: On the diameter of Cayley graphs of the symmetric group. J. Combin. Theory Ser. A **49**(1), 175–179 (1988)

74. Babai, L., Seress, Á.: On the diameter of permutation groups. Eur. J. Combin. **13**(4), 231–243 (1992)

75. Babai, L., Sós, V.T.: Sidon sets in groups and induced subgraphs of Cayley graphs. Eur. J. Combin. **6**(2), 101–114 (1985)

76. Bachoc, C., Vallentin, F.: New upper bounds for kissing numbers from semidefinite programming. J. Am. Math. Soc. **21**(3), 909–924 (2008)

77. Bader, U., Gelander, T., Monod, N.: A fixed point theorem for L^1 spaces. Invent. Math. **189**(1), 143–148 (2012)

78. Badziahin, D., Pollington, A., Velani, S.: On a problem in simultaneous Diophantine approximation: Schmidt's conjecture. Ann. Math. **174**(3), 1837–1883 (2011)

79. Bajmóczy, E.G., Bárány, I.: On a common generalization of Borsuk's and Radon's theorem. Acta Mathematica Academiae Scientiarum Hungarica **34**(3-4), 347–350 (1979)

80. Bajnok, B.: Construction of spherical t-designs. Geom. Dedicata **43**(2), 167–179 (1992)

81. Baker, R.C.: Diagonal cubic equations. II. Acta Arith. **53**(3), 217–250 (1989)

82. Baker, R.C., Harman, G.: The difference between consecutive primes. Proc. Lond. Math. Soc. **3**(2), 261–280 (1996)

83. Baker, R.C., Harman, G., Pintz, J.: The difference between consecutive primes, II. Proc. Lond. Math. Soc. **83**(3), 532–562 (2001)

84. Baker, T., Gill, J., Solovay, R.: Relativizations of the P = ?NP question. SIAM J. Comput. **4**(4), 431–442 (1975)

85. Balázs, M., Seppäläinen, T.: Order of current variance and diffusivity in the asymmetric simple exclusion process. Ann. Math. **171**(2), 1237–1265 (2010)

86. Balister, P., Bollobás, B., Morris, R.: The sharp threshold for making squares. Ann. Math. **188**(1), 49–143 (2018)

87. Balister, P., Bollobás, B., Morris, R., Sahasrabudhe, J., Tiba, M.: Flat Littlewood polynomials exist. Ann. Math. **192**(3), 977–1004 (2020)

88. Ball, K.: A lower bound for the optimal density of lattice packings. Int. Math. Res. Not. IMRN **1992**(10), 217–221 (1992)

89. Balla, I., Dräxler, F., Keevash, P., Sudakov, B.: Equiangular lines and spherical codes in Euclidean space. Invent. Math. **211**(1), 179–212 (2018)

90. Balogh, J., Morris, R., Samotij, W.: Independent sets in hypergraphs. J. Am. Math. Soc. **28**(3), 669–709 (2015)

91. Banach, S.: Sur les opérations dans les ensembles abstraits et leur application aux équations intégrales. Fund. Math. **3**(1), 133–181 (1922)

92. Banach, S.: Théorie des opérations linéaires. Monografje Matematuczne **1**, vii+254 pp. (1932)

93. Banach, S., Tarski, A.: Sur la décomposition des ensembles de points en parties respectivement congruentes. Fund. Math. **6**(1), 244–277 (1924)

94. Bang, A.S.: Taltheoretiske Undersøgelser. (Fortsat, se S. 80). Tidsskrift for mathematik **4**, 130–137 (1886)

95. Banuelos, R., Burdzy, K.: On the hot spots conjecture of J. Rauch. J. Funct. Anal. **164**(1), 1–33 (1999)

96. Baragar, A.: The exponent for the Markoff–Hurwitz equations. Pac. J. Math. **182**(1), 1–21 (1998)

97. Barak, B., Rao, A., Shaltiel, R., Wigderson, A.: 2-source dispersers for $n^{o(1)}$ entropy, and Ramsey graphs beating the Frankl–Wilson construction. Ann. Math. **176**(3), 1483–1543 (2012)
98. Bárány, I., Shlosman, S.B., Szücs, A.: On a topological generalization of a theorem of Tverberg. J. Lond. Math. Soc. **2**(1), 158–164 (1981)
99. Barlow, M.T., Murugan, M.: Stability of the elliptic Harnack inequality. Ann. Math. **187**(3), 777–823 (2018)
100. Barnet-Lamb, T., Geraghty, D., Harris, M., Taylor, R.: A family of Calabi–Yau varieties and potential automorphy II. Publ. Res. Inst. Math. Sci. **47**(1), 29–98 (2011)
101. Barnette, D.: w_v paths on 3-polytopes. J. Comb. Theory **7**(1), 62–70 (1969)
102. Bartal, Y., Linial, N., Mendel, M., Naor, A.: On metric Ramsey-type phenomena. Ann. Math. **162**(2), 643–709 (2005)
103. Barvinok, A., Woods, K.: Short rational generating functions for lattice point problems. J. Am. Math. Soc. **16**(4), 957–979 (2003)
104. Basu, S., Sombra, M.: Polynomial partitioning on varieties of codimension two and point-hypersurface incidences in four dimensions. Discrete Comput. Geom. **55**(1), 158–184 (2016)
105. Bate, D.: Purely unrectifiable metric spaces and perturbations of Lipschitz functions. Acta Math. **224**(1), 1–65 (2020)
106. Bateman, M., Katz, N.: New bounds on cap sets. J. Am. Math. Soc. **25**(2), 585–613 (2012)
107. Baudier, F., Lancien, G., Schlumprecht, T.: The coarse geometry of Tsirelson's space and applications. J. Am. Math. Soc. **31**(3), 699–717 (2018)
108. Bebendorf, M.: A note on the Poincaré inequality for convex domains. Zeitschrift für Analysis und ihre Anwendungen **22**(4), 751–756 (2003)
109. Beck, J.: An algorithmic approach to the Lovász local lemma. I. Random Struct. Algorithm. **2**(4), 343–365 (1991)
110. Behrend, F.A.: On sets of integers which contain no three terms in arithmetical progression. Proc. Natl. Acad. Sci. USA **32**(12), 331 (1946)
111. Behrend, R.E., Fischer, I., Konvalinka, M.: Diagonally and antidiagonally symmetric alternating sign matrices of odd order. Adv. Math. **315**, 324–365 (2017)
112. Bell, J., Zelmanov, E.: On the growth of algebras, semigroups, and hereditary languages. Preprint (2019). arXiv:1907.01777
113. Beltrán, C., Etayo, U., Marzo, J., Ortega-Cerdà, J.: A sequence of polynomials with optimal condition number. Preprint (2019). arXiv:1903.01356
114. Beltrán, C., Pardo, L.: Smale's 17th problem: average polynomial time to compute affine and projective solutions. J. Am. Math. Soc. **22**(2), 363–385 (2009)
115. Bender, E.A., Canfield, E.R.: The asymptotic number of rooted maps on a surface. J. Combin. Theory Ser. A **43**(2), 244–257 (1986)
116. Bender, E.A., Gao, Z., Wormald, N.C.: The number of 2-connected labelled planar graphs. Elec. J. Comb. **9**, 43 (2002)
117. Benjamini, I., Kesten, H., Peres, Y., Schramm, O.: Geometry of the uniform spanning forest: transitions in dimensions 4, 8, 12, ... Ann. Math. **160**(2), 465–491 (2004)
118. Benjamini, I., Schramm, O.: Percolation beyond $\mathbb{Z}^d$, many questions and a few answers. Electron. Commun. Probab. **1**, 71–82 (1996)
119. Benjamini, I., Schramm, O.: Percolation in the hyperbolic plane. J. Am. Math. Soc. **14**(2), 487–507 (2001)
120. Benjamini, I., Schramm, O.: Recurrence of distributional limits of finite planar graphs. Electron. J. Probab. **6**, 1–13 (2001)
121. Bennett, C.H., Gill, J.: Relative to a random oracle A, $P^A \neq NP^A \neq$ co-NP^A with probability 1. SIAM J. Comput. **10**(1), 96–113 (1981)
122. Bennett, J., Carbery, A., Tao, T.: On the multilinear restriction and Kakeya conjectures. Acta Math. **196**(2), 261–302 (2006)
123. Bennett, M.A., Dahmen, S.R.: Klein forms and the generalized superelliptic equation. Ann. Math. **177**(1), 171–239 (2013)

124. Bennett, M.A., Siksek, S.: A conjecture of Erdős, supersingular primes and short character sums. Ann. Math. **191**(2), 355–392 (2020)
125. Benoist, O.: The period-index problem for real surfaces. Publications mathématiques de l'IHÉS **130**(1), 63–110 (2019)
126. Bentkus, V., Götze, F.: Lattice point problems and distribution of values of quadratic forms. Ann. Math. **150**(3), 977–1027 (1999)
127. Beresnevich, V.: Rational points near manifolds and metric Diophantine approximation. Ann. Math. **175**(1), 187–235 (2012)
128. Beresnevich, V.: Badly approximable points on manifolds. Invent. Math. **202**(3), 1199–1240 (2015)
129. Beresnevich, V., Dickinson, D., Velani, S.: Diophantine approximation on planar curves and the distribution of rational points. Ann. Math. **166**(2), 367–426 (2007)
130. Beresnevich, V., Velani, S.: A mass transference principle and the Duffin–Schaeffer conjecture for Hausdorff measures. Ann. Math. **164**(3), 971–992 (2006)
131. van den Berg, M., Bolthausen, E., den Hollander, F.: Moderate deviations for the volume of the Wiener sausage. Ann. Math. **153**(2), 355–406 (2001)
132. van den Berg, M., Bolthausen, E., den Hollander, F.: On the volume of the intersection of two Wiener sausages. Ann. Math. **159**(2), 741–782 (2004)
133. Bergelson, V., Håland, I.J.: Sets of recurrence and generalized polynomials. In: Convergence in Ergodic Theory and Probability, vol. 5, pp. 91–110. de Gruyter Berlin (1996)
134. Bergelson, V., Leibman, A.: Polynomial extensions of van der Waerden's and Szemerédi's theorems. J. Am. Math. Soc. **9**(3), 725–753 (1996)
135. Bergelson, V., Leibman, A.: Distribution of values of bounded generalized polynomials. Acta Math. **198**(2), 155–230 (2007)
136. Bessis, D., Moussa, P., Villani, M.: Monotonic converging variational approximations to the functional integrals in quantum statistical mechanics. J. Math. Phys. **16**(11), 2318–2325 (1975)
137. Bhargava, M.: Higher composition laws III: The parametrization of quartic rings. Ann. Math. **159**(3), 1329–1360 (2004)
138. Bhargava, M.: The density of discriminants of quartic rings and fields. Ann. Math. **162**(2), 1031–1063 (2005)
139. Bhargava, M.: Higher composition laws IV: The parametrization of quintic rings. Ann. Math. **167**(1), 53–94 (2008)
140. Bhargava, M.: The density of discriminants of quintic rings and fields. Ann. Math. **172**(3), 1559–1591 (2010)
141. Bhargava, M., Gross, B.H.: The average size of the 2-Selmer group of Jacobians of hyperelliptic curves having a rational Weierstrass point. Preprint (2012). arXiv:1208.1007
142. Bhargava, M., Shankar, A.: Binary quartic forms having bounded invariants, and the boundedness of the average rank of elliptic curves. Ann. Math. **181**(1), 191–242 (2015)
143. Bhargava, M., Shankar, A.: Ternary cubic forms having bounded invariants, and the existence of a positive proportion of elliptic curves having rank 0. Ann. Math. **181**(2), 587–621 (2015)
144. Bhargava, M., Shankar, A., Taniguchi, T., Thorne, F., Tsimerman, J., Zhao, Y.: Bounds on 2-torsion in class groups of number fields and integral points on elliptic curves. J. Am. Math. Soc. **33**(4), 1087–1099 (2020)
145. Bhatia, R., Holbrook, J.: Noncommutative geometric means. Math. Intelligencer **28**(1), 32–39 (2006)
146. Bhatia, R., Holbrook, J.: Riemannian geometry and matrix geometric means. Linear Algebra Appl. **413**(2-3), 594–618 (2006)
147. Bigelow, S.: Braid groups are linear. J. Am. Math. Soc. **14**(2), 471–486 (2001)
148. Billera, L.J., Lee, C.W.: Sufficiency of Mcmullen's conditions for f-vectors of simplicial polytopes. Bull. Am. Math. Soc. **2**(1), 181–185 (1980)
149. Bilu, Y., Hanrot, G., Voutier, P.M.: Existence of primitive divisors of Lucas and Lehmer numbers. J. Reine Angew. Math. **539**(75), 122 (2001)

150. Birch, B.J., Swinnerton-Dyer, H.P.F.: Notes on elliptic curves. I. J. Reine Angew. Math. **212**(7), 25 (1963)
151. Birget, J.C., Ol'shanskii, A.Y., Rips, E., Sapir, M.V.: Isoperimetric functions of groups and computational complexity of the word problem. Ann. Math. **156**(2), 467–518 (2002)
152. Birkhoff, G.D.: A determinant formula for the number of ways of coloring a map. Ann. Math. **14**(1/4), 42–46 (1912)
153. Birkhoff, G.D.: Proof of the ergodic theorem. Proc. Natl. Acad. Sci. USA **17**(12), 656–660 (1931)
154. Bishop, C.J., Jones, P.W.: Harmonic measure, L^2-estimates and the Schwarzian derivative. J. d'Analyse Math. **62**(1), 77–113 (1994)
155. Blanc, J., Lamy, S., Zimmermann, S.: Quotients of higher dimensional Cremona groups. Preprint (2019). arXiv:1901.04145
156. Blichfeldt, H.F.: The minimum values of positive quadratic forms in six, seven and eight variables. Math. Z. **39**(1), 1–15 (1935)
157. Bloom, T.F.: A quantitative improvement for Roth's theorem on arithmetic progressions. J. Lond. Math. Soc. **93**(3), 643–663 (2016)
158. Bloom, T.F., Sisask, O.: Breaking the logarithmic barrier in Roth's theorem on arithmetic progressions. Preprint (2020). arXiv:2007.03528
159. Bobkov, S.G., Nazarov, F.L.: Large deviations of typical linear functionals on a convex body with unconditional basis. In: Stochastic Inequalities and Applications, pp. 3–13. Springer (2003)
160. Bobkov, S.G., Nazarov, F.L.: On convex bodies and log-concave probability measures with unconditional basis. In: Geometric Aspects of Functional Analysis, pp. 53–69. Springer (2003)
161. Bogachev, L.V., Molchanov, S.A., Pastur, L.A.: On the level density of random band matrices. Math. Notes **50**(6), 1232–1242 (1991)
162. Bollobás, B.: The chromatic number of random graphs. Combinatorica **8**(1), 49–55 (1988)
163. Bombieri, E.: On the large sieve. Mathematika **12**(2), 201–225 (1965)
164. Bombieri, E., De Giorgi, E., Giusti, E.: Minimal cones and the Bernstein problem. Invent. Math. **7**(3), 243–268 (1969)
165. Bombieri, E., Friedlander, J.B., Iwaniec, H.: Primes in arithmetic progressions to large moduli. Acta Math. **156**(1), 203–251 (1986)
166. Bondarenko, A., Radchenko, D., Viazovska, M.: Optimal asymptotic bounds for spherical designs. Ann. Math. **178**(2), 443–452 (2013)
167. Bondarenko, A.V., Viazovska, M.S.: Spherical designs via Brouwer fixed point theorem. SIAM J. Discrete Math. **24**(1), 207–217 (2010)
168. Bonichon, N., Gavoille, C., Hanusse, N., Poulalhon, D., Schaeffer, G.: Planar graphs, via well-orderly maps and trees. Graphs Combin. **22**(2), 185–202 (2006)
169. Borcea, J., Brändén, P.: Pólya–Schur master theorems for circular domains and their boundaries. Ann. Math. **170**(1), 465–492 (2009)
170. Borgwardt, K.H.: Untersuchungen zur Asymptotik der mittleren Schrittzahl von Simplexverfahren in der linearen Optimierung. Ph.D. thesis, Universit at Kaiserslautern (1977)
171. Böröczky, K.: Packing of spheres in spaces of constant curvature. Acta Math. Hungar. **32**(3-4), 243–261 (1978)
172. Böröczky, K.J., Lutwak, E., Yang, D., Zhang, G.: The log-Brunn–Minkowski inequality. Adv. Math. **231**(3-4), 1974–1997 (2012)
173. Borozdkin, K.G.: On a problem of Vinogradov's constant. Trudy Mat. Soc., SSSR **1**(3) (1956)
174. Borsuk, K.: Drei sätze über die n-dimensionale euklidische Sphäre. Fund. Math. **20**(1), 177–190 (1933)
175. Borwein, J., Bradley, D., Broadhurst, D., Lisoněk, P.: Special values of multiple polylogarithms. Trans. Am. Math. Soc. **353**(3), 907–941 (2001)
176. Borwein, P., Dobrowolski, E., Mossinghoff, M.J.: Lehmer's problem for polynomials with odd coefficients. Ann. Math. **166**(2), 347–366 (2007)

177. Borwein, P., Erdélyi, T., Ferguson, R., Lockhart, R.: On the zeros of cosine polynomials: solution to a problem of Littlewood. Ann. Math. **167**(3), 1109–1117 (2008)
178. Bourgain, J.: On Lipschitz embedding of finite metric spaces in Hilbert space. Isr. J. Math. **52**(1-2), 46–52 (1985)
179. Bourgain, J.: Pointwise ergodic theorems for arithmetic sets. Publications Mathématiques de l'Institut des Hautes Études Scientifiques **69**(1), 5–41 (1989)
180. Bourgain, J.: Besicovitch type maximal operators and applications to Fourier analysis. Geom. Funct. Anal. **1**(2), 147–187 (1991)
181. Bourgain, J.: On the distribution of polynomials on high dimensional convex sets. In: Geometric Aspects of Functional Analysis, pp. 127–137. Springer (1991)
182. Bourgain, J.: Periodic nonlinear Schrödinger equation and invariant measures. Comm. Math. Phys. **166**(1), 1–26 (1994)
183. Bourgain, J.: On triples in arithmetic progression. Geom. Funct. Anal. **9**(5), 968–984 (1999)
184. Bourgain, J., Chang, M.C.: On the size of k-fold sum and product sets of integers. J. Am. Math. Soc. **17**(2), 473–497 (2004)
185. Bourgain, J., Demeter, C., Guth, L.: Proof of the main conjecture in Vinogradov's mean value theorem for degrees higher than three. Ann. Math. **184**(2), 633–682 (2016)
186. Bourgain, J., Fuchs, E.: A proof of the positive density conjecture for integer Apollonian circle packings. J. Am. Math. Soc. **24**(4), 945–967 (2011)
187. Bourgain, J., Gamburd, A.: Uniform expansion bounds for Cayley graphs of $SL_2(F_p)$. Ann. Math. **167**(2), 625–642 (2008)
188. Bourgain, J., Katz, N., Tao, T.: A sum-product estimate in finite fields, and applications. Geom. Funct. Anal. **14**(1), 27–57 (2004)
189. Bourgain, J., Kontorovich, A.: On the local-global conjecture for integral Apollonian gaskets. Invent. Math. **196**(3), 589–650 (2014)
190. Bourgain, J., Kontorovich, A.: On Zaremba's conjecture. Ann. Math. **180**(1), 137–196 (2014)
191. Bourgain, J., Lindenstrauss, J., Milman, V.: Estimates related to Steiner symmetrizations. In: Geometric Aspects of Functional Analysis, pp. 264–273. Springer (1989)
192. Bourgain, J., Lindenstrauss, J., Milman, V.D.: Minkowski sums and symmetrizations. In: Geometric Aspects of Functional Analysis, pp. 44–66. Springer (1988)
193. Bourgain, J., Vu, V.H., Wood, P.M.: On the singularity probability of discrete random matrices. J. Funct. Anal. **258**(2), 559–603 (2010)
194. Bousch, T., Mairesse, J.: Asymptotic height optimization for topical IFS, Tetris heaps, and the finiteness conjecture. J. Am. Math. Soc. **15**(1), 77–111 (2002)
195. Boyd, D.W.: The sequence of radii of the Apollonian packing. Math. Comp. **39**(159), 249–254 (1982)
196. Boys, C.V.: Soap-bubbles and The Forces Which Mould Them. Course of Three Lectures. E. and JB Young and Co New York (1896)
197. Brams, S.J., Taylor, A.D.: An envy-free cake division protocol. Amer. Math. Monthly **102**(1), 9–18 (1995)
198. Brändén, P., Huh, J.: Lorentzian polynomials. Ann. Math. **192**(3), 821–891 (2020)
199. Brandes, U.: A faster algorithm for betweenness centrality. J. Math. Sociol. **25**(2), 163–177 (2001)
200. Brattka, V., Miller, J., Nies, A.: Randomness and differentiability. Trans. Am. Math. Soc. **368**(1), 581–605 (2016)
201. Braun, M., Etzion, T., Östergård, P.R.J., Vardy, A., Wassermann, A.: Existence of q-analogs of Steiner systems. Forum Math. Pi **4**, e7 (2016)
202. Braverman, M., Makarychev, K., Makarychev, Y., Naor, A.: The Grothendieck constant is strictly smaller than Krivine's bound. Forum Math. Pi **1**, e4 (2013)
203. Brendle, S.: Embedded minimal tori in S^3 and the Lawson conjecture. Acta Math. **211**(2), 177–190 (2013)
204. Brendle, S.: Embedded self-similar shrinkers of genus 0. Ann. Math. **183**(2), 715–728 (2016)

205. Brendle, S.: The isoperimetric inequality for a minimal submanifold in Euclidean space. e-prints (2019). arXiv–1907
206. Brenner, J.L., Wiegold, J., et al.: Two-generator groups. I. Mich. Math. J. **22**(1), 53–64 (1975)
207. Breuer, T., Guralnick, R.M., Kantor, W.M.: Probabilistic generation of finite simple groups, II. J. Algebra **320**(2), 443–494 (2008)
208. Breuil, C., Conrad, B., Diamond, F., Taylor, R.: On the modularity of elliptic curves over $\mathbb{Q}$: wild 3-adic exercises. J. Am. Math. Soc. **14**(4), 843–939 (2001)
209. Breuillard, E., Green, B.J., Tao, T.: Suzuki groups as expanders. Groups Geom. Dyn. **5**(2), 281–299 (2011)
210. Bridson, M.R., McReynolds, D.B., Reid, A.W., Spitler, R.: Absolute profinite rigidity and hyperbolic geometry. Ann. Math. **192**(3), 679–719 (2020)
211. Bridson, M.R., Wilton, H.: The triviality problem for profinite completions. Invent. Math. **202**(2), 839–874 (2015)
212. Brieussel, J., Zheng, T.: Speed of random walks, isoperimetry and compression of finitely generated groups. Preprint (2015). arXiv:1510.08040
213. Bringmann, K., Ono, K.: Dyson's ranks and Maass forms. Ann. Math. **171**(1), 419–449 (2010)
214. Broder, A.Z.: How hard is it to marry at random? (On the approximation of the permanent). In: Proceedings of the Eighteenth Annual ACM Symposium on the Theory of Computing, pp. 50–58 (1986)
215. Brüdern, J., Wooley, T.D.: The Hasse principle for pairs of diagonal cubic forms. Ann. Math. **166**(3), 865–895 (2007)
216. Bruin, H., van Strien, S.: Monotonicity of entropy for real multimodal maps. J. Am. Math. Soc. **28**(1), 1–61 (2015)
217. Brunn, H.: Ueber Ovale und Eiflachen: Inaugural Dissertation. Akademische Buchdruckerei von F. Straub (1887)
218. Bucur, D., Henrot, A.: Maximization of the second non-trivial Neumann eigenvalue. Acta Math. **222**(2), 337–361 (2019)
219. Buczolich, Z., Mauldin, R.D.: Divergent square averages. Ann. Math. **171**(3), 1479–1530 (2010)
220. Buczynski, J., Postinghel, E., Rupniewski, F.: On Strassen's rank additivity for small three-way tensors. SIAM J. Matrix Anal. Appl. **41**(1), 106–133 (2020)
221. Budzinski, T., Louf, B.: Local limits of uniform triangulations in high genus. Preprint (2019). arXiv–1902
222. Buff, X., Chéritat, A.: Quadratic Julia sets with positive area. Ann. Math. **176**(2), 673–746 (2012)
223. Bugeaud, Y.: Sets of exact approximation order by rational numbers II. Unif. Distrib. Theory **3**(2), 9–20 (2008)
224. Bugeaud, Y.: Multiplicative diophantine approximation. Dynamical systems and Diophantine approximation, Sémin. Congr **19**, 105–125 (2009)
225. Bugeaud, Y., Mignotte, M., Siksek, S.: Classical and modular approaches to exponential diophantine equations I. Fibonacci and Lucas perfect powers. Ann. Math. **163**(3), 969–1018 (2006)
226. Bukh, B.: Bounds on equiangular lines and on related spherical codes. SIAM J. Discrete Math. **30**(1), 549–554 (2016)
227. Bulatov, A.A.: The complexity of the counting constraint satisfaction problem. J. ACM **60**(5), 1–41 (2013)
228. Bulatov, A.A.: A dichotomy theorem for nonuniform CSPs. Preprint (2017). arXiv:1703.03021
229. Burau, W.: Über zopfgruppen und gleichsinnig verdrillte Verkettungen. Abhandlungen aus dem Mathematischen Seminar der Universität Hamburg **11**(1), 179–186 (1935)
230. Burdzy, K., Werner, W.: A counterexample to the "hot spots" conjecture. Ann. Math. **149**(1), 309–317 (1999)

231. Burgess, D.A.: On character sums and L-series. II. Proc. Lond. Math. Soc. **3**(1), 524–536 (1963)
232. Bürgisser, P., Cucker, F.: On a problem posed by Steve Smale. Ann. Math. **174**(3), 1785–1836 (2011)
233. Burness, T.C., Guralnick, R.M., Harper, S.: The spread of a finite group. Preprint (2020). arXiv:2006.01421
234. Burr, S.A., Erdős, P.: On the magnitude of generalized Ramsey numbers for graphs. Colloq. Math. Soc. Janos Bolyai **10**, 215–240 (1973)
235. Byrka, J., Aardal, K.: An optimal bifactor approximation algorithm for the metric uncapacitated facility location problem. SIAM J. Comput. **39**(6), 2212–2231 (2010)
236. de Caen, D.: Large equiangular sets of lines in Euclidean space. Electron. J. Comb. **7**, R55–R55 (2000)
237. Cai, J.Y., Chen, X.: Complexity of counting CSP with complex weights. J. ACM **64**(3), 1–39 (2017)
238. Calabi, E.: Problems in differential geometry. In: Proceedings of the United States-Japan Seminar in Differential Geometry, Kyoto, Japan, p. 170 (1965)
239. Calinescu, G., Chekuri, C., Pal, M., Vondrák, J.: Maximizing a monotone submodular function subject to a matroid constraint. SIAM J. Comput. **40**(6), 1740–1766 (2011)
240. Caluza Machado, F., de Oliveira Filho, F.M.: Improving the semidefinite programming bound for the kissing number by exploiting polynomial symmetry. Exp. Math. **27**(3), 362–369 (2018)
241. Cameron, P.J.: Generalisation of Fisher's inequality to fields with more than one element. In: Combinatorics, Lecture Note Series, pp. 9–13. London Mathematical Society (1973)
242. Candès, E., Tao, T.: The Dantzig selector: Statistical estimation when p is much larger than n. Ann. Statist. **35**(6), 2313–2351 (2007)
243. Candès, E.J., Li, X., Ma, Y., Wright, J.: Robust principal component analysis? J. ACM **58**(3), 1–37 (2011)
244. Candès, E.J., Recht, B.: Exact matrix completion via convex optimization. Found. Comput. Math. **9**(6), 717 (2009)
245. Candès, E.J., Romberg, J.K., Tao, T.: Stable signal recovery from incomplete and inaccurate measurements. Commun. Pure Appl. Math. J. Issued by the Courant Inst. Math. Sci. **59**(8), 1207–1223 (2006)
246. Candès, E.J., Strohmer, T., Voroninski, V.: Phaselift: Exact and stable signal recovery from magnitude measurements via convex programming. Comm. Pure Appl. Math. **66**(8), 1241–1274 (2013)
247. Cantarella, J., Kusner, R.B., Sullivan, J.M.: On the minimum ropelength of knots and links. Invent. Math. **150**(2), 257–286 (2002)
248. Cantat, S., Lamy, S., Cornulier, Y.: Normal subgroups in the Cremona group. Acta Math. **210**(1), 31–94 (2013)
249. Carathéodory, C., Study, E.: Zwei Beweise des Satzes, daß der Kreis unter allen Figuren gleichen Umfanges den größten Inhalt hat. Math. Ann. **68**(1), 133–140 (1909)
250. Carleman, T.: Zur theorie der Minimalflächen. Math. Z. **9**(1-2), 154–160 (1921)
251. Carlini, E., Catalisano, M.V., Geramita, A.V.: The solution to the Waring problem for monomials and the sum of coprime monomials. J. Algebra **370**, 5–14 (2012)
252. Carlson, J.A., Jaffe, A., Wiles, A.: The millennium prize problems. American Mathematical Soc. (2006)
253. Carroll, F.W., Eustice, D., Figiel, T.: The minimum modulus of polynomials with coefficients of modulus one. J. Lond. Math. Soc. **2**(1), 76–82 (1977)
254. Casazza, P.G., Fickus, M., Tremain, J.C., Weber, E.: The Kadison–Singer problem in mathematics and engineering: a detailed account. Contemp. Math. **414**, 299 (2006)
255. Catalan, E.: Note extraite d'une lettre adressée à l'éditeur par Mr. E. Catalan, Répétiteur à l'école polytechnique de Paris. J. Reine Angew. Math. **1844**(27), 192–192 (1844)
256. Chaika, J.: Every ergodic transformation is disjoint from almost every interval exchange transformation. Ann. Math. pp. 237–253 (2012)

257. Chaika, J., Masur, H.: The set of non-uniquely ergodic d-IETs has Hausdorff codimension 1/2. Invent. Math. **222**(3), 749–832 (2020)

258. Chambolle, A.: An algorithm for total variation minimization and applications. J. Math. Imaging Vision **20**(1-2), 89–97 (2004)

259. Chambolle, A., Pock, T.: A first-order primal-dual algorithm for convex problems with applications to imaging. J. Math. Imaging Vision **40**(1), 120–145 (2011)

260. Chambolle, A., Pock, T.: On the ergodic convergence rates of a first-order primal–dual algorithm. Math. Program. **159**(1), 253–287 (2016)

261. Chandrasekaran, K., Goyal, N., Haeupler, B.: Deterministic algorithms for the Lovász local lemma. SIAM J. Comput. **42**(6), 2132–2155 (2013)

262. Chang, M.C.: The Erdős–Szemerédi problem on sum set and product set. Ann. Math. **157**(3), 939–957 (2003)

263. Chao, K.: On the diophantine equation $x^2 = y^n + 1$. Acta Sci. Natur. Univ. Szechuan **2**, 57–64 (1960)

264. Charikar, M., Wirth, A.: Maximizing quadratic programs: Extending Grothendieck's inequality. In: 45th Annual IEEE Symposium on Foundations of Computer Science, pp. 54–60. IEEE (2004)

265. Chattopadhyay, E., Zuckerman, D.: Explicit two-source extractors and resilient functions. Ann. Math. **189**(3), 653–705 (2019)

266. Chawla, S., Gupta, A., Räcke, H.: Embeddings of negative-type metrics and an improved approximation to generalized sparsest cut. ACM Trans. Algorithm. (TALG) **4**(2), 1–18 (2008)

267. Chazy, J.: Sur certaines trajectoires du probleme des n corps. Bull. Astron. **35**, 321–389 (1918)

268. Chen, E.R., Engel, M., Glotzer, S.C.: Dense crystalline dimer packings of regular tetrahedra. Discrete Comput. Geom. **44**(2), 253–280 (2010)

269. Chen, W., Li, C., Ou, B.: Classification of solutions for an integral equation. Commun. Pure Appl. Math. **59**(3), 330–343 (2006)

270. Chen, X., Deng, X., Teng, S.H.: Settling the complexity of computing two-player Nash equilibria. J. ACM **56**(3), 1–57 (2009)

271. Chen, Y.: An almost constant lower bound of the isoperimetric coefficient in the KLS conjecture. Preprint (2020). arXiv:2011.13661

272. Cheng, Z., Lebowitz, J.L.: Statistics of energy levels in integrable quantum systems. Phys. Rev. A **44**(6), R3399 (1991)

273. Cheung, Y.: Hausdorff dimension of the set of nonergodic directions. Ann. Math. **158**(2), 661–678 (2003)

274. Cheung, Y.: Hausdorff dimension of the set of singular pairs. Ann. Math. **173**(1), 127–167 (2011)

275. Chiang, Y.M., Feng, S.J.: On the Nevanlinna characteristic of $f(x + \eta)$ and difference equations in the complex plane. Ramanujan J. **16**(1), 105–129 (2008)

276. Chowla, S.: The Riemann hypothesis and Hilbert's tenth problem. Gordon and Breach Science Publishers (1965)

277. Christensen, J.P.R.: Borel structures in groups and semigroups. Math. Scand. **28**(1), 124–128 (1971)

278. Christofides, N.: Worst-case analysis of a new heuristic for the travelling salesman problem. Tech. rep., Carnegie-Mellon Univ Pittsburgh PA Management Sciences Research Group (1976)

279. Chudnovsky, M., Cornuéjols, G., Liu, X., Seymour, P., Vušković, K.: Recognizing Berge graphs. Combinatorica **25**(2), 143–186 (2005)

280. Chudnovsky, M., Robertson, N., Seymour, P., Thomas, R.: The strong perfect graph theorem. Ann. Math. **164**(1), 51–229 (2006)

281. Chung, F.R.K., Füredi, Z., Graham, R.L., Seymour, P.: On induced subgraphs of the cube. J. Combin. Theory Ser. A **49**(1), 180–187 (1988)

282. Chvátal, V.: A greedy heuristic for the set-covering problem. Math. Oper. Res. **4**(3), 233–235 (1979)
283. Chvátal, V., Erdős, P.: Biased positional games. Ann. Discrete Math. **2**, 221–229 (1978)
284. Cohen, G.: Two-source dispersers for polylogarithmic entropy and improved Ramsey graphs. In: STOC'16. ACM (2019)
285. Cohen, P.J.: The independence of the continuum hypothesis. Proc. Natl. Acad. Sci. USA pp. 1143–1148 (1963)
286. Cohen, P.J.: The independence of the continuum hypothesis, II. Proc. Natl. Acad. Sci. USA **51**(1), 105 (1964)
287. Cohn, H., Elkies, N.: New upper bounds on sphere packings I. Ann. Math. **157**(2), 689–714 (2003)
288. Cohn, H., Kumar, A.: Optimality and uniqueness of the Leech lattice among lattices. Ann. Math. **170**(3), 1003–1050 (2009)
289. Cohn, H., Kumar, A., Miller, S.D., Radchenko, D., Viazovska, M.: The sphere packing problem in dimension 24. Ann. Math. **185**(3), 1017–1033 (2017)
290. Cohn, H., Kumar, A., Miller, S.D., Radchenko, D., Viazovska, M.: Universal optimality of the E_8 and Leech lattices and interpolation formulas. Preprint (2019). arXiv:1902.05438
291. Cohn, J.H.E.: On square Fibonacci numbers. J. Lond. Math. Soc. **1**(1), 537–537 (1964)
292. Coja-Oghlan, A., Panagiotou, K., Steger, A.: On the chromatic number of random graphs. J. Combin. Theory Ser. B **98**(5), 980–993 (2008)
293. Colding, T.H., Minicozzi, W.P.: The Calabi–Yau conjectures for embedded surfaces. Ann. Math. **167**(1), 211–243 (2008)
294. Conforti, M., Cornuéjols, G., Vušković, K.: Square-free perfect graphs. J. Combin. Theory Ser. B **90**(2), 257–307 (2004)
295. Conlon, D.: A new upper bound for diagonal Ramsey numbers. Ann. Math. **170**(2), 941–960 (2009)
296. Conlon, D., Ferber, A.: Lower bounds for multicolor Ramsey numbers. Preprint (2020). arXiv:2009.10458
297. Conlon, D., Fox, J., Sudakov, B.: Hypergraph Ramsey numbers. J. Am. Math. Soc. **23**(1), 247–266 (2010)
298. Conlon, D., Gowers, W.T.: Combinatorial theorems in sparse random sets. Ann. Math. pp. 367–454 (2016)
299. Constantin, P., Weinan, E., Titi, E.S.: Onsager's conjecture on the energy conservation for solutions of Euler's equation. Comm. Math. Phys. **165**(1), 207 (1994)
300. Conway, J.H., Torquato, S.: Packing, tiling, and covering with tetrahedra. Proc. Natl. Acad. Sci. USA **103**(28), 10612–10617 (2006)
301. Cook, S.A.: The complexity of theorem-proving procedures. In: Proceedings of the Third Annual ACM Symposium on the Theory of Computing, pp. 151–158. ACM (1971)
302. Coppersmith, D., Rivlin, T.J.: The growth of polynomials bounded at equally spaced points. SIAM J. Math. Anal. **23**(4), 970–983 (1992)
303. Coppersmith, D., Winograd, S.: Matrix multiplication via arithmetic progressions. J. Symb. Comput. **9**(3), 251–280 (1990)
304. van der Corput, J.G.: Über Summen von Primzahlen und Primzahlquadraten. Math. Ann. **116**(1), 1–50 (1939)
305. Corvaja, P., Zannier, U.: Finiteness of integral values for the ratio of two linear recurrences. Invent. Math. **149**(2), 431–451 (2002)
306. Couveignes, J.M.: Enumerating number fields. Ann. Math. **192**(2), 487–497 (2020)
307. Cox, D.A.: Introduction to Fermat's last theorem. Amer. Math. Monthly **101**(1), 3–14 (1994)
308. Coxeter, H.S.M.: An upper bound for the number of equal nonoverlapping spheres that can touch another of the same size. In: Convexity: Proceedings of the Seventh Symposium in Pure Mathematics of the American Mathematical Society, vol. 7, pp. 53–71. American Mathematical Soc. (1963)
309. Cramér, H.: On the order of magnitude of the difference between consecutive prime numbers. Acta Arith. **2**, 23–46 (1936)

310. Croot, E., Granville, A., Pemantle, R., Tetali, P.: On sharp transitions in making squares. Ann. Math. **175**(3), 1507–1550 (2012)
311. Croot, E., Lev, V.F., Pach, P.P.: Progression-free sets in $\mathbb{Z}_4^n$ are exponentially small. Ann. Math. **185**(1), 331–337 (2017)
312. Croot III, E.S.: On a coloring conjecture about unit fractions. Ann. Math. **157**(2), 545–556 (2003)
313. Csörnyei, M., Jones, P.: Product formulas for measures and applications to analysis and geometry (2012). www.math.sunysb.edu/Videos/dfest/PDFs/38-Jones.pdf
314. Danciger, J., Maloni, S., Schlenker, J.M.: Polyhedra inscribed in a quadric. Invent. Math. **221**(1), 237–300 (2020)
315. Dani, S.G.: Divergent trajectories of flows on homogeneous spaces and Diophantine approximation. J. Reine Angew. Math. **1985**(359), 55–89 (1985)
316. Dantzig, G.B.: Maximization of a linear function of variables subject to linear inequalities. Activity Anal. Prod. Alloc. **13**, 339–347 (1951)
317. Darbinyan, A.: Groups with decidable word problem that do not embed in groups with decidable conjugacy problem. Preprint (2017). arXiv:1708.09047
318. Darmon, H., Granville, A.: On the equations $z^m = f(x, y)$ and $ax^p + by^q = cz^r$. Bull. Lond. Math. Soc. **27**(6), 513–543 (1995)
319. Daskalakis, C., Goldberg, P.W., Papadimitriou, C.H.: The complexity of computing a Nash equilibrium. SIAM J. Comput. **39**(1), 195–259 (2009)
320. Davenport, H.: On the class-number of binary cubic forms (I). J. Lond. Math. Soc. **1**(3), 183–192 (1951)
321. Davenport, H.: Cubic forms in sixteen variables. Proc. Roy. Soc. Lond. Ser. A **272**(1350), 285–303 (1963)
322. Davenport, H., Heilbronn, H.: On the density of discriminants of cubic fields. II. Proc. Roy. Soc. Lond. Ser. A Math. Phys. Sci., 405–420 (1971)
323. Davenport, H., Schmidt, W.: Approximation to real numbers by algebraic integers. Acta Arith. **15**(4), 393–416 (1969)
324. Davenport, H., Schmidt, W.M.: Approximation to real numbers by quadratic irrationals. Acta Arith. **13**(2), 169–176 (1967)
325. Davie, A.M.: Matrix norms related to Grothendieck's inequality. In: Banach Spaces, pp. 22–26. Springer (1985)
326. Davie, A.M., Stothers, A.J.: Improved bound for complexity of matrix multiplication. Proc. Roy. Soc. Edinburgh Sect. A **143**(2), 351 (2013)
327. Davies, E.: Counting proper colourings in 4-regular graphs via the Potts model. Electron. J. Comb. **25**(4), 4–7 (2018)
328. Davies, E., Jenssen, M., Perkins, W., Roberts, B.: Extremes of the internal energy of the Potts model on cubic graphs. Random Struct. Algorithm. **53**(1), 59–75 (2018)
329. Davies, R.O.: Some remarks on the Kakeya problem. Proc. Camb. Philos. Soc. **69**(3), 417–421 (1971)
330. De Bruijn, N.G.: The roots of trigonometric integrals. Duke Math. J. **17**(3), 197–226 (1950)
331. De Giorgi, E.: Sulla proprieta isoperimetrica dell'ipersfera, nella classe degli insiemi aventi frontiera orientata di misura finita. Atti della Accademia Nazionale dei Lincei. Memorie. Classe di scienze fisiche, matematiche e naturali. Ser. 8, sez. 1a, vol. 5, fasc. 2 (1958)
332. De Giorgi, E.: Convergence problems for functionals and operators. In: Proc. Internat. Meeting on Recent Methods in Nonlinear Analysis, pp. 131–188 (1978)
333. De Guzmán, M.: Real variable methods in Fourier analysis. Elsevier (1981)
334. De Lellis, C., Székelyhidi, L.: Dissipative continuous Euler flows. Invent. Math. **193**(2), 377–407 (2013)
335. De Lellis, C., Székelyhidi, L.: Dissipative Euler flows and Onsager's conjecture. J. Eur. Math. Soc. (JEMS) **16**(7), 1467–1505 (2014)
336. Dehn, M.: Über den Rauminhalt. Math. Ann. **55**(3), 465–478 (1901)
337. Del Pino, M., Kowalczyk, M., Wei, J.: On De Giorgi's conjecture in dimension $n \geq 9$. Ann. Math. **174**(3), 1485–1569 (2011)

338. Delecroix, V., Goujard, E., Zograf, P., Zorich, A.: Enumeration of meanders and Masur–Veech volumes. Forum Math. Pi **8**, e4 (2020)

339. Delone, B.N., Faddeev, D.: The theory of irrationalities of the third degree, vol. 10. American Mathematical Society (1964)

340. Delsarte, P., Goethals, J.M., Seidel, J.J.: Spherical codes and designs. In: Geometry and Combinatorics, pp. 68–93. Elsevier (1991)

341. Delzell, C.N.: A continuous, constructive solution to Hilbert's 17-th problem. Invent. Math. **76**(3), 365–384 (1984)

342. Dembo, A., Peres, Y., Rosen, J., Zeitouni, O.: Thick points for planar Brownian motion and the Erdős–Taylor conjecture on random walk. Acta Math. **186**(2), 239–270 (2001)

343. Dembo, A., Peres, Y., Rosen, J., Zeitouni, O.: Cover times for Brownian motion and random walks in two dimensions. Ann. Math. **160**(2), 433–464 (2004)

344. Dembo, A., Poonen, B., Shao, Q.M., Zeitouni, O.: Random polynomials having few or no real zeros. J. Am. Math. Soc. **15**(4), 857–892 (2002)

345. Denjoy, A.: Sur les fonctions dérivées sommables. Bulletin de la Société Mathématique de France **43**, 161–248 (1915)

346. Denne, E., Diao, Y., Sullivan, J.M.: Quadrisecants give new lower bounds for the ropelength of a knot. Geometry Topology **10**(1), 1–26 (2006)

347. Descartes, B.: A three colour problem. Eureka **9**(21), 24–25 (1947)

348. Descartes, R., de Jonquières, E.J.P.F.: De solidorum elementis. Firmin-Didot (1890)

349. Di Nasso, M., Goldbring, I., Jin, R., Leth, S., Lupini, M., Mahlburg, K.: On a sumset conjecture of Erdős. Canad. J. Math. **67**(4), 795–809 (2015)

350. Dimitrov, V.: A proof of the Schinzel–Zassenhaus conjecture on polynomials. Preprint (2019). arXiv:1912.12545

351. Ding, J., Lee, J.R., Peres, Y.: Cover times, blanket times, and majorizing measures. Ann. Math. **175**(3), 1409–1471 (2012)

352. Ding, J., Sly, A., Sun, N.: Maximum independent sets on random regular graphs. Acta Math. **217**(2), 263–340 (2016)

353. Dinur, I., Khot, S., Kindler, G., Minzer, D., Safra, M.: On non-optimally expanding sets in Grassmann graphs. In: Proceedings of the 50th Annual ACM SIGACT Symposium on the Theory of Computing, pp. 940–951 (2018)

354. Dinur, I., Safra, S.: On the hardness of approximating minimum vertex cover. Ann. Math. **162**(1), 439–485 (2005)

355. Dinur, I., Steurer, D.: Analytical approach to parallel repetition. In: Proceedings of the Forty-Sixth Annual ACM Symposium on the Theory of Computing, pp. 624–633 (2014)

356. Dobbins, M.G.: A point in a nd-polytope is the barycenter of n points in its d-faces. Invent. Math. **199**(1), 287–292 (2015)

357. Dodziuk, J.: Difference equations, isoperimetric inequality and transience of certain random walks. Trans. Am. Math. Soc. **284**(2), 787–794 (1984)

358. Donoho, D., Tanner, J.: Counting faces of randomly projected polytopes when the projection radically lowers dimension. J. Am. Math. Soc. **22**(1), 1–53 (2009)

359. Donoho, D.L.: Compressed sensing. IEEE Trans. Inf. Theory **52**(4), 1289–1306 (2006)

360. Donsker, M.D., Varadhan, S.R.S.: Asymptotics for the Wiener sausage. Comm. Pure Appl. Math. **28**(4), 525–565 (1975)

361. Drasin, D., Pankka, P.: Sharpness of Rickman's Picard theorem in all dimensions. Acta Math. **214**(2), 209–306 (2015)

362. Drugan, G.: An immersed S^2 self-shrinker. Trans. Am. Math. Soc. **367**(5), 3139–3159 (2015)

363. Du, X., Guth, L., Ou, Y., Wang, H., Wilson, B., Zhang, R.: Weighted restriction estimates and application to Falconer distance set problem. Preprint (2018). arXiv:1802.10186

364. Du, X., Zhang, R.: Sharp L^2 estimates of the Schrödinger maximal function in higher dimensions. Ann. Math. **189**(3), 837–861 (2019)

365. Dubins, L., Hirsch, M.W., Karush, J.: Scissor congruence. Isr. J. Math. **1**(4), 239–247 (1963)

366. Dubois, O., Boufkhad, Y.: A general upper bound for the satisfiability threshold of random r-SAT formulae. J. Algorithm. **24**(2), 395–420 (1997)
367. Dudko, A., Sutherland, S.: On the Lebesgue measure of the Feigenbaum Julia set. Invent. Math., 1–36 (2020)
368. Duffin, R.J., Schaeffer, A.C.: Khintchine's problem in metric Diophantine approximation. Duke Math. J. **8**(2), 243–255 (1941)
369. Dugger, D., Isaksen, D.C.: The Hopf condition for bilinear forms over arbitrary fields. Ann. Math. **165**(3), 943–964 (2007)
370. Dujella, A.: There are only finitely many Diophantine quintuples. J. Reine Angew. Math. pp. 183–214 (2004)
371. Duminil-Copin, H., Raoufi, A., Tassion, V.: Sharp phase transition for the random-cluster and Potts models via decision trees. Ann. Math. **189**(1), 75–99 (2019)
372. Duminil-Copin, H., Smirnov, S.: The connective constant of the honeycomb lattice equals $\sqrt{2 + \sqrt{2}}$. Ann. Math. **175**(3), 1653–1665 (2012)
373. Dunnett, C.W., Sobel, M.: Approximations to the probability integral and certain percentage points of a multivariate analogue of Student's t-distribution. Biometrika **42**(1/2), 258–260 (1955)
374. Duplantier, B., Kwon, K.H.: Conformal invariance and intersections of random walks. Phys. Rev. Lett. **61**(22), 2514 (1988)
375. Dvir, Z.: On the size of Kakeya sets in finite fields. J. Am. Math. Soc. **22**(4), 1093–1097 (2009)
376. Dyer, M., Richerby, D.: An effective dichotomy for the counting constraint satisfaction problem. SIAM J. Comput. **42**(3), 1245–1274 (2013)
377. Dynnikov, I.: Arc-presentations of links: monotonic simplification. Fund. Math. **1**(190), 29–76 (2006)
378. Dyson, F.J.: Some guesses in the theory of partitions. Eureka (Cambridge) **8**(10), 10–15 (1944)
379. Eberhard, S., Green, B., Manners, F.: Sets of integers with no large sum-free subset. Ann. Math. pp. 621–652 (2014)
380. Edel, Y.: Extensions of generalized product caps. Des. Codes Cryptogr. **31**(1), 5–14 (2004)
381. Edelman, A., Kostlan, E., Shub, M.: How many eigenvalues of a random matrix are real? J. Am. Math. Soc. **7**(1), 247–267 (1994)
382. Einsiedler, M., Katok, A., Lindenstrauss, E.: Invariant measures and the set of exceptions to Littlewood's conjecture. Ann. Math. **164**(2), 513–560 (2006)
383. Eldan, R.: Thin shell implies spectral gap up to polylog via a stochastic localization scheme. Geom. Funct. Anal. **23**(2), 532–569 (2013)
384. Elekes, G., Sharir, M.: Incidences in three dimensions and distinct distances in the plane. Combin. Probab. Comput. **20**(4), 571–608 (2011)
385. Ellenberg, J.S., Gijswijt, D.: On large subsets of $\mathbb{F}_q^n$ with no three-term arithmetic progression. Ann. Math. **185**(1), 339–343 (2017)
386. Ellenberg, J.S., Venkatesh, A.: The number of extensions of a number field with fixed degree and bounded discriminant. Ann. Math. **163**(2), 723–741 (2006)
387. Elliott, P.D.T.A., Halberstam, H.: A conjecture in prime number theory. In: Symposia Mathematica, vol. 4, pp. 59–72 (1968)
388. Ellis, D., Friedgut, E., Pilpel, H.: Intersecting families of permutations. J. Am. Math. Soc. **24**(3), 649–682 (2011)
389. Elman, R., Lam, T.Y.: Quadratic forms and the u-invariant. I. Math. Z. **131**(4), 283–304 (1973)
390. Enflo, P.: On the nonexistence of uniform homeomorphisms between L_p-spaces. Arkiv för matematik **8**(2), 103–105 (1970)
391. Enflo, P.: On infinite-dimensional topological groups. Séminaire Analyse fonctionnelle (dit "Maurey–Schwartz") pp. 1–11 (1978)

392. Epstein, D.B.A., Paterson, M.S., Cannon, J.W., Holt, D.F., Levy, S.V., Thurston, W.P.: Word Processing in Groups. Jones and Bartlett Publishers, Boston, MA (1992)

393. Erdogan, M.B.: A bilinear Fourier extension theorem and applications to the distance set problem. Int. Math. Res. Not. IMRN **2005**(23), 1411–1425 (2005)

394. Erdős, P.: On the difference of consecutive primes. Q. J. Math. **6**(1), 124–128 (1935)

395. Erdős, P.: On sequences of integers no one of which divides the product of two others and on some related problems. Mitt. Forsch.-Inst. Math. Mech. Univ. Tomsk **2**, 74–82 (1938)

396. Erdős, P.: On a family of symmetric Bernoulli convolutions. Am. J. Math. **61**(4), 974–976 (1939)

397. Erdős, P.: On sets of distances of n points. Am. Math. Monthly **53**(5), 248–250 (1946)

398. Erdős, P.: Some remarks on the theory of graphs. Bull. Am. Math. Soc. **53**(4), 292–294 (1947)

399. Erdős, P.: On integers of the form $2^k + p$ and some related problems. Summa Brasil. Math. **2**, 113–123 (1950)

400. Erdős, P.: Arithmetical properties of polynomials. J. Lond. Math. Soc. **28**, 416–425 (1953)

401. Erdős, P.: Graph theory and probability. Canad. J. Math. **11**, 34–38 (1959)

402. Erdős, P.: On the representation of large integers as sums of distinct summands taken from a fixed set. Acta. Arith. **7**, 345–354 (1962)

403. Erdős, P.: Extremal problems in number theory. In: Proc. Sympos. Pure Math, pp. 181–189. Citeseer (1965)

404. Erdős, P.: Some remarks on number theory. Isr. J. Math. **3**(1), 6–12 (1965)

405. Erdős, P.: On the distribution of the convergents of almost all real numbers. J. Number Theory **2**(4), 425–441 (1970)

406. Erdős, P.: Problems and results in combinatorial number theory. Astérisque **24**(25), 295–310 (1975)

407. Erdős, P.: On the combinatorial problems which I would most like to see solved. Combinatorica **1**(1), 25–42 (1981)

408. Erdős, P., Frankl, P., Rödl, V.: The asymptotic number of graphs not containing a fixed subgraph and a problem for hypergraphs having no exponent. Graphs Combin. **2**(1), 113–121 (1986)

409. Erdős, P., Graham, R.L.: Old and new problems and results in combinatorial number theory, Monographies de L'Enseignement Mathématique, vol. 28. Universitéde Genève, Geneva (1980)

410. Erdős, P., Hajnal, A.: On Ramsey like theorems. Problems and results. In: Combinatorics (Proc. Conf. Combinatorial Math., Math. Inst., Oxford, 1972), pp. 123–140. Citeseer (1972)

411. Erdős, P., Hanani, H.: On a limit theorem in combinatorical analysis. Publ. Math. Debrecen **10**(10-13), 2–2 (1963)

412. Erdős, P., Ko, C., Rado, R.: Intersection theorems for systems of finite sets. Q. J. Math. **12**(1), 313–320 (1961)

413. Erdős, P., Lovász, L.: Problems and results on 3-chromatic hypergraphs and some related questions. In: A. Hajnal, R. Rado, V. Sos (eds.) Infinite and Finite Sets, II North-Holland,, pp. 609–627. Citeseer (1975)

414. Erdős, P., Rado, R.: Combinatorial theorems on classifications of subsets of a given set. Proc. Lond. Math. Soc. **3**(1), 417–439 (1952)

415. Erdős, P., Rényi, A.: On the evolution of random graphs. Publ. Math. Inst. Hung. Acad. Sci **5**(1), 17–60 (1960)

416. Erdős, P., Rényi, A.: On the existence of a factor of degree one of a connected random graph. Acta Math. Acad. Sci. Hungar. **17**, 359–368 (1966)

417. Erdős, P., Rubin, A.L., Taylor, H.: Choosability in graphs. In: Proc. West Coast Conf. on Combinatorics, Graph Theory and Computing, Congressus Numerantium, vol. 26, pp. 125–157 (1979)

418. Erdős, P., Selfridge, J.L.: The product of consecutive integers is never a power. Ill. J. Math. **19**, 292–301 (1975)

419. Erdős, P., Simonovits, M.: On a valence problem in extremal graph theory. Discrete Math. **5**(4), 323–334 (1973)
420. Erdős, P., Szekeres, G.: A combinatorial problem in geometry. Compos. Math. **2**, 463–470 (1935)
421. Erdős, P., Szekeres, G.: On the product $\prod_{k=1}^{n}(1 - z^{a_k})$. Publications de l'Institut Mathématique **13**, 29–34 (1959)
422. Erdős, P., Szekeres, G.: On some extremum problems in elementary geometry. In: Annales Univ. Sci. Budapest, pp. 3–4 (1960)
423. Erdős, P., Szemerédi, E.: On sums and products of integers. In: Studies in Pure Mathematics, pp. 213–218. Springer (1983)
424. Erdős, P., Taylor, S.J.: Some problems concerning the structure of random walk paths. Acta Math. Hungar. **11**(1-2), 137–162 (1960)
425. Erdős, P., Turán, P.: On some sequences of integers. J. Lond. Math. Soc. **1**(4), 261–264 (1936)
426. Erdős, P., et al.: The difference of consecutive primes. Duke Math. J. **6**(2), 438–441 (1940)
427. Erdős, P., et al.: Some unsolved problems. Mich. Math. J. **4**(3), 291–300 (1957)
428. Eremenko, A., Gabrielov, A.: Rational functions with real critical points and the B. and M. Shapiro conjecture in real enumerative geometry. Ann. Math. **155**(1), 105–129 (2002)
429. Eremenko, A.E.: On the iteration of entire functions. Banach Center Publications **23**(1), 339–345 (1989)
430. Erschler, A., Zheng, T.: Growth of periodic Grigorchuk groups. Invent. Math. **219**(3), 1069–1155 (2020)
431. Eskin, A., Mirzakhani, M.: Invariant and stationary measures for the action on moduli space. Publications Mathématiques de l'IHÉS **127**(1), 95–324 (2018)
432. Eskin, A., Mozes, S., Oh, H.: On uniform exponential growth for linear groups. Invent. Math. **160**(1), 1–30 (2005)
433. Estermann, T.: Einige sätze über quadratfreie zahlen. Math. Ann. **105**(1), 653–662 (1931)
434. Euler, L.: Elementa doctrinae solidorum. Novi commentarii academiae scientiarum Petropolitanae **4**, 109–140 (1758)
435. Euler, L.: De motu rectilineo trium corporum se mutuo attrahentium. Novi commentarii academiae scientiarum Petropolitanae **11**, 144–151 (1767)
436. Eyink, G.L.: Energy dissipation without viscosity in ideal hydrodynamics I. Fourier analysis and local energy transfer. Phys. D Nonlinear Phenomena **78**(3-4), 222–240 (1994)
437. Fagin, R., Lotem, A., Naor, M.: Optimal aggregation algorithms for middleware. J. Comput. Syst. Sci. **66**(4), 614–656 (2003)
438. Falconer, K.J.: On the Hausdorff dimensions of distance sets. Mathematika **32**(2), 206–212 (1985)
439. Fatou, P.: Sur l'itération des fonctions transcendantes entières. Acta Math. **47**(4), 337–370 (1926)
440. Feder, T., Vardi, M.Y.: Monotone monadic SNP and constraint satisfaction. In: Proceedings of the Twenty-Fifth Annual ACM Symposium on the Theory of Computing, pp. 612–622 (1993)
441. Fefferman, C., Klartag, B.: Fitting a C^m-smooth function to data, I. Ann. Math. **169**(1), 315–346 (2009)
442. Fefferman, C., Klartag, B., et al.: Fitting a C^m-smooth function to data II. Revista Matemática Iberoamericana **25**(1), 49–273 (2009)
443. Fefferman, C.L.: A sharp form of Whitney's extension theorem. Ann. Math. **161**(1), 509–577 (2005)
444. Feige, U.: A threshold of $\ln n$ for approximating set cover. J. ACM **45**(4), 634–652 (1998)
445. Feige, U., Kilian, J.: Zero knowledge and the chromatic number. J. Comput. Syst. Sci. **57**(2), 187–199 (1998)
446. Feigenbaum, M.J.: Quantitative universality for a class of nonlinear transformations. J. Stat. Phys. **19**(1), 25–52 (1978)

447. Fejes Tóth, L.: Über das kürzeste Kurvennetz das eine Kugeloberfläche in flächengleiche konvexe Teil zerlegt. Mat. Term.-tud. Értesitö **62**, 349–354 (1943)

448. Fejes Tóth, L.: Über die dichteste Kugellagerung. Math. Z. **48**, 676–684 (1943)

449. Fejes Tóth, L.: Lagerungen in der Ebene auf der kugel und im Raum. Die Grundlagen der mathematischen Wissenchaften in Einzeldarstellung. Springer (1953)

450. Feng, D.J., Hu, H.: Dimension theory of iterated function systems. Commun. Pure Appl. Math. J. Issued Courant Inst. Math. Sci. **62**(11), 1435–1500 (2009)

451. Ferenczi, S., Mauduit, C.: Transcendence of numbers with a low complexity expansion. J. Number Theory **67**(2), 146–161 (1997)

452. Ferrari, P.A., Fontes, L.R.G.: Current fluctuations for the asymmetric simple exclusion process. Ann. Probab., 820–832 (1994)

453. Filaseta, M., Ford, K., Konyagin, S., Pomerance, C., Yu, G.: Sieving by large integers and covering systems of congruences. J. Am. Math. Soc. **20**(2), 495–517 (2007)

454. Fiorini, S., Massar, S., Pokutta, S., Tiwary, H.R., De Wolf, R.: Linear vs. semidefinite extended formulations: exponential separation and strong lower bounds. In: Proceedings of the Forty-Fourth Annual ACM Symposium on the Theory of Computing, pp. 95–106 (2012)

455. Flory, P.J.: Principles of polymer chemistry. Cornell University Press (1953)

456. Foisy, J., Alfaro Garcia, M., Brock, J., Hodges, N., Zimba, J.: The standard double soap bubble in $\mathbb{R}^2$ uniquely minimizes perimeter. Pac. J. Math. **159**(1), 47–59 (1993)

457. Folkman, J.: The homology groups of a lattice. J. Math. Mech. **15**(4), 631–636 (1966)

458. Ford, K.: The distribution of integers with a divisor in a given interval. Ann. Math. **168**(2), 367–433 (2008)

459. Ford, K., Green, B., Konyagin, S., Maynard, J., Tao, T.: Long gaps between primes. J. Am. Math. Soc. **31**(1), 65–105 (2018)

460. Ford, K., Green, B., Konyagin, S., Tao, T.: Large gaps between consecutive prime numbers. Ann. Math., 935–974 (2016)

461. Fouvry, É., Klüners, J.: On the negative Pell equation. Ann. Math. **172**(3), 2035–2104 (2010)

462. Fouvry, É., Klüners, J.: The parity of the period of the continued fraction of $\sqrt{d}$. Proc. Lond. Math. Soc. **101**(2), 337–391 (2010)

463. Fox, J.: A new proof of the graph removal lemma. Ann. Math., 561–579 (2011)

464. Fox, R.H.: A quick trip through knot theory. In: Topology of 3-manifolds and Related Topics, pp. 120–167. Prentice-Hall (1962)

465. Fraenkel, A.: Uber die Zermelosche Begründung der Mengenlehre. Jahresbericht der Deutschen Mathematiker-Vereinigung **30**(2), 45–46 (1921)

466. Frank, R.L., Lenzmann, E., Silvestre, L.: Uniqueness of radial solutions for the fractional Laplacian. Comm. Pure Appl. Math. **69**(9), 1671–1726 (2016)

467. Frankl, P., Deza, M.: On the maximum number of permutations with given maximal or minimal distance. J. Combin. Theory Ser. A **22**(3), 352–360 (1977)

468. Frankl, P., Wilson, R.M.: Intersection theorems with geometric consequences. Combinatorica **1**(4), 357–368 (1981)

469. Frantzikinakis, N., Host, B.: Higher order Fourier analysis of multiplicative functions and applications. J. Am. Math. Soc. **30**(1), 67–157 (2017)

470. Frantzikinakis, N., Host, B.: The logarithmic Sarnak conjecture for ergodic weights. Ann. Math. **187**(3), 869–931 (2018)

471. Freedman, M.: The topology of four-dimensional manifolds. J. Differ. Geom. **17**(3), 357–453 (1982)

472. Freeman, L.C.: A set of measures of centrality based on betweenness. Sociometry **40**(1), 35–41 (1977)

473. Freiman, G.A.: Foundations of a structural theory of set addition. Kazan, 1966 (in Russian). English translation in: Translations of Mathematical Monographs, vol. 37. Amer. Math. Soc., Providence, RI (1973)

474. Freiman, G.A.: Diofantovy priblizheniya i geometriya chisel (zadacha Markova). Diophantine approximations and the geometry of numbers (Markov's problem)] Kalinin. Gosudarstv. Univ., Kalinin (1975)

475. Frick, F.: Counterexamples to the topological Tverberg conjecture. Preprint (2015). arXiv:1502.00947

476. Friedlander, J., Iwaniec, H.: The polynomial $x^2 + y^4$ captures its primes. Ann. Math. **148**, 945–1040 (1998)

477. Friedman, J.: A proof of Alon's second eigenvalue conjecture and related problems, Memoirs of the American Mathematical Society, vol. 195. American Mathematical Soc. (2008)

478. Friedman, J.: Sheaves on graphs, their homological invariants, and a proof of the Hanna Neumann conjecture: with an appendix by Warren Dicks, Memoirs of the American Mathematical Society, vol. 233. American Mathematical Society (2015)

479. Fuglede, B.: Commuting self-adjoint partial differential operators and a group theoretic problem. J. Funct. Anal. **16**(1), 101–121 (1974)

480. Füredi, Z., Hajnal, P.: Davenport–Schinzel theory of matrices. Discrete Math. **103**(3), 233–251 (1992)

481. Fürer, M.: Faster integer multiplication. SIAM J. Comput. **39**(3), 979–1005 (2009)

482. Furstenberg, H.: Disjointness in ergodic theory, minimal sets, and a problem in Diophantine approximation. Math. Syst. Theory **1**(1), 1–49 (1967)

483. Furstenberg, H.: Intersections of Cantor sets and transversality of semigroups. In: Problems in Analysis (Sympos. Salomon Bochner, Princeton Univ., Princeton, NJ, 1969), pp. 41–59. Princeton University Press (1970)

484. Furstenberg, H.: Ergodic behavior of diagonal measures and a theorem of Szemerédi on arithmetic progressions. J. d'Analyse Math. **31**(1), 204–256 (1977)

485. Furstenberg, H., Katznelson, Y.: An ergodic Szemerédi theorem for commuting transformations. J. d'Analyse Math. **34**(1), 275–291 (1978)

486. Furstenberg, H., Katznelson, Y.: A density version of the Hales–Jewett theorem. J. d'Analyse Math. **57**(1), 64–119 (1991)

487. Fusco, N., Maggi, F., Pratelli, A.: The sharp quantitative isoperimetric inequality. Ann. Math. **168**(3), 941–980 (2008)

488. Galvin, D.: Maximizing H-colorings of a regular graph. J. Graph Theory **73**(1), 66–84 (2013)

489. Galvin, D., Tetali, P.: On weighted graph homomorphisms. DIMACS Ser. Discrete Math. Theor. Comput. Sci. **63**, 97–104 (2004)

490. Gamburd, A., Magee, M., Ronan, R.: An asymptotic formula for integer points on Markoff–Hurwitz varieties. Ann. Math. **190**(3), 751–809 (2019)

491. Gandolfi, A., Keane, M.S., Newman, C.M.: Uniqueness of the infinite component in a random graph with applications to percolation and spin glasses. Probab. Theory Related Fields **92**(4), 511–527 (1992)

492. Gardner, R.J.: Geometric Tomography. Cambridge University Press (1995)

493. Gauss, C.F.: Untersuchungen über die Eigenschaften der positiven ternären quadratischen Formen von Ludwig August Seeber. Göttingische gelehrte Anzeigen (1831)

494. Gelfond, A.: Sur les nombres qui ont des propriétés additives et multiplicatives données. Acta Arith. **13**(3), 259–265 (1968)

495. Gentry, C.: Fully homomorphic encryption using ideal lattices. In: Proceedings of the Forty-First Annual ACM Symposium on the Theory of Computing, pp. 169–178 (2009)

496. Ghioca, D., Tucker, T.J., Zieve, M.E.: Intersections of polynomial orbits, and a dynamical Mordell–Lang conjecture. Invent. Math. **171**(2), 463–483 (2008)

497. Ghoussoub, N., Gui, C.: On a conjecture of De Giorgi and some related problems. Math. Ann. **311**(3), 481–491 (1998)

498. Giménez, O., Noy, M.: Asymptotic enumeration and limit laws of planar graphs. J. Am. Math. Soc. **22**(2), 309–329 (2009)

499. Gluskin, E.D.: Diameter of the Minkowski compactum is approximately equal to n. Funktsional'nyi Analiz i ego Prilozheniya **15**(1), 72–73 (1981)

500. Goddard, W., Lyle, J.: Dense graphs with small clique number. J. Graph Theory **66**(4), 319–331 (2011)

501. Gödel, K.: Über formal unentscheidbare Sätze der Principia Mathematica und verwandter Systeme I. Monatshefte für Mathematik und Physik **38**(1), 173–198 (1931)

502. Goemans, M.X.: Semidefinite programming in combinatorial optimization. Math. Program. **79**(1-3), 143–161 (1997)

503. Goldston, D.A., Pintz, J., Yalçin Yıldırım, C.: Primes in tuples I. Ann. Math. **170**(2), 819–862 (2009)

504. Goldston, D.A., Pintz, J., Yalçin Yıldırım, C.: Primes in tuples II. Acta Math. **204**(1), 1–47 (2010)

505. Gorban, A.N., Golubkov, A., Grechuk, B., Mirkes, E.M., Tyukin, I.Y.: Correction of AI systems by linear discriminants: Probabilistic foundations. Information Sciences **466**, 303–322 (2018)

506. Gotsman, C., Linial, N.: The equivalence of two problems on the cube. J. Combin. Theory Ser. A **61**(1), 142–146 (1992)

507. Götze, F.: Lattice point problems and values of quadratic forms. Invent. Math. **157**(1), 195–226 (2004)

508. Gowers, W.T.: A new proof of Szemerédi's theorem. Geom. Funct. Anal. **11**(3), 465–588 (2001)

509. Gowers, W.T.: An infinite Ramsey theorem and some Banach-space dichotomies. Ann. Math. **156**(3), 797–833 (2002)

510. Gowers, W.T.: Hypergraph regularity and the multidimensional Szemerédi theorem. Ann. Math. **166**(3), 897–946 (2007)

511. Gowers, W.T.: Quasirandom groups. Combin. Probab. Comput. **17**(3), 363–387 (2008)

512. Grabowski, Ł., Máthé, A., Pikhurko, O.: Measurable circle squaring. Ann. Math. **185**(2), 671–710 (2017)

513. Grafakos, L., Li, X.: Uniform bounds for the bilinear Hilbert transforms, I. Ann. Math. **159**(3), 889–933 (2004)

514. Graham, R.L., Lagarias, J.C., Mallows, C.L., Wilks, A.R., Yan, C.H.: Apollonian circle packings: number theory. J. Number Theory **100**(1), 1–45 (2003)

515. Granville, A., Soundararajan, K.: The spectrum of multiplicative functions. Ann. Math. **153**(2), 407–470 (2001)

516. Granville, A., Soundararajan, K.: Large character sums: pretentious characters and the Pólya–Vinogradov theorem. J. Am. Math. Soc. **20**(2), 357–384 (2007)

517. Granville, A., Soundararajan, K.: An uncertainty principle for arithmetic sequences. Ann. Math. **165**(2), 593–635 (2007)

518. Gravel, S., Elser, V., Kallus, Y.: Upper bound on the packing density of regular tetrahedra and octahedra. Discrete Comput. Geom. **46**(4), 799–818 (2011)

519. Gray, R.D.: Undecidability of the word problem for one-relator inverse monoids via right-angled Artin subgroups of one-relator groups. Invent. Math. **219**(3), 987–1008 (2020)

520. Grechuk, B.: Theorems of the 21st Century: Volume I. Springer (2019)

521. Grechuk, B., Gorban, A.N., Tyukin, I.Y.: General stochastic separation theorems with optimal bounds. Neural Networks **138**, 33–56 (2021)

522. Green, B.: Roth's theorem in the primes. Ann. Math. **161**(3), 1609–1636 (2005)

523. Green, B., Tao, T.: The primes contain arbitrarily long arithmetic progressions. Ann. Math. **167**(2), 481–547 (2008)

524. Green, B., Tao, T.: Linear equations in primes. Ann. Math. **171**(3), 1753–1850 (2010)

525. Green, B., Tao, T.: The Möbius function is strongly orthogonal to nilsequences. Ann. Math. **175**(2), 541–566 (2012)

526. Green, B., Tao, T., Ziegler, T.: An inverse theorem for the Gowers $U^{s+1}[N]$-norm. Ann. Math. **176**(2), 1231–1372 (2012)

527. Greenfeld, R., Lev, N.: Fuglede's spectral set conjecture for convex polytopes. Anal. PDE **10**(6), 1497–1538 (2017)
528. de Grey, A.D.N.J.: The chromatic number of the plane is at least 5. Geombinatorics **28**(1), 5–18 (2018)
529. Griffin, M., Ono, K., Rolen, L., Zagier, D.: Jensen polynomials for the Riemann zeta function and other sequences. Proc. Natl. Acad. Sci. USA **116**(23), 11103–11110 (2019)
530. Grigorchuk, R., Medynets, K.: Topological full groups are locally embeddable into finite groups. Preprint (2012). arXiv:1105.0719 v3
531. Grigorchuk, R.I.: Burnside problem on periodic groups. Funktsional'nyi Analiz i ego Prilozheniya **14**(1), 53–54 (1980)
532. Grigorchuk, R.I.: Degrees of growth of finitely generated groups, and the theory of invariant means. Izv. Ross. Akad. Nauk Ser. Mat. **48**(5), 939–985 (1984)
533. Grohe, M., Marx, D.: Structure theorem and isomorphism test for graphs with excluded topological subgraphs. SIAM J. Comput. **44**(1), 114–159 (2015)
534. Gromov, M.: Structures métriques pour les variétés riemanniennes. Textes Mathématiques [Mathematical Texts], vol. 1. Cedic (1981)
535. Gromov, M., Milman, V.D.: A topological application of the isoperimetric inequality. Am. J. Math. **105**(4), 843–854 (1983)
536. Gromov, M., et al.: Filling Riemannian manifolds. J. Differ. Geom. **18**(1), 1–147 (1983)
537. Grothendieck, A.: Resume de la theorie metrique des produits tensoriels topologiques. Bol. Soc. Mat. Sao Paulo **8**, 1–79 (1953)
538. Grötschel, M., Lovász, L., Schrijver, A.: Geometric algorithms and combinatorial optimization. Algorithms Combin. **2**, 65–84 (1988)
539. Grove, K., Karcher, H.: How to conjugate C^1-close group actions. Math. Z. **132**(1), 11–20 (1973)
540. Grünbaum, B.: Higher-dimensional analogs of the four-color problem and some inequalities for simplicial complexes. J. Comb. Theory **8**(2), 147–153 (1970)
541. Gupta, S.D., Eaton, M.L., Olkin, I., Perlman, M., Savage, L.J., Sobel, M.: Inequalities on the probability content of convex regions for elliptically contoured distributions. In: Proceedings of the Sixth Berkeley Symposium on Mathematical Statistics and Probability (Univ. California, Berkeley, Calif., 1970/1971), vol. 2, pp. 241–265 (1972)
542. Guralnick, R.M., Kantor, W.M.: Probabilistic generation of finite simple groups. J. Algebra **234**(2), 743–792 (2000)
543. Guralnick, R.M., Liebeck, M.W., O'Brien, E.A., Shalev, A., Tiep, P.H.: Surjective word maps and Burnside's $p^a q^b$ theorem. Invent. Math. **213**(2) (2018)
544. Gurarii, V.I.: Subspaces and bases in spaces of continuous functions. Doklady Akademii Nauk **167**(5), 971–973 (1966)
545. Gurel-Gurevich, O., Nachmias, A.: Recurrence of planar graph limits. Ann. Math. **177**(2), 761–781 (2013)
546. Guth, L.: The endpoint case of the Bennett–Carbery–Tao multilinear Kakeya conjecture. Acta Math. **205**(2), 263–286 (2010)
547. Guth, L.: Restriction estimates using polynomial partitioning II. Acta Math. **221**(1), 81–142 (2018)
548. Guth, L., Iosevich, A., Ou, Y., Wang, H.: On Falconer's distance set problem in the plane. Invent. Math. **219**(3), 779–830 (2020)
549. Guth, L., Katz, N.H.: On the Erdős distinct distances problem in the plane. Ann. Math. **181**(1), 155–190 (2015)
550. Haantjes, J.: Equilateral point-sets in elliptic two-and three-dimensional spaces. Nieuw Arch. Wiskunde (2) **22**, 355–362 (1948)
551. Hadwiger, H.: Einfache herleitung der isoperimetrischen ungleichung für abgeschlossene Punktmengen. Math. Ann. **124**(1), 158–160 (1951)
552. Hadwiger, H.: Vorlesungen über Inhalt, Oberfldche und Isoperimetrie. Berlin-Göttingen-Heidelberg, Springer-Verlag (1957)

553. Häggström, O., Peres, Y., Steif, J.E.: Dynamical percolation. Ann. Inst. Henri Poincaré, Probab. Stat. **33**(4), 497–528 (1997)
554. Hales, A.W., Jewett, R.I.: Regularity and positional games. Trans. Am. Math. Soc. **106**(2), 222–229 (1963)
555. Hales, T., McLaughlin, S.: The dodecahedral conjecture. J. Am. Math. Soc. **23**(2), 299–344 (2010)
556. Hales, T.C.: The honeycomb conjecture. Discrete Comput. Geom. **25**(1), 1–22 (2001)
557. Hales, T.C.: A proof of the Kepler conjecture. Ann. Math. **162**(3), 1065–1185 (2005)
558. Hales, T.C., Adams, M., Bauer, G., Dang, T.D., Harrison, J., Le Truong, H., Kaliszyk, C., Magron, V., McLaughlin, S., Nguyen, T.T., et al.: A formal proof of the Kepler conjecture. Forum Math. Pi **5**, e2 (2017)
559. Hall, R.R.: Proof of a conjecture of Heath-Brown concerning quadratic residues. Proc. Edinb. Math. Soc. **39**(3), 581–588 (1996)
560. Hampton, M., Moeckel, R.: Finiteness of relative equilibria of the four-body problem. Invent. Math. **163**(2), 289–312 (2006)
561. Hans-Gill, R.J., Raka, M., Sehmi, R.: On conjectures of Minkowski and Woods for $n = 7$. J. Number Theory **129**(5), 1011–1033 (2009)
562. Hans-Gill, R.J., Raka, M., Sehmi, R.: On conjectures of Minkowski and Woods for $n = 8$. Acta Arith. **147**, 337–385 (2011)
563. Hara, T., Slade, G.: Mean-field critical behaviour for percolation in high dimensions. Comm. Math. Phys. **128**(2), 333–391 (1990)
564. Hara, T., et al.: Decay of correlations in nearest-neighbor self-avoiding walk, percolation, lattice trees and animals. Ann. Probab. **36**(2), 530–593 (2008)
565. Hardy, G.H., Littlewood, J.E.: Some problems of Diophantine approximation: A remarkable trigonometrical series. Proc. Natl. Acad. Sci. USA **2**(10), 583 (1916)
566. Hardy, G.H., Littlewood, J.E., et al.: Some problems of 'partitio numerorum'; III: On the expression of a number as a sum of primes. Acta Math. **44**, 1–70 (1923)
567. Harper, A.J.: Sharp conditional bounds for moments of the Riemann zeta function. Preprint (2013). arXiv:1305.4618
568. Harper, A.J.: Moments of random multiplicative functions, I: Low moments, better than squareroot cancellation, and critical multiplicative chaos. Forum Math. Pi **8**, e1 (2020)
569. Hartmann, A.K., Weigt, M.: Phase transitions in combinatorial optimization problems, vol. 67. Wiley Online Library (2005)
570. Harvey, D., van der Hoeven, J.: Faster integer multiplication using short lattice vectors. The Open Book Series **2**(1), 293–310 (2019)
571. Harvey, D., Van Der Hoeven, J.: Integer multiplication in time $O(n \log n)$ (online preprint) (2019)
572. Hass, J., Lagarias, J.: The number of Reidemeister moves needed for unknotting. J. Am. Math. Soc. **14**(2), 399–428 (2001)
573. Hass, J., Schlafly, R.: Double bubbles minimize. Ann. Math. **151**(2), 459–515 (2000)
574. Hasse, H.: Zur Theorie der quadratischen Formen, insbesondere ihrer Darstellbarkeitseigenschaften im Bereich der rationalen Zahlen und ihrer Einteilung in Geschlechter. Ph.D. thesis, University of Marburg (1921)
575. Hasse, H.: Zur Theorie der abstrakten elliptischen Funktionenkörper III. Die Struktur des Meromorphismenrings. Die Riemannsche Vermutung. J. Reine Angew. Math. **1936**(175), 193–208 (1936)
576. Håstad, J.: Almost optimal lower bounds for small depth circuits. In: Proceedings of the Eighteenth Annual ACM Symposium on the Theory of Computing, pp. 6–20 (1986)
577. Håstad, J.: Clique is hard to approximate within $n^{1-\varepsilon}$. Acta Math. **182**, 105–142 (1999)
578. Håstad, J.: Some optimal inapproximability results. J. ACM **48**(4), 798–859 (2001)
579. Håstad, J., Rossman, B., Servedio, R.A., Tan, L.Y.: An average-case depth hierarchy theorem for boolean circuits. J. ACM **64**(5), 1–27 (2017)
580. Hausdorff, F.: Summen von $\aleph_1$ Mengen. Fund. Math. **26**, 241–255 (1936)

581. He, B., Togbé, A., Ziegler, V.: There is no Diophantine quintuple. Trans. Am. Math. Soc. **371**(9), 6665–6709 (2019)

582. He, X., Wigderson, Y.: Hedetniemi's conjecture is asymptotically false. J. Combin. Theory Ser. B (2020)

583. Heath-Brown, D.R.: The fourth power moment of the Riemann zeta function. Proc. Lond. Math. Soc. **3**(3), 385–422 (1979)

584. Heath-Brown, D.R.: Primes represented by $x^3 + 2y^3$. Acta Math. **186**(1), 1–84 (2001)

585. Heath-Brown, D.R.: The density of rational points on curves and surfaces. Ann. Math. **155**(2), 553–598 (2002)

586. Heath-Brown, D.R.: Cubic forms in 14 variables. Invent. Math. **170**(1), 199–230 (2007)

587. Heckel, A.: Non-concentration of the chromatic number of a random graph. Preprint (2019). arXiv–1906

588. Hedetniemi, S.T.: Homomorphisms of graphs and automata. Tech. rep., University of Michigan (1966)

589. Hefetz, D., Stich, S.: On two problems regarding the Hamiltonian cycle game. Electron. J. Comb. **16**(1) (2009). Article no. R28

590. Heintz, J.: Definability and fast quantifier elimination in algebraically closed fields. Theor. Comput. Sci. **24**(3), 239–277 (1983)

591. Helfgott, H., Venkatesh, A.: Integral points on elliptic curves and 3-torsion in class groups. J. Am. Math. Soc. **19**(3), 527–550 (2006)

592. Helfgott, H.A.: Growth and generation in $SL_2(\mathbb{Z}/p\mathbb{Z})$. Ann. Math. **167**(2), 601–623 (2008)

593. Helfgott, H.A.: The ternary Goldbach conjecture is true. Preprint (2013). arXiv:1312.7748

594. Helfgott, H.A., Platt, D.J.: Numerical verification of the ternary Goldbach conjecture up to $8.875 \cdot 10^{30}$. Exp. Math. **22**(4), 406–409 (2013)

595. Helfgott, H.A., Seress, Á.: On the diameter of permutation groups. Ann. Math. **179**(2), 611–658 (2014)

596. Helfgott, H.A., et al.: Square-free values of $f(p)$, f cubic. Acta Math. **213**(1), 107–135 (2014)

597. Helson, H.: Hankel forms. Studia Math. **1**(198), 79–84 (2010)

598. Helton, J.W.: "positive" noncommutative polynomials are sums of squares. Ann. Math. **156**(2), 675–694 (2002)

599. Hermon, J., Hutchcroft, T.: Supercritical percolation on nonamenable graphs: Isoperimetry, analyticity, and exponential decay of the cluster size distribution. Preprint (2019). arXiv:1904.10448

600. Hertweck, M.: A counterexample to the isomorphism problem for integral group rings. Ann. Math. **154**(1), 115–138 (2001)

601. Heule, M.: Schur number five. In: Proceedings of the Thirty-Second AAAI Conference on Artificial Intelligence (2018)

602. Heule, M.J.H., Kullmann, O., Marek, V.W.: Solving and verifying the boolean pythagorean triples problem via cube-and-conquer. In: International Conference on Theory and Applications of Satisfiability Testing, pp. 228–245. Springer (2016)

603. Hiary, G.A.: Fast methods to compute the Riemann zeta function. Ann. Math. **174**(2), 891–946 (2011)

604. Higman, G.: The units of group-rings. Proc. Lond. Math. Soc. **2**(1), 231–248 (1940)

605. Hilbert, D.: Mathematical problems. Bull. Am. Math. Soc. **8**(10), 437–479 (1902)

606. Hilbert, D.: Beweis für die Darstellbarkeit der ganzen Zahlen durch eine feste Anzahln ter Potenzen (Waringsches Problem). Math. Ann. **67**(3), 281–300 (1909)

607. Hindman, N.: Finite sums from sequences within cells of a partition of $\mathbb{N}$. J. Combin. Theory Ser. A **17**(1), 1–11 (1974)

608. Hochman, M.: On self-similar sets with overlaps and inverse theorems for entropy. Ann. Math. **180**, 773–822 (2014)

609. Hochman, M., Meyerovitch, T.: A characterization of the entropies of multidimensional shifts of finite type. Ann. Math. **171**(3), 2011–2038 (2010)

610. Hochman, M., Shmerkin, P.: Local entropy averages and projections of fractal measures. Ann. Math. **175**(3), 1001–1059 (2012)
611. Hodgson, C.D., Rivin, I., Smith, W.D.: A characterization of convex hyperbolic polyhedra and of convex polyhedra inscribed in the sphere. Bull. Am. Math. Soc. **27**(2), 246–251 (1992)
612. Hoeffding, W.: Probability inequalities for sums of bounded random variables. J. Amer. Statist. Assoc. **58**(301), 13–30 (1963)
613. Hoffman, D., Meeks III, W.H.: Embedded minimal surfaces of finite topology. Ann. Math. **131**(1), 1–34 (1990)
614. Hoffman, D., Traizet, M., White, B.: Helicoidal minimal surfaces of prescribed genus. Acta Math. **216**(2), 217–323 (2016)
615. Hoggar, S.G.: Chromatic polynomials and logarithmic concavity. J. Combin. Theory Ser. B **16**(3), 248–254 (1974)
616. Hoheisel, G.: Primzahlprobleme in der Analysis. Sitz. Preuss. Akad. Wiss. **33**, 3–11 (1930)
617. Holmsen, A.F., Nassajian, M.H., Pach, J., Tardos, G.: Two extensions of the Erdős–Szekeres problem. J. Eur. Math. Soc. (JEMS) (2020)
618. Holroyd, A.E., Liggett, T.M.: Finitely dependent coloring. Forum Math. Pi **4**, e9 (2016). https://doi.org/10.1017/fmp.2016.7
619. Hopf, H.: Ein toplogischer Beitrag zur reellen Algebra. Comment. Math. Helv. **13**(1), 219–239 (1941)
620. Host, B., Kra, B.: Nonconventional ergodic averages and nilmanifolds. Ann. Math. **161**(1), 397–488 (2005)
621. Hough, B.: Solution of the minimum modulus problem for covering systems. Ann. Math., 361–382 (2015)
622. Howson, A.G.: On the intersection of finitely generated free groups. J. Lond. Math. Soc. **1**(4), 428–434 (1954)
623. Huang, H.: Induced subgraphs of hypercubes and a proof of the Sensitivity Conjecture. Ann. Math. **190**(3), 949–955 (2019)
624. Huang, R., Rota, G.C.: On the relations of various conjectures on Latin squares and straightening coefficients. Discrete Math. **128**(1-3), 225–236 (1994)
625. Huang, S.: An improvement to Zaremba's conjecture. Geom. Funct. Anal. **25**(3), 860–914 (2015)
626. Hubbard, J., Schleicher, D., Sutherland, S.: How to find all roots of complex polynomials by Newton's method. Invent. Math. **146**(1), 1–33 (2001)
627. Huh, J.: Milnor numbers of projective hypersurfaces and the chromatic polynomial of graphs. J. Am. Math. Soc. **25**(3), 907–927 (2012)
628. Huh, J., Katz, E.: Log-concavity of characteristic polynomials and the Bergman fan of matroids. Math. Ann. **354**(3), 1103–1116 (2012)
629. Huh, J., Schröter, B., Wang, B.: Correlation bounds for fields and matroids. Preprint (2018). arXiv:1806.02675
630. Huisken, G.: Asymptotic-behavior for singularities of the mean-curvature flow. J. Differ. Geom. **31**(1), 285–299 (1990)
631. Hurwitz, A.: Über die angenäherte Darstellung der Irrationalzahlen durch rationale Brüche. Math. Ann. **39**(2), 279–284 (1891)
632. Hutchcroft, T.: Nonuniqueness and mean-field criticality for percolation on nonunimodular transitive graphs. J. Am. Math. Soc. **33**(4), 1101–1165 (2020)
633. Hutchings, M., Morgan, F., Ritoré, M., Ros, A.: Proof of the double bubble conjecture. Ann. Math. **155**(2), 459–489 (2002)
634. Hyde, J.: The group of boundary fixing homeomorphisms of the disc is not left-orderable. Ann. Math. **190**(2), 657–661 (2019)
635. Hyde, J., Lodha, Y.: Finitely generated infinite simple groups of homeomorphisms of the real line. Invent. Math. **218**(1), 83–112 (2019)
636. Ibragimov, I.A., Linnik, J.V.: Nezavisimye stalionarno svyazannye velichiny. M: Nauka **16**, 102–111 (1965)

637. Iosevich, A., Katz, N., Tao, T.: The Fuglede spectral conjecture holds for convex planar domains. Math. Res. Lett. **10**(5), 559–569 (2003)
638. Iosevich, A., Kolountzakis, M.N.: Periodicity of the spectrum in dimension one. Anal. PDE **6**(4), 819–827 (2013)
639. Isaacs, I.M., Liebeck, M.W., Navarro, G., Tiep, P.H.: Fields of values of odd-degree irreducible characters. Adv. Math. **354**, 106757 (2019)
640. Isett, P.: A proof of Onsager's conjecture. Ann. Math. **188**(3), 871–963 (2018)
641. Isett, P.J.: Hölder continuous Euler flows with compact support in time. Princeton University, Princeton, NJ (2013)
642. Ivanisvili, P., van Handel, R., Volberg, A.: Rademacher type and Enflo type coincide. Ann. Math. **192**(2), 665–678 (2020)
643. Iwaniec, H.: Primes represented by quadratic polynomials in two variables. Acta Arith. **5**(24), 435–459 (1974)
644. Izhboldin, O.T.: Fields of u-invariant 9. Ann. Math. **154**(3), 529–587 (2001)
645. Jackson, S., Mauldin, R.: On a lattice problem of H. Steinhaus. J. Am. Math. Soc. **15**(4), 817–856 (2002)
646. Jadbabaie, A., Lin, J., Morse, A.S.: Coordination of groups of mobile autonomous agents using nearest neighbor rules. IEEE Trans. Automatic Control **48**(6), 988–1001 (2003)
647. Jaeger, F.: Flows and generalized coloring theorems in graphs. J. Combin. Theory Ser. B **26**(2), 205–216 (1979)
648. Jain, K., Mahdian, M., Saberi, A.: A new greedy approach for facility location problems. In: Proceedings of the Thiry-Fourth Annual ACM Symposium on the Theory of Computing, pp. 731–740 (2002)
649. Jain, V., Sah, A., Sawhney, M.: Singularity of discrete random matrices II. Preprint (2020). arXiv:2010.06554
650. James, R.: New pentagonal tiling reported by Martin Gardner. Sci. Am. **1975**, 117–118 (1975)
651. Jaye, B., Tolsa, X., Villa, M.: A proof of Carleson's ε^2-conjecture. Preprint (2019). arXiv:1909.08581
652. Jensen, I.: Improved lower bounds on the connective constants for two-dimensional self-avoiding walks. J. Phys. A **37**(48), 11521 (2004)
653. Jensen, I., Guttmann, A.J.: Self-avoiding walks, neighbour-avoiding walks and trails on semiregular lattices. J. Phys. A **31**(40), 8137 (1998)
654. Jerison, D., Levine, L., Sheffield, S.: Logarithmic fluctuations for internal DLA. J. Am. Math. Soc. **25**(1), 271–301 (2012)
655. Jerison, D., Levine, L., Sheffield, S.: Internal DLA in higher dimensions. Electron. J. Probab. **18**(98), 1–14 (2013)
656. Jerrum, M., Sinclair, A., Vigoda, E.: A polynomial-time approximation algorithm for the permanent of a matrix with nonnegative entries. J. ACM **51**(4), 671–697 (2004)
657. Jerrum, M.R., Valiant, L.G., Vazirani, V.V.: Random generation of combinatorial structures from a uniform distribution. Theor. Comput. Sci. **43**, 169–188 (1986)
658. Jessen, B., Wintner, A.: Distribution functions and the Riemann zeta function. Trans. Am. Math. Soc. **38**(1), 48–88 (1935)
659. Ji, Z., Natarajan, A., Vidick, T., Wright, J., Yuen, H.: MIP* = RE. Preprint (2020). arXiv:2001.04383
660. Johansson, A., Kahn, J., Vu, V.: Factors in random graphs. Random Struct. Algorithm. **33**(1), 1–28 (2008)
661. Johnson, W.B.: Banach spaces all of whose subspaces have the approximation property. Séminaire Analyse fonctionnelle (dit "Maurey–Schwartz") pp. 1–11 (1979)
662. Johnson, W.B., Odell, E.: The diameter of the isomorphism class of a Banach space. Ann. Math. **162**(1), 423–437 (2005)
663. Johnson, W.B., Szankowski, A.: Hereditary approximation property. Ann. Math. **176**(3), 1987–2001 (2012)

664. Johnstone, I.M.: On the distribution of the largest eigenvalue in principal components analysis. Ann. Statist. pp. 295–327 (2001)
665. Jorge, L.P.d.M., Xavier, F.: A complete minimal surface in $\mathbb{R}^3$ between two parallel planes. Ann. Math., 203–206 (1980)
666. Judge, C., Mondal, S.: Euclidean triangles have no hot spots. Ann. Math. **191**(1), 167–211 (2020)
667. Juschenko, K., Monod, N.: Cantor systems, piecewise translations and simple amenable groups. Ann. Math. **178**(2), 775–787 (2013)
668. Kabatyanskiı, G.A., Levenshteın, V.I.: Bounds for packings on a sphere and in space. Probl. Inf. Transm. **95**, 148–158 (1974)
669. Kabatyanskiı, G.A., Levenshteın, V.I.: On bounds for packings on a sphere and in space. Problemy Peredachi Informatsii **14**(1), 3–25 (1978)
670. Kadison, R.V., Singer, I.M.: Extensions of pure states. Am. J. Math. **81**(2), 383–400 (1959)
671. Kahane, J.P.: Sur les polynomes a coefficients unimodulaires. Bull. Lond. Math. Soc. **12**(5), 321–342 (1980)
672. Kahn, J., Komlós, J., Szemerédi, E.: On the probability that a random ±1-matrix is singular. J. Am. Math. Soc. **8**(1), 223–240 (1995)
673. Kalai, G.: The diameter of graphs of convex polytopes and f-vector theory. In: Applied Geometry and Discrete Mathematics, pp. 387–412 (1990)
674. Kalai, G., Kleitman, D.J.: A quasi-polynomial bound for the diameter of graphs of polyhedra. Bull. Am. Math. Soc.(NS) **26**(math. MG/9204233), 315–316 (1992)
675. Kalkowski, M., Karoński, M., Pfender, F.: Vertex-coloring edge-weightings: towards the 1-2-3-conjecture. J. Combin. Theory Ser. B **100**(3), 347–349 (2010)
676. Kallus, Y., Elser, V., Gravel, S.: Dense periodic packings of tetrahedra with small repeating units. Discrete Comput. Geom. **44**(2), 245–252 (2010)
677. Kaluba, M., Kielak, D., Nowak, P.W.: On property (T) for Aut(F_n) and SL$_n(\mathbb{Z})$. Ann. Math. **193**(2), 539–562 (2021)
678. Kaluba, M., Nowak, P.W., Ozawa, N.: Aut(F_5) has property (T). Math. Ann. **375**(3-4), 1169–1191 (2019)
679. Kannan, R.: Lattice translates of a polytope and the Frobenius problem. Combinatorica **12**(2), 161–177 (1992)
680. Kaplansky, I.: Quadratic forms. J. Math. Soc. Jpn. **5**(2), 200–207 (1953)
681. Karatsuba, A.: Multiplication of multidigit numbers on automata. Sov. Phys. Doklady **7**, 595–596 (1963)
682. Karlin, A.R., Klein, N., Gharan, S.O.: A (slightly) improved approximation algorithm for metric TSP. Preprint (2020). arXiv:2007.01409
683. Karmarkar, N.: A new polynomial-time algorithm for linear programming. Combinatorica **4**(4), 373–395 (1984)
684. Karoński, M., Łuczak, T., Thomason, A.: Edge weights and vertex colours. J. Comb. Theory Ser. B **91**(1), 151–157 (2004)
685. Karp, R.M., Upfal, E., Wigderson, A.: Constructing a perfect matching is in random NC. Combinatorica **6**(1), 35–48 (1986)
686. Karpinski, M., Lampis, M., Schmied, R.: New inapproximability bounds for TSP. J. Comput. Syst. Sci. **81**(8), 1665–1677 (2015)
687. Kassabov, M.: Symmetric groups and expander graphs. Invent. Math. **170**(2), 327–354 (2007)
688. Kassabov, M., Lubotzky, A., Nikolov, N.: Finite simple groups as expanders. Proc. Natl. Acad. Sci. USA **103**(16), 6116–6119 (2006)
689. Kassabov, M., Pak, I.: Groups of oscillating intermediate growth. Ann. Math. **177**(3), 1113–1145 (2013)
690. Kathuria, L., Raka, M.: On conjectures of Minkowski and Woods for $n = 9$. Proc. Math. Sci. **126**(4), 501–548 (2016)
691. Kathuria, L., Raka, M.: On conjectures of Minkowski and Woods for $n = 10$. Preprint (2020). arXiv:2009.09992

692. Katok, A.: Interval exchange transformations and some special flows are not mixing. Isr. J. Math. **35**(4), 301–310 (1980)
693. Katz, N., Zahl, J.: An improved bound on the Hausdorff dimension of Besicovitch sets in $\mathbb{R}^3$. J. Am. Math. Soc. **32**(1), 195–259 (2019)
694. Katz, N.H., Tardos, G.: A new entropy inequality for the Erdős distance problem. Contemp. Math. **342**, 119–126 (2004)
695. Kaveh, K., Khovanskii, A.G.: Newton–Okounkov bodies, semigroups of integral points, graded algebras and intersection theory. Ann. Math. **176**(2), 925–978 (2012)
696. Kazhdan, D.A.: Connection of the dual space of a group with the structure of its close subgroups. Funct. Anal. Appl. **1**(1), 63–65 (1967)
697. Kearns, M., Littman, M.L., Singh, S.: Graphical models for game theory. In: Proceedings of the Seventeenth Conference on Uncertainty in Artificial Intelligence, pp. 253–260 (2001)
698. Kedlaya, K.S.: Large product-free subsets of finite groups. J. Combin. Theory Ser. A **77**(2), 339–343 (1997)
699. Keevash, P.: The existence of designs. Preprint (2014). arXiv:1401.3665
700. Keith, S., Zhong, X.: The Poincaré inequality is an open ended condition. Ann. Math. **167**(2), 575–599 (2008)
701. Kempe, D., Kleinberg, J., Tardos, É.: Maximizing the spread of influence through a social network. Theory Comput. **11**(4), 105–147 (2015)
702. Kepler, J.: Narratio de obseruatis a se quatuor Iouis satellitibus erronibus, quos Galilaeus Galilaeus Mathematicus Florentinus iure inuentionis Medicaea sidera nuncupauit : cum adiuncta Dissertatione de nuncio sidereo nuper ad mortales misso. Sumptibus Zachariae Palthenii (1611)
703. Kerckhoff, S., Masur, H., Smillie, J.: Ergodicity of billiard flows and quadratic differentials. Ann. Math. **124**(2), 293–311 (1986)
704. Kershner, R.B.: On paving the plane. Amer. Math. Monthly **75**(8), 839–844 (1968)
705. Kesten, H., Sidoravicius, V.: A shape theorem for the spread of an infection. Ann. Math. **167**(3), 701–766 (2008)
706. Kesten, H., et al.: The critical probability of bond percolation on the square lattice equals 1/2. Comm. Math. Phys. **74**(1), 41–59 (1980)
707. Khachiyan, L.G.: A polynomial algorithm in linear programming. Doklady Akademii Nauk **244**(5), 1093–1096 (1979)
708. Khalfalah, A., Szemerédi, E.: On the number of monochromatic solutions of $x + y = z^2$. Comb. Probab. Comput. **15**(1-2), 213 (2006)
709. Khintchine, A.: Einige Sätze über Kettenbrüche, mit Anwendungen auf die Theorie der diophantischen Approximationen. Math. Ann. **92**(1-2), 115–125 (1924)
710. Khorunzhiy, O.: Estimates for moments of random matrices with gaussian elements. In: Séminaire de probabilités XLI, pp. 51–92. Springer (2008)
711. Khot, S.: On the power of unique 2-prover 1-round games. In: Proceedings of the Thiry-Fourth Annual ACM Symposium on the Theory of Computing, pp. 767–775. ACM (2002)
712. Khot, S., Kindler, G., Mossel, E., O'Donnell, R.: Optimal inapproximability results for MAX-CUT and other 2-variable CSPs? SIAM J. Comput. **37**(1), 319–357 (2007)
713. Khot, S., Minzer, D., Safra, M.: Pseudorandom sets in grassmann graph have near-perfect expansion. In: 2018 IEEE 59th Annual Symposium on the Foundations of Computer Science (FOCS), pp. 592–601. IEEE (2018)
714. Khot, S., Regev, O.: Vertex cover might be hard to approximate to within $2 - \varepsilon$. J. Comput. Syst. Sci. **74**(3), 335–349 (2008)
715. Khot, S.A., Vishnoi, N.K.: The unique games conjecture, integrality gap for cut problems and embeddability of negative-type metrics into l_1. J. ACM **62**(1), 1–39 (2015)
716. Khukhro, E.I., Mazurov, V.D.: Unsolved problems in group theory. The Kourovka notebook. Preprint (2014). arXiv:1401.0300
717. Ki, H., Kim, Y.O., Lee, J.: On the de Bruijn–Newman constant. Adv. Math. **222**(1), 281–306 (2009)

718. Klartag, B.: $5n$ Minkowski symmetrizations suffice to arrive at an approximate Euclidean ball. Ann. Math. **156**(3), 947–960 (2002)

719. Klartag, B.: On convex perturbations with a bounded isotropic constant. Geom. Funct. Anal. **16**(6), 1274–1290 (2006)

720. Klartag, B.: A central limit theorem for convex sets. Invent. Math. **168**(1), 91–131 (2007)

721. Klartag, B., Milman, V.D.: Isomorphic Steiner symmetrization. Invent. Math. **153**(3), 463–485 (2003)

722. Klazar, M.: The Füredi–Hajnal conjecture implies the Stanley–Wilf conjecture. In: Formal Power Series and Algebraic Combinatorics, pp. 250–255. Springer (2000)

723. Klee, V.: Is a body spherical if its HA-measurements are constant? Am. Math. Monthly **76**(5), 539–542 (1969)

724. Klee, V.: Is every polygonal region illuminable from some point? Am. Math. Monthly **76**(2), 180–180 (1969)

725. Klee, V., Minty, G.J.: How good is the simplex algorithm? Inequalities **3**(3), 159–175 (1972)

726. Kneser, H.: Die Topologie der Mannigfaltigkeiten. Jahresbericht der Deut. Math. Verein **34**, 1–13 (1926)

727. Kochol, M.: An equivalent version of the 3-flow conjecture. J. Combin. Theory Ser. B **83**(2), 258–261 (2001)

728. Koenigsmann, J.: Defining $\mathbb{Z}$ in $\mathbb{Q}$. Ann. Math. **183**, 73–93 (2016)

729. Kohayakawa, Y., Łuczak, T., Rödl, V.: On K^4-free subgraphs of random graphs. Combinatorica **17**(2), 173–213 (1997)

730. Kolountzakis, M.N.: Non-symmetric convex domains have no basis of exponentials. Ill. J. Math. **44**(3), 542–550 (2000)

731. Kolountzakis, M.N., Matolcsi, M.: Complex Hadamard matrices and the spectral set conjecture. Collect. Math. **57**(1), 281–291 (2006)

732. Komlós, J.: On determinant of $(0, 1)$ matrices. Studia Sci. Math. Hungarica **2**, 7–21 (1967)

733. Komlós, J., Major, P., Tusnády, G.: An approximation of partial sums of independent RV's, and the sample DF. I. Zeitschrift für Wahrscheinlichkeitstheorie und verwandte Gebiete **32**(1), 111–131 (1975)

734. Konev, B., Lisitsa, A.: Computer-aided proof of Erdős discrepancy properties. Artificial Intelligence **224**, 103–118 (2015)

735. Kőnig, D.: Graphs and matrices. Matematikai és Fizikai Lapok **38**, 116–119 (1931)

736. Kontorovich, A., Oh, H.: Apollonian circle packings and closed horospheres on hyperbolic 3-manifolds. J. Am. Math. Soc. **24**(3), 603–648 (2011)

737. Korec, I.: A density estimate for the $3x + 1$ problem. Mathematica Slovaca **44**(1), 85–89 (1994)

738. Korevaar, J., Meyers, J.L.H.: Spherical Faraday cage for the case of equal point charges and Chebyshev-type quadrature on the sphere. Integral Transforms Spec. Funct. **1**(2), 105–117 (1993)

739. Korkine, A., Zolotareff, G.: Sur les formes quadratiques. Math. Ann. **6**(3), 366–389 (1873)

740. Korkine, A., Zolotareff, G.: Sur les formes quadratiques positives. Math. Ann. **11**(2), 242–292 (1877)

741. Koukoulopoulos, D., Maynard, J.: On the Duffin–Schaeffer conjecture. Ann. Math. **192**(1), 251–307 (2020)

742. Koutis, I., Miller, G.L., Peng, R.: A nearly-$m \log n$ time solver for sdd linear systems. In: 2011 IEEE 52nd Annual Symposium on the Foundations of Computer Science, pp. 590–598. IEEE (2011)

743. Kővári, P., Sós, V., Turán, P.: On a problem of Zarankiewicz. In: Colloquium Mathematicum, vol. 3, pp. 50–57. Polska Akademia Nauk (1954)

744. Kozlovski, O., Shen, W., van Strien, S.: Density of hyperbolicity in dimension one. Ann. Math. **166**(1), 145–182 (2007)

745. Kozma, G., Nachmias, A.: Arm exponents in high dimensional percolation. J. Am. Math. Soc. **24**(2), 375–409 (2011)

746. Kozma, G., Nitzan, S.: Combining Riesz bases. Invent. Math. **199**(1), 267–285 (2015)

747. Kozma, G., Nitzan, S.: Combining Riesz bases in $\mathbb{R}^d$. Revista matemática iberoamericana **32**(4), 1393–1406 (2016)
748. Krammer, D.: The braid group b_4 is linear. Invent. Math. **142**(3), 451–486 (2000)
749. Krammer, D.: Braid groups are linear. Ann. Math. **155**(1), 131–156 (2002)
750. Krivelevich, M.: The critical bias for the Hamiltonicity game is $(1 + o(1))n/\log n$. J. Am. Math. Soc. **24**(1), 125–131 (2011)
751. Krivine, J.L.: Sur la constante de Grothendieck. CR Acad. Sci. Paris Ser. AB **284**(8), A445–A446 (1977)
752. Kuperberg, G.: Symmetry classes of alternating-sign matrices under one roof. Ann. Math. **156**(3), 835–866 (2002)
753. Lackenby, M.: A polynomial upper bound on Reidemeister moves. Ann. Math. **182**(2) (2015)
754. Laczkovich, M.: Equidecomposability and discrepancy; a solution of Tarski's circle-squaring problem. J. Reine Angew. Math. **1990**(404), 77–117 (1990)
755. Ladner, R.E.: On the structure of polynomial time reducibility. J. ACM **22**(1), 155–171 (1975)
756. Lagarias, J.C., Wang, Y.: The finiteness conjecture for the generalized spectral radius of a set of matrices. Linear Algebra Appl. **214**, 17–42 (1995)
757. Lagrange, J.L.: Essai sur le probleme des trois corps. Prix de l'académie royale des Sciences de Paris **9**, 292 (1772)
758. Lairez, P.: A deterministic algorithm to compute approximate roots of polynomial systems in polynomial average time. Found. Comput. Math. **17**(5), 1265–1292 (2017)
759. Lairez, P.: Rigid continuation paths I. Quasilinear average complexity for solving polynomial systems. J. Am. Math. Soc. **33**(2), 487–526 (2020)
760. Laishram, S., Shorey, T.N.: Perfect powers in arithmetic progressions. J. Comb. Number Theory **7**(2), 95 (2015)
761. Landim, C., Quastel, J., Salmhofer, M., Yau, H.T.: Superdiffusivity of asymmetric exclusion process in dimensions one and two. Comm. Math. Phys. **244**(3), 455–481 (2004)
762. Langevin, M.: Quelques applications de nouveaux résultats de van der Poorten. Séminaire Delange-Pisot-Poitou. Théorie des nombres **17**(2), G1–G11 (1976)
763. Larsen, M., Shalev, A.: Word maps and Waring type problems. J. Am. Math. Soc. **22**(2), 437–466 (2009)
764. Larsen, M., Shalev, A., Tiep, P.H.: The Waring problem for finite simple groups. Ann. Math. **174**(3), 1885–1950 (2011)
765. Larsen, M., Shalev, A., Tiep, P.H.: Probabilistic Waring problems for finite simple groups. Ann. Math. **190**(2), 561–608 (2019)
766. Lasserre, J.B.: Global optimization with polynomials and the problem of moments. SIAM J. Optim. **11**(3), 796–817 (2001)
767. Lawler, G.F.: On the covering time of a disc by simple random walk in two dimensions. In: Seminar on Stochastic Processes, 1992, pp. 189–207. Springer (1993)
768. Lawler, G.F., Bramson, M., Griffeath, D.: Internal diffusion limited aggregation. Ann. Probab. pp. 2117–2140 (1992)
769. Lawler, G.F., Puckette, E.E.: The intersection exponent for simple random walk. Combin. Probab. Comput. **9**(5), 441–464 (2000)
770. Lawler, G.F., Schramm, O., Werner, W.: Values of Brownian intersection exponents, II: Plane exponents. Acta Math. **187**(2), 275–308 (2001). https://doi.org/10.1007/BF02392619
771. Lawson, H.B.: Complete minimal surfaces in S^3. Ann. Math. **92**, 335–374 (1970)
772. Lawson, H.B.: The unknottedness of minimal embeddings. Invent. Math. **11**(3), 183–187 (1970)
773. Lawson, J., Lim, Y.: Monotonic properties of the least squares mean. Math. Ann. **351**(2), 267–279 (2011)
774. Lebesgue, V.A.: Sur l'impossibilité en nombres entiers de l'équation $x^m = y^2 + 1$, Nouv. Ann. Math. **9**(9), 178–181 (1850)
775. Lee, C.: Ramsey numbers of degenerate graphs. Ann. Math. pp. 791–829 (2017)

776. Lee, J.R., Peres, Y.: Harmonic maps on amenable groups and a diffusive lower bound for random walks. Ann. Probab. **41**(5), 3392–3419 (2013)
777. Lee, Y.T., Vempala, S.S.: Eldan's stochastic localization and the KLS hyperplane conjecture: an improved lower bound for expansion. In: 2017 IEEE 58th Annual Symposium on the Foundations of Computer Science (FOCS), pp. 998–1007. IEEE (2017)
778. Lee, Y.T., Vempala, S.S.: The Kannan–Lovász–Simonovits conjecture. In: Current Developments in Mathematics, pp. 1–36. International Press of Boston (2017)
779. Legendre, A.M.: Éléments de Géométrie. Imprimerie Firmin Didot, Pére et Fils (1794)
780. Lehmer, D.H.: Factorization of certain cyclotomic functions. Ann. Math. **34**(3), 461–479 (1933)
781. Leighton, T., Rao, S.: Multicommodity max-flow min-cut theorems and their use in designing approximation algorithms. J. ACM **46**(6), 787–832 (1999)
782. Lelievre, S., Monteil, T., Weiss, B.: Everything is illuminated. Geometry Topology **20**(3), 1737–1762 (2016)
783. Lemmens, P.W.H., Seidel, J.J.: Equiangular lines. J. Algebra **24**(3), 494–512 (1973)
784. Leuzinger, E., Young, R.: Filling functions of arithmetic groups. Preprint (2017). arXiv:1710.00732
785. Lev, N., Matolcsi, M.: The Fuglede conjecture for convex domains is true in all dimensions. Preprint (2019). arXiv:1904.12262
786. Levenshtein, V.I.: On bounds for packings in n-dimensional Euclidean space. Doklady Akademii Nauk **245**(6), 1299–1303 (1979)
787. LeVeque, W.J.: On the equation $y^m = f(x)$. Acta Arith. **9**(3), 209–219 (1964)
788. Levine, L., Pegden, W., Smart, C.K.: The Apollonian structure of integer superharmonic matrices. Ann. Math. **186**(1), 1–67 (2017)
789. Lewko, M.: An improved upper bound for the sum-free subset constant. J. Integer Seq. **13**(2), 3 (2010)
790. Li, P., Schoen, R., Yau, S.T.: On the isoperimetric inequality for minimal surfaces. Annali della Scuola Normale Superiore di Pisa-Classe di Scienze **11**(2), 237–244 (1984)
791. Li, P., Yau, S.T.: A new conformal invariant and its applications to the Willmore conjecture and the first eigenvalue of compact surfaces. Invent. Math. **69**(2), 269–291 (1982)
792. Li, S.: A 1.488 approximation algorithm for the uncapacitated facility location problem. Inform. Comput. **222**, 45–58 (2013)
793. Li, X.: Non-malleable extractors and non-malleable codes: partially optimal constructions. In: Proceedings of the 34th Computational Complexity Conference, pp. 1–49 (2019)
794. Lieb, E.H.: Proof of Wehrl's entropy conjecture. Comm. Math. Phys. **62**, 35–40 (1978)
795. Lieb, E.H.: Sharp constants in the Hardy–Littlewood–Sobolev and related inequalities. Ann. Math. **118**(2), 349–374 (1983)
796. Liebeck, M., O'Brien, E., Shalev, A., Tiep, P.: Products of squares in finite simple groups. Proc. Am. Math. Soc. **140**(1), 21–33 (2012)
797. Liebeck, M.W., O'Brien, E.A., Shalev, A., Tiep, P.H.: The Ore conjecture. J. Eur. Math. Soc. (JEMS) **12**(4), 939–1008 (2010)
798. Liebeck, M.W., Shalev, A.: Diameters of finite simple groups: sharp bounds and applications. Ann. Math. **154**(2), 383–406 (2001)
799. Lindenstrauss, J., Preiss, D.: On Fréchet differentiability of Lipschitz maps between Banach spaces. Ann. Math. **157**(1), 257–288 (2003)
800. Linial, N.: Finite metric spaces – combinatorics, geometry and algorithms. In: Proceedings of the International Congress of Mathematicians III. Citeseer (2002)
801. Linnik, U.V.: On the least prime in an arithmetic progression. I. The basic theorem. Mat. Sbornik NS **15**(2), 139–178 (1944)
802. van Lint, J.H., Seidel, J.J.: Equilateral point sets in elliptic geometry. Indag. Math **28**(3), 335–34 (1966)
803. Lipton, R.J., Markakis, E., Mehta, A.: Playing large games using simple strategies. In: Proceedings of the 4th ACM Conference on Electronic Commerce, pp. 36–41 (2003)

804. Litherland, R.A., Simon, J., Durumeric, O., Rawdon, E.: Thickness of knots. Topology Appl. **91**(3), 233–244 (1999)
805. Littlewood, J.E.: On polynomials $\sigma^n \pm z^m$, $\sigma^n e^{\alpha_m i} z^m$, $z = e^{\theta i}$. J. Lond. Math. Soc. **1**(1), 367–376 (1966)
806. Littlewood, J.E.: Some problems in real and complex analysis. DC Heath (1968)
807. Liu, H., Pikhurko, O., Staden, K.: The exact minimum number of triangles in graphs with given order and size. Forum Math. Pi **8**, e8 (2020)
808. Liu, M.C., Wang, T.Z.: On the Vinogradov bound in the three primes Goldbach conjecture. AcAri **105**(2), 133–175 (2002)
809. Ljunggren, W.: On the diophantine equation $x^2 + 4 = ay^2$. Norske Vid. Selsk. Forh., Trondheim **24**, 82–84 (1951)
810. Logunov, A.: Nodal sets of Laplace eigenfunctions: proof of Nadirashvili's conjecture and of the lower bound in Yau's conjecture. Ann. Math. **187**(1), 241–262 (2018)
811. London, H., Finkelstein, R.: On Fibonacci and Lucas numbers which are perfect powers. Fibonacci Quart. **7**(5), 476–481 (1969)
812. Lonjou, A.: Non simplicité du groupe de Cremona sur tout corps. Annales de l'Institut Fourier **66**(5), 2021–2046 (2016)
813. Lovász, L.: A characterization of perfect graphs. J. Combin. Theory Ser. B **13**(2), 95–98 (1972)
814. Lovász, L., Simonovits, M.: On the number of complete subgraphs of a graph II. In: Studies in Pure Mathematics, pp. 459–495. Springer (1983)
815. Lovász, L.M., Thomassen, C., Wu, Y., Zhang, C.Q.: Nowhere-zero 3-flows and modulo k-orientations. J. Combin. Theory Ser. B **103**(5), 587–598 (2013)
816. Lubotzky, A.: Discrete groups, expanding graphs and invariant measures. Springer Science & Business Media (2010)
817. Lubotzky, A., Weiss, B.: Groups and expanders. In: Expanding Graphs (Princeton, NJ, 1992), vol. 10, pp. 95–109. American Mathematical Society (1993)
818. Łuczak, T.: The chromatic number of random graphs. Combinatorica **11**(1), 45–54 (1991)
819. Luczak, T., Rucinńki, A.: Tree-matchings in graph processes. SIAM J. Discrete Math. **4**(1), 107–120 (1991)
820. Łuczak, T., Thomassé, S.: Coloring dense graphs via VC-dimension. Preprint (2010). arXiv:1007.1670
821. Ludwig, M.: Minkowski valuations. Trans. Am. Math. Soc. **357**(10), 4191–4213 (2005)
822. Lutz, E.: Sur l'equation $y^2 = x^3 - ax - b$ dans les corps p-adic. J. Reine Angew. Math. **177**, 238–247 (1937)
823. Lyapunov, A.M.: A general proposition of probability theory. CR Acad. Sci. Paris **132**, 814–815 (1901)
824. Lyubich, M.: Almost every real quadratic map is either regular or stochastic. Ann. Math. **156**(1), 1–78 (2002)
825. Maderna, E., Venturelli, A.: Viscosity solutions and hyperbolic motions: A new PDE method for the n-body problem. Ann. Math. **192**(2), 499–550 (2020)
826. Magnus, W.: Das Identitätsproblem für Gruppen mit einer definierenden Relation. Math. Ann. **106**(1), 295–307 (1932)
827. Mahajan, M., Varadarajan, K.R.: A new NC-algorithm for finding a perfect matching in bipartite planar and small genus graphs. In: Proceedings of the Thirty-Second Annual ACM Symposium on the Theory of Computing, pp. 351–357 (2000)
828. Maharam, D.: An algebraic characterization of measure algebras. Ann. Math. **48**(1), 154–167 (1947)
829. Mahler, K.: Zur Approximation algebraischer Zahlen. III. Acta Math. **62**(1), 91–166 (1933)
830. Maier, H., Pomerance, C.: Unusually large gaps between consecutive primes. Trans. Am. Math. Soc. **322**(1), 201–237 (1990)
831. Maier, H., et al.: Primes in short intervals. Mich. Math. J. **32**(2), 221–225 (1985)
832. Maier, H., et al.: Small differences between prime numbers. Mich. Math. J. **35**(3), 323–344 (1988)

833. Malle, G., Späth, B.: Characters of odd degree. Ann. Math. **184**(3), 869–908 (2016)
834. Malliaris, M., Shelah, S.: Cofinality spectrum theorems in model theory, set theory, and general topology. J. Am. Math. Soc. **29**(1), 237–297 (2016)
835. Mankiewicz, P.: On Lipschitz mappings between Fréchet spaces. Studia Math. **3**(41), 225–241 (1972)
836. Mann, C., McLoud-Mann, J., Von Derau, D.: Convex pentagons that admit i-block transitive tilings. Geom. Dedicata **194**(1), 141–167 (2018)
837. Manners, F.: A solution to the pyjama problem. Invent. Math. **202**(1), 239–270 (2015)
838. Manolescu, C.: Pin (2)-equivariant Seiberg–Witten Floer homology and the triangulation conjecture. J. Am. Math. Soc. **29**(1), 147–176 (2016)
839. Mantel, W.: Vraagstuk XXVIII. Wiskundige Opgaven **10**, 60–61 (1907)
840. Marcus, A., Tardos, G.: Excluded permutation matrices and the Stanley–Wilf conjecture. J. Combin. Theory Ser. A **107**(1), 153–160 (2004)
841. Marcus, A.W., Spielman, D.A., Srivastava, N.: Interlacing families I: Bipartite Ramanujan graphs of all degrees. Ann. Math. **182**(1), 307–325 (2015)
842. Marcus, A.W., Spielman, D.A., Srivastava, N.: Interlacing families II: Mixed characteristic polynomials and the Kadison–Singer problem. Ann. Math. **182**(1), 327–350 (2015)
843. Marcus, A.W., Spielman, D.A., Srivastava, N.: Interlacing families IV: Bipartite Ramanujan graphs of all sizes. SIAM J. Comput. **47**(6), 2488–2509 (2018)
844. Margulis, G.A.: Explicit constructions of concentrators. Problemy Peredachi Informatsii **9**(4), 71–80 (1973)
845. Marklof, J.: Pair correlation densities of inhomogeneous quadratic forms. Ann. Math. **158**(2), 419–471 (2003)
846. Marks, A.S., Unger, S.T.: Borel circle squaring. Ann. Math. **186**(2), 581–605 (2017)
847. Marques, F.C., Neves, A.: Min-Max theory and the Willmore conjecture. Ann. Math. **179**(2), 683–782 (2014)
848. Marszalek, R.: On the product of consecutive elements of an arithmetic progression. Monatshefte für Mathematik **100**(3), 215–222 (1985)
849. Martinez, C., Zelmanov, E.: Products of powers in finite simple groups. Isr. J. Math. **96**(2), 469–479 (1996)
850. Mason, J.H.: Matroids: unimodal conjectures and Motzkin's theorem. Combinatorics (Proc. Conf. Comb. Math. Math. Inst. Oxford, 1972) **5**, 49 (1972)
851. Masur, H.: Interval exchange transformations and measured foliations. Ann. Math. **115**(1), 169–200 (1982)
852. Masur, H., et al.: Hausdorff dimension of the set of nonergodic foliations of a quadratic differential. Duke Math. J. **66**(3), 387–442 (1992)
853. Matiyasevich, Y.V.: The Diophantineness of enumerable sets. Doklady Akademii Nauk **191**(2), 279–282 (1970)
854. Matomäki, K., Radziwiłł, M.: Multiplicative functions in short intervals. Ann. Math. pp. 1015–1056 (2016)
855. Matomäki, K., Radziwiłł, M., Tao, T.: Sign patterns of the Liouville and Möbius functions. Forum Math. Sigma **4**, e14 (2016)
856. Matomäki, K., Radziwiłł, M., Tao, T.: Fourier uniformity of bounded multiplicative functions in short intervals on average. Invent. Math. **220**(1), 1–58 (2020)
857. Matomäki, K., Radziwiłł, M., Tao, T., et al.: An averaged form of Chowla's conjecture. Algebra Number Theory **9**(9), 2167–2196 (2015)
858. Matui, H.: Some remarks on topological full groups of Cantor minimal systems. Int. J. Math. **17**(02), 231–251 (2006)
859. Mauduit, C., Rivat, J.: Sur un probleme de Gelfond: la somme des chiffres des nombres premiers. Ann. Math. **171**(3), 1591–1646 (2010)
860. Maynard, J.: Small gaps between primes. Ann. Math. **181**(1), 383–413 (2015)
861. Maynard, J.: Large gaps between primes. Ann. Math. **183**(3), 915–933 (2016)
862. Maynard, J.: Primes with restricted digits. Invent. Math. **217**(1), 127–218 (2019)
863. Maynard, J.: Primes represented by incomplete norm forms. Forum Math. Pi **8**, e3, (2020)

864. McCool, J.: A faithful polynomial representation of Out F_3. Math. Proc. Camb. Philos. Soc. **106**(2), 207–213 (1989)
865. McKay, J.: Irreducible representations of odd degree. J. Algebra **20**(2), 416–418 (1972)
866. McMullen, C.: Minkowski's conjecture, well-rounded lattices and topological dimension. J. Am. Math. Soc. **18**(3), 711–734 (2005)
867. Meakin, P., Deutch, J.M.: The formation of surfaces by diffusion limited annihilation. J. Chem. Phys. **85**(4), 2320–2325 (1986)
868. Meeks, W.H., Pérez, J., Ros, A.: Bounds on the topology and index of minimal surfaces. Acta Math. **223**(1), 113–149 (2019)
869. Meeks III, W.H., Pérez, J., Ros, A.: Properly embedded minimal planar domains. Ann. Math. **181**(2), 473–546 (2015)
870. Meeks III, W.H., Rosenberg, H.: The uniqueness of the helicoid. Ann. Math. **161**(2), 727–758 (2005)
871. Megretski, A.: Relaxations of quadratic programs in operator theory and system analysis. In: Systems, Approximation, Singular Integral Operators, and Related Topics, pp. 365–392. Springer (2001)
872. Mehta, M.: Random Matrices and the Statistical Theory of Energy Levels. Academic Press, New York (1967)
873. Melas, A.: On the centered Hardy–Littlewood maximal operator. Trans. Am. Math. Soc. **354**(8), 3263–3273 (2002)
874. Melas, A.D.: The best constant for the centered Hardy–Littlewood maximal inequality. Ann. Math. **157**(2), 647–688 (2003)
875. Menarguez, M.T., Soria, F.: Weak type $(1, 1)$ inequalities of maximal convolution operators. Rendiconti del Circolo Matematico di Palermo **41**(3), 342–352 (1992)
876. Mendel, M., Naor, A.: Euclidean quotients of finite metric spaces. Adv. Math. **189**(2), 451–494 (2004)
877. Mendel, M., Naor, A.: Ramsey partitions and proximity data structures. J. Eur. Math. Soc. (JEMS) **9**(2), 253–275 (2007)
878. Mendel, M., Naor, A.: Metric cotype. Ann. Math. **168**(1), 247–298 (2008)
879. Mendelson, S., Vershynin, R.: Entropy and the combinatorial dimension. Invent. Math. **152**(1), 37–55 (2003)
880. Menger, K.: Zur allgemeinen kurventheorie. Fund. Math. **10**(1), 96–115 (1927)
881. Menshikov, M.V.: Coincidence of critical points in percolation problems. Sov. Math. Doklady **33**, 856–859 (1986)
882. Merkur'ev, A.S.: Kaplansky conjecture in the theory of quadratic forms. J. Sov. Math. **57**(6), 3489–3497 (1991)
883. Meshulam, R.: On subsets of finite abelian groups with no 3-term arithmetic progressions. J. Combin. Theory Ser. A **71**(1), 168–172 (1995)
884. Meyer, A.: Mathematische Mittheilungen. Vierteljahrschrift der Naturforschenden Gesellschaft in Zürich **29**, 209–222 (1884)
885. Meyniel, H.: A new property of critical imperfect graphs and some consequences. Eur. J. Combin. **8**(3), 313–316 (1987)
886. Mézard, M., Parisi, G.: Replicas and optimization. J. de Physique Lettres **46**(17), 771–778 (1985)
887. Mézard, M., Parisi, G.: Mean-field equations for the matching and the travelling salesman problems. EPL (Europhys. Lett.) **2**(12), 913 (1986)
888. Mézard, M., Parisi, G.: A replica analysis of the travelling salesman problem. J. de Physique **47**(8), 1285–1296 (1986)
889. Mihailescu, P.: Primary cyclotomic units and a proof of Catalan's conjecture. J. Reine Angew. Math. **2004**(572), 167–195 (2004)
890. Mihăilescu, P.: A class number free criterion for catalan's conjecture. J. Number Theory **99**(2), 225–231 (2003)
891. Miller, G.L.: Riemann's hypothesis and tests for primality. J. Comput. Syst. Sci. **13**(3), 300–317 (1976)

892. Miller, G.L., Naor, J.: Flow in planar graphs with multiple sources and sinks. SIAM J. Comput. **24**(5), 1002–1017 (1995)
893. Miller III, C.F., et al.: On Group-theoretic Decision Problems and Their Classification. 68. Princeton University Press (1971)
894. Mills, W.H., Robbins, D.P., Rumsey Jr, H.: Alternating sign matrices and descending plane partitions. J. Combin. Theory Ser. A **34**(3), 340–359 (1983)
895. Milman, E.: On the role of convexity in isoperimetry, spectral gap and concentration. Invent. Math. **177**(1), 1–43 (2009)
896. Milnor, J.: Remarks on iterated cubic maps. Exp. Math. **1**(1), 5–24 (1992)
897. Milnor, J., Tresser, C.: On entropy and monotonicity for real cubic maps. Comm. Math. Phys. **209**(1), 123–178 (2000)
898. Mineyev, I.: Submultiplicativity and the Hanna Neumann conjecture. Ann. Math. **175**, 393–414 (2012)
899. Minkowski, H.: Geometrie der zahlen (2 vol.). Teubner (1896)
900. Moakher, M.: A differential geometric approach to the geometric mean of symmetric positive-definite matrices. SIAM J. Matrix Anal. Appl. **26**(3), 735–747 (2005)
901. Moise, E.E.: Affine structures in 3-manifolds: V. The triangulation theorem and Hauptvermutung. Ann. Math. **56**, 96–114 (1952)
902. Monasson, R., Zecchina, R.: Statistical mechanics of the random K-satisfiability model. Phys. Rev. E **56**(2), 1357 (1997)
903. Montgomery, H.L.: The pair correlation of zeros of the zeta function. In: Proc. Symp. Pure Math, vol. 24, pp. 181–193 (1973)
904. Moody, J.A., et al.: The Burau representation of the braid group B_n is unfaithful for large n. Bull. Am. Math. Soc. **25**(2), 379–384 (1991)
905. Moore, E.H.: On the reciprocal of the general algebraic matrix. Bull. Am. Math. Soc. **26**, 394–395 (1920)
906. Mordell, L.J.: On the rational resolutions of the indeterminate equations of the third and fourth degree. Proc. Camb. Phil. Soc. **21**, 179–192 (1922)
907. Moreira, C.G.: Geometric properties of the Markov and Lagrange spectra. Ann. Math. **188**(1), 145–170 (2018)
908. Moreira, C.G.T.d.A., Yoccoz, J.C.: Stable intersections of regular Cantor sets with large Hausdorff dimensions. Ann. Math. **154**(1), 45–96 (2001)
909. Moreira, J.: Monochromatic sums and products in $\mathbb{N}$. Ann. Math. **185**(3), 1069–1090 (2017)
910. Moreira, J., Richter, F.K., Robertson, D.: A proof of a sumset conjecture of Erdős. Ann. Math. **189**(2), 605–652 (2019)
911. Moser, J.: On Harnack's theorem for elliptic differential equations. Comm. Pure Appl. Math. **14**(3), 577–591 (1961)
912. Moser, J.: A sharp form of an inequality by N. Trudinger. Ind. Univ. Math. J. **20**(11), 1077–1092 (1971)
913. Moser, R.A., Tardos, G.: A constructive proof of the general Lovász local lemma. J. ACM **57**(2), 1–15 (2010)
914. Mossel, E., O'Donnell, R., Oleszkiewicz, K.: Noise stability of functions with low influences: Invariance and optimality. Ann. Math. **171**(1), 295–341 (2010)
915. Motzkin, T.S.: The arithmetic-geometric inequality. Inequalities (Proc. Sympos. Wright-Patterson Air Force Base, Ohio, 1965) pp. 205–224 (1967)
916. Mukhin, E., Tarasov, V., Varchenko, A.: The B. and M. Shapiro conjecture in real algebraic geometry and the Bethe ansatz. Ann. Math. **170**(2), 863–881 (2009)
917. Musin, O.R.: The kissing number in four dimensions. Ann. Math. **168**(1), 1–32 (2008)
918. Nadirashvili, N.: Hadamard's and Calabi–Yau's conjectures on negatively curved and minimal surfaces. Invent. Math. **126**(3), 457–466 (1996)
919. Nadirashvili, N.: Geometry of nodal sets and multiplicity of eigenvalues. In: Current Developments in Mathematics, pp. 231–235. International Press of Boston (1997)
920. Nagell, T.: Solution de quelques problèmes dans la théorie arithmétique des cubiques planes du premier genre. Wid. Akad. Skrifter Oslo I **1**, 1–25 (1935)

921. Naor, A.: Discrete Riesz transforms and sharp metric x_p inequalities. Ann. Math. **184**(3), 991–1016 (2016)

922. Naor, A., Schechtman, G.: Metric X_p inequalities. Forum Math. Pi **4**, e3, 1–81 (2016)

923. Naor, A., Young, R.: Vertical perimeter versus horizontal perimeter. Ann. Math. **188**(1), 171–279 (2018)

924. Nash, J.F., et al.: Equilibrium points in n-person games. Proc. Natl. Acad. Sci. USA **36**(1), 48–49 (1950)

925. Nathanson, M.B.: Sumsets contained in infinite sets of integers. J. Comb. Theory Ser. A **28**(2), 150–155 (1980)

926. Navarro, G., Tiep, P.H.: The fields of values of characters of degree not divisible by p. Forum Math. Pi **9**, e2 (2021). https://doi.org/10.1017/fmp.2021.1

927. Nazarov, F., Ryabogin, D., Zvavitch, A.: An asymmetric convex body with maximal sections of constant volume. J. Am. Math. Soc. **27**(1), 43–68 (2014)

928. Nekrashevych, V.: Palindromic subshifts and simple periodic groups of intermediate growth. Ann. Math. **187**(3), 667–719 (2018)

929. Nemirovski, A., Roos, C., Terlaky, T.: On maximization of quadratic form over intersection of ellipsoids with common center. Math. Program. **86**(3), 463–473 (1999)

930. Neuen, D.: Isomorphism testing for graphs excluding small topological subgraphs. Preprint (2020). arXiv:2011.14730

931. Neumann, B.: Sharing ham and eggs. Iota, Manchester University, pp. 14–18 (1959)

932. Neumann, H.: On the intersection of finitely generated free groups. Addendum. Publ. Math. Debrecen **5**(128), 58 (1957)

933. von Neumann, J.: Zusatz zur Arbeit. Fund. Math. **13**(1), 333–116 (1929)

934. von Neumann, J.: Collected Works: Volume V – Design of Computers, Theory of Automata and Numerical Analysis. Pergamon Press (1963)

935. Neumann, W.D.: On intersections of finitely generated subgroups of free groups. In: Groups – Canberra 1989, pp. 161–170. Springer (1990)

936. Newman, C.M.: Fourier transforms with only real zeros. Proc. Am. Math. Soc. **61**(2), 245–251 (1976)

937. Nielsen, P.P.: A covering system whose smallest modulus is 40. J. Number Theory **129**(3), 640–666 (2009)

938. Nienhuis, B.: Exact critical point and critical exponents of $O(n)$ models in two dimensions. Phys. Rev. Lett. **49**(15), 1062 (1982)

939. Nikiforov, V.: The number of cliques in graphs of given order and size. Trans. Am. Math. Soc. **363**(3), 1599–1618 (2011)

940. Nisan, N., Szegedy, M.: On the degree of Boolean functions as real polynomials. Comput. Complexity **4**(4), 301–313 (1994)

941. Nitsche, M.: Computer proofs for property (T), and SDP duality. Preprint (2020). arXiv:2009.05134

942. Novikov, P.S.: Algorithmic unsolvability of the word problem in group theory. Trydu Mat. Inst. Stelkov **44**, 3–143 (1955)

943. Odlyzko, A.M.: On subspaces spanned by random selections of ± 1 vectors. J. Combin. Theory Ser. A **47**(1), 124–133 (1988)

944. Odlyzko, A.M., Sloane, N.J.A.: New bounds on the number of unit spheres that can touch a unit sphere in n dimensions. J. Combin. Theory Ser. A **26**(2), 210–214 (1979)

945. O'Donnell, R., Saks, M., Schramm, O., Servedio, R.A.: Every decision tree has an influential variable. In: 46th Annual IEEE Symposium on Foundations of Computer Science (FOCS'05), pp. 31–39. IEEE (2005)

946. Oesterlé, J.: Nouvelles approches du "théoreme" de Fermat. Astérisque **161**(162), 165–186 (1988)

947. Okada, S.: Enumeration of symmetry classes of alternating sign matrices and characters of classical groups. J. Algebraic Combin. **23**(1), 43–69 (2006)

948. Oliver, R.J.L., Thorne, F.: Upper bounds on number fields of given degree and bounded discriminant. Preprint (2020). arXiv:2005.14110

949. Ol'shanskii, A.Y.: On the problem of the existence of an invariant mean on a group. Russ. Math. Surv. **35**(4), 180 (1980)

950. Ol'shanskii, A.Y., Sapir, M.V.: Non-amenable finitely presented torsion-by-cyclic groups. Publications Mathématiques de l'Institut des Hautes Études Scientifiques **96**(1), 43–169 (2003)

951. Ono, K.: Distribution of the partition function modulo m. Ann. Math. **151**(1), 293–307 (2000)

952. Onsager, L.: Statistical hydrodynamics. Il Nuovo Cimento (1943–1954) **6**(2), 279–287 (1949)

953. Ordentlich, O., Regev, O., Weiss, B.: New bounds on the density of lattice coverings. Preprint (2020). arXiv:2006.00340

954. Ore, O.: Some remarks on commutators. Proc. Am. Math. Soc. **2**(2), 307–314 (1951)

955. Osajda, D.: Small cancellation labellings of some infinite graphs and applications. Acta Math. **225**(1), 159–191 (2020)

956. Oseledets, I.V.: Tensor-train decomposition. SIAM J. Sci. Comput. **33**(5), 2295–2317 (2011)

957. Osin, D.: Small cancellations over relatively hyperbolic groups and embedding theorems. Ann. Math. **172**(1), 1–39 (2010)

958. Ozawa, N.: Noncommutative real algebraic geometry of Kazhdan's property (T). J. Inst. Math. Jussieu **15**(1), 85–90 (2016)

959. Ozaydin, M.: Equivariant maps for the symmetric group (1987). Unpublished preprint

960. Painlevé, P.: Leçons, sur la théorie analytique des équations différentielles: professées à Stockholm (septembre, octobre, novembre 1895) sur l'invitation de SM le roi de Suède et de Norwège. A. Hermann (1897)

961. Paley, R.E.A.C.: Some theorems on abstract spaces. Bull. Am. Math. Soc. **42**(4), 235–240 (1936)

962. Paley, R.E.A.C., Wiener, N.: Fourier transforms in the complex domain, Colloquium Publications, vol. 19. American Mathematical Soc. (1934)

963. Palis, J.: Homoclinic orbits, hyperbolic dynamics and dimension of Cantor sets. Contemp. Math. **58**(26), 203–216 (1987)

964. Paouris, G.: Concentration of mass and central limit properties of isotropic convex bodies. Proc. Am. Math. Soc. **133**(2), 565–575 (2005)

965. Paouris, G.: Concentration of mass on convex bodies. Geom. Funct. Anal. **16**(5), 1021–1049 (2006)

966. Papadimitriou, C.H.: On the complexity of the parity argument and other inefficient proofs of existence. J. Comput. Syst. Sci. **48**(3), 498–532 (1994)

967. Pardon, J.: On the distortion of knots on embedded surfaces. Ann. Math. **174**(1), 637–646 (2011)

968. Pastur, L.A.: On the spectrum of random matrices. Theor. Math. Phys. **10**(1), 67–74 (1972)

969. Payne, L.E., Weinberger, H.F.: An optimal Poincaré inequality for convex domains. Arch. Ration. Mech. Anal. **5**(1), 286–292 (1960)

970. Peluse, S.: Bounds for sets with no polynomial progressions. Forum Math. Pi **8**, e16, 1–55 (2020)

971. Pemantle, R.: Choosing a spanning tree for the integer lattice uniformly. Ann. Probab. **19**(4), 1559–1574 (1991)

972. Penrose, R.: A generalized inverse for matrices. Math. Proc. Camb. Philos. Soc. **51**(3), 406–413 (1955)

973. Perelman, G.: The entropy formula for the Ricci flow and its geometric applications. Preprint (2002). arXiv:math/0211159

974. Perelman, G.: Finite extinction time for the solutions to the Ricci flow on certain three-manifolds. Preprint (2003). arXiv:math/0307245

975. Perelman, G.: Ricci flow with surgery on three-manifolds. preprint (2003). arXiv:math/0303109

976. Perkins, E.A., Taylor, S.J.: Uniform measure results for the image of subsets under Brownian motion. Probab. Theory Relat. Fields **76**(3), 257–289 (1987)

977. Petersen, C.L., Zakeri, S.: On the Julia set of a typical quadratic polynomial with a Siegel disk. Ann. Math. **159**(1), 1–52 (2004)
978. Petrow, I., Young, M.P.: The fourth moment of Dirichlet L-functions along a coset and the Weyl bound. Preprint (2019). arXiv:1908.10346
979. Petrow, I., Young, M.P.: The Weyl bound for Dirichlet L-functions of cube-free conductor. Ann. Math. **192**(2), 437–486 (2020)
980. Pfister, A.: On quadratic forms and abelian varieties over function fields. Contemp. Math. **8**, 249–264 (1982)
981. Piccirillo, L.: The Conway knot is not slice. Ann. Math. **191**(2), 581–591 (2020)
982. Pietsch, A.: Theorie der Operatorenideale: Zusammenfassung. Friedrich-Schiller Universität (1972)
983. Pikhurko, O., Razborov, A.: Asymptotic structure of graphs with the minimum number of triangles. Combin. Probab. Comput. **26**(1), 138–160 (2017)
984. Pila, J.: Density of integral and rational points on varieties. Astérisque **228**, 183–187 (1995)
985. Pisier, G.: Probabilistic methods in the geometry of Banach spaces. In: Probability and Analysis, pp. 167–241. Springer (1986)
986. Pisier, G.: Weak Hilbert spaces. Proc. Lond. Math. Soc. **3**(3), 547–579 (1988)
987. Podewski, K.P., Steffens, K.: Injective choice functions for countable families. J. Combin. Theory Ser. B **21**(1), 40–46 (1976)
988. Poincaré, H.: Sur les équations aux dérivées partielles de la physique mathématique. Am. J. Math. pp. 211–294 (1890)
989. Poincaré, H.: Cinquième complément à l'analysis situs. Rendiconti del Circolo Matematico di Palermo (1884–1940) **18**(1), 45–110 (1904)
990. Pokrovskiy, A.: Rota's basis conjecture holds asymptotically. Preprint (2020). arXiv:2008.06045
991. Pólya, G.: Über eine Aufgabe der Wahrscheinlichkeitsrechnung betreffend die Irrfahrt im Straßennetz. Math. Ann. **84**(1-2), 149–160 (1921)
992. Pólya, G.: Mathematics and plausible reasoning. Vol. 1: Induction and analogy in mathematics. Princeton University Press (1954)
993. Pólya, G., Jensen, J.L.W.V.: Über die algebraisch-funktionentheoretische Untersuchungen von JLWV Jensen. AF Høst (1927)
994. Pólya, G., Schur, I.: Über zwei Arten von Faktorenfolgen in der Theorie der algebraischen Gleichungen. J. Reine Angew. Math. **144**, 89–113 (1914)
995. Pólya, G., Szegő, G.: Isoperimetric inequalities in mathematical physics. Princeton University Press (1951)
996. Polymath, D.H.J.: A new proof of the density Hales–Jewett theorem. Ann. Math. pp. 1283–1327 (2012)
997. Polymath, D.H.J.: Variants of the Selberg sieve, and bounded intervals containing many primes. Res. Math. Sci. **1**(1), 12 (2014)
998. Polymath, D.H.J.: Effective approximation of heat flow evolution of the Riemann ξ function, and a new upper bound for the de Bruijn–Newman constant. Res. Math. Sci. **6**(3), 31 (2019)
999. Pomerance, C.: The role of smooth numbers in number theoretic algorithms. In: Proceedings of the International Congress of Mathematicians, pp. 411–422. Springer (1995)
1000. Pönitz, A., Tittmann, P.: Improved upper bounds for self-avoiding walks in $\mathbb{Z}^d$. Electron. J. Comb. **7**, R21–R21 (2000)
1001. Poonen, B.: Characterizing integers among rational numbers with a universal-existential formula. Am. J. Math. **131**(3), 675–682 (2009)
1002. Poonen, B., Stoll, M.: The Cassels–Tate pairing on polarized abelian varieties. Ann. Math. **150**(3), 1109–1149 (1999)
1003. Poonen, B., Stoll, M.: Most odd degree hyperelliptic curves have only one rational point. Ann. Math. **180**(3), 1137–1166 (2014)
1004. van der Poorten, A.J.: Solution de la conjecture de pisot sur le quotient de Hadamard de deux fractions rationnelles. CR Acad. Sci. Paris **306**(97), 102 (1988)

1005. van der Poorten, A.J.: Some facts that should be better known, especially about rational functions. In: Number Theory and Applications (Banff, AB, 1988), vol. 265, pp. 497–528. Kluwer Academic Publishers (1989)

1006. Pournin, L.: The diameter of associahedra. Adv. Math. **259**, 13–42 (2014)

1007. Preiss, D., Speight, G.: Differentiability of Lipschitz functions in Lebesgue null sets. Invent. Math. **199**(2), 517–559 (2015)

1008. Procaccia, A.D.: Thou shalt covet thy neighbor's cake. In: Twenty-First International Joint Conference on Artificial Intelligence (2009)

1009. Prunescu, M.: The exponential diophantine problem for $\mathbb{Q}$. J. Symb. Log. **85**(2), 671–672 (2020)

1010. Puder, D.: Primitive words, free factors and measure preservation. Isr. J. Math. **201**(1), 25–73 (2014)

1011. Puder, D., Parzanchevski, O.: Measure preserving words are primitive. J. Am. Math. Soc. **28**(1), 63–97 (2015)

1012. Pyber, L., Szabó, E.: Growth in finite simple groups of Lie type. J. Am. Math. Soc. **29**(1), 95–146 (2016)

1013. Rabin, M.O.: Probabilistic algorithm for testing primality. J. Number Theory **12**(1), 128–138 (1980)

1014. Radchenko, D., Viazovska, M.: Fourier interpolation on the real line. Publications mathématiques de l'IHÉS **129**(1), 51–81 (2019)

1015. Rado, R.: Studien zur kombinatorik. Math. Z. **36**(1), 424–470 (1933)

1016. Radó, T.: Uber den Begriff der Riemannschen Flache. Acta Litt. Sci. Szeged **2**, 101–121 (1925)

1017. Radon, J.: Mengen konvexer Körper, die einen gemeinsamen Punkt enthalten. Math. Ann. **83**(1-2), 113–115 (1921)

1018. Raghavendra, P.: Optimal algorithms and inapproximability results for every CSP? In: Proceedings of the Fortieth Annual ACM Symposium on the Theory of Computing, pp. 245–254 (2008)

1019. Rakhmanov, E.A.: Bounds for polynomials with a unit discrete norm. Ann. Math. **165**(1), 55–88 (2007)

1020. Ramachandra, K.: Some remarks on the mean value of the Riemann zeta-function and other Dirichlet series 1. Hardy–Ramanujan J. **1**, 1–15 (1978)

1021. Ramanujan, S.: Congruence properties of partitions. Math. Z. **9**(1-2), 147–153 (1921)

1022. Ramsey, F.P.: On a problem of formal logic. Proc. Lond. Math. Soc. **2**(1), 264–286 (1930)

1023. Ran, A.C.M., Reurings, M.C.B.: A fixed point theorem in partially ordered sets and some applications to matrix equations. Proc. Am. Math. Soc. **132**(5), 1435–1443 (2004)

1024. Ranestad, K., Schreyer, F.O.: On the rank of a symmetric form. J. Algebra **346**(1), 340–342 (2011)

1025. Rankin, R.A.: The difference between consecutive prime numbers. J. Lond. Math. Soc. **s1-13**(4), 242–247 (1938). https://doi.org/10.1112/jlms/s1-13.4.242. https://londmathsoc. onlinelibrary.wiley.com/doi/abs/10.1112/jlms/s1-13.4.242

1026. Rao, M.: Exhaustive search of convex pentagons which tile the plane. Preprint (2017). arXiv:1708.00274

1027. Rauch, J.: Five problems: An introduction to the qualitative theory of partial differential equations. In: Partial Differential Equations and Related Topics, pp. 355–369. Springer (1975)

1028. Raz, R., Tal, A.: Oracle separation of BQP and PH. In: Proceedings of the 51st Annual ACM SIGACT Symposium on the Theory of Computing, pp. 13–23 (2019)

1029. Razborov, A.: A product theorem in free groups. Ann. Math. **179**(2), 405–429 (2014)

1030. Razborov, A.A.: On the minimal density of triangles in graphs. Comb. Probab. Comput. **17**(4), 603–618 (2008)

1031. Razborov, A.A., et al.: Flag algebras. J. Symb. Log. **72**(4), 1239–1282 (2007)

1032. Razumov, A.V., Stroganov, Y.G.: Enumeration of quarter-turn-symmetric alternating-sign matrices of odd order. Theor. Math. Phys. **149**(3), 1639–1650 (2006)

1033. Razumov, A.V., Stroganov, Y.G.: Enumerations of half-turn-symmetric alternating-sign matrices of odd order. Theor. Math. Phys. **148**(3), 1174–1198 (2006)

1034. Read, R.C.: An introduction to chromatic polynomials. J. Comb. Theory **4**(1), 52–71 (1968)

1035. Regev, O., Shapira, U., Weiss, B.: Counterexamples to a conjecture of Woods. Duke Math. J. **166**(13), 2443–2446 (2017)

1036. Reid, M.: Superflip required 20 face turns (1995). http://cube20.org/cubelovers/CL15/002.txt

1037. Reid, W.T.: The isoperimetric inequality and associated boundary problems. J. Math. Mech. **8**(6), 897–905 (1959)

1038. Reidemeister, K.: Elementare begründung der knotentheorie. Abhandlungen aus dem Mathematischen Seminar der Universität Hamburg **5**(1), 24–32 (1927)

1039. Reiher, C.: The clique density theorem. Ann. Math., 683–707 (2016)

1040. Reingold, O.: Undirected connectivity in log-space. J. ACM **55**(4), 17 (2008)

1041. Reingold, O., Vadhan, S., Wigderson, A.: Entropy waves, the zig-zag graph product, and new constant-degree expanders. Ann. Math. **155**(1), 157–187 (2002)

1042. Reinhardt, K.: Uber die Zerlegung der Ebene in Polygone. Inaugural Dissertation, Univ. Frankfurt a. M. (1918)

1043. Ribet, K.A.: On modular representations of $\mathrm{Gal}(\bar{Q}/q)$ arising from modular forms. Invent. Math. **100**(1), 431–476 (1990)

1044. Rickman, S.: On the number of omitted values of entire quasiregular mappings. J. d'Analyse Math. **37**(1), 100–117 (1980)

1045. Rickman, S.: The analogue of Picard's theorem for quasiregular mappings in dimension three. Acta Math. **154**(3-4), 195–242 (1985)

1046. Riemann, B.: Ueber die Anzahl der Primzahlen unter einer gegebenen Grosse. Ges. Math. Werke und Wissenschaftlicher Nachlaß **2**, 145–155 (1859)

1047. Rivest, R.L., Adleman, L., Dertouzos, M.L.: On data banks and privacy homomorphisms. Found. Secure Comput. **4**(11), 169–180 (1978)

1048. Rivoire, O.: Phases vitreuses, optimisation et grandes déviations. Ph.D. thesis, Université Paris Sud-Paris XI (2005)

1049. Robbins, D.P.: The story of 1, 2, 7, 42, 429, 7436, ... Math. Intell. **13**(2), 12–19 (1991)

1050. Robertson, N., Seymour, P.D.: Graph minors. I. Excluding a forest. J. Combin. Theory Ser. B **35**(1), 39–61 (1983)

1051. Robertson, N., Seymour, P.D.: Graph minors. XX. Wagner's conjecture. J. Combin. Theory Ser. B **92**(2), 325–357 (2004)

1052. Robinson, J.: Definability and decision problems in arithmetic. J. Symb. Log. **14**(2), 98–114 (1949)

1053. Roche-Newton, O., Rudnev, M., Shkredov, I.D.: New sum-product type estimates over finite fields. Adv. Math. **293**, 589–605 (2016)

1054. Rodgers, B., Tao, T.: The De Bruijn–Newman constant is non-negative. Forum Math. Pi **8**, e6, 1–62 (2020)

1055. Rödl, V.: On a packing and covering problem. Eur. J. Combin. **6**(1), 69–78 (1985)

1056. Rogers, C.A.: Lattice coverings of space. Mathematika **6**(1), 33–39 (1959)

1057. Rokicki, T., Kociemba, H., Davidson, M., Dethridge, J.: The diameter of the Rubik's cube group is twenty. SIAM Review **56**(4), 645–670 (2014)

1058. Rónyai, L., Babai, L., Ganapathy, M.: On the number of zero-patterns of a sequence of polynomials. J. Am. Math. Soc. **14**(3), 717–735 (2001)

1059. Rosendal, C.: Continuity of universally measurable homomorphisms. Forum Math. Pi **7**, e5, 1–20 (2019)

1060. Rota, G.C.: Combinatorial theory, old and new. In: Actes du Congres International des Mathématiciens (Nice, 1970), vol. 3, pp. 229–233 (1970)

1061. Roth, K.F.: On certain sets of integers. J. Lond. Math. Soc. **1**(1), 104–109 (1953)

1062. Rothberger, F.: On some problems of Hausdorff and of Sierpiński. Fund. Math. **1**(35), 29–46 (1948)

1063. Rothvoss, T.: The matching polytope has exponential extension complexity. J. ACM **64**(6), 1–19 (2017)

1064. Rottenfusser, G., Rückert, J., Rempe, L., Schleicher, D.: Dynamic rays of bounded-type entire functions. Ann. Math., 77–125 (2011)

1065. Roy, D.: Approximation to real numbers by cubic algebraic integers. II. Ann. Math. **158**(3), 1081–1087 (2003)

1066. Royen, T.: A simple proof of the Gaussian correlation conjecture extended to multivariate gamma distributions. Far East J. Theor. Stat. **48**(2), 139–145 (2014)

1067. Rudelson, M.: Invertibility of random matrices: norm of the inverse. Ann. Math. **168**(2), 575–600 (2008)

1068. Rudelson, M., Vershynin, R.: Combinatorics of random processes and sections of convex bodies. Ann. Math. **164**(2), 603–648 (2006)

1069. Rudelson, M., Vershynin, R.: The Littlewood–Offord problem and invertibility of random matrices. Adv. Math. **218**(2), 600–633 (2008)

1070. Rudelson, M., Vershynin, R.: Invertibility of random matrices: unitary and orthogonal perturbations. J. Am. Math. Soc. **27**(2), 293–338 (2014)

1071. Rudin, W.: Some theorems on Fourier coefficients. Proc. Am. Math. Soc. **10**(6), 855–859 (1959)

1072. Ruf, B.: A sharp Trudinger–Moser type inequality for unbounded domains in $\mathbb{R}^2$. J. Funct. Anal. **219**, 340–367 (2005)

1073. Sah, A.: Diagonal Ramsey via effective quasirandomness. Preprint (2020). arXiv:2005.09251

1074. Sah, A., Sawhney, M., Stoner, D., Zhao, Y.: A reverse Sidorenko inequality. Invent. Math. pp. 1–47 (2020)

1075. Šajna, M.: Cycle decompositions III: complete graphs and fixed length cycles. J. Combin. Des. **10**(1), 27–78 (2002)

1076. Samet, B., Vetro, C., Vetro, P.: Fixed point theorems for α-ψ-contractive type mappings. Nonlinear Anal. Theory Methods Appl. **75**(4), 2154–2165 (2012)

1077. Sanders, T.: On Roth's theorem on progressions. Ann. Math. **174**(1), 619–636 (2011)

1078. Santos, F.: A counterexample to the Hirsch conjecture. Ann. Math. **176**(1), 383–412 (2012)

1079. Sapir, M.V., Birget, J.C., Rips, E.: Isoperimetric and isodiametric functions of groups. Ann. Math. **156**(2), 345–466 (2002)

1080. Sarkaria, K.S.: Shifting and embeddability of simplicial complexes. Max-Planck-Institut für Mathematik (1992)

1081. Sárközy, A.: On difference sets of sequences of integers. I. Acta Mathematica Academiae Scientiarum Hungarica **31**(1-2), 125–149 (1978)

1082. Savin, O.: Regularity of flat level sets in phase transitions. Ann. Math. **169**(1), 41–78 (2009)

1083. Saxton, D., Thomason, A.: Hypergraph containers. Invent. Math. **201**(3), 925–992 (2015)

1084. Schacht, M.: Extremal results for random discrete structures. Ann. Math. **184**(2), 333–365 (2016)

1085. Schäffer, J.J.: Geometry of spheres in normed spaces. Dekker (1976)

1086. Schattschneider, D.: Tiling the plane with congruent pentagons. Mathematics Magazine **51**(1), 29–44 (1978)

1087. Schinzel, A., Tijdeman, R.: On the equation $y^m = p(x)$. Acta Arith. **31**(2), 199–204 (1976)

1088. Schinzel, A., Zassenhaus, H., et al.: A refinement of two theorems of Kronecker. Mich. Math. J **12**, 81–85 (1965)

1089. Schmidt, W.M.: Open problems in diophantine approximation. Diophantine Approximations and Transcendental numbers (Luminy, 1982), Progr. Math **31**, 271–287 (1983)

1090. Schmidt, W.M.: Integer points on curves of genus 1. Compos. Math. **81**(1), 33–59 (1992)

1091. Schmidt, W.M.: Number fields of given degree and bounded discriminant. Astérisque **228**(4), 189–195 (1995)

1092. Schmidt, W.M.: Diophantine approximation, Lecture Notes in Mathematics, vol. 785. Springer Science & Business Media (1996)

1093. Schnorr, C.P.: Zufälligkeit und Wahrscheinlichkeit. Eine algorithmische Begründung der Wahrscheinlichkeitstheorie, Lecture Notes in Mathematics, vol. 218. Springer, Heidelberg (1971)

1094. Schoen, T.: Improved bound in Roth's theorem on arithmetic progressions. Preprint (2020). arXiv:2005.01145

1095. Schönhage, A.: Numerik analytischer Funktionen und Komplexität. Jahresber. Deutsch. Math.-Verein 92(1), 1–20 (1990)

1096. Schönhage, A., Strassen, V.: Schnelle multiplikation grosser zahlen. Computing 7(3-4), 281–292 (1971)

1097. Schramm, O., Steif, J.E.: Quantitative noise sensitivity and exceptional times for percolation. Ann. Math. 171(2), 619–672 (2010)

1098. Schur, I.: Über die kongruenz $x^m + y^m \equiv z^m \pmod{p}$. Jahresbericht der Deutschen Mathematikervereinigung 25, 114–117 (1916)

1099. Schütte, K., van der Waerden, B.L.: Das problem der dreizehn Kugeln. Math. Ann. 125(1), 325–334 (1952)

1100. Schwartz, R.E.: Unbounded orbits for outer billiards I. J. Mod. Dyn. 1(3), 371–424 (2007)

1101. Schwartz, R.E.: Obtuse triangular billiards II: One hundred degrees worth of periodic trajectories. Exp. Math. 18(2), 137–171 (2009)

1102. Serdyukov, A.: On some extremal walks in graphs. Upravlyaemye systemy 17(76-79), 1 (1978)

1103. Sernau, L.: Graph operations and upper bounds on graph homomorphism counts. J. Graph Theory 87(2), 149–163 (2018)

1104. Seymour, P.D.: Nowhere-zero 6-flows. J. Combin. Theory Ser. B 30(2), 130–135 (1981)

1105. Shakan, G.: On higher energy decompositions and the sum–product phenomenon. Math. Proc. Camb. Philos. Soc. 167(3), 599–617 (2019)

1106. Shalev, A.: Word maps, conjugacy classes, and a noncommutative Waring-type theorem. Ann. Math. 170(3), 1383–1416 (2009)

1107. Shalev, A.: Some results and problems in the theory of word maps. In: Erdős Centennial, pp. 611–649. Springer (2013)

1108. Shamir, A.: IP = PSPACE. J. ACM 39(4), 869–877 (1992)

1109. Shamir, E., Spencer, J.: Sharp concentration of the chromatic number on random graphs $g_{n,p}$. Combinatorica 7(1), 121–129 (1987)

1110. Shannon, C.E.: A mathematical theory of communication. Bell Syst. Tech. J. 27(3), 379–423 (1948)

1111. Shannon, C.E.: Communication in the presence of noise. Proc. IRE 37(1), 10–21 (1949)

1112. Shannon, C.E.: Probability of error for optimal codes in a Gaussian channel. Bell Syst. Tech. J. 38(3), 611–656 (1959)

1113. Shannon, C.E., Weaver, W.: The Mathematical Theory of Communication. University of Illinois Press (1949)

1114. Shapira, U.: A solution to a problem of Cassels and Diophantine properties of cubic numbers. Ann. Math. 173(1), 543–557 (2011)

1115. Shapiro, H.S.: Extremal problems for polynomials and power series. Ph.D. thesis, Massachusetts Institute of Technology (1952)

1116. Shitov, Y.: Counterexamples to Hedetniemi's conjecture. Ann. Math. 190(2), 663–667 (2019)

1117. Shitov, Y.: Counterexamples to Strassen's direct sum conjecture. Acta Math. 222(2), 363–379 (2019)

1118. Shmerkin, P.: On the exceptional set for absolute continuity of Bernoulli convolutions. Geom. Funct. Anal. 24(3), 946–958 (2014)

1119. Shmerkin, P.: On Furstenberg's intersection conjecture, self-similar measures, and the L^q norms of convolutions. Ann. Math. 189(2), 319–391 (2019)

1120. Shmoys, D.B., Tardos, É., Aardal, K.: Approximation algorithms for facility location problems. In: Proceedings of the Twenty-Ninth Annual ACM Symposium on the Theory of Computing, pp. 265–274 (1997)

1121. Shnirelman, A.: On the nonuniqueness of weak solution of the Euler equation. Commun. Pure Appl. Math. J. Issued by the Courant Inst. Math. Sci. **50**(12), 1261–1286 (1997)

1122. Shnirelman, A.: Weak solutions with decreasing energy of incompressible Euler equations. Comm. Math. Phys. **210**(3), 541–603 (2000)

1123. Shor, P.W.: Algorithms for quantum computation: discrete logarithms and factoring. In: Proceedings of the 35th Annual Symposium on the Foundations of Computer Science, pp. 124–134. IEEE (1994)

1124. Shorey, T.N.: Perfect powers in products of arithmetical progressions with fixed initial term. Indag. Math. (N.S.) **7**(4), 521–525 (1996)

1125. Shub, M., Smale, S.: Complexity of Bezout's theorem II: Volumes and probabilities. In: Computational Algebraic Geometry, pp. 267–285. Springer (1993)

1126. Shub, M., Smale, S.: Complexity of Bézout's theorem III: Condition number and packing. J. Complexity **9**(1), 4–14 (1993)

1127. Siegel, C.L.: Über einige Anwendungen diophantischer Approximationen. Akad. de Gruyter in Komm. (1929)

1128. Siegel, C.L.: The average measure of quadratic forms with given determinant and signature. Ann. Math. **45**(4), 667–685 (1944)

1129. Sierpiński, W.: Sur un probleme de H. Steinhaus concernant les ensembles de points sur le plan. Fund. Math. **2**(46), 191–194 (1958)

1130. Simmons, H.: The word problem for absolute presentations. J. Lond. Math. Soc. **2**(2), 275–280 (1973)

1131. Simons, J.: Minimal varieties in Riemannian manifolds. Ann. Math. **88**, 62–105 (1968)

1132. Siudeja, B.: Hot spots conjecture for a class of acute triangles. Math. Z. **280**(3-4), 783–806 (2015)

1133. Skolem, T.: Diophantische gleichungen. J. Springer (1938)

1134. Skubenko, B.F.: A proof of Minkowski's conjecture on the product of n linear inhomogeneous forms in variables for $n \le 5$. J. Sov. Math. **6**(6), 627–650 (1976)

1135. Sleator, D.D., Tarjan, R.E., Thurston, W.P.: Rotation distance, triangulations, and hyperbolic geometry. J. Am. Math. Soc. **1**(3), 647–681 (1988)

1136. Smale, S.: The generalized Poincaré conjecture in higher dimensions. Bull. Am. Math. Soc. **66**(5), 373–375 (1960)

1137. Smale, S.: Mathematical problems for the next century. Math. Intell. **20**(2), 7–15 (1998)

1138. Smith, A.: The congruent numbers have positive natural density. Preprint (2016). arXiv:1603.08479

1139. Smith, A.: 2-selmer groups, 2-class groups, and Goldfeld's conjecture. Preprint (2017). arXiv:1702.02325

1140. Smullyan, R.M.: Undecidability and recursive inseparability. In: Recursion Theory for Metamathematics. Oxford University Press (1958)

1141. Sodin, S.: The spectral edge of some random band matrices. Ann. Math. **172**(3), 2223–2251 (2010)

1142. Solomyak, B.: On the random series $\sigma \pm \lambda^n$ (an Erdős problem). Ann. Math. **142**(3), 611–625 (1995)

1143. Solymosi, J.: Bounding multiplicative energy by the sumset. Adv. Math. **222**(2), 402–408 (2009)

1144. Solymosi, J., Tao, T.: An incidence theorem in higher dimensions. Discrete Comput. Geom. **48**(2), 255–280 (2012)

1145. Solynin, A.Y., Zalgaller, V.A.: An isoperimetric inequality for logarithmic capacity of polygons. Ann. Math. **159**(1), 277–303 (2004)

1146. Soundararajan, K.: Moments of the Riemann zeta function. Ann. Math. **170**(2), 981–993 (2009)

1147. Spielman, D.A., Teng, S.H.: Smoothed analysis of algorithms. In: Proceedings of the International Congress of Mathematicians, vol. I, pp. 597–606. Beijing (2002)

1148. Spielman, D.A., Teng, S.H.: Smoothed analysis of algorithms: Why the simplex algorithm usually takes polynomial time. J. ACM **51**(3), 385–463 (2004)

1149. Spielman, D.A., Teng, S.H.: Nearly linear time algorithms for preconditioning and solving symmetric, diagonally dominant linear systems. SIAM J. Matrix Anal. Appl. **35**(3), 835–885 (2014)

1150. Spitzer, F.: Electrostatic capacity, heat flow, and Brownian motion. Zeitschrift für Wahrscheinlichkeitstheorie und Verwandte Gebiete **3**(2), 110–121 (1964)

1151. Stahl, H.R.: Proof of the BMV conjecture. Acta Math. **211**(2), 255–290 (2013)

1152. Stam, A.J.: Some inequalities satisfied by the quantities of information of Fisher and Shannon. Inf. Control **2**(2), 101–112 (1959)

1153. Stanley, R.P.: The number of faces of a simplicial convex polytope. Adv. Math. **35**(3), 236–238 (1980)

1154. Stein, E.M.: Some problems in harmonic analysis. In: Harmonic analysis in Euclidean spaces, Proceedings of the Symposium in Pure Mathematics of the Amer. Math. Soc., Williams College, Mass, Proc. Sympos. Pure Math., XXXV Part I, 1979, pp. 3–20. Amer. Math. Soc. (1979)

1155. Stein, E.M., Strömberg, J.O.: Behavior of maximal functions in $\mathbb{R}^n$ for large n. Arkiv för matematik **21**(1-2), 259–269 (1983)

1156. Stein, R.: A new pentagon tiler. Mathematics Magazine **58**, 308 (1985)

1157. Steinberg, R.: Generators for simple groups. Canad. J. Math. **14**, 277–283 (1962)

1158. Steiner, J.: Systematisch Entwickelung der Abhängikeit geometrischer Gestalten von Einander, mit Berücksichtigung der Arbeiten alter und neuer Geometer über Porismen, Projections-Methoden, Geometrie der Lage, etc. Erster Theil. Finke (1832)

1159. Steiner, J.: Einfache Beweise der isoperimetrischen Hauptsätze. J. Reine Angew. Math. **1838**(18), 281–296 (1838)

1160. Steinitz, E.: Über isoperimetrische Probleme bei konvexen Polyedern. J. Reine Angew. Math. **1928**(159), 133–143 (1928)

1161. Stevenhagen, P.: The number of real quadratic fields having units of negative norm. Exp. Math. **2**(2), 121–136 (1993)

1162. Stewart, C.L., et al.: On divisors of Lucas and Lehmer numbers. Acta Math. **211**(2), 291–314 (2013)

1163. Stone, A.: On the isoperimetric inequality on a minimal surface. Calc. Var. Part. Differ. Equ. **17**(4), 369–391 (2003)

1164. Stone, A.H., Tukey, J.W., et al.: Generalized "sandwich" theorems. Duke Math. J. **9**(2), 356–359 (1942)

1165. Strassen, V.: Gaussian elimination is not optimal. Numer. Math. **13**(4), 354–356 (1969)

1166. Strassen, V.: Vermeidung von divisionen. J. Reine Angew. Math. **264**, 184–202 (1973)

1167. Sudakov, V.N.: Typical distributions of linear functionals in finite-dimensional spaces of higher dimension. Doklady Akademii Nauk **243**(6), 1402–1405 (1978)

1168. Suk, A.: On the Erdős–Szekeres convex polygon problem. J. Am. Math. Soc. **30**(4), 1047–1053 (2017)

1169. Suzuki, M.: A new type of simple groups of finite order. Proc. Natl. Acad. Sci. USA **46**(6), 868 (1960)

1170. Sviridenko, M.: An improved approximation algorithm for the metric uncapacitated facility location problem. In: International Conference on Integer Programming and Combinatorial Optimization, pp. 240–257. Springer (2002)

1171. Sydler, J.P.: Conditions nécessaires et suffisantes pour l'équivalence des polyèdres de l'espace euclidien à trois dimensions. Comment. Math. Helv. **40**(1), 43–80 (1965)

1172. Szegő, G.: Inequalities for certain eigenvalues of a membrane of given area. J. Ration. Mech. Anal. **3**, 343–356 (1954)

1173. Szekeres, G., Peters, L.: Computer solution to the 17-point Erdős–Szekeres problem. ANZIAM J. **48**(2), 151–164 (2006)

1174. Szemerédi, E.: On sets of integers containing no k elements in arithmetic progression. Acta Arith. **27**, 299–345 (1975)

1175. Szemerédi, E., Trotter, W.T.: Extremal problems in discrete geometry. Combinatorica **3**(3-4), 381–392 (1983)

1176. Szemerédi, E., Vu, V.H.: Finite and infinite arithmetic progressions in sumsets. Ann. Math. **163**(1), 1–35 (2006)

1177. Talagrand, M.: Regularity of Gaussian processes. Acta Math. **159**(1), 99–149 (1987)

1178. Talagrand, M.: Isoperimetry, logarithmic Sobolev inequalities on the discrete cube, and Margulis' graph connectivity theorem. Geom. Funct. Anal. **3**(3), 295–314 (1993)

1179. Talagrand, M.: Maharam's problem. Ann. Math. **168**(3), 981–1009 (2008)

1180. Tao, T.: Fuglede's conjecture is false in 5 and higher dimensions. Math. Res. Lett. **11**(2), 251–258 (2004)

1181. Tao, T.: The Erdős discrepancy problem. Discrete Anal. p. 27 (2016)

1182. Tao, T.: The logarithmically averaged Chowla and Elliott conjectures for two-point correlations. Forum Math. Pi **4**, e8, 1–36 (2016)

1183. Tao, T.: Almost all orbits of the Collatz map attain almost bounded values. Preprint (2019). arXiv:1909.03562

1184. Tao, T., Vu, V.: On random ± 1 matrices: singularity and determinant. Random Struct. Algorithm. **28**(1), 1–23 (2006)

1185. Tao, T., Vu, V.: On the singularity probability of random Bernoulli matrices. J. Am. Math. Soc. **20**(3), 603–628 (2007)

1186. Tao, T., Vu, V.: Random matrices: Universality of ESDs and the circular law. Ann. Probab. **38**(5), 2023–2065 (2010)

1187. Tao, T., Vu, V.: Random matrices: universality of local eigenvalue statistics. Acta Math. **206**(1), 127–204 (2011)

1188. Tao, T., Vu, V.: Random matrices: universality of local spectral statistics of non-Hermitian matrices. Ann. Probab. **43**(2), 782–874 (2015)

1189. Tao, T., Vu, V.H.: Inverse Littlewood–Offord theorems and the condition number of random discrete matrices. Ann. Math. **169**(2), 595–632 (2009)

1190. Tao, T., Ziegler, T.: The primes contain arbitrarily long polynomial progressions. Acta Math. **201**(2), 213–305 (2008)

1191. Tao, T., Ziegler, T.: Polynomial patterns in the primes. Forum Math. Pi **6**, e1, 1–60 (2018)

1192. Tao, Z.: On the representation of large odd integer as a sum of three almost equal primes. Acta Mathematica Sinica **7**(3), 259–272 (1991)

1193. Tate, J., Shafarevich, I.R.: The rank of elliptic curves. Doklady Akademii Nauk **175**(4), 770–773 (1967)

1194. Terras, R.: A stopping time problem on the positive integers. Acta Arith. **3**(30), 241–252 (1976)

1195. Thomason, A.: Pseudo-random graphs. In: North-Holland Mathematics Studies, vol. 144, pp. 307–331. Elsevier (1987)

1196. Thomassen, C.: On the chromatic number of triangle-free graphs of large minimum degree. Combinatorica **22**(4), 591–596 (2002)

1197. Thomassen, C.: The weak 3-flow conjecture and the weak circular flow conjecture. J. Combin. Theory Ser. B **102**(2), 521–529 (2012)

1198. Thompson, J.G.: Invariants of finite groups. J. Algebra **69**(1), 143–145 (1981)

1199. Thue, A.: Über die dichteste Zusammenstellung von kongruenten Kreisen in einer Ebene. Christiania: Dybwad in Komm. (1910)

1200. Thunder, J.L.: Decomposable form inequalities. Ann. Math. **153**(3), 767–804 (2001)

1201. Thurston, W.P.: Three dimensional manifolds, Kleinian groups and hyperbolic geometry. Bull. Am. Math. Soc. **6**(3), 357–381 (1982)

1202. Tiep, P.H.: The α-invariant and Thompson's conjecture. Forum Math. Pi **4**, e5, 1–28 (2016)

1203. Tijdeman, R.: On the equation of Catalan. Acta Arith. **29**, 197–209 (1976)

1204. Tikhomirov, K.: Singularity of random Bernoulli matrices. Ann. Math. **191**(2), 593–634 (2020)

1205. Tokarsky, G.W.: Polygonal rooms not illuminable from every point. Am. Math. Monthly **102**(10), 867–879 (1995)

1206. Tolsa, X.: Painlevé's problem and the semiadditivity of analytic capacity. Acta Math. **190**(1), 105–149 (2003). https://doi.org/10.1007/BF02393237

1207. Toom, A.L.: The complexity of a scheme of functional elements realizing the multiplication of integers. Sov. Math. Doklady **3**(4), 714–716 (1963)

1208. Torquato, S., Jiao, Y.: Analytical constructions of a family of dense tetrahedron packings and the role of symmetry. Preprint (2009). arXiv:0912.4210

1209. Tóth, C.D.: The Szemerédi–Trotter theorem in the complex plane. Combinatorica **35**(1), 95–126 (2015)

1210. Tracy, C.A., Widom, H.: Level-spacing distributions and the Airy kernel. Phys. Lett. B **305**, 115–118 (1993)

1211. Tracy, C.A., Widom, H.: Level-spacing distributions and the Airy kernel. Comm. Math. Phys. **159**(1), 151–174 (1994)

1212. Tracy, C.A., Widom, H.: On orthogonal and symplectic matrix ensembles. Comm. Math. Phys. **177**(3), 727–754 (1996)

1213. Tropp, J.A.: User-friendly tail bounds for sums of random matrices. Found. Comput. Math. **12**(4), 389–434 (2012)

1214. Trudinger, N.S.: On imbeddings into Orlicz spaces and some applications. J. Math. Mech. **17**(5), 473–483 (1967)

1215. Tunnell, J.B.: A classical Diophantine problem and modular forms of weight 3/2. Invent. Math. **72**(2), 323–334 (1983)

1216. Turán, P.: Egy gráfelméleti szélsoértékfeladatról. Mat. Fiz. Lapok **48**(3), 436 (1941)

1217. Turán, P.: Eine extremalaufgabe aus der graphentheorie. Mat. Fiz. Lapok **48**(436-452), 61 (1941)

1218. Turing, A.M.: On computable numbers, with an application to the Entscheidungsproblem. Proc. Lond. Math. Soc. **2**(1), 230–265 (1937)

1219. Tutte, W.T.: On the imbedding of linear graphs in surfaces. Proc. Lond. Math. Soc. **2**(1), 474–483 (1949)

1220. Tverberg, H.: A generalization of Radon's theorem. J. Lond. Math. Soc. **1**(1), 123–128 (1966)

1221. Ulmer, D.: Elliptic curves with large rank over function fields. Ann. Math. **155**(1), 295–315 (2002)

1222. Vaaler, J.D.: On the metric theory of Diophantine approximation. Pac. J. Math. **76**(2), 527–539 (1978)

1223. Valiant, L.G.: The complexity of computing the permanent. Theor. Comput. Sci. **8**(2), 189–201 (1979)

1224. Varjú, P.: Absolute continuity of Bernoulli convolutions for algebraic parameters. J. Am. Math. Soc. **32**(2), 351–397 (2019)

1225. Varjú, P.P.: On the dimension of Bernoulli convolutions for all transcendental parameters. Ann. Math. **189**(3), 1001–1011 (2019)

1226. Vaserstein, L.: Polynomial parametrization for the solutions of Diophantine equations and arithmetic groups. Ann. Math. **171**(2), 979–1009 (2010)

1227. Vaughan, R.C.: On pairs of additive cubic equations. Proc. Lond. Math. Soc. **3**(2), 354–364 (1977)

1228. Vaughan, R.C., Velani, S.: Diophantine approximation on planar curves: the convergence theory. Invent. Math. **166**(1), 103–124 (2006)

1229. Vazirani, V.V.: NC algorithms for computing the number of perfect matchings in $K_{3,3}$-free graphs and related problems. Inf. Comput. **80**(2), 152–164 (1989)

1230. Veech, W.A.: Gauss measures for transformations on the space of interval exchange maps. Ann. Math. **115**(2), 201–242 (1982)

1231. Veech, W.A.: The metric theory of interval exchange transformations I. Generic spectral properties. Am. J. Math. **106**(6), 1331–1359 (1984)

1232. Venkov, B.A.: On a class of Euclidean polyhedra. Vestnik Leningrad. Univ. Ser. Mat. Fiz. Him. **9**(2), 11–31 (1954)

1233. Viazovska, M.S.: The sphere packing problem in dimension 8. Ann. Math. **185**(3), 991–1015 (2017)

1234. Vicsek, T., Czirók, A., Ben-Jacob, E., Cohen, I., Shochet, O.: Novel type of phase transition in a system of self-driven particles. Phys. Rev. Lett. **75**(6), 1226 (1995)

1235. Vinogradov, A.I.: On the density conjecture for Dirichlet L-series. Izv. Akad. Nauk SSSR Ser. Mat. **29**, 903–934 (1965)

1236. Vinogradov, I.M.: Representation of an odd number as a sum of three primes. In: Goldbach Conjecture, pp. 61–64. World Scientific (1984)

1237. Vitushkin, A.G.: The analytic capacity of sets in problems of approximation theory. Russ. Math. Surv. **22**, 139–200 (1967)

1238. Vivaldi, F., Shaidenko, A.V.: Global stability of a class of discontinuous dual billiards. Comm. Math. Phys. **110**(4), 625–640 (1987)

1239. Vlăduţ, S.: Lattices with exponentially large kissing numbers. Mosc. J. Comb. Number Theory **8**(2), 163–177 (2019)

1240. van der Waerden, B.: Beweis einer baudetschen vermutung. Nieuw Arch. Wisk. **19**, 212–216 (1927)

1241. Wagner, G., Volkmann, B.: On averaging sets. Monatshefte für Mathematik **111**(1), 69–78 (1991)

1242. Wagner, K.: Über eine Eigenschaft der ebenen Komplexe. Math. Ann. **114**(1), 570–590 (1937)

1243. Wagner, K.: Graphentheorie, vol. 248/248a. BJ Hochschultaschenbucher, Mannheim, p. 61 (1970)

1244. Wagon, S.: The Banach–Tarski paradox, volume 24 of. Encyclopedia of Mathematics and its Applications (1985)

1245. Walsh, M.N.: The polynomial method over varieties. Invent. Math. **222**(2), 469–512 (2020)

1246. Wang, T., Yu, Q.: On vertex-coloring 13-edge-weighting. Front. Math. China **3**(4), 581–587 (2008)

1247. Wästlund, J.: The mean field traveling salesman and related problems. Acta Math. **204**(1), 91–150 (2010)

1248. Wästlund, J.: Replica symmetry of the minimum matching. Ann. Math. **175**(3), 1061–1091 (2012)

1249. Weaire, D., Phelan, R.: A counter-example to Kelvin's conjecture on minimal surfaces. Phil. Mag. Lett. **69**(2), 107–110 (1994)

1250. Weber, M., Hoffman, D., Wolf, M.: An embedded genus-one helicoid. Ann. Math. **169**(2), 347–448 (2009)

1251. Weierstrass, K.: Über continuirliche functionen eines reellen arguments, die für Keinen Werth des letzteren einen bestimmten Differentailqutienten besitzen. In: Mathematische Werke, vol. II, pp. 71–74. Mayer & Müller, Berlin (1872)

1252. Weinberger, H.F.: An isoperimetric inequality for the n-dimensional free membrane problem. J. Ration. Mech. Anal. **5**(4), 633–636 (1956)

1253. Westzynthius, E.J.: Über die Verteilung der Zahlen: die zu den n ersten Primzahlen teilerfremd sind. Ph.D. thesis, Helsingfors (1931)

1254. Weyl, H.: Über die gleichverteilung von zahlen mod. eins. Math. Ann. **77**(3), 313–352 (1916)

1255. Whitney, H.: A logical expansion in mathematics. Bull. Am. Math. Soc. **38**(8), 572–579 (1932)

1256. Whitney, H.: Analytic extensions of differentiable functions defined in closed sets. Trans. Am. Math. Soc. **36**(1), 63–89 (1934)

1257. Whittaker, E.T.: On the functions which are represented by the expansion of interpolating theory. Proc. Roy. Soc. Edinburgh **35**, 181–194 (1915)

1258. Wigderson, A.: Mathematics and Computation: A Theory Revolutionizing Technology and Science. Princeton University Press (2019)

1259. Wigderson, Y.: An improved lower bound on multicolor Ramsey numbers. e-print (2020). arXiv–2009

1260. Wigner, E.: Characteristic vectors of bordered matrices with infinite dimensions. Ann. Math. **63**(3), 548–564 (1955)

1261. Wigner, E.P.: On the distribution of the roots of certain symmetric matrices. Ann. Math. **67**, 325–327 (1958)

1262. Wiles, A.: Modular elliptic curves and Fermat's last theorem. Ann. Math. **141**(3), 443–551 (1995)

1263. Wilf, H.S.: On crossing numbers, and some unsolved problems. Comb. Geom. Probab. pp. 557–562 (1993)

1264. Wilker, J.B.: Sizing up a solid packing. Period. Math. Hungar. **8**(2), 117–134 (1977)

1265. Willmore, T.J.: Note on embedded surfaces. An. Sti. Univ. "Al. I. Cuza" Iasi Sect. I a Mat.(NS) B **11**, 493–496 (1965)

1266. Wilson, J.: First-order group theory. In: Infinite Groups 1994, pp. 301–314. De Gruyter (2017)

1267. Wilson, J.S.: On exponential growth and uniformly exponential growth for groups. Invent. Math. **155**(2), 287–303 (2004)

1268. Wilson, R.M.: An existence theory for pairwise balanced designs, III: Proof of the existence conjectures. J. Combin. Theory Ser. A **18**(1), 71–79 (1975)

1269. Winkler, P., Zuckerman, D.: Multiple cover time. Random Struct. Algorithm. **9**(4), 403–411 (1996)

1270. Wolff, T.: Decay of circular means of Fourier transforms of measures. Int. Math. Res. Not. IMRN **1999**(10), 547–567 (1999)

1271. Wolff, T.: Recent work connected with the Kakeya problem. In: Prospects in Mathematics (Princeton, NJ, 1996), pp. 129–162. Princeton University Press (1999)

1272. Wolff, T.H.: An improved bound for Kakeya type maximal functions. Revista Matemática Iberoamericana **11**(3), 651–674 (1995)

1273. Woods, A.C.: Covering six space with spheres. J. Number Theory **4**(2), 157–180 (1972)

1274. Wooley, T.D.: Vinogradov's mean value theorem via efficient congruencing. Ann. Math. **175**(3), 1575–1627 (2012)

1275. Wooley, T.D.: The cubic case of the main conjecture in Vinogradov's mean value theorem. Adv. Math. **294**, 532–561 (2016)

1276. Wu, M.: A proof of Furstenberg's conjecture on the intersections of p- and q-invariant sets. Ann. Math. **189**(3), 707–751 (2019)

1277. Xia, Z.: The existence of noncollision singularities in Newtonian systems. Ann. Math. **135**(3), 411–468 (1992)

1278. Xue, J.: Non-collision singularities in a planar 4-body problem. Acta Math. **224**(2), 253–388 (2020)

1279. Xylouris, T.: On the least prime in an arithmetic progression and estimates for the zeros of Dirichlet L-functions. Acta Arith. **150**(1), 65–91 (2011)

1280. Yang, Y., Zhang, D.: The log-Brunn–Minkowski inequality in $\mathbb{R}^3$. Proc. Am. Math. Soc. **147**(10), 4465–4475 (2019)

1281. Yau, H.T.: $(\ln t)^{2/3}$ law of the two dimensional asymmetric simple exclusion process. Ann. Math. **159**(1), 377–405 (2004)

1282. Yau, S.T.: Problem section, in 'Seminar on Differential Geometry'. Ann. Math. Study **102**, 669–706 (1982)

1283. Yau, S.T.: Review of geometry and analysis. In: Mathematics: Frontiers and Perspectives, pp. 353–401. Amer. Math. Soc., Providence, RI (2000)

1284. Young, M.P.: The fourth moment of Dirichlet L-functions. Ann. Math. **173**(1), 1–50 (2011)

1285. Young, R.: The Dehn function of $\mathrm{SL}_n(\mathbb{Z})$. Ann. Math. **177**(3), 969–1027 (2013)

1286. Zagier, D.: Values of zeta functions and their applications. In: First European Congress of Mathematics Paris, July 6–10, 1992, pp. 497–512. Springer (1994)

1287. Zagier, D.: Evaluation of the multiple zeta values $\zeta(2, \ldots, 2, 3, 2, \ldots, 2)$. Ann. Math. **175**(2), 977–1000 (2012)

1288. Zaremba, S.K.: La méthode des "bons treillis" pour le calcul des intégrales multiples. In: Applications of Number Theory to Numerical Analysis, pp. 39–119. Elsevier (1972)

1289. Zeilberger, D.: Proof of the alternating sign matrix conjecture. Electron. J. Combin. **3**(2), R13 (1996)

1290. Zermelo, E.: Untersuchungen über die Grundlagen der Mengenlehre. I. Math. Ann. **65**(2), 261–281 (1908)

1291. Zhang, Y.: Bounded gaps between primes. Ann. Math. **179**(3), 1121–1174 (2014)

1292. Zhuk, D.: A proof of the CSP dichotomy conjecture. J. ACM **67**(5), 1–78 (2020)

1293. Zsigmondy, K.: Zur theorie der potenzreste. Monatshefte für Mathematik und Physik **3**(1), 265–284 (1892)

1294. Zuckerman, D.: Linear degree extractors and the inapproximability of max clique and chromatic number. Theory Comput. **3**, 103–128 (2007)

1295. Zykov, A.A.: On some properties of linear complexes. Matematicheskii sbornik **66**(2), 163–188 (1949)

Author Index

Subject Index

Printed in the United States
by Baker & Taylor Publisher Services